INTERNATIONAL SEMINAR ON NUCLEAR WAR AND PLANETARY EMERGENCIES 44th Session

THE ROLE OF SCIENCE IN THE THIRD MILLENNIUM

World Energy Crisis - Global Nuclear Energy Issues after Fukushima - Energy Efficiency / Energy and Pollution - Unconventional Natural Gas: Benefits and Risks / Climate - Cosmic Rays and Climatic Processes / Water and Pollution - Water Scarcity and Pollution - Contaminants of Emerging Concern / Information Security - The Role of Science in Information Technologies and Internet Tools in Developing Countries Food, Soil & Medicine - Greenhouse Gases Consequences - Evidence-Based Third Millennium Medicine Lectio Magistralis – The Role of Science in the Third Millennium

THE SCIENCE AND CULTURE SERIES
Nuclear Strategy and Peace Technology*

Series Editor: Antonino Zichichi

2007 — International Seminar on Nuclear War and Planetary Emergencies — 38th Session: World Energy Crisis; Managing Climate Change; Mitigation of Greenhouse Gases; Geoengineering & Adaptation; Theoretical Alternatives to Climate Modelling; US Missile Defence Shield; Global Monitoring of the Planet; Life Cycle Nuclear Energy Environmental Issues; The Epidemic of Alzheimer; Infectious Agents and Cancer

2008 — International Seminar on Nuclear War and Planetary Emergencies — 40th Session: Energy: Nuclear Power Present and Future; Sustainability of Biofuels; Resolving the Nuclear Waste – Climatology Model and Statistics; Ozone and Climate Change Interaction; Spatio-Temporal Field of Atmospheric CO_2; Forest Policies – Medicine: Vector-Borne Diseases; Screening Technology – Pollution: Air-Borne Particulates – Global Monitoring of the Planet: Disarmament and Non-Proliferation Regime; The Crisis in Internet Security; The Northern Sea Route – The Erice Science for Peace Award Scientific Session

2009 — International Seminar on Nuclear War and Planetary Emergencies — 42nd Session: World Energy Crisis—Energy & Pollution: Essential Technologies for Managing the Coupled Challenges of Climate Change and Energy Security, Energy, Water, Climate, Pollution & Limits of Development in Asian Countries; Global Monitoring of the Planet—Sensitivity of Climate to Additional CO_2 as Indicated by Water Cycle Feedback Issues, Climate Uncertainties Addressed by Satellites, The Basic Mathematics Needed for all Models; Pollution and Medicine—The Revolution in the Environmental Health Sciences and the Emergence of Green Chemistry; Information Security—Cyber Conflict and Cyber Stability: Finding a Path to Cyber Peace; Cultural Pollution —The Erice Science for Peace Award Scientific Session

2010 — International Seminar on Nuclear War and Planetary Emergencies — 43rd Session: The Role of Science in the Third Millennium Energy: Nuclear Power Plants—Sustainable Systems / Pollution: Oil Spills – Air, Water and Waste Criteria / Climate: Paleoclimate Natural Variations and Implications – Sensitivity, Climategate, USA Legislation / Cosmic Objects: Countermeasures / Mitigation of Terrorist Acts: Scientific Contributions to Biosecurity / Lectio Magistralis: Science in Everyday Life / Information Security: Cyber Conflicts and Stability / The Erice Science for Peace Prize Award

2011 — International Seminar on Nuclear War and Planetary Emergencies — 44th Session: The Role of Science in the Third Millennium: World Energy Crisis - Global Nuclear Energy Issues after Fukushima - Energy Efficiency / Energy and Pollution - Unconventional Natural Gas: Benefits and Risks / Climate - Cosmic Rays and Climatic Processes / Water and Pollution - Water Scarcity and Pollution - Contaminants of Emerging Concern / Information Security - The Role of Science in Information Technologies and Internet Tools in Developing Countries Food, Soil & Medicine - Greenhouse Gases Consequences - Evidence-Based Third Millennium Medicine Lectio Magistralis - The Role of Science in the Third Millennium

*For the complete list of titles in this series, please go to:
http://worldscibooks.com/series/scsnsp_series.shtml

THE SCIENCE AND CULTURE SERIES

Nuclear Strategy and Peace Technology

INTERNATIONAL SEMINAR ON NUCLEAR WAR AND PLANETARY EMERGENCIES

44th Session

THE ROLE OF SCIENCE IN THE THIRD MILLENNIUM

World Energy Crisis - Global Nuclear Energy Issues after Fukushima - Energy Efficiency / Energy and Pollution - Unconventional Natural Gas: Benefits and Risks / Climate - Cosmic Rays and Climatic Processes / Water and Pollution - Water Scarcity and Pollution - Contaminants of Emerging Concern / Information Security - The Role of Science in Information Technologies and Internet Tools in Developing Countries Food, Soil & Medicine - Greenhouse Gases Consequences - Evidence-Based Third Millennium Medicine Lectio Magistralis – The Role of Science in the Third Millennium

"E. Majorana" Centre for Scientific Culture

Erice, Italy, 19 – 24 August 2011

Series Editor and Chairman: A. Zichichi

Edited by R. Ragaini

World Scientific

NEW JERSEY • LONDON • SINGAPORE • BEIJING • SHANGHAI • HONG KONG • TAIPEI • CHENNAI

Published by

World Scientific Publishing Co. Pte. Ltd.
5 Toh Tuck Link, Singapore 596224
USA office: 27 Warren Street, Suite 401-402, Hackensack, NJ 07601
UK office: 57 Shelton Street, Covent Garden, London WC2H 9HE

British Library Cataloguing-in-Publication Data
A catalogue record for this book is available from the British Library.

INTERNATIONAL SEMINAR ON NUCLEAR WAR AND PLANETARY EMERGENCIES — 44TH SESSION: THE ROLE OF SCIENCE IN THE THIRD MILLENNIUM: WORLD ENERGY CRISIS - GLOBAL NUCLEAR ENERGY ISSUES AFTER FUKUSHIMA - ENERGY EFFICIENCY / ENERGY AND POLLUTION - UNCONVENTIONAL NATURAL GAS: BENEFITS AND RISKS / CLIMATE - COSMIC RAYS AND CLIMATIC PROCESSES / WATER AND POLLUTION - WATER SCARCITY AND POLLUTION - CONTAMINANTS OF EMERGING CONCERN / INFORMATION SECURITY - THE ROLE OF SCIENCE IN INFORMATION TECHNOLOGIES AND INTERNET TOOLS IN DEVELOPING COUNTRIES FOOD, SOIL & MEDICINE - GREENHOUSE GASES CONSEQUENCES - EVIDENCE-BASED THIRD MILLENNIUM MEDICINE LECTIO MAGISTRALIS - THE ROLE OF SCIENCE IN THE THIRD MILLENNIUM

ISBN-13 978-981-4415-00-2
ISBN-10 981-4415-00-6

Printed in Singapore by B & Jo Enterprise Pte Ltd

CONTENTS

SESSION 1

OPENING SESSION

SCIENCE, CULTURE AND THE PLANETARY EMERGENCIES

ANTONINO ZICHICHI
INFN and University of Bologna, Italy, CERN, Geneva, Switzerland, Enrico Fermi Centre, Rome, Italy, World Federation of Scientists, Beijing, Geneva, Moscow, New York

Dear Colleagues, Ladies and Gentlemen,

Let me first of all welcome you to this 44th Session of the International Seminars on Nuclear War and Planetary Emergencies. My lecture will be based on three points:

- **Introduction**
  Brief recall of our activities, started nearly half a century ago.
- **Analysis of the so-called Modern Culture**
  Why? Because of the new Science called "Complexity".
- **Where are we going next**
  Why? Because we want an "Effective Action" to be implemented.

INTRODUCTION

Brief Recall of Our Activities, Started Nearly Half a Century Ago.

The Planetary Emergencies have been, since a long time, at the centre of our attention and here we are again, one hundred scientists from 39 countries, gathered in Erice to analyse a series of crucial multidisciplinary scientific issues.

- World Energy Crisis
- Global Monitoring of the Planet
- Pollution
- Water
- Information Security
- Climate

They are all part of the 63 Planetary Emergencies identified by the World Federation of Scientists 25 years ago. As we have many participants coming for the first time, let me very briefly recall our past activities that started nearly half a century ago.

1. This year is the first of the three years devoted to the celebrations of the 50th Anniversary of the Ettore Majorana Centre for Scientific Culture, which I started in 1961 with John Bell at CERN and formally established (EMFCSC) in 1962 with Bell, Blackett, Weisskopf and Rabi in Geneva (CERN); the first activity at Erice being, in June 1963, the Subnuclear Physics School. These three events are the reason why the celebrations will last three years. During these past five decades, 109,250 scientists (145 Nobel, of which 76 were awarded after their being at the EMFCSC) from 932 laboratories and research

centres of 140 nations have taken part in post-university activities to promote a Science without secrets and without borders.

2 This is the basis of our successful policy in imposing the idea of a "Science without secrets and without frontiers", East-West and North-South, first through the Erice Statement, and next by opening World Laboratory branches in Russia and various Eastern and newly emerging countries.

3 Our activity has also been devoted to the defusing of the USA–USSR Nuclear confrontation, through the International Seminars for Nuclear War and Planetary Emergencies in the early 80's, thus contributing to the avoidance of a nuclear holocaust.

4 In 1986, we launched the International Centre for Scientific Culture—World Laboratory, based on scientific volunteering. This has allowed us to implement 100 pilot projects in all domains of science, in 50 developing countries, while granting thousands of scholarships to their young scientists.

5 Our continuing fight against an environmental holocaust was started many decades ago, and numerous recommendations have been issued to governments and international organisations.

6 And now some good news, which I am pleased to give to you. An exceptional financial support has been granted for the three years celebration of the 50th Anniversary. These funds will be devoted each year—in 2011, 2012, 2013—to the selection of excellent fellows in the NSP (National Scholarship Programme) of the WFS.

We all hoped that by disseminating our findings and experience, based on rigorous scientific method, we would show the way and help governments and international institutions take the lead in coming up with an "effective action" to solve the problems of Planetary Emergencies. We are nowhere near that point, as we will see at the end of this lecture.

And now the two basic points:

FIRST: an analysis of the so-called "*Modern Culture*".
SECOND: a proposal based on "*Where we are going next*". Here is the "Effective Action".

FIRST POINT—ANALYSIS OF THE SO-CALLED MODERN CULTURE

<u>Why? Because of the new Science called "Complexity".</u>

I would like to invite all of you to think about a problem which is debated the world over.

Reductionism versus **Holism**.

<u>Why?</u>

Because the **Culture of our time** claims that the Science of Complexity will win, allowing Modern Society to solve all problems which go from the study of Social and Economic Systems to the behaviours of Financial Markets to the Traffic Flux, etc.

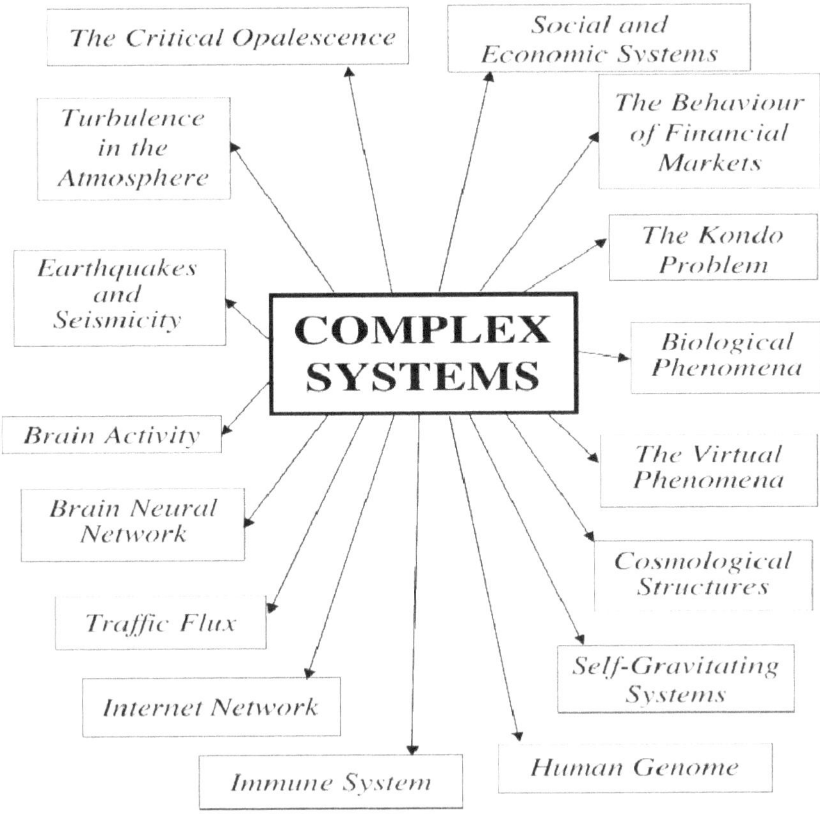

Fig. 1.

People claim that if we go on with the Reductionistic method we will never be able to understand the Logic of Nature for the very simple reason that our research work ignores the interaction with the environment, i.e., the world which is around us. From the **Basic Quantities** up to the constituents of our body (p, n, e), atoms and molecules, not one of these elements could care less about the environment, as illustrated in Figure 2.

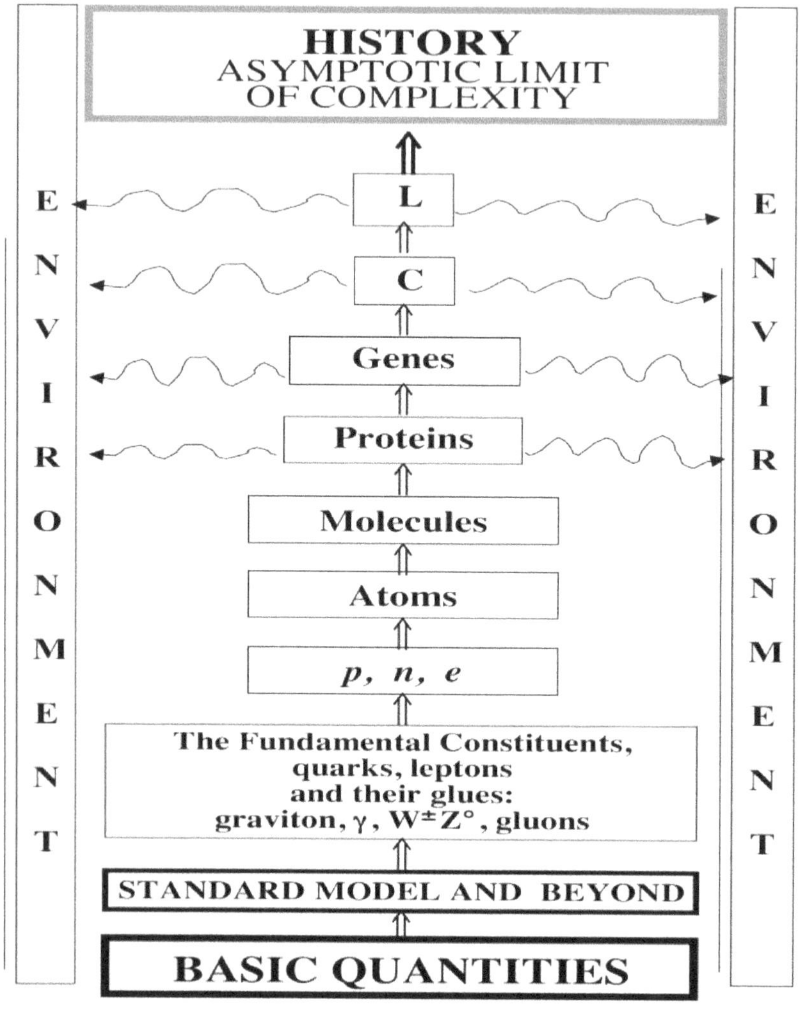

Fig. 2.

From proteins up to History, which needs the living form of matter, L, i.e., us, for which it is necessary the most elementary form of life, the so called "minimal life cell", C, for which genes and proteins are needed; from proteins up to History **everything interacts with the environment**.

This is why these fields of Science where Holism is needed must have priority. Reductionism needs to become Holism.

It is generally accepted that *Simplicity* is the outcome of *Reductionism*, while *Complexity* is the result of *Holism*.

The most celebrated example of ***Simplicity*** is ***Science***, while the most celebrated example of ***Complexity*** is ***History***.

Let us have a look at **History** and **Science**.

A common characteristic to both Science and History is *Evolution*. In fact, *Science* can be identified with the Evolution of our Basic Understanding of the Laws governing the World in its Structures (**EBUS**), and *History* with the Evolution of the World in Real Life (**EWRL**).

The real world is characterized by two basic features, which appear to be one the opposite of the other:

Simplicity and *Complexity*.

The existence of Complexity in the real world emerges from two experimentally well-established basic elements:

1) The **Anderson-Feynman-Beethoven**-type phenomena (**AFB**), i.e., phenomena whose Laws and Regularities ignore the existence of the fundamental Laws of Nature from which they originate;
2) The Sarajevo-type effects, i.e., **Unexpected Events** of quasi irrelevant magnitude, which produce **Enormous Consequences** (**UEEC**).

The definition of "Sarajevo-type" event is due to our friends working in History. Following our discovery of "**Virtual Physics**", they have invented "**Virtual History**".

The two "Virtual" effects have nothing in common.

The first fellow engaged in the calculation of the first virtual effect in physics is Victor Weisskopf, who computed the so called "vacuum polarization" contribution in the energy level of the Hydrogen. This was found to be negative and small. But in 1947, Lamb and Rutherford discovered that the experimental result was positive and big. The reason being that the electron in Hydrogen can emit and absorb a photon, as described by the very simple Feynman diagram:

while the process computed by Weisskopf was:

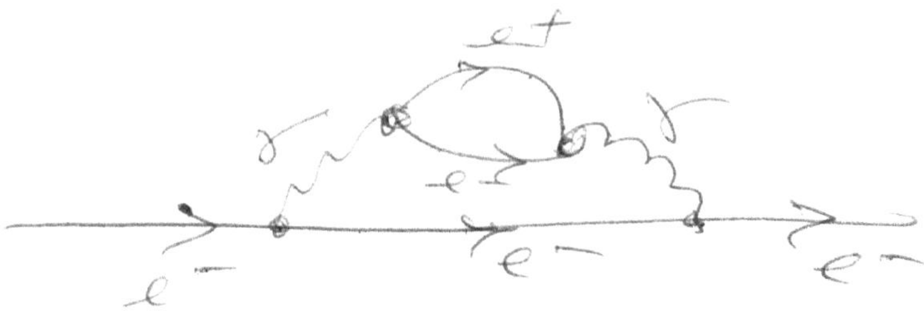

All these virtual effects obey the Logic of Nature.

"**Virtual History**" would produce a world totally different as illustrated in Figure 3. "What if?", i.e., what would our word be if in Sarajevo, the Austrian Archduke had not been assassinated. This totally unexpected event produced the 1st World War and the 2nd World War.

The world would not be as it is if Lenin had been killed during his travelling through Germany. And so on, as illustrated in Figure 3, "What if?"

Virtual History would be totally different if any or few UEEC events did not occur.

On the contrary, "**Virtual Physics**" is exactly as it should be since the phenomena studied obey the Logic of Nature.

Only if "destiny" existed, virtual history would be exactly like real history since the occurrence of a **UEEC** event would be dictated by destiny.

Let us go back to the two basic experimental evidences which characterize Complexity, namely the **AFB** phenomenology and the ***Sarajevo-type Events***. They are present in **History** as well as in **Science**, as reported in Figure 3.

'WHAT IF?'

	In History ≡ EWRL		In Science ≡ EBUS
I	What if Julius Caesar had been assassinated many years before?	*I*	What if Galileo Galilei had not discovered that $F = mg$?
II	What if Napoleon had not been born?	*II*	What if Newton had not discovered that $F = G\frac{m_1 \cdot m_2}{R_{12}^2}$?
III	What if America had been discovered a few centuries later?	*III*	What if Maxwell had not discovered the unification of electricity, magnetism and optical phenomena, which allowed him to conclude that light is a vibration of the EM field?
IV	What if Louis XVI had been able to win against the 'Storming of the Bastille'?	*IV*	What if Planck had not discovered that $h \neq 0$?
V	What if the 1908 Tunguska Comet had fallen somewhere in Europe instead of Tunguska in Siberia?	*V*	What if Lorentz had not discovered that space and time cannot both be real?
VI	What if the killer of the Austrian Archduke Francisco Ferdinand had been arrested the day before the Sarajevo event?	*VI*	What if Einstein had not discovered the existence of time-like and space-like real worlds? Only in the time-like world, simultaneity does not change, with changing observer.
VII	What if Lenin had been killed during his travelling through Germany?	*VII*	What if Rutherford had not discovered the nucleus?
VIII	What if Hitler had not been appointed Chancellor by the President of the Republic of Weimar Paul von Hindenburg?	*VIII*	What if Hess had not discovered cosmic rays?
IX	What if the first nuclear weapon had been built either by Japan before Pearl Harbour (1941) or by Hitler in 1942 or by Stalin in 1943?	*IX*	What if Dirac had not discovered his equation, which opens new horizons, including the existence of the antiworld?
X	What if Nazi Germany had defeated the Soviet Union?	*X*	What if Fermi had not discovered weak forces?
XI	What if Karol Wojtyla had not been elected Pope, thus becoming John Paul II?	*XI*	What if Fermi and Dirac had not discovered the Fermi–Dirac statistics?
XII	What if the USSR had not collapsed?	*XII*	What if the 'strange particles' had not been discovered in the Blackett Lab?

Fig. 3.

To prove that Complexity exists at all levels, including the elementary one, it is necessary to study:

1. The experimental evidence which allows us to identify the existence of Complexity;
2. The Standard Model including its extensions;
3. The Platonic Simplicity.

The result is that Complexity exists at the Fundamental Level and permeates all ways of life from quarks and gluons to molecular biology and life in general.

Let me go back to Figure 2.

The work of thousands of physicists, engineers and specialized technicians, in big Labs in Europe, USA and former Soviet Union, during five decades were needed to reach the knowledge which goes from the "Basic Quantities to the "Standard Model and Beyond".

Is this knowledge needed to improve the quality of our life**?** The answer is **yes**.

If it was not for our unprecedented knowledge of the **Basic Quantities** acquired, as mentioned above, during the last five decades in our Labs, my friend **David Scott**— the Commander of Apollo XV—would not have been able **to go to the Moon** and perform the Galilean experiment of the "feather" and the "hammer", which is the proof that "inertial" and "gravitational" masses are the same "mass".

Here is the **Commander of Apollo XV in the Moon** with the "**feather**" and the "**hammer**" (realized in the course of the NASA mission from the 26 July to 7 August 1971 on the 380th Anniversary of Galilean experiment).

This discovery was celebrated in China with millions of postcards printed in 1991, on the 400th Anniversary of the **Pisa-Tower experiment,** when Galilei discovered that $\mathbf{m_i = m_g}$.

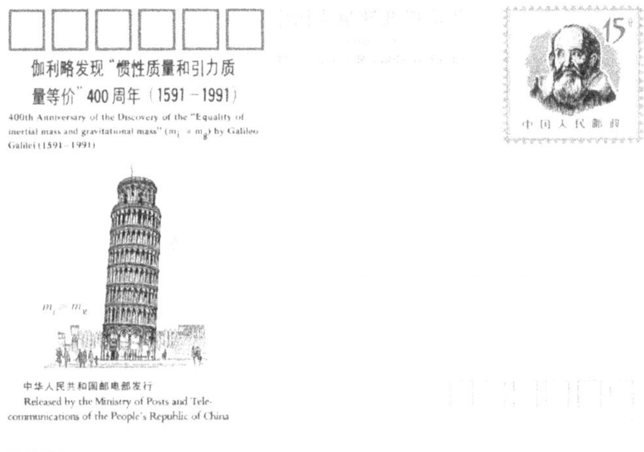

The celebration in China of the experiment that Galilei carried out by dropping two identical balls from the Leaning Tower of Pisa, one made of stone and the other of lead. Galilei discovered that they fell in exactly the same way.

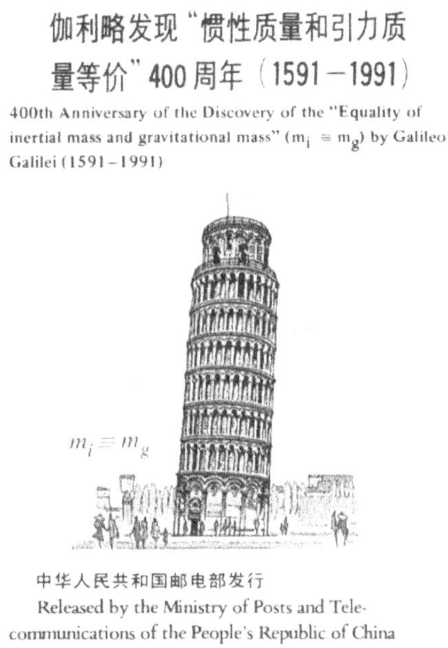

China celebrated a great Galilean discovery.

The Chinese government paid tribute to Galilei by printing postcards to commemorate his discovery regarding the falling feather and hammer, namely the equality between inertial mass (m_i) and gravitational mass (m_g).

Erice 2001—Celebration of the 30th Anniversary of David Scott Galilean experiment in the Moon.

David Scott and Antonino Zichichi (Erice 2001).

If we now go from Space technology to Medical technology, the **Story repeats itself**: if the human expectation for life is over 80 years, the reason is in our understanding of the **Basic Quantities**. Let me try to summarize in a few slides our present knowledge of the **Basic Quantities**.

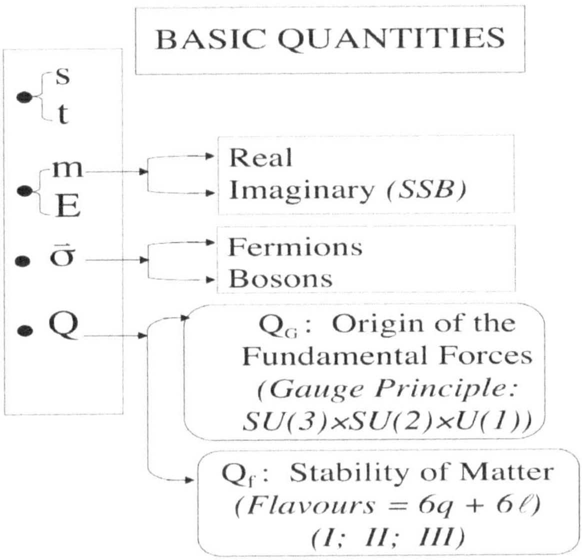
BASIC QUANTITIES
s
t
m
E
σ⃗
Q
Real
Imaginary (SSB)
Fermions
Bosons
Q_G : Origin of the Fundamental Forces
(Gauge Principle: SU(3)×SU(2)×U(1))
Q_f : Stability of Matter
(Flavours = 6q + 6ℓ)
(I; II; III)

Fig. 4.

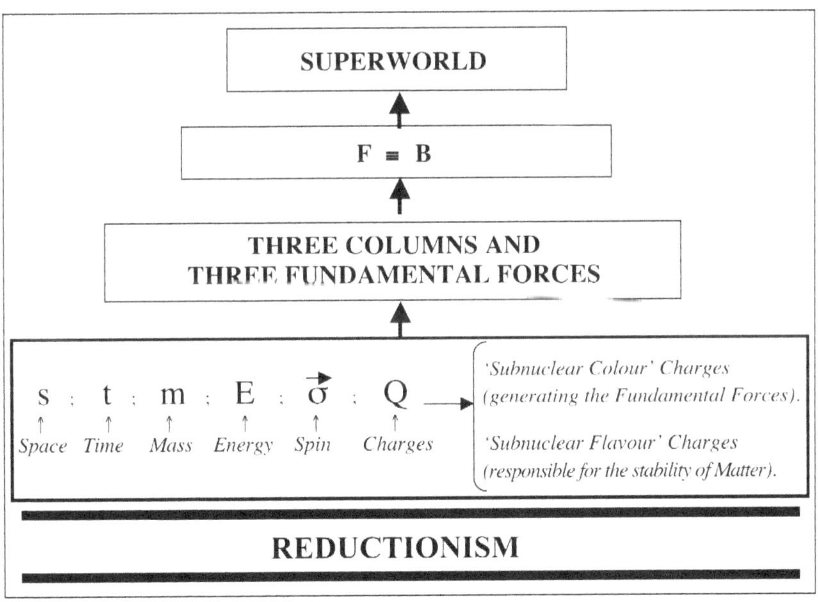
SUPERWORLD
F ≡ B
THREE COLUMNS AND
THREE FUNDAMENTAL FORCES
s : t : m : E : σ⃗ : Q
Space Time Mass Energy Spin Charges
'Subnuclear Colour' Charges
(generating the Fundamental Forces).
'Subnuclear Flavour' Charges
(responsible for the stability of Matter).
REDUCTIONISM

Fig. 5.

- **Three Columns** = *Families of quarks and leptons*
- **Three Forces** = *Electroweak, Subnuclear Strong,*
- **43** *Space-Times Dimensions*

The Three Columns

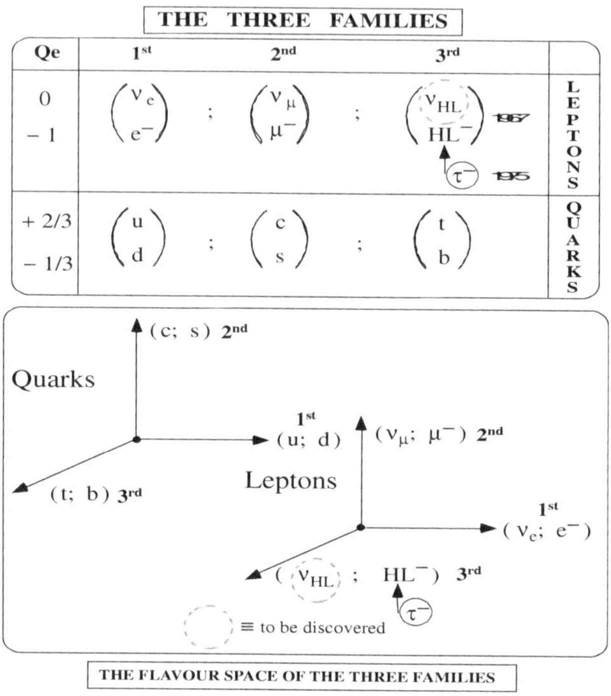

Fig. 6.

The upper panel of Figure 6 shows the three families of quarks and leptons, together with the respective values of their electric charge, Q_e.

The lower panel shows the spaces of the so-called "subnuclear flavour" charges (with additively conserved quantum numbers).

Notice that the third family lepton τ^- was originally (1967) named HL^- by the BCF (Bologna-CERN-Frascati) Group.

Its partner ν_{HL}, the existence of which was postulated in 1967 in the experimental proposal by the BCF group to search for heavy leptons at Frascati, would have been directly observed only recently.

The Three Fundamental Forces

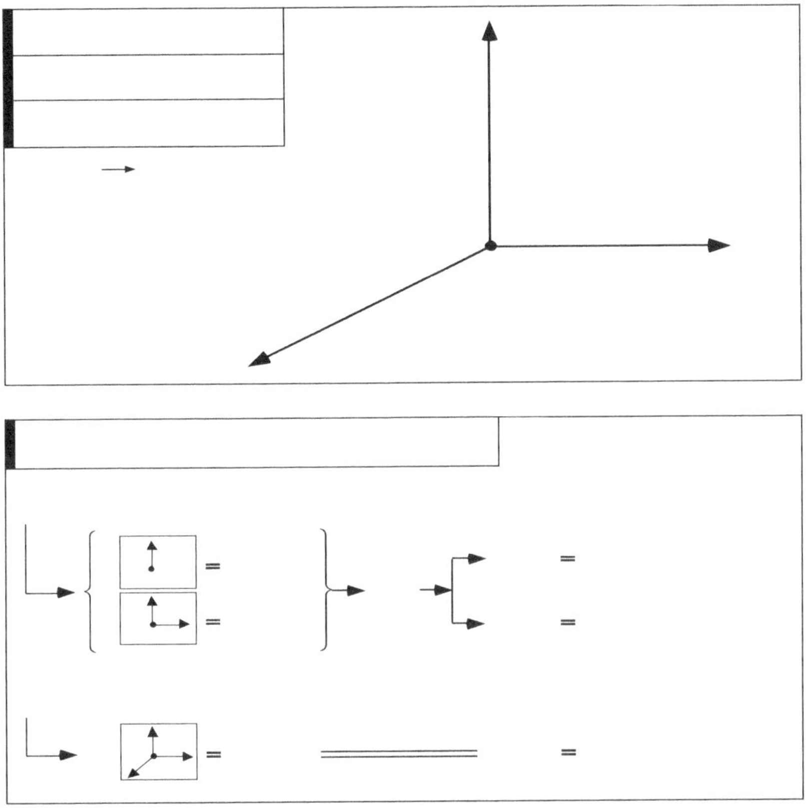

Fig. 7.

The upper panel of Figure 7 shows a standard Lorentz Space-Time to illustrate how gravitational Forces originate from the fact that, in any point $(i\bar{x}, t)$ of such a Space, a change of reference system is possible without altering any physical result.

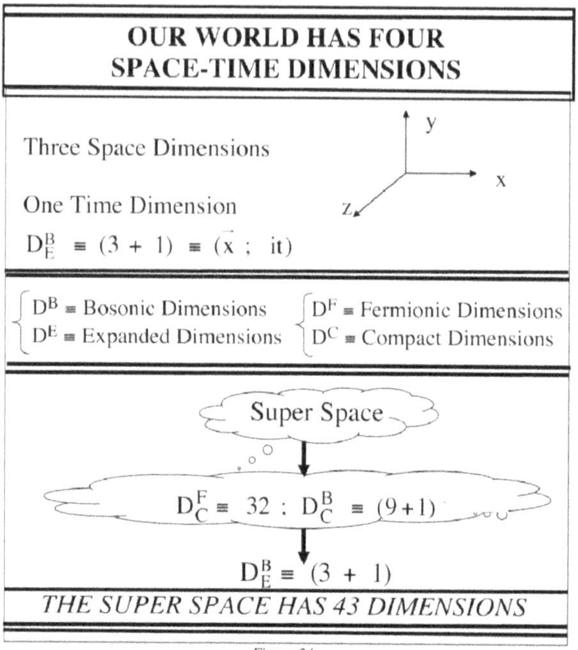

Figure 34

Fig. 8.

In the lower panel, the origin of the other forces, the [SU(2) × U(1)] electroweak one and the [SU(3)] strong one, is sketched.

These forces derive from the possibility to operate in fictitious Spaces with **one** [U(1)], **two** [SU(2)] and **three** [SU(3)] **complex dimensions**, under the condition of invariance of the physics results. Why do columns have half-integer spin and why do forces have integer spin? Can columns with integer spin and forces with half-integer spin exist?

<u>Why do we need the Superworld?</u>

There are fundamental reasons making the Superworld a need.

1. The two energy scales must be kept separate: 10^{19} GeV (Planck) and 10^2 GeV (Fermi).
2. The gravitational attraction of light must be prevented from being infinite. Otherwise we could see neither the light from Stars, nor the light of our Sun. The "gravitino" (Supergravity: F ≡) allows the gravitational attraction of light to be finite. For this to happen the Supersymmetry Law (F ≡ B) must exist.

3. Gravitational attraction is powerful but it cannot be infinite. We would be stuck to the Sun. Space would not exist between Stars and Galaxies. Cosmic expansion would not exist. In order to have a finite gravitational attraction, theories are needed in which the Euclidean concept of point is abandoned. The point is replaced by a string. No more Pointlike Theories but String Theories. These theories must be supersymmetric: the already quoted Supersymmetry Law (F ≡ B) must be valid. Otherwise "tachions" would appear. This is the origin of Relativistic Quantum String Theory (RQST).
4. If we want the Unification of all fundamental phenomena—the synthesis of which is provided by three "gauge couplings", $\alpha_1 \alpha_2 \alpha_3$, running with the energy (Renormalization Group Equations)—the Supersymmetry Law (F ≡ B) must necessarily be introduced.
5. An interesting detail: no-Scale-Supergravity is an Infrared solution of RQST. This might allow us to understand the extremely small value of the Cosmological Constant.
6. Finally: why Three Columns and Three Forces? The answer to this question should come once we will be able from the compactification of the 43-dimensional Super Space to go to our present world with (3+1) Space-Time dimensions.
7. Problem. Supersymmetry does not show up at our energy scale. Hence the problem arises to compute the energy above which the (F ≡ B) Law starts to act. Thanks to the EGM effect, this energy level is 700 times more accessible than thought so far.

Where is the SUPERWORLD?

Where could be the proof that we come from the Superworld? Our Galaxy rotates too fast! The ashes of the Superworld (the so-called neutralinos) could explain **THE COMPACTNESS OF OUR GALAXY** (and of all Galaxies in the Universe).

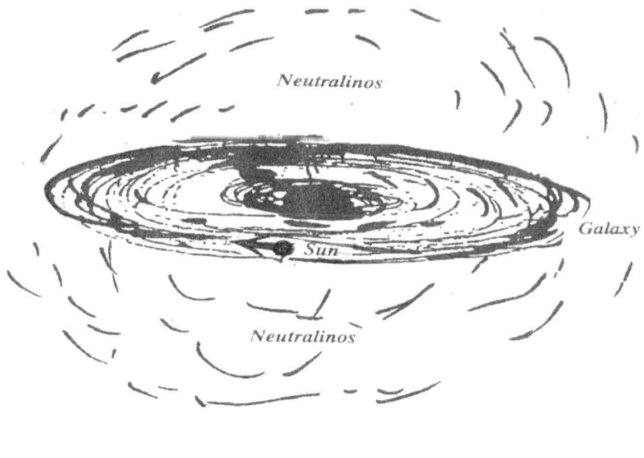

Fig. 9.

Neutralinos cannot aggregate into Stars since, being neutral, they lose little energy. This would allow neutralinos to remain in a sphere concentric with our Galactic centre.

Even if they aggregated into Stars, neutralinos would not emit light, like ordinary Stars do.

Fire needs the plasma of protons and electrons. This is why super Stars cannot exist. **The neutralinos** $\tilde{\chi}_1^0$ belong to what we call the **Standard Model & Beyond (SM&B).** The **SM&B** is the **greatest synthesis** of **all times** in the study of **NATURE** and of its **LOGIC.**

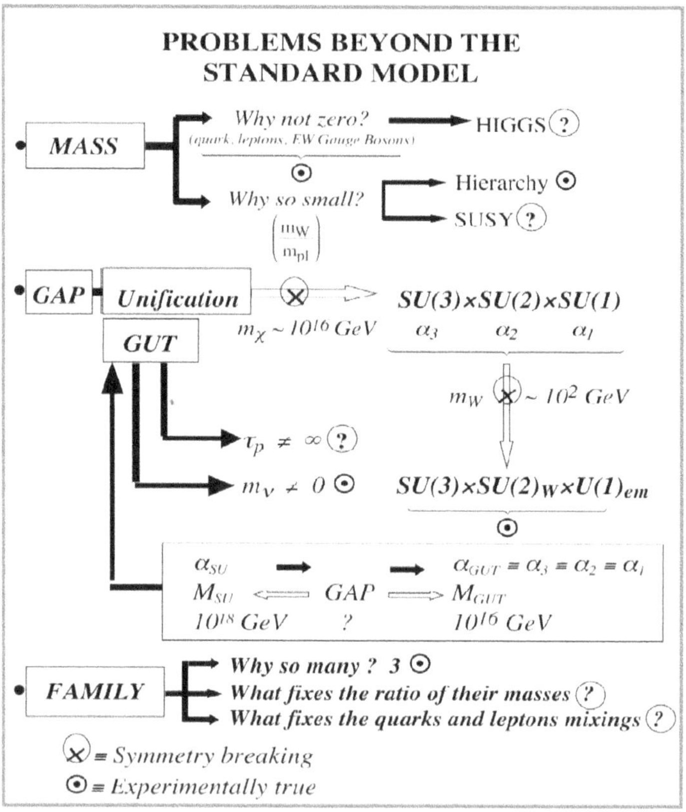

Figure 16

Fig. 10.

The dots indicate the problems whose nature is experimentally checked. The question marks are open problems. Let us see how we got it.

The Standard Model And Beyond

← **RGEs** (α_i $(i \equiv 1, 2, 3)$; m_j $(j \equiv q, l, G, H)$): $f(k^2)$.
  λ GUT ($\alpha_{GUT} \cong 1/24$) & GAP (10^{16} 10^{18}) GeV.
  λ SUSY (to stabilize $m_F/m_P \cong 10^{17}$).
λ RQST (to quantize Gravity).

↑ **Gauge Principle (hidden and expanded dimensions).**
— How a Fundamental Force is generated: SU(3); SU(2) × U(1) and Gravity.

→ **The Physics of Imaginary Masses: SSB.**
— The Imaginary Mass in SU(2) × U(1) produces masses ($m_{W^\pm}$; m_{Z^0} ; m_q; m_l), including $m_\gamma = 0$.
— The Imaginary Mass in SU(5) ⇒ SU(3) × SU(2) × U(1) or in any higher Symmetry Group (not containing U(1)) ⇒ SU(3) × SU(2) × U(1) produces Monopoles.
— The Imaginary Mass in $SU(3)_c$ generates Confinement.

↓ **Flavour Mixings & C ≠ , P ≠ , CP ≠ , T ≠ .**
— No need for it but it is there.

° **Anomalies & Instantons.**
— Basic Features of all non-Abelian Forces.

Note:
q ≡ quark and squark;
l ≡ lepton and slepton;
G ≡ Gauge boson and Gaugino;
H ≡ Higgs and Shiggs;
RGEs ≡ Renormalization Group Equations;
GUT ≡ Grand Unified Theory;
SUSY ≡ Supersymmetry;
RQST ≡ Relativistic Quantum String Theory;
SSB ≡ Spontaneous Symmetry Breaking.
m_F ≡ Fermi mass scale;
m_P ≡ Planck mass scale;
k ≡ quadrimomentum;
C ≡ Charge Conjugation;
P ≡ Parity;
T ≡ Time Reversal;
≠ ≡ Breakdown of Symmetry Operators

Fig. 11: The five basic steps in our understanding of nature.

← The renormalization group equations (RGEs) imply that the gauge couplings (α_i) and the masses (m_j) all run with k^2. It is this running which allows GUT, suggests SUSY and produces the need for a non-point-like description (RQST) of physics processes, thus opening the way to quantize gravity.

↑ All forces originate in the same way: the gauge principle.

→ Imaginary masses play a central role in describing nature.

↓ The mass-eigenstates are mixed when the Fermi forces come in.

° The abelian force QED has lost its role of being the guide for all fundamental forces. The non-abelian gauge forces dominate and have features which are not present in QED.

We have reached this extremely "simple" understanding of the fundamental natural phenomena via an incredible series of ABF (**Anderson-Feynman-Beethoven**) and UEEC (**Sarajevo-type**) **events**.

It turns out that (**AFB**) phenomenology and **UEEC** events exist in the Real World, no matter the Space-Time-Energy Scale considered.

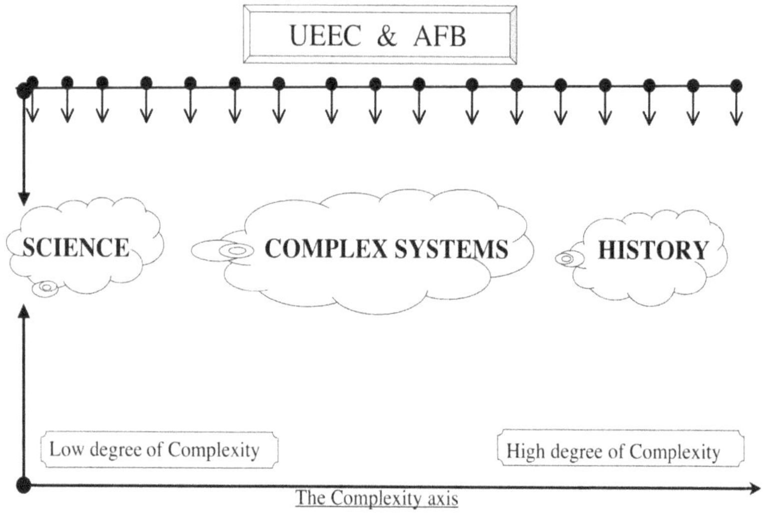

Fig. 12.

We do not know what will be discovered in the near future. What we do know is that all these attempts to understand the Logic of Nature appear to be very "*simple*" when compared with the "*Complexity*" of the world we are part of.

We show in Figure 13 the various components of our knowledge which make up the apparently very different domains of our world.

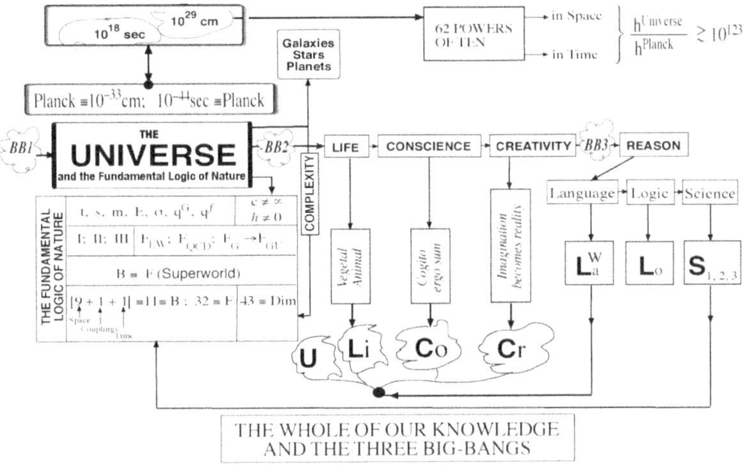

Fig. 13.

Latest News: Two Schools of Thought about Complexity

- Complexity & Big Numbers
- Complexity & Controllability

A Remark on Big Numbers: Is Big Numbers the way to understand the origin of Complexity?

Experimentally known **Big Numbers**.

Big Numbers come from the ratio of Space, Time, Mass-Energy and Action, as given below:

$$\boxed{\text{Space}} \equiv\rangle \quad \frac{S^{Universe}}{Planck\ \ Length} \cong 10^{62}$$

$$\boxed{\text{Time}} \equiv\rangle \quad \frac{\tau_p}{Planck\ \ Time} = \frac{10^{33} \cdot \pi \times 10^{7}\ \text{sec}}{10^{-43}\ \text{sec}} =$$

$$= \frac{10^{40}\ \text{sec}}{10^{-43}\ \text{sec}} \cong 10^{83}$$

$$\boxed{\text{Mass-Energy}} \equiv\rangle \quad \frac{M^{Universe}}{M_{\nu}} \cong \frac{10^{80}\, GeV \cdot 10^{2}}{10^{-4}\, eV} =$$

$$= \frac{10^{91}\, eV}{10^{-4}\, eV} = 10^{95}$$

$$\boxed{\text{Action}} \equiv\rangle \quad \frac{h^{Universe}}{h_{Planck}} = 10^{123}$$

Fig. 14.

With these Big Numbers compete those coming from the Mathematical Model of our Brain. According to some models, in order to produce new ideas the number of possible combinations is much larger than

$$\mathbf{10^{123}}.$$

[Donald GLASER (Bubble Chamber)]

Complexity and Controllability

Selected Proposal for the Totally Unexpected Discovery (TUD)

Proposal: a competition worldwide where young fellows start to think about these problems.

Why? Because the Culture of our Time, called modern, is in fact pre-Aristotelian: neither **rigorous theoretical logic** (**Math**) nor **rigorous experimental logic** (**Science**) have the role they should have.

Let me remind you that **the funds needed** to study the **Planetary Emergencies** compete with other projects like **man going to Mars**.

In order to allow decision making leaders to focus their attention on Planetary Emergencies, we need to convince them that **the Science of Complexity is in our hands**.

This is the reason why the World Federation of Scientists through the National Scholarship Programmes has established a competition for all young fellows to think about TUD.

A Totally Unexpected Discovery (TUD)

TUD needs a motivation; of course the motivation must be outside all possible expectations.

We have a competition; in all fields of Science, with a prize of 3.000 CHF for each TUD

WORLD FEDERATION OF SCIENTISTS

THE PRESIDENT

Professor Valery P. Kukhar
Institute for Bioorganic Chemistry and Petrochemistry
Laboratory of Fine Organic Synthesis
Kiev, Ukraine

Geneva, 20 June 2011

The President Office: CERN, CH-1211 GENEVE 23, Switzerland
Tel 0041 22.7679958 — e-mail: claude.manoli@federationofscientists.org
Tel 0039.0923.869133 — e-mail: fiorella.ruggiu@ccsem.infn.it
Tel 0041.22.7677755 — e-mail: esther.laperriere@cern.ch

Dear Professor Kukhar,

The three-year celebration of the 50th Anniversary the Ettore Majorana Foundation and Centre for Scientific Culture starts in 2011.

One of the highlights will be the remittal, during the August 2011 International Seminar, of a 3'000 Swiss Francs prize to a certain number of highly distinguished fellows from the WFS National Scholarship Programmes (NSPs). Each NSP representative is kindly requested to designate, out of all the fellows who undertook the NSP over the years, one he considers to be really outstanding. During the Seminar, the prize will be remitted to the NSP National Representative, who will then forward it to the designated fellow.

As the WFS Representative of your WFS National Scholarship Programme, if you feel that one of the beneficiaries of the Programme, past or present, has really proved to have an outstanding level, I would appreciate your sending me his name along with the motivation (not exceeding one page) for his excellence.

With best regards.

Professor Dr. Antonino Zichichi
President

Main centres:
ESTONIA – EE0010 TALLIN – 10 Ravala Boulevard – Tel +372/2/441304 – Fax: +372/2/440640 – E-mail: rebanek@park.tartu.ee
GEORGIA – 380008 TBILISI – 52 Rustaveli Avenue – Tel. +995/32/998891 – Fax: +995/32/998823 – E-mail: lado@acad.acnet.ge
LITHUANIA – 232600 VILNIUS – 3 Gedimino Avenue – Tel: +3/702/613651 – Fax: +3/702/618464 – E-mail: smtc.pllr@wllb.lt
MEDITERRANEAN area – FCCSEM, 191016 ERICE – 26 Via Guarnotta – Tel: +39/0923/869133 – Fax: +39/0923/869226 – E-mail: fiorella.ruggiu@ccsem.infn.it
PEOPLE'S REPUBLIC OF CHINA – BEIJING – 52 Sanlihe Road – Tel: +86/1/8597701 – Fax: +86/1/8511095
RUSSIA – 123182 MOSCOW – Ulitsa Kurchatova 46 – Tel: +7/095/1969241 – Fax: +7/095/9430074
UKRAINE – 252054 KIEV – 32-A Str. Turgenevskaya – Tel: +380/44/2167012 – Fax: +380/44/2167012
USA – Columbia University in the City of New York, NY 10027 NEW YORK – 538 West 120th Street – Tel: +1/212/8543339 – Fax: +1/212/9323169
CH-1211 GENEVA 23, Switzerland – Tel. +41/22/7677755 – Fax: +41/22/7850219 – E-mail: claude.manoli@federationofscientists.org
Sections in the following countries:
ALGERIA • ARGENTINA • BELARUS • BELGIUM • BULGARIA • BURKINA FASO • CANADA • CHILE • CHINA • COLUMBIA • CZECH REPUBLIC • DENMARK • EGYPT • ESTONIA • FINLAND • FRANCE • GERMANY • GREECE • HUNGARY • INDIA • ISRAEL • ITALY • JAPAN • JORDAN • KENYA • KOREA • LITHUANIA • NIGERIA • PAKISTAN • POLAND • PORTUGAL • REPUBLIC OF GEORGIA • ROMANIA • RUSSIA • SLOVENIA • SPAIN • SWEDEN • SWITZERLAND • THE NETHERLANDS • TUNISIA • TURKEY • UK • UKRAINE • USA • VIETNAM

Something totally unexpected. Something which no one has been able so far to imagine.

Conclusion of the First Point:

The Logic of Nature allows the existence of Science (the asymptotic limit of Simplicity) and of History (the asymptotic limit of Complexity). The present status of the Reductionistic achievements is the **Standard Model** and its extension which predicts GUT (the Grand Unification Theory), the existence of the Superworld and the resolution of the quantum-gravity problem via the powerful theoretical structure of RQST (Relativistic Quantum String Theory).

Since Nature shows structures which are considered as being complex, the general trend is that, in order to understand their roots, we must go from Reductionism to Holism whose frontier is Complexity. But "**Complexity**" is ill-defined; nevertheless people speak of "Complexity" as a source of new insights in Physics, Biology, Geology, Cosmology, Social Sciences and in all intellectual activities which look at the world through the lens of a standard analysis in terms of either Simplicity or Complexity.
We have seen that the necessity to abandon the Reductionism in favour of the Holistic approach is the existence of phenomena whose Laws and Regularities have apparently no links with the Fundamental Laws of Nature from which they originate: **AFB** (**A**nderson-**F**eynman-**B**eethoven) phenomena.

The other reason to introduce the Holistic approach, and therefore Complexity, is the appearance of Unexpected Events, apparently marginal, but producing Enormous Consequences: the **UEEC** events.

These two basic elements, **AFB** and **UEEC**—which are at the origin of Complexity with its Holistic consequences permeating all our existence, from molecular biology to life in all its innumerable forms up to our own, including History—**do exist at the Fundamental Level**.

The fundamental proof comes from the sequence of **AFB** and **UEEC** (totally unexpected discoveries) which brought us to the Standard Model and its extension (**SM&B**), which is the most spectacular synthesis of Rigorous Science to understand the **Logic of Nature**.

The final Conclusion of this First Point is that, if we want to understand Complexity, the only way is to continue with the Reductionistic analysis of the phenomena, no matter the Mass-Energy and Space-Time scales involved.

WHERE ARE WE GOING NEXT

Why?

Because we want an "Effective Action" to be implemented.

A project to study the Planetary Emergencies needs through rigorous Galilean reductionism.

Project for a Centre where the Planetary Emergencies are studied: **CSEW** (Centre for the Study of Emergencies Worldwide).

1. **Nowhere in the world can one find a Centre dedicated to the Study of Emergencies in the World (CSEW),** whose main objective is aimed at the solution of the problems that are the source of **Planetary Emergencies**, making it thus possible to provide to the political powers the means to increase the level of the quality of life, not only in it's components and material structures but also taking into consideration the importance of it's values in everyday's life. To attain such objectives, the three components most important to the development of a civil society, in step with its great achievements, must be reunited: Science, Technology and Culture.
2. **Scientific Research** and its applications (better known as Technological Research) must be focalised on a common priority goal: increase the level of

the quality of life, whether to resolve the daily problems with their multiple complex requirements or to give to all the capacity—thanks to Culture—to choose the supreme values of our existence. Culture has the obligation to remind us always that we are the only form of living matter endowed with reason.

3. The **CSEW "Project" has also as an objective to confront, study and resolve the problems inherent to the actual system of world equilibrium between nations**, where the various multipolar components that had emerged, once the danger of the Nuclear Holocaust from the USA-USSR confrontation was overcome, continue to prevail with the main motivation of confronting the enemy and the resulting arms race, as if the Wall which was felled down in Berlin had been reconstructed in each nation of the multipolar system.
4. The **CSEW Project** would be pure utopia unless one could rely on the results obtained through the existing structures involved in scientific, technological and cultural activities, which have demonstrated their effective potential of innovation to promote new activities related to the common priority goal of increasing the level of the quality of life.

<u>These structures do exist</u>

They are:

- **The World Federation of Scientists (WFS)**, the largest association of scientists based on "scientific volunteering". The WFS has studied the Planetary Emergencies and classified them into 15 groups for a total of 63 emergencies. The WFS has implemented dozens of pilot projects in dozens of countries throughout the world, to prove to the governments of this world that, if there is a political will, it is possible to fight and overcome the Planetary Emergencies.
- **The "Ettore Majorana" Centre for Scientific Culture in Erice**, which succeeded in giving life to a new spirit of international scientific collaboration, for a "Science without secrets and without frontiers", involving in it's activities in Erice, over the last nearly half century, 109'250 scientists from 932 Laboratories and Universities and 140 countries. The Centre has had, since its foundation, three fundamental pillars: scientific excellence, meritocracy and the scientific volunteering. All with a minimum of bureaucracy.
- **CERN in Geneva**, the largest laboratory in the world of a Galilean inspiration. CERN represents the scientific unification of Europe – based on the political actions of great European Physicists – and is an example of a structure based on meritocracy and scientific excellence, with a minimum of bureaucracy.
- **The new "Enrico Fermi" Centre for Studies and Research**, which should act as the originating and co-ordinating centre for scientific, technological and cultural activities of the **CSEW "Project"**.

Let me close with an overview of all Planetary Emergencies.

THE WORLD FEDERATION OF SCIENTISTS NATIONAL SCHOLARSHIP PROGRAMME

BEST FELLOW PRIZE—2011

To Celebrate the 50th Anniversary of the Ettore Majorana Foundation and Centre for Scientific Culture

COUNTRY	**LAUREATE**	**REPRESENTATIVE**
Azerbaijan	SEVDA RZAEVA	Dr. Adalat Hasanov
China	LIANGJIANG WEN	Prof. Tsung Dao Lee
Georgia	IRAKLI TITVINIDZE	Prof. Merab Eliashvili
Iran	KATAYOUN DOROUD	Dr. Crispin Williams
Jordan	ABEER LABABNEH	Dr. Muheeb Awawdeh
Kenya	JOEL L. BARGUL	Dr. Daniel Masiga
Kyrgyzstan	YULIA LOMTEVA	Dr. Askar Kutanov
Romania	CIPRIAN CHIRILA	Prof. G. Stratan
Senegal	FATOU NDOYE	Dr. M'Barek Diop
Slovenia	URBAN BREN	Dr. Edward Kobal
Tajikistan	RABAJOVA LUTFYA	Prof. Nusrat Rajabov
Ukraine	ANDREY SIZOV	Prof. Valery Kukhar
Vietnam	PHAM NGOC DIEP	Prof. Pierre Darriulat

For more details see Appendices and References

APPENDIX I: SEVEN DEFINITIONS OF COMPLEXITY
APPENDIX II: AFB PHENOMENA FROM BEETHOVEN TO THE SUPERWORLD
APPENDIX III: EXAMPLES OF UEEC EVENTS IN THE CONSTRUCTION OF THE STANDARD MODEL AND BEYOND: A PERSONAL EXPERIENCE
APPENDIX IV: A FEW WORDS ON HOW IS THE WORLD MADE AND THE SUPERWORLD
APPENDIX V: THE PLATONIC GRAND UNIFICATION
APPENDIX VI: THE PLATONIC SUPERSYMMETRY
APPENDIX VII: THE TIME-SEQUENCE OF LANGUAGE—LOGIC—SCIENCE AND THE PRESENT STATUS OF SCIENCE
APPENDIX VIII: JOHN PAUL II AND ENRICO FERMI

1. *Complexity at the Fundamental Level*
   A. Zichichi, presented at the 42nd Course of the International School of Subnuclear Physics, Erice, 29 Aug–7 Sept 2004, Vol. 42, p. 251, World Scientific (2007).
2. *Rigorous Logic in the Theory of Evolution*
   A. Zichichi, Plenary Session on "*Scientific Insights into the Evolution of the Universe and of Life*", 31 October–4 November 2008, Vatican City, The Pontifical Academy of Sciences—in Proceedings of the Plenary Sessions, Acta 20, pages 101-178, Vatican City (2009).

APPENDIX I
SEVEN DEFINITIONS OF COMPLEXITY

People speak of "Complexity" as a source of new insights in physics, biology, geology, cosmology, social sciences, evolution of the human species and in all intellectual activities which look at the world through the lens of a standard analysis in terms of either Simplicity or Complexity. **But "Complexity"** is ill-defined, as shown by the existence of at least seven definitions of Complexity.

DEFINITION NUMBER 1

Complexity is a property of systems that are somewhere in between a completely random and a completely regular state, often described by a highly non linear set of equations but sometimes not describable by equations at all.

DEFINITION NUMBER 2

Bad ones:

1. Chaos.
2. The need for lengthy calculations.
3. The need for many distinct variables.

Better ones:

4. Unexpected difficulty when attempting to describe something in a precisely formulated theory.
5. What is left over after all systematic approaches failed.

But it could also be that: Complexity is an excuse for sloppy thinking.

DEFINITION NUMBER 3

The Complexity of a theory (problem) is the minimum amount of computer time and storage required to simulate (solve) it to a specified level of precision.

DEFINITION NUMBER 4

If we admit that biological or linguistic evolution, or financial dynamics are complex phenomena, then their typical dynamics are somehow between strong chaos (i.e., positive Lyapunov exponents) and simple orbits (i.e., negative Lyapunov exponents). In other words, Complexity (or at least some form of it) is deeply related to the edge of chaos (i.e., vanishing maximal Lyapunov exponent). Since the edge of chaos appears to be related paradigmatically to an entropy index "q" different from unity, there must be some deep connection between Complexity and generalized entropies such as "S_q".

DEFINITION NUMBER 5

From the mathematical point of view:

- A problem can be polynomial, which means that it is not to hard to predict surprises.

- A problem can be NP or NP-complete, which represent different degrees of difficulty in predicting surprises.
- Surprises means: UEEC event (see later).
- That degree of difficulty can be associated with the level of Complexity.

DEFINITION NUMBER 6

A system is "complex" when it is no longer useful to describe it in terms of its fundamental constituents.

DEFINITION NUMBER 7

The simplest definition of Complexity: "*Complexity is the opposite of Simplicity*". This is why we have studied the platonic Grand Unification (Appendix V) and the Platonic Supersymmetry (Appendix VI), in order to show that Nature does not follow Platonic Simplicity.

APPENDIX II
AFB PHENOMENA FROM BEETHOVEN TO THE SUPERWORLD

Beethoven and the laws of acoustics.

Beethoven could compose superb masterpieces of music without any knowledge of the laws governing acoustic phenomena.

But these masterpieces could not exist if the laws of acoustics were not there.

The living cell and QED.

To study the mechanisms governing a living cell, we do not need to know the laws of electromagnetic phenomena whose advanced formulation is QED.

All mechanisms needed for life are, to a great extent, examples of electromagnetic processes. If QED was not there, Life could not exist.

Nuclear physics and QCD.

Proton and neutron interactions appear as if a fundamental force of nature is at work: the nuclear force, with its rules and its regularities.

These interactions ignore that protons and neutrons are made with quarks and gluons.

Nuclear physics does not appear to care about the existence of Quantum ChromoDynamics (QCD), the fundamental force acting between quarks and gluons at the heart of the subnuclear world.

Nuclear physics ignores QCD but all phenomena occurring in nuclear physics have their roots in the interactions of quarks and gluons.

In other words, protons and neutrons behave like Beethoven: they interact and build up nuclear physics without "knowing" the laws governing QCD.

The world and the Superworld. The most recent example of an Anderson-Feynman-Beethoven-type phenomenology is **the World which could not care less about the existence of the Superworld**. The world needs only 4 dimensions (3 Space + 1 Time); the Superworld 43 dimensions.

APPENDIX III
EXAMPLES OF UEEC EVENTS IN THE CONSTRUCTION OF THE STANDARD MODEL AND BEYOND: A PERSONAL EXPERIENCE

There are many UEEC events in the construction of the Standard Model and Beyond (SM&B). In some of them I have been directly involved. They are summarized in Figure 3.1.

Each **UEEC** event (except the last one) is coupled with a **despite**, in order to emphasize the reason why the event is unexpected. The no. 7 event has only the unexpected details.

UEEC EVENTS IN THE CONSTRUCTION OF THE SM&B = MY PERSONAL EXPERIENCE

- **The 3rd lepton, HL** (now called τ) with its own neutrino, ν_{HL} (now called ν_τ),
  despite *the abundance of neutrinos:* ν_e *and* ν_μ

- **Antimatter**
  despite *S-matrix and C, P, CP, T breakings.*

- **Nucleon Time-like** EM structure
  despite *S-matrix*

- **No quarks** in violent (pp) collisions
  despite *scaling.*

- **Meson mixings**
  $\theta_V \neq \theta_{PS}$: $(51^o) \neq (10^o) \neq 0$ ***despite*** $SU(3)_{uds}$.

- **Effective energy:** the Gribov QCD-light
  despite *QCD.*

⑦ **The running** of $\alpha_1 \alpha_2 \alpha_3$ versus **energy**:
the EGM effect, the GAP between E_{GUT} *and* E_{SU}*, and the absence of the Platonic straight line convergence.*

Fig. 3.1.

APPENDIX IV
A FEW WORDS ON HOW IS THE WORLD MADE AND THE SUPERWORLD

A flower, the sea, the air we breathe, the Moon, the Stars, the Sun itself, whatever we call World is made of Fermions and Bosons.

- Fermions: (with 'half-integer' spin like the electron). These particles—called quarks and leptons—are the 'bricks', of our material existence (shown in Figure 4.1): more precisely, the tiny 'tops' with minimum amount of "intrinsic angular momentum": ½ $\hbar$.
- Bosons: (with 'integer' spin like the photon). Particles having integer 'spin' value are the 'glues' of the fundamental forces (shown in Figure 4.2), acting between 'bricks'. The 'glues' are also tiny 'tops'. The 'electroweak' and the subnuclear forces have the minimum value of integer spin: one in Planck unit (1 $\hbar$), the 'gravitational' glue has twice this integer value of the spin: (2 $\hbar$).

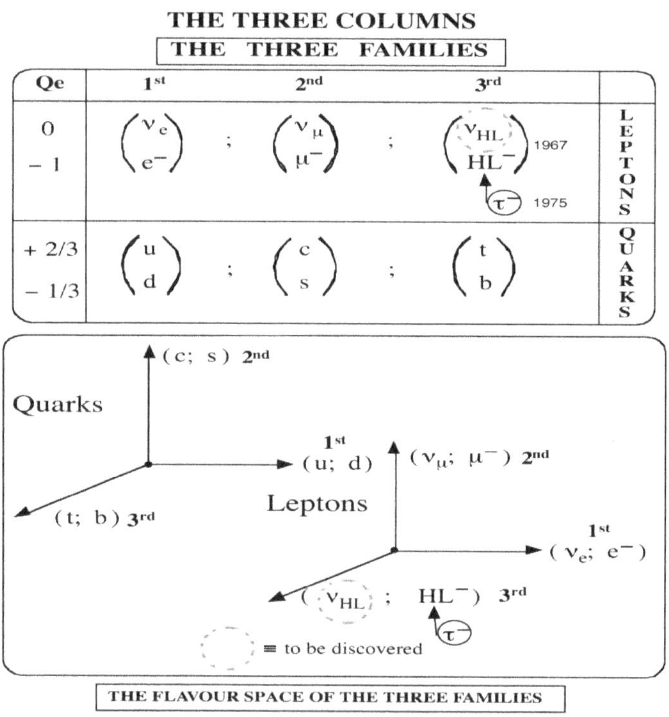

Fig. 4.1.

The upper panel of Figure 4.1 shows the three families of quarks and leptons, together with the respective values of their electric charge, Q_e.

The lower panel shows the spaces of the so-called 'subnuclear flavour' charges (with additively conserved quantum numbers).

Notice that the third family lepton τ^- was originally (1967) named HL^- by the BCF (Bologna-CERN-Frascati) Group [8–11]. Its partner ν_{HL}, the existence of which was postulated in 1967 in the experimental proposal by the BCF group to search for heavy leptons at Frascati, would have been directly observed only recently.

The upper panel of Figure 4.2 shows a standard Lorentz Space-Time to illustrate how gravitational Forces originate from the fact that, in any point $(i\vec{x}, t)$ of such a Space, a change of reference system is possible without altering any physical result. In the lower panel the origin of the other forces, the [SU(2) × U(1)] electroweak one and the [SU(3)] strong one, is sketched.

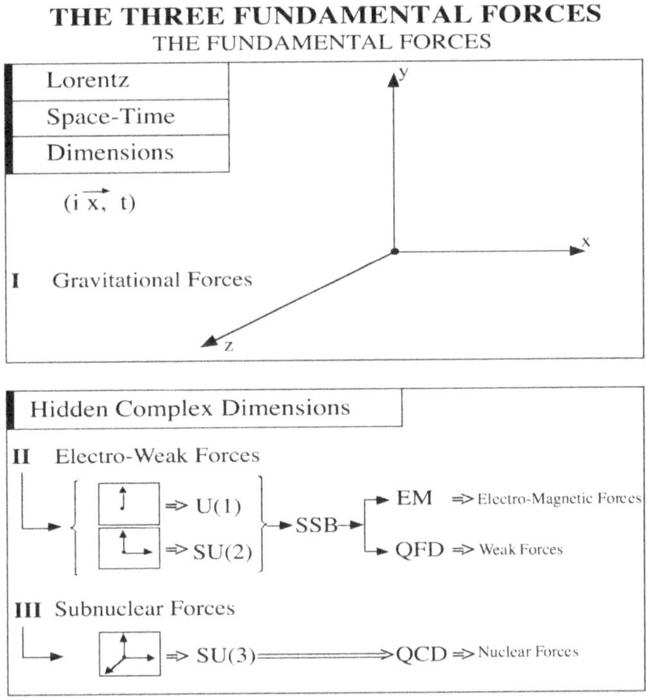

Fig. 4.2.

These forces originate from the possibility to operate in fictitious Spaces with one [U(1)], two [SU(2)] and three [SU(3)] complex dimensions, under the condition of local invariance of the physics results. This is the reason why the glues have integer spin.

Why do the columns have half-integer spin we do not know: it is an experimental unexpected discovery? Can columns with integer spin and forces with half-integer spin exist? The answer is Yes if we introduce supersymmetry whose foundation is superspace (Figure 4.3).

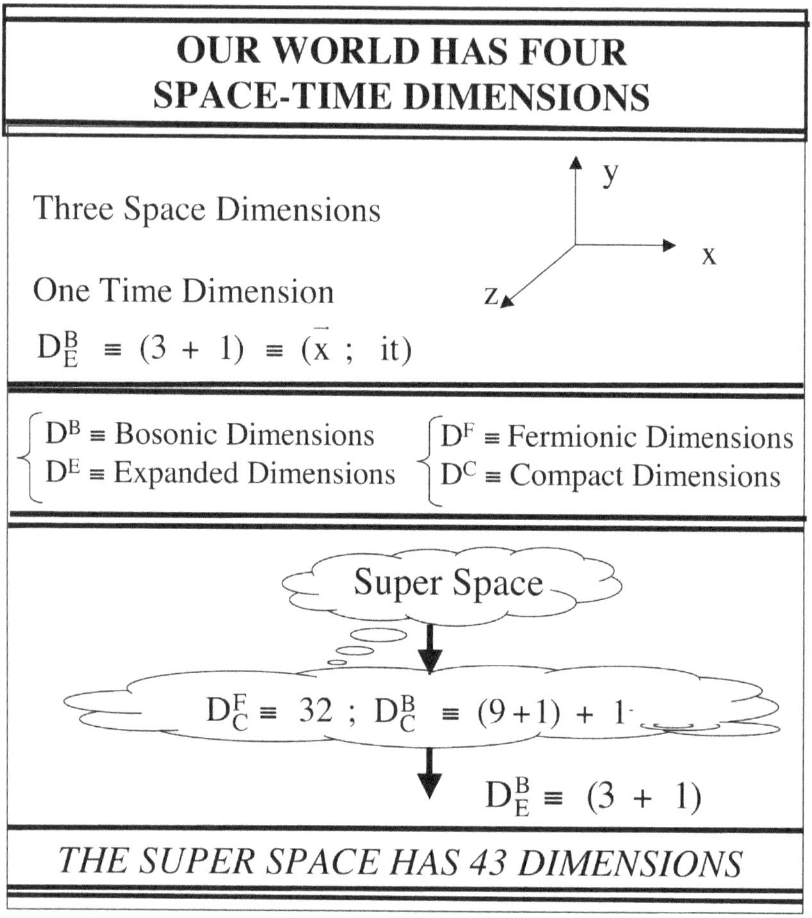

Fig. 4.3.

Our space-time has (3+1) bosonic dimensions; superspace has [(9+1) + 1] bosonic and 32 fermionic dimensions.

The (9+1) dimensions are 9 for Euclidian space and 1 for time, while the additional + 1 is the compactified dimension which produces the gravitational gauge coupling.

APPENDIX V
THE PLATONIC GRAND UNIFICATION

Figure 5.1 is the best example of Platonic Grand Unification. The points have a sequence of 100 GeV in energy. The last point where the "ideal" platonic straight line intercepts the theoretical prediction is at the energy of the Grand Unification. This corresponds to $E_{GU} = 10^{16.2}$ GeV. Other detailed information on the theoretical inputs: the number of fermionic families, N_F, is 3; the number of Higgs particles, N_H, is 2. The input values of the gauge couplings at the Z^0-mass is $\alpha_3\ (M_Z) = 0.118 \pm 0.008$; the other input is the ratio of weak and electromagnetic couplings also measured at the Z^0-mass value: $\sin^2\theta_W\ (M_Z) = 0.2334 \pm 0.0008$.

The Platonic Grand Unification should be along the straight line, small dots (blue), but Nature seems to follow the big dots (red).

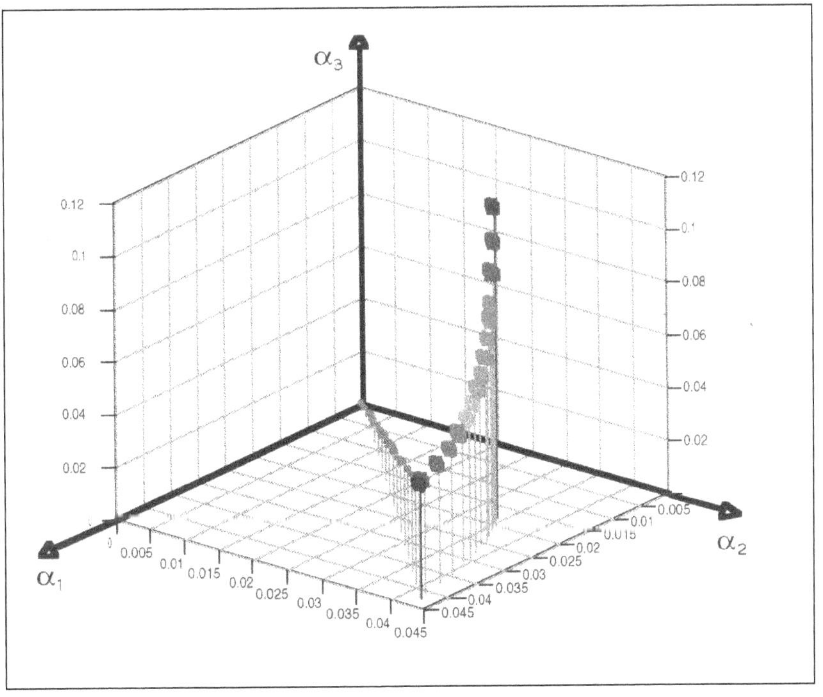

Fig. 5.1.

APPENDIX VI
THE PLATONIC SUPERSYMMETRY

THE PLATONIC CONCEPT OF SUPERSYMMETRY

The **Gauge Principle** should generate a
Gauge Force ⟶ **Gauge Bosons**
If *NATURE* was platonically *SUPERSYMMETRIC*
Supersymmetry Transformation should generate **Gauginos**

1st **DEVIATION FROM PLATONIC SIMPLICITY**
OUR FERMIONS ARE NOT THE GAUGINOS

2nd **DEVIATION FROM PLATONIC SIMPLICITY**
THE FUNDAMENTAL FERMIONS ARE OF TWO DIFFERENT CLASSES: LEPTONS AND QUARKS

3rd **DEVIATION FROM PLATONIC SIMPLICITY**
THERE IS NOT ONLY ONE BUT THREE FAMILIES OF FUNDAMENTAL FERMIONS

4th **DEVIATION FROM PLATONIC SIMPLICITY**
THE FUNDAMENTAL FERMIONS BECOME MIXED WHEN THE WEAK FORCES ARE SWITCHED ON: MIXINGS EXIST

5th **DEVIATION FROM PLATONIC SIMPLICITY**
THERE ARE DIFFERENT MIXINGS

Fig. 6.1.

APPENDIX VII

THE TIME-SEQUENCE OF LANGUAGE—LOGIC—SCIENCE AND THE PRESENT STATUS OF SCIENCE

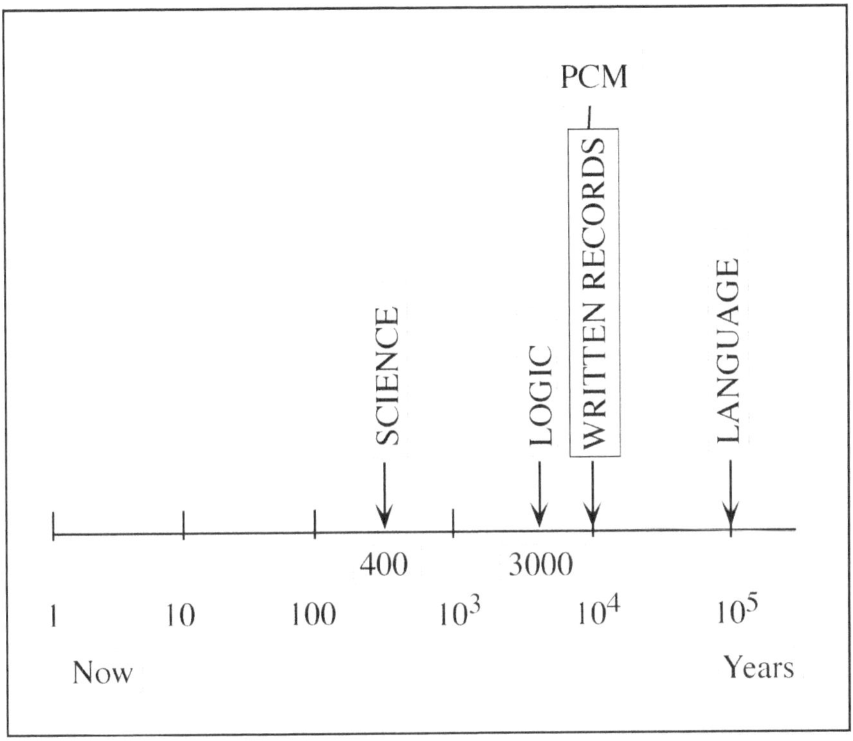

Fig. 7.1.

After 400 years of Reductionism, the present trend is to go from **Reductionism** to **Holism** and the Future could be dominated by **Science being NP-complete**.

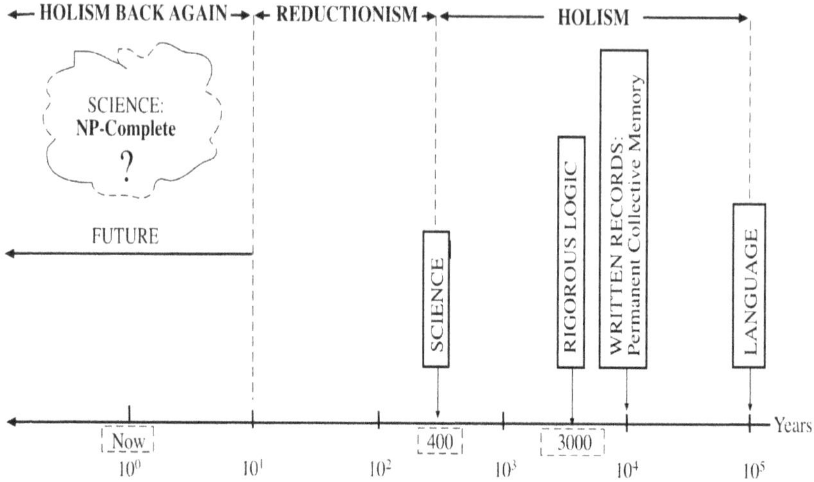

Fig. 7.2.

It turns out that Complexity in the real world exists, no matter the mass-energy and space-time scales considered [1–4], as illustrated in Figure 7.3.

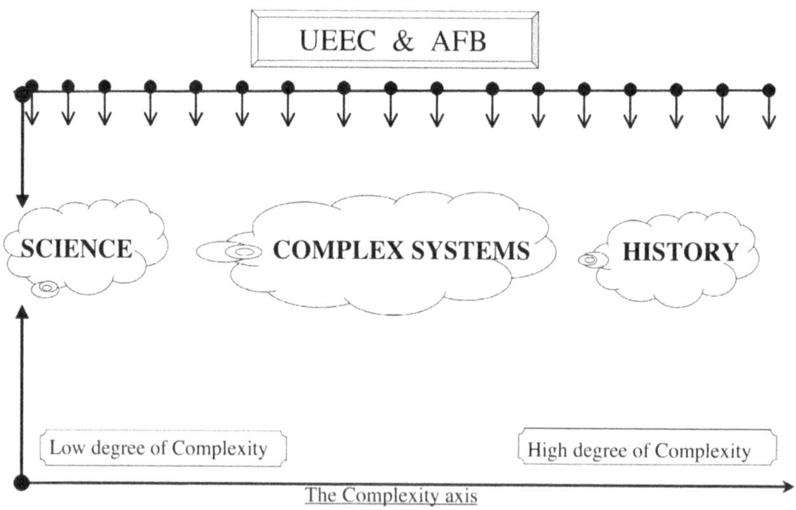

Fig. 7.3.

It is thanks to this field of Science that it has been possible to identify the Basic Quantities needed to build the world where we live, as shown in Figure 7.4.

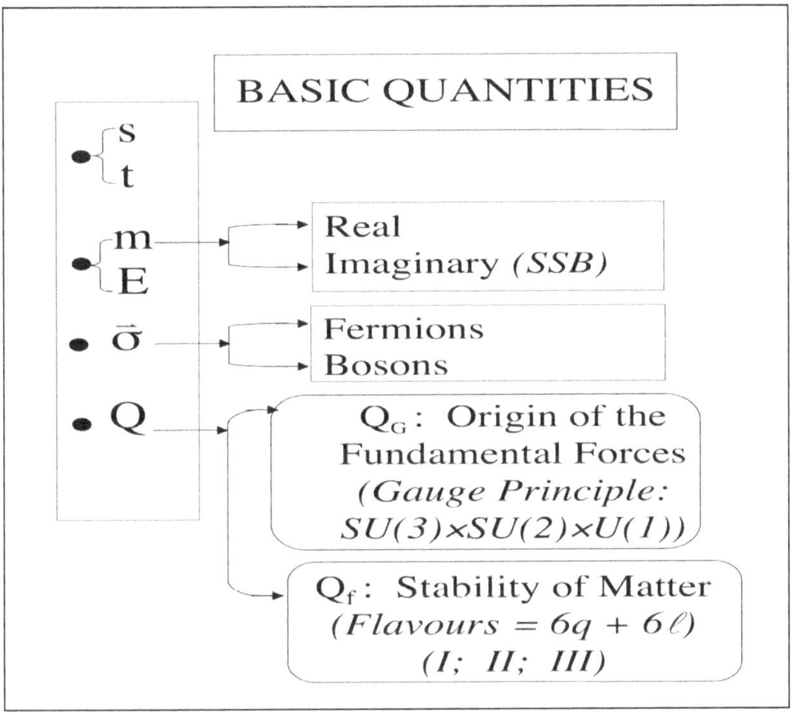

Fig. 7.4.

The conceptual understanding of the constituents of the Universe after 400 years of Reductionism are shown in Figure 7.5.

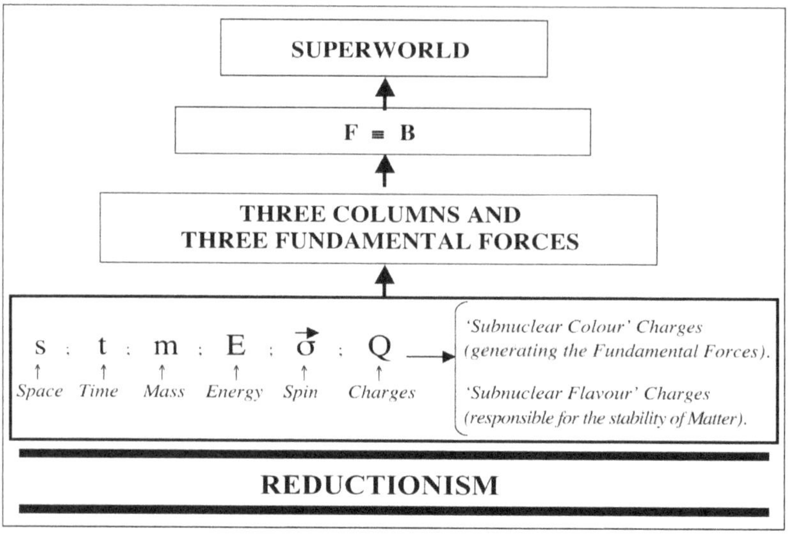

Fig. 7.5

The key point is that the vacuum we are looking for must have the properties shown in Figure 7.5.

APPENDIX VIII
JOHN PAUL II AND ENRICO FERMI

—∗—∗—∗—

The **Project** has, as its main reference points H.H. John Paul II who, for the **CSEW Project**, was the supreme moral and cultural guide, and Enrico Fermi, the greatest Galilean of the 20th Century, whose teachings in Italy and the world were the scientific and technological driving force of the twentieth century.

NUCLEAR POWER IN RUSSIA—25 YEARS AFTER CHERNOBYL

PROFESSOR EUGENI VELIKHOV
Russian Research Centre, "Kurchatov Institute"
Moscow, Russia

Experts from the Kurchatov Institute have taken part, since the day after the accident, in works on liquidation of its consequences ("Operative group"). In 1987, "Complex Expedition" was created at Kurchatov Institute. Within 1992, the Kurchatov Institute carried out a scientific management of department of nuclear and radiation safety in ISTC "Shelter" (Ukrainian Academy of Science).

<u>26 April 1986</u>

The world's worst nuclear accident in the history of atomic industry took place at the Chernobyl Nuclear Power Plant. As a result of the reactor explosion, its core was completely destroyed; the reactor building, de-aeration stack, turbine hall and some other constructions were damaged. Protection barriers and safety systems were destroyed.

<u>26 April to 06 May 1986—Active Stage of Accident</u>

Millions of curies of radioactive particles were released. From the first days after accident, specialists of the Kurchatov Institute took part in works on liquidation of its consequences.

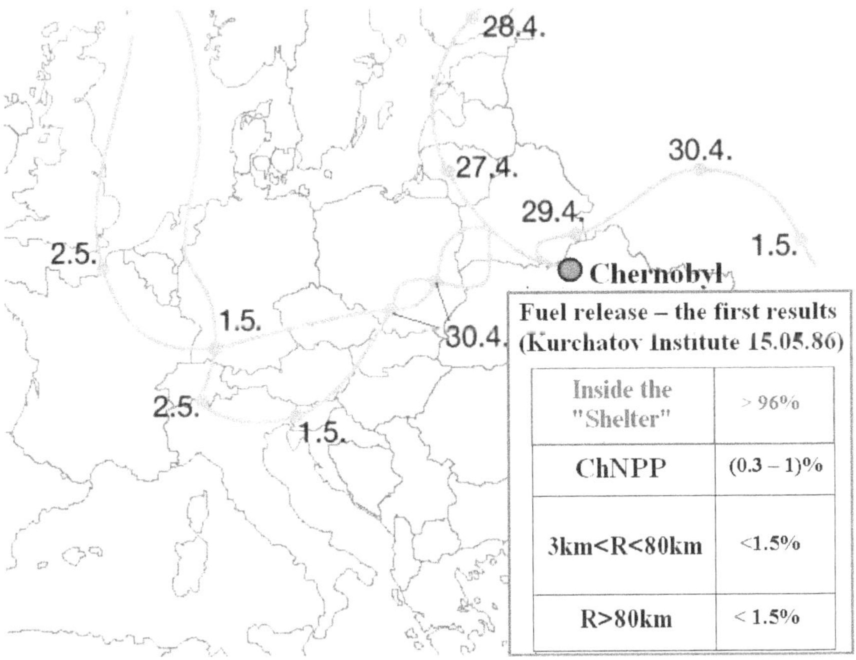

Fuel release – the first results (Kurchatov Institute 15.05.86)

Inside the "Shelter"	> 96%
ChNPP	(0.3 – 1)%
3km<R<80km	<1.5%
R>80km	< 1.5%

May to November 1986—Building of the Shelter

The Kurchatov Institute was the scientific supervisor for creation of the Shelter.

August to November 1986—The Program "BUOY"

Composed of:

1. Thermal flux detectors.
2. Thermometers.
3. Dose rate of gamma radiation detectors.
4. Anemometers.
5. Panels of communications.

The main conclusion: more than 90% of the fuel had remained inside the 4th reactor unit.

August – November. The program "BUOY"

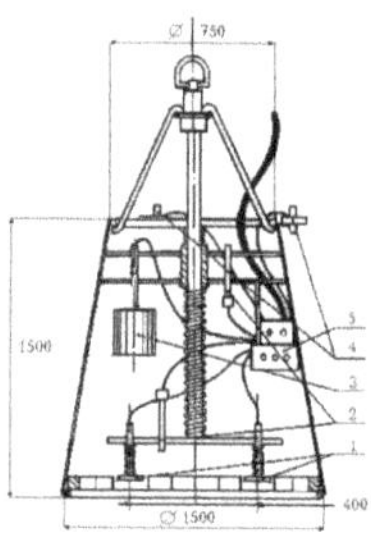

1. Detectors of a thermal flux.
2. Thermometers.
3. Detectors of a dose rate of gamma radiation.
4. Anemometers.
5. Panels of communications.

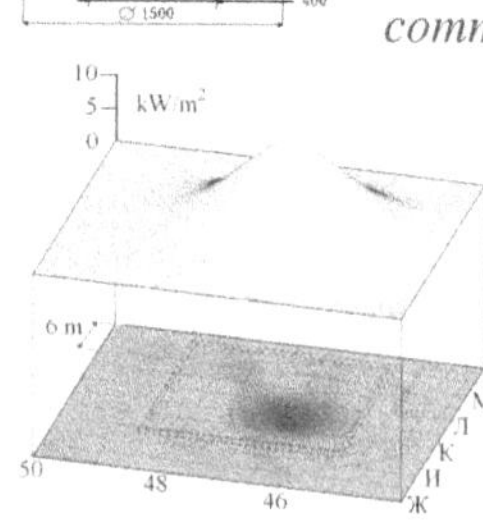

The main conclusion - more than 90 % of fuel remain in rooms of 4th Unit

30 November 1986—The Shelter is Completed.

1987 – 1992—The Kurchatov Institute's Complex Expedition

Highly radioactive fields, debris from destroyed construction material and concrete was poured into the building during the Shelter's creation and created barriers to the access to rooms close to the reactor.

It was necessary to estimate and reduce nuclear, radiation, and thermal hazards. For this purpose, new methods and equipment had to be developed which would allow the study of main fuel masses.

In 1987, the program of accessing inside Unit 4 was prepared by scientists and engineers of the Kurchatov Institute. In December 1987, for the implementation of this program, the Kurchatov Institute's Complex Expedition was created.

The Complex Expedition consisted of a comparatively small scientific division and a strong complement for procurement and implementation of building structures. In the "rush hours", the personnel of the Complex Expedition consisted of about 3,500 people.

1988 – 1991

In accordance with the programme, the rooms to the west and the south of the reactor unit were cleaned up, drilling equipment was installed and boreholes were drilled through concrete and steel, to potential fuel locations. These boreholes became a main source of information about nuclear fuel in the destroyed unit.

FCM parameters were measured with the help of special thermal, gamma and neutron detectors.

The full number of bore holes ~150.

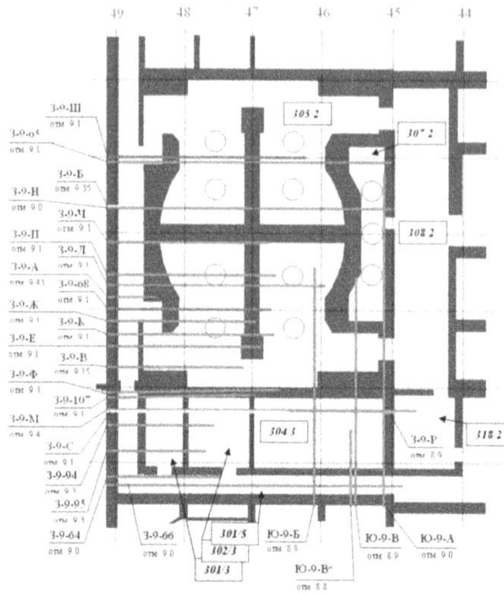

These bore holes became a main source of the information about nuclear fuel and the destroyed Unit.

Bore holes on marks (8.8 - 9.4)m (in under reactor rooms)

FCM parameters were measured with the help of special thermal, gamma and neutron detectors

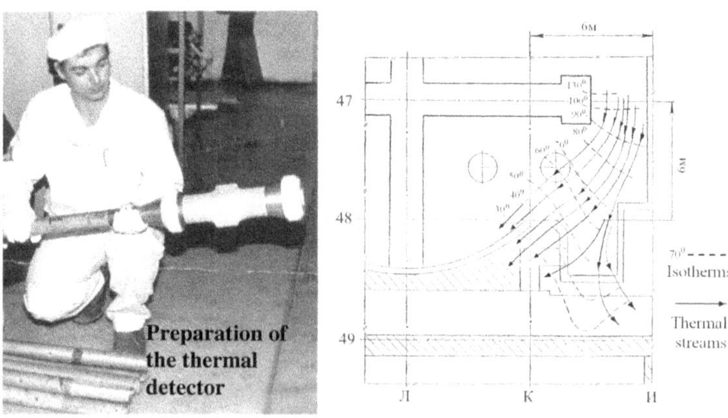

Measurement of temperature and thermal streams by means of bore holes in a basic plate of the reactor

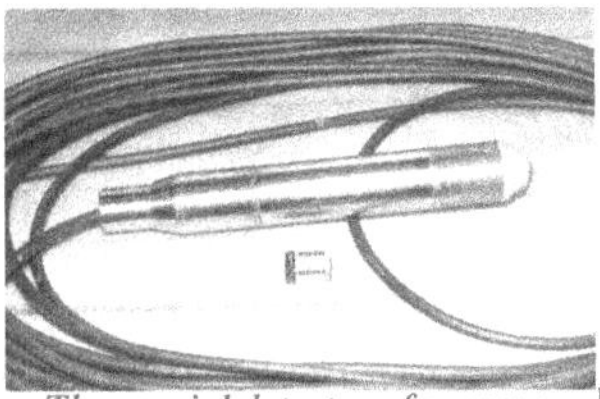

The special detector of gamma radiation

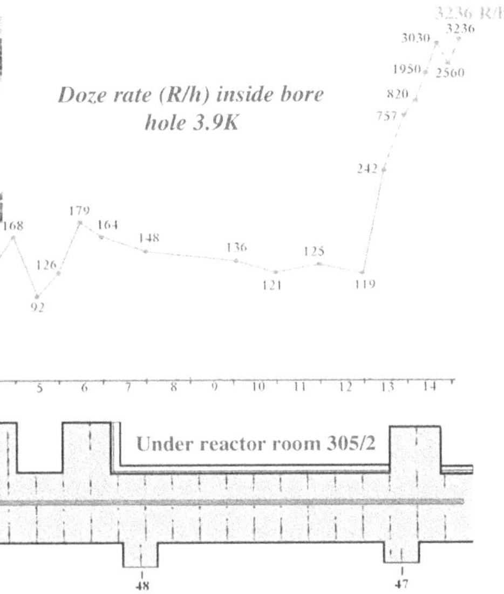

Measurements of neutron flax

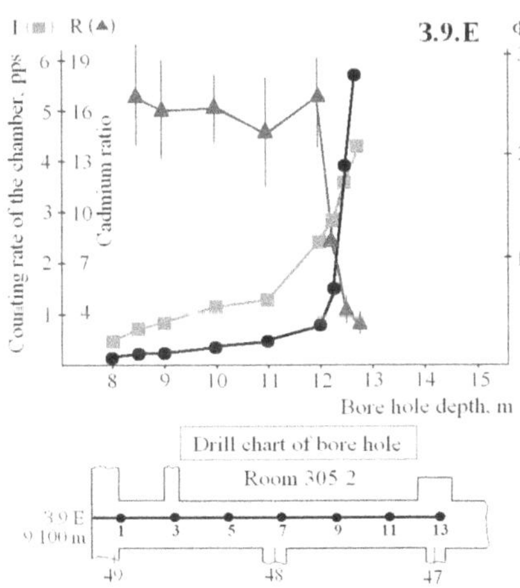

Investigations in 87-91 years have allowed reconstructing geometry of reactor space after accident

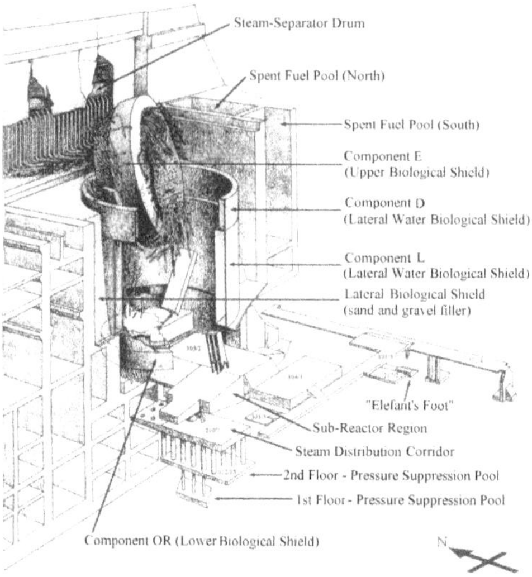

After the end of the USSR, the Ukraine took the responsibility for the Chernobyl Zone, the Chernobyl Nuclear Power Plant and the Shelter.

On the 4th of February 1992, an Inter-sectorial Scientific-Technical Centre “Shelter” (ISTC) was formed on the basis of the Complex Expedition and the Institute of Nuclear Research of the Ukrainian Academy of Sciences. The Department of Nuclear and Radioactive Safety, whose scientific supervisor is the Russian Research Centre “Kurchatov Institute”, was also included in ISTC.

The Situation of the Nuclear Fuel inside the Shelter—Lava-like Fuel Containing Materials (LFCM)

Nuclear Fuel in the Shelter

1. Central Hall, other upper rooms. Core fragments, fuel dust.
2. Southern spent fuel pool. Fuel assemblies.
3. The most part of the premises. Fuel dust.
4. Rooms 305/2 +307/2 + debris on the Scheme "OR" in the reactor vault. LFCM, core fragments.
5. Steam distribution corridor LFCM
6. Pressure suppression pool (2 floor) LFCM
7. Pressure suppression pool (1 floor) LFCM
8. 304/3, 303/3, 301/5, 301/6, Elephant Foot. LFCM (don't presented at the picture)
9. Reactor compartment, Turbine Hall, etc. Water with uranium salts (don't presented at the picture.)
10. Under the Cascade wall Core fragments (don't presented at the picture.)
11. The fuel at the site. Core fragments, fuel dust

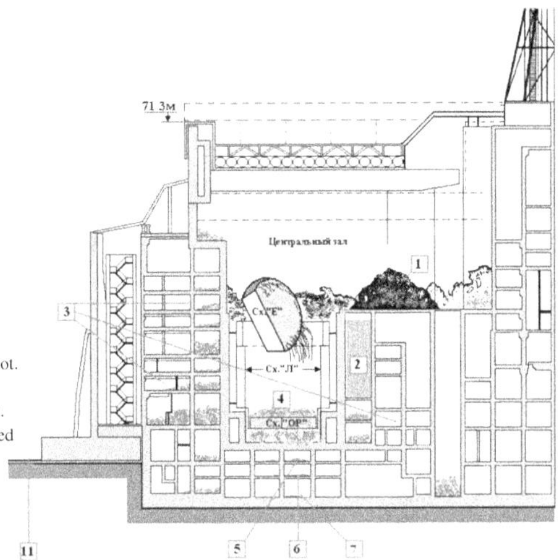

Physical and chemical properties:

- Fuel content in LFCM: Black ceramics contains (5.5 ± 1.5) % of fuel, brown ceramics (10 ± 2) %.
- Burning up: Brown ceramics - (12.6 ± 0.4), black - (12.5 ± 0.5) MW per a day/kg (U).
- Radio Nuclides Content: LFCM suffered from the impact of high temperature (higher than 1500°C). Under such temperature impact, volatile fission products including Cesium should be released from the fuel to the atmosphere.

Lava - like fuel containing materials (LFCM). Physical and chemical properties (elements, %).

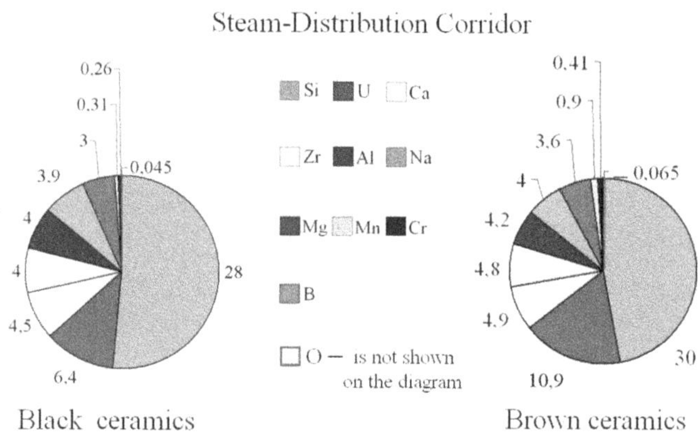

Lava-like Fuel Containing Masses (LFCM) Spreading—Radioactive Dust

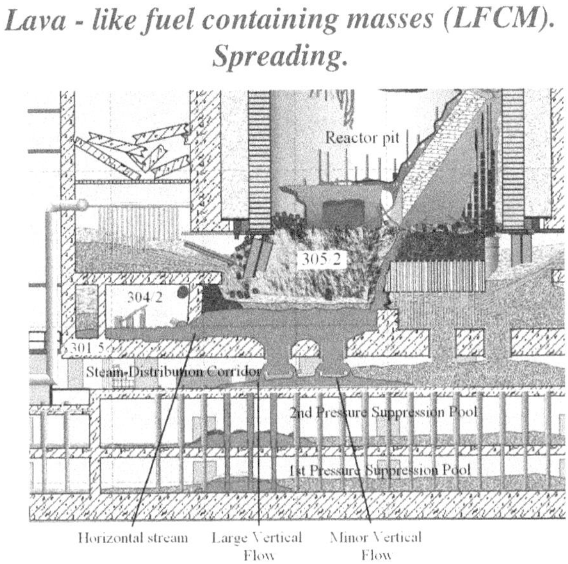

The fuel dust is dangerous in terms of human internal exposure and in environmental contamination.

The aerosol diameters of the Shelter fuel dust range from 0.1 to 20 µm. Primary radioactive particles (generated during the accident) contained in the Shelter are divided into two general types—condensation and fuel particles (hot fuel particles).

As shown, the majority of the particles are concentrated in the micron domain. The contribution of submicron particles to radioactive transport in the Shelter is insignificant.

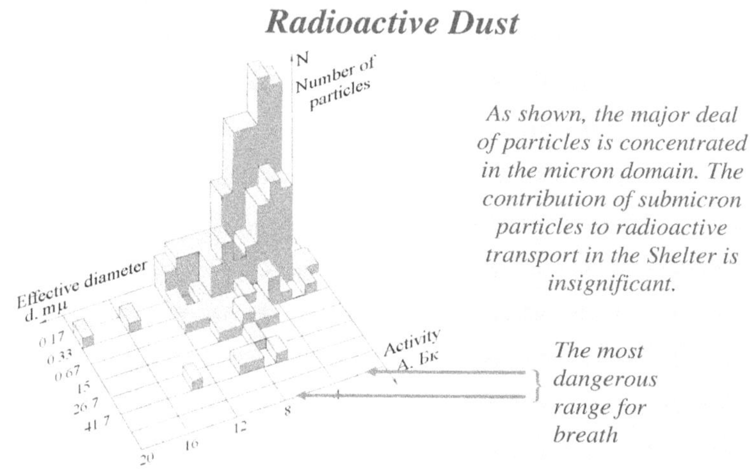

Distribution of aerosol particles by sizes (µm) and β-activity (Bq)

Assessment of Dust Amount and Surface Contamination Density inside the Shelter.

From 1996 to 1998, a number of additional investigations of radioactive dust characteristics were undertaken. Characteristics of the Shelter dust were studied using different techniques and methodologies, both standard and specially developed for the Shelter conditions.

The integrated stock of fuel dust in the central hall was about 5 tons (on uranium) from which about 100 kg can be carried to removable contamination.

Water in the Shelter Rooms

Water penetrating into the Shelter is a source of several hazards, the main being the following:

- Water gets into FCM accumulations and leads to increase in K (effect.);
- Water destroys FCM;

- Water increases potential nuclear risk of the Shelter by transporting dissolved salts of enriched uranium.

Therefore, regular investigations of water accumulations in rooms are carried out at the object.

Balance of water of the Shelter (2007)

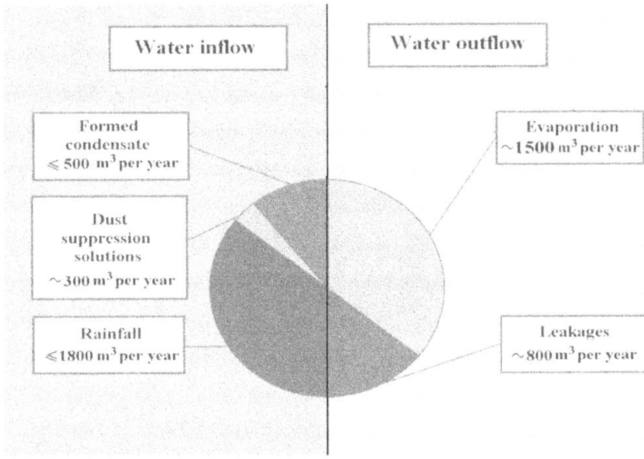

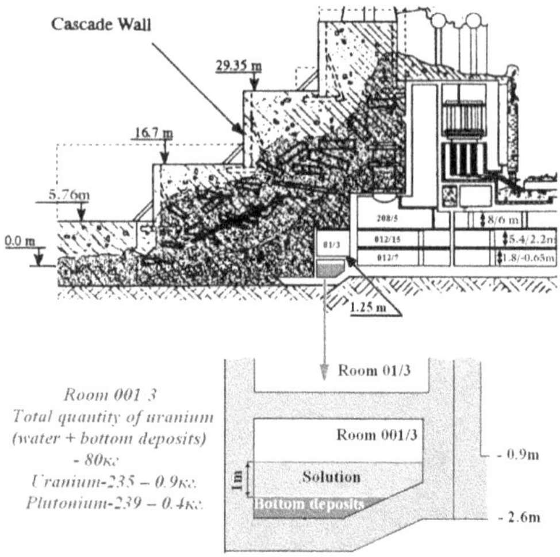

The biggest accumulation of water in the Shelter:

300 l (solution) and 100 l (bottom deposits)

<ins>Value Assessments of Fuel Allocated in the Shelter Rooms</ins>

Value assessments of fuel allocated in the Shelter rooms

Name (number) of Rooms	FCM Modifications in Room	Elevation, m	Note
Central Hall (914/2)	The Core fragments	More than 21	There is LFCM availability possible
Southern Cooling pond (505/3)	The Core fragments	14.8	129 cassettes with spent fuel. LFCM availability possible
All the upper rooms including the CH	Fuel dust	~30	
304/3	LFCM	6 ± 2	"Lava-Like Horizontal Flow".
301/5+301/6+303/3	LFCM	4.5 ± 2.5	"Lava-Like Horizontal Flow"
217/2	LFCM	0.4 ± 0.2	"Elephant Foot", "Stalactites"
305/2 and 504/2	The Core fragments, LFCM, dust	85 ± 25	It's the beginning of all the flows of Lava-like FCM.
SDC (210/5+210/6+210/7)	LFCM	25 ± 11	"Large Vertical Flow" and "Little Vertical Flow"
PB-2 (012/14+012/15+012/16)	LFCM	8 ± 3	
PB-1 (012/5+012/6+012/7)	LFCM	1.5 ± 0.7	
Water in all the reactor compartment rooms	Soluble Uranium salts, suspensions	<10 kg	
Fuel at the Shelter site	The Core fragments, dust	0.75 ± 0.25	

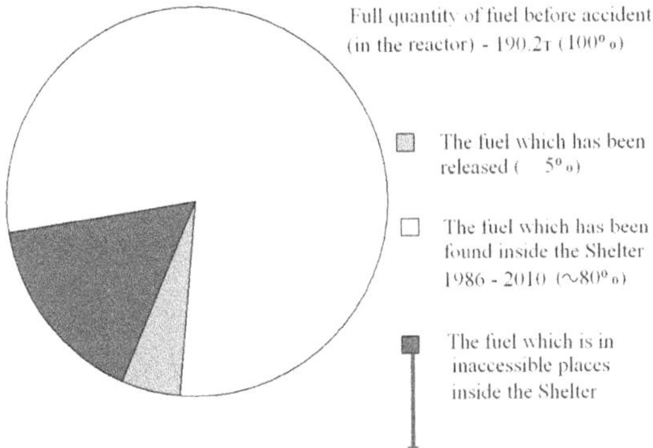

Till present no reliable information is obtained about many places (inside the "Shelter") where sufficient quantity of fuel might be allocated. Now a lack of data of fuel allocated in the following places is the most "blank spots":

- ***in the Central Hall under dropped materials;***
- ***under Cascade Wall***
- ***in Room 305/2, inside large sizeable aggregations***
- ***in Room 307/2.***

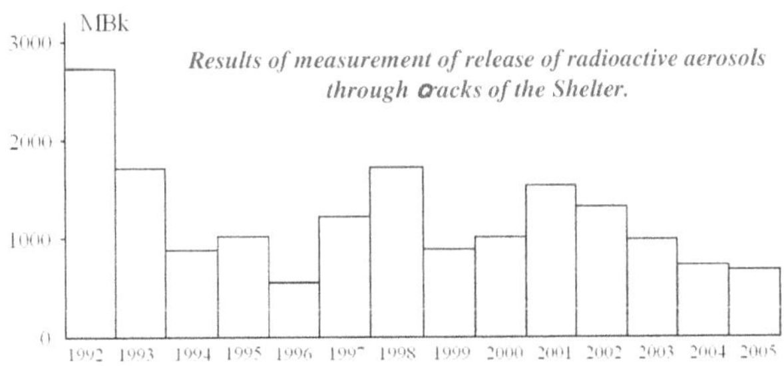

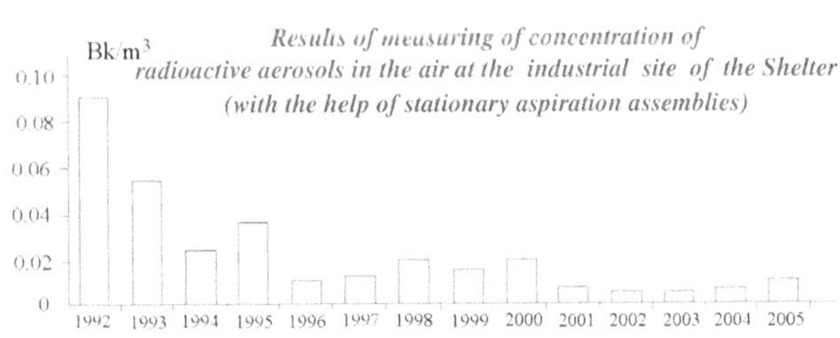

Nuclear safety of lava

System: ($UO_2 + SiO_2 + H_2O$)

It is completely safe for "a pure" lava

Nuclear safety of system—(lava + fragments of active zone) in under-reactor room No 305/2.

In the event that water fills in system (lava + fragments of active zone), it can become dangerous.

Though nuclear danger is only a hypothesis, we can't consider the Shelter as a nuclear safety facility. However, the initiation of a fission chain reaction may be dangerous only for the staff working inside the Shelter.

Nuclear Safety of Water Accumulations

What do we know about enriched uranium in the water in the rooms of the Shelter? Danger concentration is more than 90g /l . Real concentrations are less than 0.3g /l .

This information makes us sure that nuclear danger is absent but that permanent control is necessary.

Radiation Safety

Before stabilization, the main danger presented by the Shelter was the release of a radioactive dust as a result of a collapse of the roof or its elements. Earthquake, tornado, wind, etc., could have caused a collapse.

Radiation safety

Probabilities of Failure of Major Confining Structures

Structures, areas, loads	Probabilities of failure, year^{-1} (Before Stabilization)
Western area	
Supporting areas of beams B1 and B2 on axis 50:	
- vertical load	0,13
- horizontal load	0,16
Wall on axis 50 along with adjacent frame	0,16
Western buttress wall	0,09
Southern area	
Deaerator stack	0,16
- Horizontal load	0,16
Southern shielding - hockey sticks	0,13
Northern and Eastern area	
Northern hockey sticks	0,04

Conversion of the Shelter into the Ecologically Safe System

Shelter—a potentially dangerous object and its danger increases with time, with the destruction of structures affected by the accident. Therefore, in the past 25 years, Russian and Ukrainian specialists were constantly watching the Shelter and have been taking all possible measures to reduce risks. At the same time, it should be noted that all these years, the shelter was never exposed to extreme environmental conditions, such as a 5 magnitude earthquake, which probability is estimated to 1×10 -2/year for Chernobyl.

The conservative approach to Shelter facility safety forced "Kurchatov Institute" even in 1989 to propose the construction of Shelter-2. It was proposed to erect a leak-tight and strong structure (Shelter-2) over the existing "Shelter" facility which could stand for many decades and, in due course, to create technologies to disassemble the destroyed unit, to take out and dispose of fuel under protection of Shelter-2.

It became clear that such Shelter conversion was a very expensive effort which would demand international assistance. A lot of efforts and time were demanded prior to the decision of the Government of Ukraine to create the International Chernobyl fund.

For this purpose "Shelter Implementation Plan" (SIP) was developed in 1996 – 1997 by the international group on the basis of the previous research performed by Russian and Ukrainian experts. Implementation of the Plan was considered to start in 1998.

Conversion of the Shelter

For this purpose "Shelter Implementation Plan" (SIP) was developed in 1996–1997 by the international group on the basis of the previous research performed by Russian and Ukrainian experts. Implementation of the Plan was considered to start in 1998.

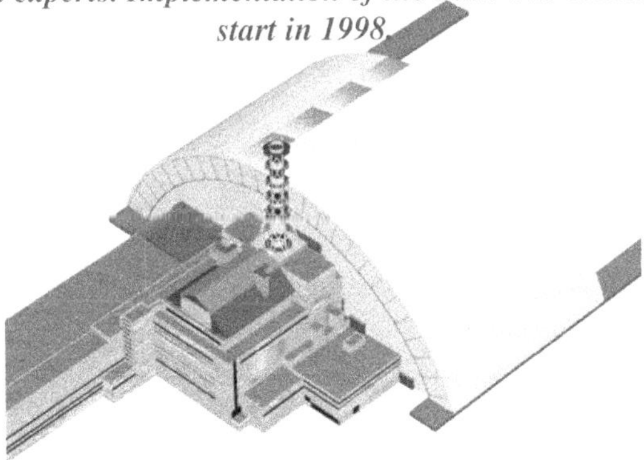

In 1996, the Shelter's current "safety analysis" was issued as a Kurchatov Institute document. It is the major document for maintenance and conversions of the Shelter. In 2001, a new version of the Analysis was created in KI.

Major work phases to convert Shelter facility into safe state:

1998 – 2003	Preparatory phase
2004 – 2008	Stabilization of the structures to guarantee the facility being intact within 10 years
2008 – 2012	Design and construction of new safe confinement (NSC) in the form of Arch. Arch sliding over existing Shelter facility
2012 – 20...	Dismantle of the structures, removal of FCM and RW.

The discussions of Shelter conversion, options, preparation and commencement of operations generated a variety of new issues, areas and variants of actions, but as a whole, the conversion strategy was in agreement with the one proposed by the Kurchatov Institute twenty years ago.

<u>Conversion of the Shelter (2004 – 2008)</u>
Stabilization measures of the Shelter building structures

Measures to stabilize the structures are now accomplished.

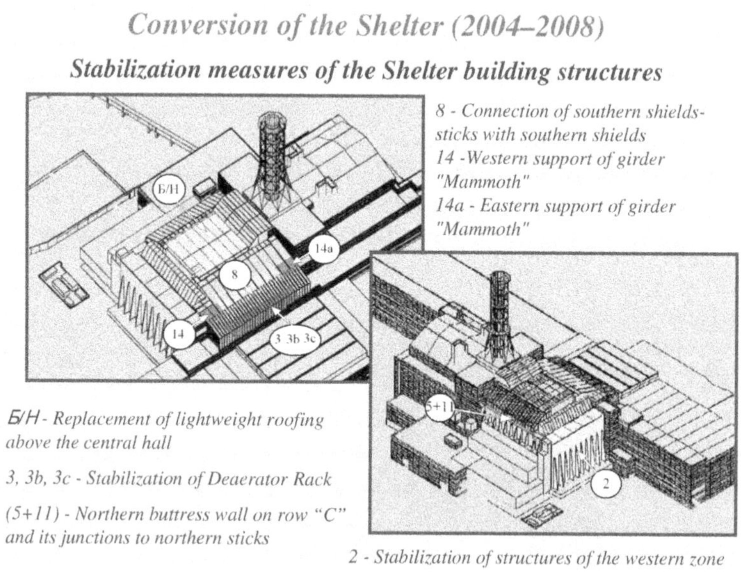

Now the priority is to construct new safe confinement (Arch) sliding over existing Shelter and partial dismantling of the destroyed structures.

Conversion of the Shelter

Now the priority is to construct new safe confinement (Arch) sliding over existing Shelter and partial dismantling of the destroyed structures.

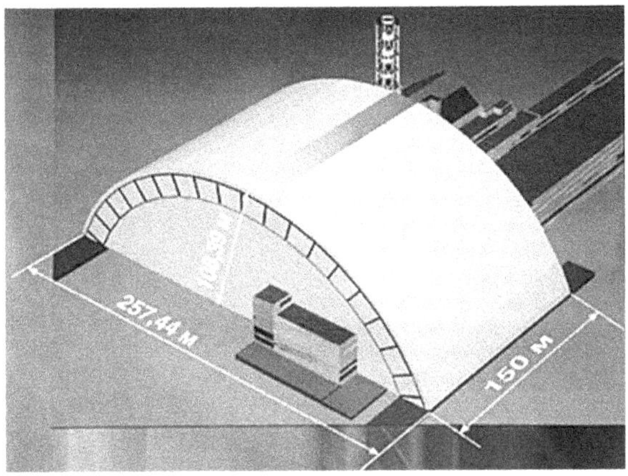

Today, the Arch project is already finished and works on Arch foundations are conducted. The main difficulty: a lack of money for building (approximately 500 million dollars). Ukraine will address again for the international help.

SIP activities have come to an end, but plans to convert the facility are still on the agenda. Removal of nuclear fuel and radioactive materials and final disposal thereof should become the next and the most difficult step forward. According to the existing estimations, preparation and performance of these operations may take many tens of years. All this time FCM will "be stored" under NSC protection.

The "Kurchatov Institute" continues to conduct works on the Shelter's problems in Moscow. The first part of a database on scientific and technical problems of accident on ChNPP has been created at the "Kurchatov Institute". This part of a database contains articles, tables, drawings, schemes and photos. The information total amount makes more than 60 megabyte.

A device for melt localization has been designed and several have been built. The first in the world industrial devices are installed at the Tianwan Nuclear Power Plant in China and at the Kudankulam Nuclear Power Plant, being constructed in India.

Safety requirements for hydrogen energy facilities were developed as well as for non-proliferation, remediation and physical protection of nuclear materials. Prospects for Floating Nuclear Power Plants overseas projects, in full compliance with present requirements of nuclear safety regulations, are being studied.

THE POTENTIAL POSITIVE IMPACT OF THE IRRATIONAL ON THE SUCCESS OF SCIENCE

THE LORD ALDERDICE FRCPsych
House of Lords
London UK

I found myself having to address the impact of the irrational on group activities many years ago when I began to try to explore the causes of terrorism in my home region of Northern Ireland. While it was clear to me from the start that terrorism was a tactic of asymmetric warfare, it also became apparent that it was not simply a rational choice of tactic as a result of a personal socio-economic cost-benefit analysis. At an individual level, the terrorist very rarely gets much personal socio-economic benefit from his actions. This is most striking in the case of suicide bombers of course, but it is also true more generally of those who engage in more standard terrorism, if less catastrophically so. In one way or another they sacrifice their well-being to the cause.

As I further explored the matter in other regions where there had also been long-term politically motivated terrorist campaigns, it increasingly became apparent to me that, in most circumstances, the community on whose behalf the terrorist claimed to act suffered the most from their actions. Whether in Northern Ireland, Peru, Nepal or elsewhere it was not the British Army, the Peruvian armed forces or the Royal Nepalese Army that suffered the most, but rather the ordinary Catholic Nationalists in Northern Ireland, the Quechua-speaking indigenous people of Peru and the rural hill village communities in Nepal. Whatever the motivation for the adoption of this tactic of terrorism, it was not the outcome of a cold, rational, realistic, socio-economic, cost-benefit analysis either for the individual or for the group, because both suffered dreadfully and with very limited success in terms of outcome.

I therefore began to explore the non-rational and even irrational elements in individual and group decision-making processes that led to such an outcome in such communities. As a psychiatrist, I was already familiar with trying to understand the thinking and communications of those judged by others to be disturbed in their thinking, experience or behaviour, including those who were frankly mad. I was aware that because of the evolutionary structure of the brain and of the mind it was possible as a psychoanalytically-informed psychiatrist to come to understand something of what was going on in the thinking of a person with a neurosis, a psychosis or a psychopathic disturbance of personality, if one could understand the form of that disturbed thinking process. So I started to apply the same approach to decipher the apparently non-rational form of the thinking of individuals and groups involved with terrorism.

In my paper to this meeting last year, I described how the common factor in the adoption of the tactic of terrorism seemed to be a group phenomenon, not an individual one. Even where it seems to be an individual act, such as in the recent horrible attacks in Norway, the terrorist makes clear that he believes himself to be acting on behalf of a community or group—virtual or actual.

Where a group feels that its interests and concerns are disregarded, in a manner which is deeply unjust and humiliatingly disrespectful, and has concluded that there is no other way to resolve the problem in the face of overwhelming odds, it may resort to terrorist violence. This violence is of course criminal but it is also not to the immediate socio-economic benefit of the group. However it is understood to be in the service of righting a terrible, historic wrong suffered by the group. In other words, the decision to adopt such a tactic is based on a 'morally-informed' decision rather than a materially-informed one. We may not share their measure of morality but we must understand that it is perceived to be a lesser evil undertaken to address a greater evil. In this sense the decision is not a rational one, at least in material terms, since it leads to socio-economic disbenefit. In last year's paper I also tried to show how such non-rational and even irrational outcomes occur in other circumstances such as where countries find themselves unable to adapt quickly to massive socio-economic and cultural change, as in sub-Saharan Africa, or where the wish to make money has led to the adoption of a pseudo-scientific approach to economics in the financial market-place with disastrous results in recent years.

Since last year's presentation, members of the Permanent Monitoring Panel on the Motivations for Terrorism have continued to explore this different set of decision-making principles in the arena of politically-motivated violence. The term 'sacred values' has been adopted to describe the non-material principles upon which these decisions seem to be based. These are not necessarily religious values, but do transcend merely material questions—for example I will angrily reject the suggestion that I will exchange the life of my child for a financial consideration because the life of my child represents a transcendent or 'sacred value' for me. It has become clear through experiments conducted by Scott Atran and other colleagues that this process of decision-making is not consequentialist, based on instrumental or material values, but is deontological, based on moral values or rules and indeed there are some indications from experiments using functional Magnetic Resonance Imaging (fMRI) that this kind of thinking may take place in parts of the brain which tend to be used for rules-based decision-making, as distinct from those parts which would engage in, for example, outcome-based weighing up of a socio-economic calculus.

At the end of my paper last year I noted that while I was focusing on adverse outcomes of such non-rational or irrational thinking, especially for the success of science, I was not implying that all such outcomes were adverse and some months afterwards President Zichichi asked that I pay attention to this question and address the possible positive aspects of this irrationality.

At one level, the past year has given us a powerful non-terrorist political example—the so-called Arab Spring. Many people, particularly young people, have come out on to the streets of countries across the Arab world and in some cases have already achieved the downfall of authoritarian regimes. We do not know what the outcomes will be, and they will certainly be different in different countries. What is clear however is that the apparent motivation was not a simple personal socio-economic cost-benefit analysis. Those who came out on to the streets did so at enormous personal risk and many thousands have lost their lives. Scott Atran has also been analysing key words in the tweets in the Arab Spring countries and it seems that words such as *freedom* and *democracy* come up much more frequently than words such as *food* and *price* in the tweets. This supports the view that it is these 'sacred values' rather than simple socio-economic themes which are critical motivators.

Most of you here will take the view that the American Revolution was also a positive development and there is evidence that the revolutionaries of the time would not have made a solid case for their actions had they simply judged it on historical or military analyses—they did not have the numbers or power on their side—however they had the passionate conviction that right was on their side and they succeeded. This is not infrequently the case even in a football game—the psychological posture is absolutely key to the outcome and so-called 'devoted actors' can bring about unexpected results when compared with so-called 'rational actors'. We must not therefore assume that non-rational decision-making will necessarily bring about a negative outcome or is only operative in regressive and ultimately damaging causes.

Darwin became aware of the issue of the non-rational imperative having a positive evolutionary benefit. He was puzzled that the brave warrior who risked his life for others and only very rarely achieved any personal benefit should nevertheless survive as an entity in evolutionary terms. He was forced to conclude that natural selection must have the propensity to select for moral virtue, that is to say individual self-sacrifice for the benefit of the group. In other words, groups survive as groups because they have individuals who make judgements based on the welfare of the group not just their own socio-economic benefit. If this is the case then it is a mistake to believe that evolution will remove the need for non-rational decision-making. On the contrary, faith-based decision-making or other morally informed deontological processes—deciding on the basis of a commitment to a political programme or a belief in the moral imperative of a scientific search for truth—may have evolutionary benefit for the survival of the group, and a group which uses exclusively consequentialist decision-making based on personal socio-economic cost-benefit analysis will have less evolutionary survival capacity. The non-rational is therefore no mere hangover from a more primitive time but rather a crucial component for communal survival.

In an analogous way, young people may devote themselves to scientific work, or to religious vocations or political struggle, with a commitment of time and energy that is way out of line with any expected personal socio-economic benefit. They may happily sacrifice personal relationships, timely relaxation and economic well-being, and not because this devotion will bring them particular socio-economic benefit. Science, and society as a whole may benefit from their devotion to the cause, but that is a very different matter from a realistic, personal, cost-benefit analysis. It is nevertheless a positive contribution to the success of science.

There are other elements of the non-rational or even irrational that contribute to scientific success.

In last year's paper I described how, in a climate of fear, an individual or a group may regress into primary process thinking in which the normal appreciation of the boundaries of time and space are suspended. We use the words neurosis and psychosis to describe various types of such disturbance. In these states the psychological division between what I remember from the past, what I perceive in the present and what I wish for the in the future are not properly differentiated, nor is there a proper perceptual boundary between what is going on inside my head and what I perceive from outside—so what I perceive to be persecutory 'voices' are actually thoughts in my own mind. Of course when an individual is captured by such irrational or non-reality orientated thinking it is to their profound disadvantage. However to be able voluntarily and transiently to slip into a similar state may be creative and helpful. Faced with a difficult problem it may be useful to fall into a state of reverie, to day-dream about the problem, to stop paying direct attention to it for a while, or to sleep on it. The solution to the problem may then become apparent, not because of an intense logical focus on the issue, but through the mind operating on the problem in a different way. The issue may be reframed or construed differently by the non-conscious mind 'thinking about it' using a different set of rules—the so-called primary process thinking which does not hold to the rules of tempero-spatial differentiation. You may well be familiar with having had such an experience yourself. Professor Ujita made passing reference to something similar yesterday at the Energy PMP Workshop when he said that some of the problems we experience are due to a 'lack of imagination'. Just as children prepare themselves for real adult life by playing and living in a fantasy world—though we should not imagine that they do not appreciate what is reality and what is fantasy—so we need the capacity to step aside from reality-orientated thinking to solve some problems.

Another element of this thinking is to allow the past to intrude transiently on the present. When I meet a new person, I do not start from the position of knowing nothing about them. I immediately call up a previous experience of someone who looked similar, dressed similarly, spoke in a similar way, came from the same country or whatever. Psychoanalysts call this transference, because we are transferring an earlier experience on to the present. I then subsequently more or less correct this template with more reality orientated perceptions from the present, but I do not start with a blank sheet. We may see something similar in visual pattern recognition. My wife is a histo- and cyto-pathologist and she must make diagnoses on the basis of her examination of many microscope slides each day. If she were to painstakingly go through a protocol of ticking boxes for each observation before making a diagnosis, she would get very few slides covered each day. Instead in many cases she will *immediately* recognize a visual pattern where she observes substantial similarities with a previous known case she has seen, despite there being some differences. She will then in a more logical way check out whether there are critical differences despite the similarities she has observed. As a psychiatrist (and indeed as a politician) I have a similar response of situational (rather than visual) pattern recognition. I immediately recognize similarities in the patient's description of their life situation (or a political picture) and will then, if I am wise, check out with a more logically-informed scrutiny how far there are also crucial differences despite the similarities. Martin Gladwell in his book 'Blink' described well this initial flash of recognition which is highly complex but almost immediate. Our PMP colleague Bob Axelrod has been exploring how far previous experience is applied, helpfully and otherwise, in the way we address political situations and phenomena.

Another way of thinking about these different modes of thinking is to see one as an artistic way of thinking and the other as a scientific way of thinking—the artist is always transgressing the boundaries of reality and the scientist is trying to map out those very boundaries, however I think in truth scientific progress is made not only through following through a logical process of analysis, though this is critical. However when a challenging problem arises, it made be necessary to dispense with what are generally assumed to be the proper boundaries of tempero-spatial 'reality' and permit our minds to play with the problem and allow a new formulation to occur to us. As our President Niño Zichichi has emphasized many times, while we live with a general assumption of three dimensions of space and one of time, reality has many more dimensions that we can appreciate mathematically, if not easily perceive in terms of mental models.

One key difference between these uses of non-rational thinking processes and mental disturbance is that we can voluntarily step back from them into a more 'reality-orientated' way of functioning. The mentally ill individual and the profoundly disturbed community are not able to do so—they are enthralled to these ways of thinking, at least for a time.

In summary, while non-rational ways of thinking may have destructive outcomes when they dominate the thinking of an individual or a group, they may also be useful and in some cases necessary to permit genuine creativity and the reframing of scientific and practical and social questions in life. This beneficial capacity however requires that they are transiently used and the outcomes are also subjected to some element of more analytical thinking. There is no reason to believe that evolution will remove these non-rational capacities because they are likely to have selective advantage for the survival of the group, and at least in some cases for the artistic or scientific successfulness of the individual.

LEARNING MORE FROM FUKUSHIMA DAI-ICHI

RICHARD L. GARWIN
Thomas J. Watson Research Center, IBM Research Division
Yorktown Heights, USA

ABSTRACT

A tsunami caused by the magnitude 9.0 Tohoku Earthquake of 11 March 2011 on the northeast coast of Japan drowned the emergency diesel generators of three boiling-water reactors at Fukushima Dai-ichi, resulting in station blackout and the meltdown of the three reactor cores. Although emergency injection of seawater was improvised to remove the decay heat from the reactors, it was too late to avoid boiling off of much of the water in the reactor pressure vessels and the reaction with steam of the zirconium alloy "clad" of the fuel rods in the reactor with the evolution of hydrogen, which in turn over pressurized the massive concrete containment of the reactor and compelled venting of the hydrogen and some of the radioactive material from the reactor. This paper reports the course of events, the resulting contamination of the environment and the evacuation of 180,000 inhabitants, the efforts to prevent further damage to the reactors and spent fuel pools, and how the world can learn to prevent such accidents and to better alert and inform the public of the hazards and how they can protect themselves.

On March 11, 2011, an intense earthquake off the northeast coast of Japan, ranked as magnitude 9.0, severely tested Japan's earthquake code and discipline. The country and its buildings passed with flying colors. Some 16 nuclear reactors shut down instantly, as planned, as the earthquake strong motion exceeded on the order of 0.5 g, as measured by strong-motion seismometers within the buildings themselves. The reactors, including three operating reactors at Fukushima Dai-ichi (1F), Units 1, 2, and 3, automatically inserted their control rods into the reactor core, terminating the neutron chain reaction, and reducing the thermal power output to that of the "decay heat" of the fission products themselves. With so much power generation suddenly off line, transmission line power to the 1F site was lost, and emergency diesel generators (EDG) took over.[1]

[1] e.g., http://allthingsnuclear.org/post/4609790167/what-happened-at-fukushima-dai-ichi

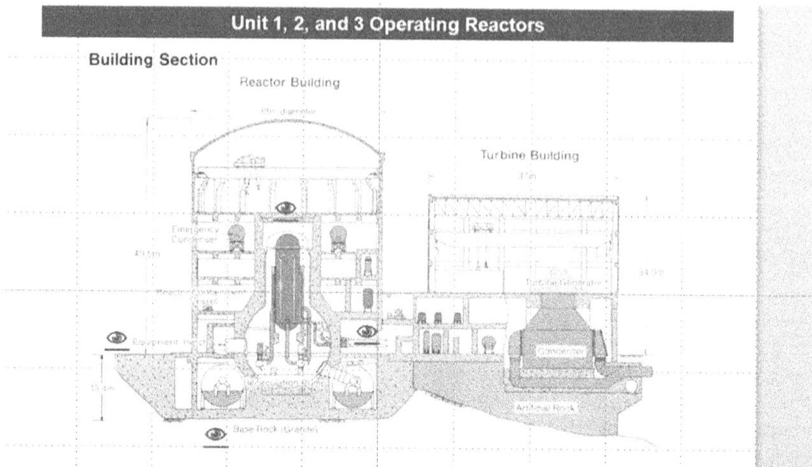

The Mark I primary containment features a drywell (the inverted lightbulb) and a torus (the donut) connected by eight vent pipes. The reactor building surrounds the primary containment. Steam and feedwater piping passes through the containment walls and reactor building to the turbine building.

For the early boiling water reactors (BWR) of 1F with power outputs on the order of 500 MWe (hence thermal power on the order of 1600 MWt), electrically powered circulating pumps continued to cool the reactor pressure vessels and to reject the heat through heat exchangers to seawater. About an hour later, an enormous tsunami swept west into Japan, striking Fukushima Dai-ichi with an inundation about 15 m above mean sea level. This was about a "500-yr" tsunami, but the seawall at 1F had been built to repel a 100-yr tsunami and it did nothing to retard the enormous flow of water that flooded far inland. In particular, the water almost instantly stopped the EDG which were supplying Units 1-4 with power in the absence of grid power. With the grid and EDG power both gone, these reactors were dependent on their sets of lead-acid batteries, like those in automobiles, which were specified to power the instrumentation (valves and lights and meters) of the site for at least 4 hrs. After that, there was no light or instrumentation power, and under these conditions of "station blackout" (SBO) the circulation of water and the cooling of the reactor pressure vessel (RPV) ceased. But fission-product decay heat continued to boil water in the RPVs.

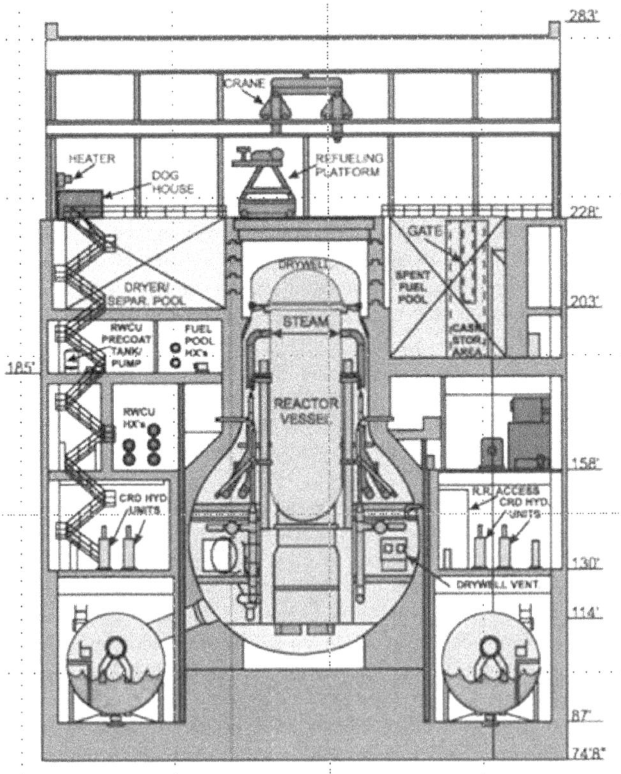

Although steam from the RPV in reactors 1-3 no longer went to the main turbine and alternator, it did flow either to an emergency condenser (Unit 1) or, in Units 2 and 3, to drive a small auxiliary turbine that itself directly powered a recirculating pump to cool the reactor core. With the batteries exhausted, the valves to the emergency turbine could no longer be opened and with continued decay heat accumulating, the pressure in the RPVs rose until steam began to emerge from the overpressure safety valve near the top of each RPV.

The decay heat, initially at the level of about 10 MWt, sufficed to evaporate water at its latent heat on the order of 2 gigajoule per ton, or 18 tons per hour.

There was no control from the control room, because there was no electrical power.

Working under the greatest difficulty, Tokyo Electric Power Company (TEPCO) staff rigged high-pressure injection of water from external pumps into the RPV, using a "feed and bleed" approach in order to replace water lost to steam and thus, they hoped, to maintain the water level above the top of the heating zone in the 5-m-long fuel rods.

Unfortunately, the only water quickly available in these amounts was seawater, which is highly corrosive to the reactors, and contains about 1.3% sodium chloride (NaCl). Thus, at a rate that would stabilize at about 7 tons per hour or 168 tons per day into each of reactors 1-3, each RPV accumulated 2 tons of salt per day, initially in the form of brine and ultimately in the form of salt cake encrusting hot portions of the reactor and the bottom regions of the RPV, or so it is thought.

Before seawater cooling could be established, the water level in the RPVs fell low enough that the zirconium alloy (zircaloy) tubes cladding the uranium oxide pellets in the fuel overheated and burst, reacting with the water to form hydrogen, which escaped with the vented steam but, unlike the steam, did not condense in the large water-filled torus shown in the figure. The hydrogen pressure increased in the "drywell," which, in turn, needed to be vented to prevent its bursting. Unfortunately, although there are valves installed to conduct the vented gas to the tall stack, the valves seem not to have retained their position after the power failure, and hydrogen invaded the reactor buildings, including Building 4. There was no fuel at all in Unit 4's RPV because it had been shut down for refueling, with its fuel all transferred to the elevated spent fuel pool in Unit 4, where it was residing deep underwater, as was the case with the old spent fuel in Units 1-3.

There was much concern and uncertainty in Japan and in the world technical nuclear community that water had leaked or was leaking from the spent fuel pools, which were without circulating water cooling. If they had not leaked, there would have been several days of safety before the cooling by evaporation to the atmosphere would uncover the top of active fuel in the stored fuel elements. Major concern was warranted because the spent fuel pool every year receives 25% of a full core-load of spent fuel, so that there is about 6 (ck) times as much long-lived fission product content (e.g., 30-yr half-live Cs-137) as in an operating reactor core.

In the attempt to cool the spent-fuel pools, futile efforts were made to provide water by helicopter, until finally a low-tech solution was implemented in the form of ordinary "giraffes" used on high-rise building sites to deliver concrete to higher floors of the buildings under construction. Of course, the giraffes delivered water and not concrete, but there was continuing uncertainty about the temperature and the water level in the spent fuel pools. Unfortunately, initially only seawater was available and was used to prevent further disaster, but at the cost of an as yet unassessed corrosion problem of the fuel rods and assemblies.

THE PROBLEM OF WATER ACCUMULATION

For weeks some 20 tons per hour (480 tons per day) of water was pumped to the reactor pressure vessels to remove the decay heat and to prevent further evolution of hydrogen and fission products. Additional water was pumped to the spent fuel pools until circulation could be restored. This water input (including much of the condensed steam from the RPVs) accumulated in the reactor buildings and the turbine buildings and in trenches among the buildings. It was apparent that this highly radioactive water would spill into the sea unless it was pumped into holding tanks. It is no small matter to dispose of 120,000 tons of such intensely radioactive water. Its radiation destroys seals and limits the technology that can be used. Some tanks at 1F have been emptied to accommodate some of this water. Large barge-tank reservoirs have been brought in, and TEPCO contracted with AREVA for ion-exchange decontamination of the water, with a stage of reverse osmosis (not usually required in the spent-fuel reprocessing facilities) in order to remove most of the salt from the water that is then decontaminated by a factor 100,000 or so. Much of the fission product radioactivity is taken out as highly radioactive resin, but the salt has been left as concentrated brine.

More recently, TEPCO has planned to introduce an evaporator to convert the brines to solid waste for eventual burial as high-level waste, but that will take much research to provide a suitable containment system. It seems also that TEPCO may plan to skip the reverse osmosis process and move directly to evaporation of the water; the problem will be to ensure that the content of radioactivity is low enough to release the steam to the atmosphere.

The damage done by the hydrogen explosion is evident in the figures, with massive destruction of Buildings 1, 3, and 4. Building 2 was spared, perhaps because staff preemptively cut a large ventilation panel in the exterior paneling.

With the venting of the hydrogen, fission products escaped into the environment as well, and the government of Japan (GOJ) on March 12 called for mandatory evacuation of a 10-km radius, and later that day a 20-km radius—a region home to 78,000 people (including those within 10 km of an undamaged set of TEPCO reactors at Fukushima Dai-ichi). A "stay-in-house" order was issued to those living within between 20 and 30 km of 1F, many of whose residents evacuated voluntarily. Under the conditions of severe tsunami damage in the region and disruption of normal local government services, there was little attention as to living conditions of those who were ordered to stay-in-house. Regions where aerial survey jointly performed by GOJ and the U.S. government (USG) showed contamination above the Protective Action Guide (PAG) have also been subject to evacuation.

The EPA Protective Action Guide, cited in the box in the figure reads, "If a person is in danger of receiving an external radiation dose greater than R Rem over Y years, the EPA recommends relocation until radiation levels decrease." It should be noted that this is not an urgent action because the dose is received over Y years. Note also, that the figure charts the dose received from ground contamination after May 9, 2011. Because each sievert (Sv) of adult exposure corresponds to an added probability of death from cancer of about 5%,[2] the 2-Rem (0.02 Sv) exposure over the first year for a person staying within the edge of the red boundary would connote about 0.1% lifetime additional probability of death by cancer for that one year of exposure. Per 100,000 people exposed, this would be about 100 additional deaths. Normal background radiation from rocks, cosmic rays, radon, and medical diagnostics contribute about 0.4 Rem per year to each person's exposure.

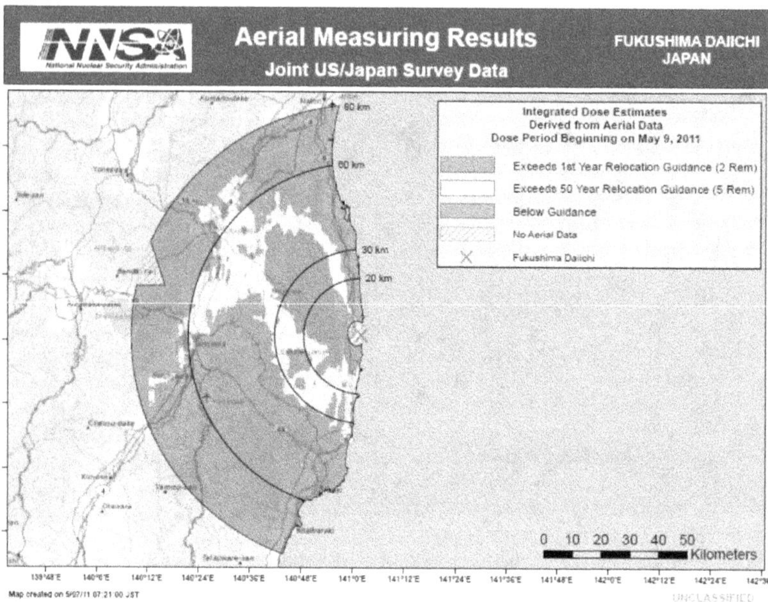

But a larger and broader tragedy had already occurred. Some 22,000 souls had been killed by the tsunami, many of them drowned on land or swept to sea. This is totally unlike the earlier reactor accidents at Windscale in England (1957), at Three-Mile Island in the United States (1979), or at Chernobyl in the Ukraine (1986), which had occurred without any external disruption. The details of the 1979 and 1986 accidents are detailed for instance in my 2005 book with Georges Charpak and Venance Journé.[3]

2 Board on Effects of Ionizing Radiation, BEIR-7, National Academies Press, 2006.

3 "De Tchernobyl en tchernobyls," by G. Charpak, R.L. Garwin, and V. Journé, (Odile Jacob, 2005).

At Three-Mile Island, the reactor core of Unit 2 (Unit 1 was unaffected) melted, and much of its load of radioactivity escaped from the reactor pressure vessel into the reactor building housing the pressurized water reactor (PWR) and into an auxiliary building. But only radioactive xenon and some iodine escaped from the reactor stack, and almost none was deposited on the ground. It is estimated that exposure from the passing cloud to the population at large totaled 20-40 person-sieverts (p-Sv) which at the rate of 0.05 lethal cancer deaths per p-Sv corresponds to an expected cancer death toll of one or two, among the millions of natural cancer deaths expected within the lifetime of those exposed to the small amount of radioactivity in the passing cloud. At Chernobyl, the world exposure has been documented as some 600,000 p-Sv, corresponding to a lethal cancer death toll of some 30,000 people. I provide here for reference a figure from a 2005 paper on comparative risks of various energy technologies.[4]

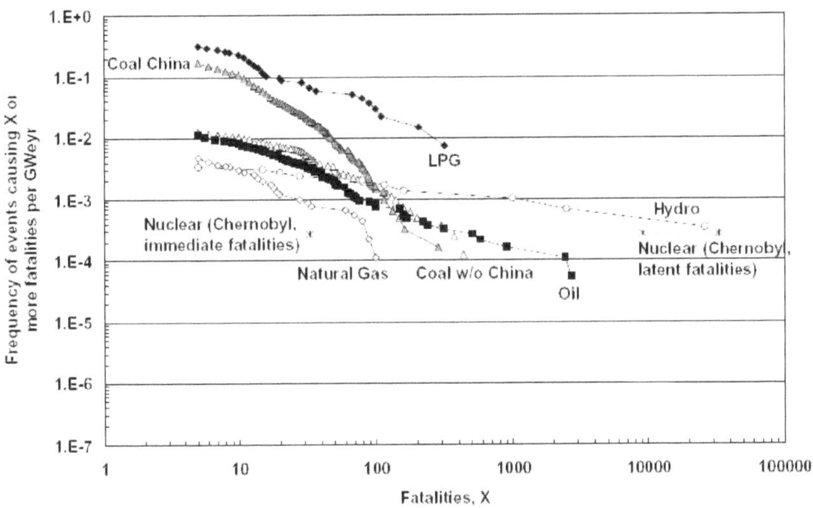

Figure 4. Comparison of frequency-consequence curves for full energy chains in non-OECD countries with partial reallocation for the period 1969-2000. The curves for coal w/o China, coal China, oil, natural gas, LPG and hydro are based on historical accidents and show <u>immediate</u> fatalities. For the nuclear chain, the <u>immediate</u> fatalities are represented by one point (Chernobyl); for the estimated Chernobyl-specific <u>latent</u> fatalities lower and upper bounds are given.

[4] "Accident Risks in the Energy Sector: Comparison of Damage Indicators and External Costs" by S. Hirschberg, P. Burgherr, A. Hunt.

At every level, the earthquake and tsunami must have impaired the response to the accident at 1F. For instance, the workers must have been concerned not only about their own safety but especially about the fate of their families. Almost all of the personal dosimeters were lost to the tsunami. And government at the local level was largely nonfunctional. At the national level, the concerns were understandably dominated by the earthquake and tsunami, and not by some possible damage to reactors.

In this brief talk, I can only touch on another urgent need, as mentioned in my June 2010 article.[5] This is the analog of a Geographic Information System (GIS) in wide use for multiple layers of information about urban utilities ranging from the location of gas pipes, to electricity nets, to telephone numbers and street addresses. Another layer should be added, which would be the local radiation exposure rate, in mSv per year (normal background is 3 mSv per year), and also the level of ground contamination, closely related to the radiation level.

In addition, without waiting for an accident, approximate shielding factors against local radiation should be calculated and recorded for easy retrieval in case of need. Access would be by the ubiquitous Web browser available not only on PCs but on smart phones and other handheld computer systems.

Feeding such an information system following a disaster (and practiced often before!) should be an automated Airborne Monitoring System (AMS) which could consist of a substantial number of 10-kg drone aircraft carrying crude radiation detectors such as NaI crystals viewed by a detector, that from an altitude of 500 m in open terrain (1000 m in high-rise cities) could readily map the averaged contamination on the ground and the derived average dose. See footnote.[6]

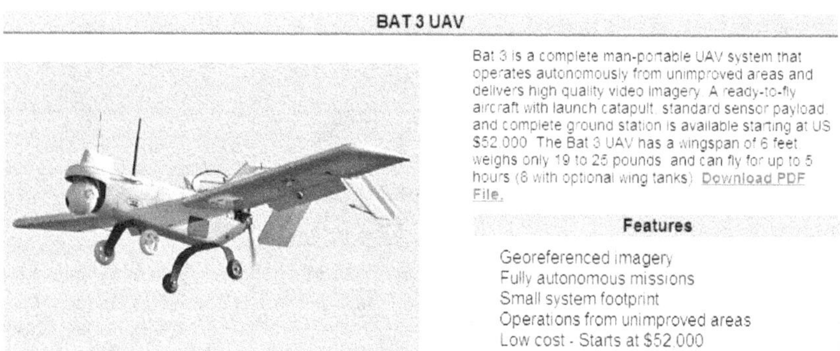

The Bat 3 is a 20-lb class UAV that seems to fit the description. These could be used for routine patrol of the Fukushima area at a non-interfering altitude of 1 km or 2 km, or at considerably lower altitude in order to resolve individual reactor sources.

---

[5] R.L. Garwin, "A Nuclear Explosion in a City or an Attack on a Nuclear Reactor," *The Bridge*, Vol. 40, No. 2, pp 20-27, Summer 2010, http://www.nae.edu/File.aspx?id=19815.

[6] http://www.spyplanes.com/pages_new/products.htm

Ultimately, they could be used to patrol populated areas such as Tokyo in order to provide a contamination map that would in large part reassure the public and guide any precautionary measures. The aircraft provide "georeferenced imagery" or (presumably) georeferenced counts of nuclear radiation. How this would actually be used depends on the database to which it is supplied and the access of authorities to that database.

Similar detectors carried on buses or taxi cabs or street-cleaning machines could supplement this over days or weeks (except in the case of nuclear explosion) with information on each and every street to provide refined data to guide decisions by individuals and by the authorities.

From the beginning of the Fukushima Dai-ichi crisis, U.S. government entities asked the National Atmospheric Release Advisory Center (NARAC) at LLNL for projections of the radiation exposure from assumed releases at 1F. Because the radiation measurements at 1F were unavailable because of lack of electrical power, absolute estimates were not available, but relative projections could be made.

The GOJ had a tool for this purpose as well, SPEEDI (System for Prediction of Environmental Emergency Dose Information), but it was not initially used because there was no agreement on the source term—the amount and type of radioactive material released from 1F. It wasn't until May 3 that the SPEEDI results were publicly available.

Finally, we may hope to bring some rationality into such blunt tools as mandatory evacuation, which itself is said to be as traumatic as a divorce or the death of a loved one, which costs should be quantified and balanced against the reduction of damage due to radiation in case one remains in the contaminated area. Taking the "50-yr" committed dose of 5 Rem at the outer edge of the yellow, which corresponds to 0.25% additional probability of death by radiation-induced lethal cancer, one needs to compare with the 20% probability of death by cancer in the same population not exposed to radiation. With the exception of the Nuclear Regulatory Commission, the U.S. government for planning purposes values a premature death at $5 million, so that an exposure to 5 Rem would indicate a cost of 0.25% x $5,000,000 = $12,500. Surely the population should have the opportunity to judge whether they would prefer to remain in the contaminated area and receive compensation in this amount, or be relocated, which would cost the state far more.

On the other hand, the government of Japan for a short time relaxed the environmental exposure limit for children from 1 to 20 mSv/yr (2 Rem/yr), in order that they should be able to continue to receive schooling at sites that would otherwise require evacuation. This edict was soon reversed after public protest and the resignation of Prof. Toshio Kosako, a radiation expert and an appointee of Prime Minister Naoto Kan. The government also stated that it would pay for the cleanup of the soil on school grounds.

COMMENT ON THE EVENTS OF FUKUSHIMA DAI-ICHI

In describing the course of the accident at Fukushima Dai-ichi, and some of its consequences thus far, it is apparent that the situation was much exacerbated by the general disaster in that area of the earthquake and tsunami.

The GOJ did provide much information on the Internet, although not in readily accessible form, either by public safety specialists or by the general public. The job was complicated by power failures and by the fact that many residents did not have their computers. In Chapter 9 of a major document[7] published by the Office of the Prime Minister, "kantei," the GOJ expresses itself on communications and messaging:

> *"While monitoring data has been quickly publicized, we need to come up with some ways to promptly communicate necessary information to the sufferers who want to obtain information but do not have access to the Internet due to power failure in such a case as combined emergency with natural disaster."*

and,

> *"The main channel of information provision has been through the mass media, which has transmitted press conferences and press releases to residents in the surrounding area, general public in Japan and international community. Hence, it is important to identify the needs of the mass media in addition to adequately communicate what people want to know. For example, when a hydrogen explosion occurred at reactor building of Units 1 and 3, television broadcast it almost real-time. The mass media strongly requested the ERC right after the explosion for an explanation of the accident by someone with appropriate knowledge in front of the camera about what really happened there and how the explosions would affect the reactors and so on. However, because it took time to verify the related facts, their needs were not always satisfied. As this issue is liable to be involved with trade-off between swiftness and accuracy, it would have been appropriate to develop a manual to respond to such situations in advance."*

I have quoted only two paragraphs, one concerned with the mechanical problem of reception of information, and the other with both substance and presentation.

In addition to the Internet, urgent messages such as the demand for evacuation were pushed by telephone, but much of the telephone service was disrupted, both to residents and to public officials. And, recall, the public was faced with clear and present disaster from the tsunami, with an uncertain number of tens of thousands of people lost at the time—not whether there might be tens of thousands of deaths eventually, but how many had actually occurred.

In previous documents, the PMP-MTA has discussed in great detail the formulation of messages in regard to pandemics ("Personal Protective Measures" in particular), the testing of those messages for understandability and effectiveness, and also for the degree to which they can be adopted.

As for format and presentation, for the most part the mechanisms envisaged by the PMP were Web based, in order to have the information available when needed, without having it lost or outdated in homes or offices.

As technology has advanced, it is now feasible to guard against loss of communications (but not necessarily of power); or the latest version of the emergency communications could be routinely downloaded into the home and office PCs, so that it

---

[7] http://www.kantei.go.jp/foreign/kan/topics/201106/pdf/chapter_ix.pdf Also, http://www.kantei.go.jp/foreign/kan/topics/201106/iaea_houkokusho_e.html

would be available if communication lines were down or overloaded. To guard against temporary loss of power, the same capability should be and is available via smart phones and tablets.

Major questions are whether families will indeed have emergency kits and whether they will practice access to the emergency communication sites.

IMPACT OF FUKUSHIMA DAI-ICHI ON THE WORLD NUCLEAR POWER PROGAM

An example of potential response is a "90-day study" by the staff of the U.S. Nuclear Regulatory Commission which draws lessons for improving the safety of U.S. nuclear reactors.[8] The Chairman of the NRC has called for rapid implementation of most of the recommendations, but the other four NRC commissioners in testimony to the U.S. Congress seem to advocate extended delay.

Non-governmental organizations in the U.S. had also provided their views, and now their response to the 90-day NRC study. In particular, the Union of Concerned Scientists (UCS) which provided timely and substantive information as the 1F crisis evolved, takes issue technically with some of the detailed recommendations[9] and calls attention to its own broader set of issues,[10] including that the NRC set similar priorities on its own performance in regard to safety as it does in regard to timely action on management and contract issues.

APPENDIX

CONTINUING GROUND RADIATION SURVEY by a fleet of GPS-guided aircraft with data telemetered to populate a Geographic Information Service (GIS) database:

In more detail, within the Permanent Monitoring Panel on Mitigation of Terrorist Acts (PMP-MTA) we have long discussed a capability to use small UAVs to monitor from the air radioactive contamination of the ground, to guide public health measures such as evacuation, "remove your shoes when entering the house", and the like. This would have been very useful in case of an incident involving a radiological dispersal device (RDD), and it will surely be valuable in the aftermath of Fukushima. The attenuation length in air[11] for a typical gamma ray of 1-MeV energy is 15.7 grams per sq. cm, and at sea level density of 1.3 mg/cc, an aircraft patrolling 1 km above ground level is shielded by 130 g/sq cm, or 8.3 mean free paths (MFP). So a signal from a 2-inch cube of NaI radiation detector might appear to be attenuated by a factor 3900 from what it would be at ground level. A contamination level of 1 R/yr at ground level[12] is 100 erg/g-yr and for a 1-MeV gamma ray (1.6×10^{-6} erg/MeV) this corresponds to 66×10^{6} MeV/g-

---

8 http://pbadupws.nrc.gov/docs/ML1118/ML111861807.pdf

9 http://www.ucsusa.org/assets/documents/nuclear_power/UCS-Response-to-NRC-90-day-recs-8-1-11.pdf

10 http://www.ucsusa.org/assets/documents/nuclear_power/ucs-rpt-nuclear-safety-recs.pdf

11 http://physics.nist.gov/PhysRefData/XrayMassCoef/ComTab/air.html

12 For comparison, "First Year DRL (Derived Response Level): If a person is in danger of receiving an external radiation dose greater than 2 Rem during the first year, the EPA recommends relocation until radiation levels decrease. This is not an urgent action because the dose is received over a full year. "

yr. Per second this is 2.20 MeV/s-g and for a 2-inch NaI cube, we need to multiply by about 250 to find that we have about 550/s counting rate from the contamination at ground level. At flight altitudes, this is much suppressed.

For instance, at 1000-m above ground level (AGL) the suppression would appear to be by a factor 3900, to about 0.14 c/s; at 500-m altitude, the attenuation could be only $e^{-4.15}$ or 62, so the count rate would be about 8.7c/s. At a mapping altitude of 500 m, the UAV would need to be warned about the location of radio transmitting towers.

In reality, for a non-collimated detector with poor energy resolution, we should use the energy-absorption coefficient, for 1-MeV photons given by NIST [6] as only 28 g/sq-cm, so that the count rates would be much higher at altitude—specifically, about 5.5 c/s at 1000-m and 55 c/s at 500-m AGL (above ground level).

Because the detector is essentially non-directional, there is an averaging over an area of linear size comparable with the patrol altitude. And the extrapolation to ground level depends on not only the altitude of the aircraft but the altitude of the terrain below, which is readily available from Google Earth.

LESSONS FROM HISTORY OF RADIATION USE AND NUCLEAR ACCIDENTS, PARTICULARLY FUKUSHIMA

RICHARD WILSON
Department of Physics, Harvard University
Cambridge, USA

THE IMPORTANCE OF UNDERSTANDING RADIATION ISSUES FOR IMMEDIATE ANALYSIS AND IMMEDIATE ACTION

Effects on Public Health

I will focus on the public health aspect of the Japanese, and indeed the world, reaction to the Fukushima crisis. Firstly I will summarize those facts that I consider to be important and explain why I believe the reaction was incorrect.

Medical Use of X-Rays.

Very soon after Roentgen's discovery of x-rays in 1895, physicians used them for diagnostic purposes. Although very early on it was realized that they caused skin and other lesions, the fantastic ability to see within the body was so important that physicians correctly argued that the benefits of x-ray use overshadowed any harm. But this only addressed one part of the risk-benefit calculation. Others, physicists in particular, pointed out that the same benefit could be achieved with far less harm by more careful use: shielding, more sensitive film and so forth.

In the 1920s there was more interest in controlling the use. In 1927 the International Commission on Radiological Protection (ICRP) was formed. This is a non governmental body but most governments heed its recommendations. But the advice of ICRP and physicists was not fully heeded till about 1970. In 1961, for example, I had an (unnecessary) chest x-ray at Stanford University and measured my dose, which was about 1 Rem. Now, that same x-ray would take 7 milliRem. But a CAT scan today is nearly 1 Rem.

Hiroshima and Nagasaki

Starting in August 1945, physicists have been involved with extensive nuclear activities. Although there is a common public misconception that 200,000 persons each died because of radiation exposure at Hiroshima and Nagasaki, most of the deaths were due to the blast and radiation accounted for only a few percent. Nonetheless, an unprecedented research activity took place. The Atomic Bomb Casualty Commission (ABCC) (now the Radiation Effects Research Foundation—RERF) was jointly funded by USA and Japanese governments to study radiation effects. The UN started the United Nations Subcommittee on the Effects of Atomic Radiation (UNSCEAR) which lists over 100,000 reports and papers on the subject. Also the U.S. National Academy of Sciences has issued a useful set of reports, "Biological Effects of Ionizing Radiation", which are more readable.

Distinction between Acute Problems and Chronic Problems

The studies find a crucial distinction between the results of radiation exposure in a short period (integrated over a week or two) and the acute effects that it causes, and radiation over a long period of a few years and the chronic effects that it causes. The acute effect of Acute Radiation Sickness (http://en.wikipedia.org/wiki/Acute_radiation_syndrome) starts with a reduction in white blood cell counts and can then lead to tissue damage. It is generally accepted that this occurs at radiation levels above 100 Rems (1 Sv) with an LD50 (least dose at which 50% of people die) if 400 Rems (4 Sv), (formerly believed to be 250 Rems) which can be extended to 500 Rems (5 Sv) by a blood transfusion. The first major example of a death from Acute Radiation Sickness was Dr. Harry Daghlian who was exposed on August 21, 1945 and died 25 days later as a result of a nuclear criticality accident. It is not always realized, but prompt evacuation is only needed to avoid Acute Radiation Sickness.

Hiroshima and Nagasaki provide the data from which effects of radiation are usually determined. As occurs with all chronic effects, they are determined at a high radiation level and a model is used to describe what happens at the lower level. At low levels, the usual model suggests low dose linearity. This comes from the realization that if a medical outcome of a pollutant or action is indistinguishable from one that occurs naturally, any addition to natural incidence is proportional to the dose at low doses. (See: Crump, K.S., Hoel, D.G., Langley, C.H., Peto, R. (1976) "Fundamental carcinogenic processes and their implications for low dose risk assessment." *Cancer Res.* 36:2973-2979, and Guess, H., Crump, K., and R. Peto. (1977) "Uncertainty estimates for low-dose-rate extrapolation of animal carcinogenicity data." *Cancer Res.* 37:3475-3483.)

But there are assumptions and approximations. In the above sentence, I used the word "indistinguishable". They must be biologically indistinguishable and not merely that a pathologist cannot distinguish. There is only one paper, to my knowledge, on this fundamental point. Cancers that occur after radiation therapy have a different DNA structure. There seems to be no interest in exploring this further either for radiation cancers or chemically produced cancers. The coefficient of the linear term is determined from data at high doses. Also the dose in Hiroshima and Nagasaki was over a short period and it is probable that doses over a long period produce smaller effects. There are animal studies that suggest a factor of 2-10 but only two data sets: the occupational doses at Ozerk in 1948 as the Russians were rushing to make a bomb before the wicked Americans killed them, and the Russians exposed at Techa River after the waste pond overflowed.

According to the above theoretical model, if someone gets a dose just below the LD50 he can still get chronic problems of which the most important is cancer. At an integrated dose of 200 Rems, there is a 10% to 20% increase in cancer probability. This depends upon a dose integrated over a long time—of the order of years. It can, therefore, rise well above 200 Rems without causing Acute Radiation Sickness. The natural incidence of fatal cancers is about 20% so no one who gets less than 100 Rems will double his natural incidence and he cannot rightly claim that it is "more likely than not" that his cancer is due to radiation. These numbers are deliberately a little higher than my best belief at the moment and do not account for a dose rate reduction, but they are at the upper end (more pessimistic) of the fraction in the most recent U.S. National Academy Report BEIR VII Table 1. At 100 millSievert, 0.1 Sv or 10 Rems, the increase in fatal cancer probability is 4% or 20% of the natural fatal cancer rate. The number of digits in the number of each entry is high but only the first has any significance. Alas, there is no easily available table where age is broken out.

Table 1.

	All Solid Cancer		Leukemia	
Excess cases (including non-fatal cases) from exposure to 100 mSv	Males	Females	Males	Females
	800 (400–1600)	1300 (690–2500)	100 (30–300)	70 (20–250)
Number of cases in the absence of exposure	45.500	36.900	830	590
Excess deaths from exposure to 100 mSv	410 (200–830)	610 (300–1200)	70 (20–220)	50 (10–190)
Number of deaths in the absence of exposure	22.100	17.500	710	530

Table 1. The table shows the estimated number of cancer cases and deaths expected to result in 100.000 persons (with an age distribution similar to that of the entire U.S. population) exposed to 100 mSv. The estimates are accompanied by 95% subjective confidence intervals shown in parentheses that reflect the most important uncertainty sources including statistical variation, uncertainty in adjusting risk for exposure at low doses and dose rates, and uncertainty in the method of transporting data from a Japanese to a U.S. population. For comparison, the number of expected cases and deaths in the absence of exposure is listed.

2Approximately 42 cancers per 100 individuals calculated from Table 12-4 in Chapter 12 of the BEIR VII report.

3 In special cases, such as in utero exposure, some evidence suggests excess cancers can be detected as low as 10 mSv.

The radionuclides that are produced by nuclear fission are well known, as are their melting points and boiling points. A listing can be found, for example in Table B3 of the report of a study Severe Accidents at Nuclear Power Plants that was carried out for the American Physical Society and is on this website at http://phys4.harvard.edu/~wilson/publications/pp341.pdf and Reviews of Modern Physics, 57, 3, pt. II, July 1985. This is reproduced here as Table 2.

The last (6th) column is for the dose integrated for the first 7 days after the accident.

Table 2.

TABLE II.B.3. Radioactive inventories and whole-body dose conversion factors (3200 MW_t—PWR).

Radionuclides	Half-life $t_{1/2}$ (days)	Shutdown inventory (10^6 Ci)	Cloud D_c/χ_0 (rem m^3/Ci s)	Inhalation, κ_i (rem/Ci inhaled) (0–50 yr)	Ground D_g/S (rem m^2/Ci)
Noble gases					
Kr-85	3950.0	0.56	$0.475E-03$	$0.310E+00$	
Kr-85*m*	0.183	24.0	$0.364E-01$	$0.260E+00$	
Kr-87	0.0528	47.0	$0.181E+00$	$0.100E+01$	
Kr-88	0.117	68.0	$0.467E+00$	$0.230E+01$	
Xe-133	5.28	170.0	$0.906E-02$	$0.700E+00$	
Xe-135	0.384	34.0	$0.567E-01$	$0.120E+01$	
Iodines					
I-131	8.05	85.0	$0.872E-01$	$0.600E+03$	$0.708E+03$
I-132	0.0958	120.0	$0.511E+00$	$0.700E+02$	$0.107E+03$
I-133	0.875	170.0	$0.154E+00$	$0.200E+03$	$0.311E+03$
I-134	0.0366	190.0	$0.533E+00$	$0.300E+02$	$0.414E+02$
I-135	0.280	150.0	$0.419E+00$	$0.150E+03$	$0.285E+03$
Cesiums and rubidiums					
Cs-134	750.0	7.5	$0.350E+00$	$0.470E+05$	$0.369E+04$
Cs-136	13.0	3.0	$0.478E+00$	$0.590E+04$	$0.410E+04$
Cs-137	11000.0	4.7	$0.122E+00$	$0.360E+05$	$0.131E+04$
Rb-86	18.7	0.026	$0.207E-01$	$0.660E+04$	$0.185E+03$
Telluriums and antimony					
Te-127	0.391	5.9	$0.936E-03$	$0.340E+02$	$0.813E+00$
Te-127*m*	109.0	1.1	$0.110E-02$	$0.240E+04$	$0.584E+02$
Te-129	0.048	31.0	$0.147E-01$	$0.980E+01$	$0.198E+01$
Te-129*m*	0.340	5.3	$0.783E-02$	$0.300E+04$	$0.246E+03$
Te-131*m*	1.25	13.0	$0.314E+00$	$0.550E+03$	$0.960E+03$
Te-132	3.25	120.0	$0.475E-01$	$0.150E+04$	$0.308E+04$
Sb-127	3.88	6.1	$0.151E+00$	$0.790E+03$	$0.920E+03$
Sb-129	0.179	33.0	$0.268E+00$	$0.110E+03$	$0.104E+03$

I call attention to the isotopes of iodine and of cesium. The former is normally gaseous and is easily released and the latter, although normally solid, is soon evaporated in an accident. Only in the high temperature of a nuclear explosion would it be likely to emit large quantities of strontium, uranium or plutonium.

Nuclear Power—Normal Operation

Physicists and engineers have been urging careful use of radioactive materials. A modern nuclear power station emits very little radioactivity. Indeed, it is often stated (correctly) that a coal-fired power station in its particulate emissions emits more. Also the exposure to the plant workers can be kept low without sacrificing performance. They (as health physicists) have set standards which are low and can be met with little cost. The benefit of a low radiation exposure is not limited by a high cost to the consumer of electricity.

But when the situation in a power plant is not normal all changes. The habits, rules, customs about radiation exposure should change accordingly and the change should be automatic and instantaneous and therefore prepared in advance. This did not happen at Fukushima. The need to balance risks is similar to the physicians' situation in 1900-1970.

Windscale, TMI and Chernobyl

There were three reactor accidents from which lessons can be learned. At Windscale in 1957, a plutonium production reactor caught fire and iodine was released. Short lived radioactive iodine (I 131 with 10 day half life) can make the major immediate hazard with a well known chain. Iodine can fall to the ground and be eaten by cows where it concentrates in the milk and babies drink the milk and concentrate the iodine in the thyroid. This has been realized for 60 years and at the Windscale accident in the UK in 1957, the government impounded and bought all milk for a couple of months. (Curiously the cows produced twice as much as usual, although this increase is not usually attributed to radiation!)

No one knows exactly how much iodine was ingested at Chernobyl, but a lot. 2,000 children got thyroid cancer of which 20 have died. No one need have gotten thyroid cancer if it were not for secrecy. There are anecdotes (which I believe) that a school teacher near Hohnichi (Belarus) and an Army general in eastern Ukraine were reprimanded by the KGB for advising children not to drink milk for a month (the half life of the iodine is 10 days or so) and thereby causing a panic. This was, and is, far less likely to happen in an open society in Japan.

There is disagreement about the effects of potassium iodide. If ingested before radioactive iodine exposure, it can reduce the ingestion of the radioactive substance. But, there are suggestions that, if taken after exposure to radioactive iodine, it can lock in the radioactive iodine already taken. Moreover, there are other side effects particularly for pregnant women so it is wise not to take it unnecessarily.

At TMI in 1979, there was a partial meltdown but mostly contained. I can find no report of what happened to the iodine, but believe that it combined with water to form HI which was pumped out of the containment into the turbine building where it stayed quietly on the floor.

After Chernobyl in 1986, we confirmed that the important releases for the long-term effects are Cesium 134 (2 year half life) and cesium 137 (30 year half life), and that the radiation from the ground deposition is the important pathway with ingestion only about 25%. Evacuation from Pripyat was delayed 36 hours and Chistallogovka 3-4 days. But prompter evacuation would not have changed the dose much. Only at Chernobyl did anyone in the plant get Acute Radiation Sickness. No one in the general public did.

In the preceding paragraphs, I note that exposure to Cs 134 and Cs 137 as the dominant long term problems. The measurements of radioactivity deposition confirms that deposition and, therefore, emission of strontium 90 and the transuranic elements was much less, even though the initial explosion dispersed them locally, and the subsequent graphite fire must have reached thousands of degrees and almost all the cesium was evaporated. That is an important observation because strontium and plutonium in particular are "bone seekers" as they enter the body giving long term irradiation to various organs.

One important feature is that the effects on health of these low levels of radioactivity are calculated, and not measured. The calculated number is too small to be directly measured. Indeed, the number might be zero if the radiation cancers are not identical with those occurring naturally or might even be less than 1 if hormesis exists. There is no hormesis for those washed out to sea in the Tsunami. Those who wish to dramatize the effect tend to stress the total number of calculated fatal cancers. Typical numbers discussed are 4,000–8000 in the USSR countries (Ukraine, Belarus and European Russia) and 20,000 world-wide. The latter is to be compared with the billion or so naturally occurring cancers in the world in that time period. But for discussions of how to manage an accident more direct information is appropriate.

Fukushima

Armed with this information, I looked at the measurements from Fukushima dai-ichi. I looked at the data on the radiation spike measured at the gate of the Fukushima complex and integrated it.

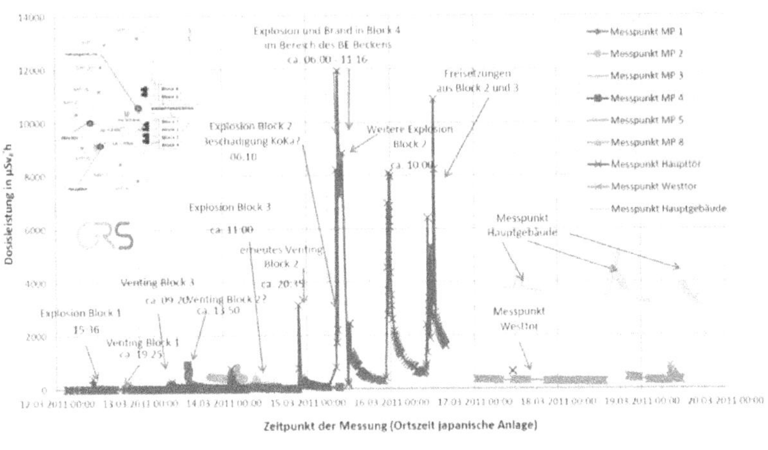

Fig. 1: (This particular figure comes from a German website.)

The releases on Friday, Saturday, Sunday and Monday were not serious. The big doses were on Tuesday, March 15th and Wednesday, March 16th. The spikes were probably doses from noble gases. The integrated dose was large—0.02 Sv (or 2 Rems). But this is less than 1 year of normal occupational dose and it should not have prevented a radiation worker from going to or being near the plant. Indeed the report to IAEA states that the average for all power plant workers as of May 23rd was only 7.7 mSv or 770 mRems, which is about the amount of a CAT scan.

Starting on Thursday, March 16th, the reactors and the spent fuel pools were being cooled by sea water, and there has been no comparable release since that time. Taking the usual decay of Cs 134 and 137 into account, one would expect an immediate drop, and my estimate would be for about 0.06 Sv (6 Rems) at the main gate for the first year and falling more slowly thereafter. At 3 miles (5 Km), this would be down a factor of 10.

Adverse Effects on Health of Dislocation or Evacuation

It has been noted in the medical community for many years that there are stresses and problems associated with relocation that can lead, by themselves, to adverse effects on public health. In 1975, I saw figures of a 5% increase in cancer probability. I note that in an accident situation this would only be a calculated increase but in that sense is directly comparable to any increase in cancer rate due to radiation. It is hard to find good numbers and I merely refer to a recent review: "Health effects of relocation following disaster: a systematic review of the literature." Uscher-Pines, L. *Disasters*. 2009 Mar:33(1):1-22. She opined:

> "Despite the frequency of post-disaster relocation and evidence of its effect on psychological morbidity, there is a relative paucity of studies; the few examples in the literature reveal weak study designs, inconsistent results, and inattention to physical health impacts and the challenges facing vulnerable populations. Further research guided by theory is needed to inform emergency preparedness and recovery policy."

In the 1980s, Dr. Crouch and I looked at the unexposed "control" rats and mice from the U.S. National Toxicology program. The rate of cancer varied many percent. For example, we found that, in some experiments, the lights in the cages were on continuously and these were rodents with an elevated "control" rate. In the large study of 30,000 mice at the National Center for Toxicological Research (NCTR), the ED01 or "megamouse" study, there was a variation in response according to where the cages were. Those on the top shelf got tumors later than those on lower shelves. I presume that is because their stress was less. But what counts as stress for a mouse is unclear.

Similarly stress from divorce and separation, or even merely lack of a partner, has been blamed for a 15% increase in the probability of cancer fatalities.

Oystein, Kravdal. "The Impact of Marital Status on Cancer Survival." *Social Science and Medicine* (2001) 32:357-368. I take 5% increase in cancer probability from relocation as a reasonable "lower limit" to the increase of cancer with the stress of evacuation.

The "official" Kemeny report after the Three Mile Island accident stated (inter alia):

"We conclude that the most serious health effect of the accident (for any reason) was severe mental stress, which was short-lived. The highest levels of distress were found among those living within 5 miles of TMI and in families with preschool children."

Japanese have noted unexpected deaths in the elderly who have been evacuated. Significant numbers of the elderly in shelters have died unexpectedly. Maybe the calorie intake is below starvation level and not all have three meals a day. Lack of hot food, running water, crowding, poor toilet facilities and lack of water for cleaning people and locations, lack of fuel and lack of hospitals to accept admissions, ambulances or medical services except what appear to be medical personnel who are themselves local victims. Yet there is no indication that these were considered by those ordering an evacuation! A simple calculation shows that this can far exceed any benefit evacuation may bring.

<u>Doses After Fukushima</u>

I first looked at doses in various locations listed by the Japanese Atomic Industrial Forum.

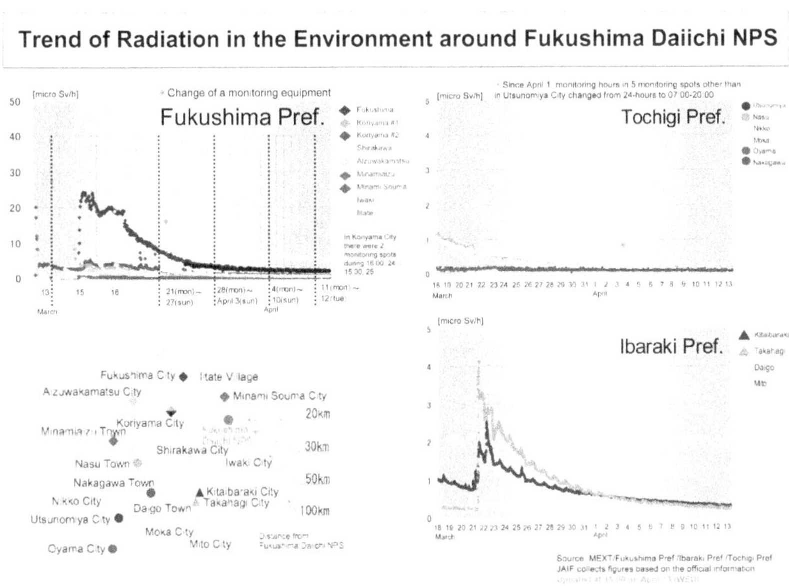

Fig. 2: <u>http://www.jaif.or.jp/english/news_images/pdf/ENGNEWS01_1302054182P.pdf</u>

Indeed the newspapers emphasized the doses in the Ibaraki region, on the way to

Tokyo. The abscissa is microSievert per hour. But the dose seems much smaller when the doses in the Ibaraki prefecture are plotted with a different abscissa and ordinate.

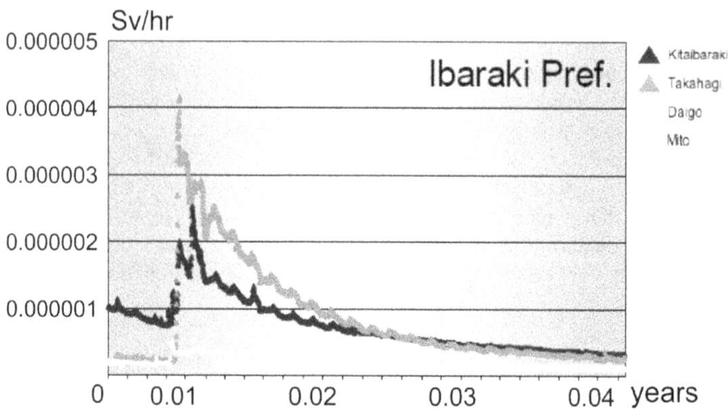

Fig. 3.

One microSievert per hour, kept up for a year, would give 8760 microSievert, or 8.76 milliSv or 876 milliRems. What does this mean?

Many actions can give anyone a dose of 876 milliRems:

- A single chest x-ray in a major hospital as late as 1960.
- A CAT scan today.
- 7 months allowable occupational dose.
- 1/25 of what a Chernobyl clean up worker got.
- 1/100 of an astronaut's allowed dose.
- About the dose I got in 1991 from a day at Chernobyl, mostly inside the sarcophagus.

Any serious student should evaluate his own lifetime dose for comparison. I have been officially a "radiation worker" since 1946. Yet my integrated dose is almost all due to medical x-rays—and I have no record of these for the first 20 years. The American Nuclear Society has a website which enables a good estimate to be made on line: Radiation Dose Chart: http://www.new.ans.org/pi/resources/dosechart/. Ideally there would be such a program with the fatal cancer rates at different ages.

In the last year, I received 2.4 Rem (0.02 Sv). I believe it is absurd to evacuate to avoid this small a dose. Certainly evacuation should not be mandatory. Of course if we go further and consider age, evacuating an 85 year old for anything except Acute Radiation Sickness is really stupid. The graphs suggest that no one in the Tochiki or Ibaraki prefectures should be concerned, but naturally they will be interested.

For the area NW of the plant, the best description of the doses comes from the Ministry of Education, Culture, Sports, Science & Technology (MEXT) in Japan. (http://www.mext.go.jp/english/)

Their map (Figure 4) shows that the major deposition was to the NW of the plant toward Fukushima itself. MEXT integrated the doses using the usual model, shown on Figure 4, and it seems that the highest dose was 0.15 Sv (15 Rems) integrated to March 2012. This is for continued occupancy in the open. It can clearly be reduced by being indoors and avoiding high dose hot spots.

For comparison, this was the official dose allowed for the "liquidators" (clean up workers) after Chernobyl. According to the studies of effects of radiation, this can add 1% or so to cancer probability. This is well below the variation in natural cancer incidence (up to 30%) from unknown causes.

There is no indication of any large deposition of either strontium or transuranic elements, suggesting that the internal deposition by these elements can be ignored.

Dose Readings Map (as of April 24, 2011)

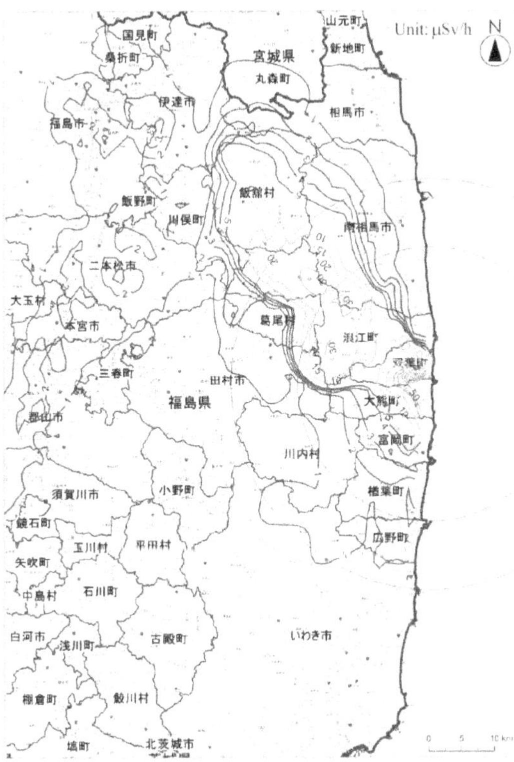

Based on the latest values indicated in the press releases at 10:00 and 13:00, April 24.
Locations monitored previously: Based on the values converted on April 24 using the ratios to the value of Location No.32.

Fig. 4: Shows a radiation map of the area NW of the plant, in April 2011.

From this, MEXT use a theory to estimate the integrated dose to March 2011 shown in Table 3. This table is a selection from the MEXT data showing the highest figure in all their measurements and a few low ones.

Table 3. Estimated Dose Integrated to March 2012 at Each Monitoring Location based on Measured Values (MEXT 2011).

	From Fukushima Dai-ichi NPP		Estimates of Integrated Dose	Latest Readings (Average)	Estimates of Integrated Dose as of March 11, 2012 (mSv)
Location	**Direction**	**Distance**	**mSv**	**mSv/h**	
Akogi Kunugidaira, Namie Town, Futaba County	Northwest	24km	68.2	0.0374	224.9
Akogi, Ishigoya, Namie Town, Futaba County	Northwest	30km	37.1	0.0158	103.1
Akogi Teshichiro, Namie Town, Futaba County	Northwest	31km	32.9	0.0163	101.0
Onami Takinoiri, Fukushima City	Northwest	56km	3.8	0.0018	11.4
Aza Kitaaramaki, Hisanohama Town, Iwaki City	South	31km	1.0	0.0002	1.6
Takahagi, Ogawa Town, Iwaki City	South-southwest	36km	0.6	0.0002	1.4

The Evacuation Decision

The question the Japanese faced was how much to evacuate. I believe they flubbed. They had not thought in advance. Then they panicked. It is unclear whether the evacuation was ordered by the government or merely suggested. But it was without analysis. But this is clearly forgivable given the history of radiation effects in Japan and the failure of the world community to provide guidance. Indeed for 30 years the world community has set guidelines which, I argue, are stupid.

They should have asked the questions:

- Is there an immediate reason to evacuate to avoid Acute Radiation Sickness? The measured doses give the answer NO.
- Would there be an appreciable increase in long term radiation dose by waiting a few days to analyze? Again the realization that Cesium was the problem would demand the answer NO.
- In retrospect was it sensible to evacuate people beyond 3 miles from the reactor, bearing in mind the competitive risks? I submit that here again the answer is NO.
- Would it have been wise to inform everyone and prepare for a VOLUNTARY evacuation for those who wished it, which preparation could avoid the chaos that occurred in New Orleans after Katrina? Here the answer is definitely YES.
- Is there an adverse effect on health in evacuation? Here the answer is definitely YES although often ignored.

I submit that the whole world nuclear power and safety community, including semi-political agencies like IAEA and politicians themselves, should ponder the above.

American friends of the Japanese people should ask themselves the following questions:

- What is the role of friends who believe they are experts?
- Careful analysis along the lines of the early part of this report?
- Off the cuff remarks at a Senate budget hearing?

Dr. Gregory Jaczko, Chairman of the Nuclear Regulatory Commission gave the following testimony Jaczko to the U.S. Congress on March 17, 2011:

> "Recently, the NRC made a recommendation that based upon the available information that we have, that for a comparable situation in the United States, we would recommend an evacuation to a much larger radius than has been currently been provided in Japan. As a result of this recommendation, the ambassador in Japan has issued a statement to American citizens that we believe it is appropriate to evacuate to a larger distance up to approximately 50 miles."

This was repeated by President Obama on Thursday, March 17, 2011. President Obama made remarks from the White House Rose Garden on the nuclear crisis in Japan shortly after paying an unannounced visit to the Japanese Embassy. After expressing condolences to the Japanese people, the president confirmed calling for an evacuation of U.S. citizens within a 50 mile radius of the reactors in northeastern Japan.

These were outlandish and not the remarks of a true friend. I so stated in a formal fax to Dr. Jaczko as soon as I saw the transcript. In my carefully considered view, Dr. Jaczko and President Obama should visit Japan and apologize to the Japanese people, perhaps by bowing deeply to the Japanese legislature. This would be politically difficult for President Obama in his present travails, but may be possible the day after the 2012 election—whichever way the electoral decision goes.

I have been told that Jaczko was merely following an NRC rule: keep the dose less than 500 mRem in the immediate accident and < 2 Rem over the first year. I argue that the events at Fukushima demonstrate clearly how stupid, and counterproductive to public health, that rule is, and it becomes a matter of urgency to modify it.

I note that in an emergency, someone in NRC has to be empowered to act without waiting for a Commission, or committee meeting. Chairman Jaczko claimed, and still claims, that power. But it was never an emergency in the USA, and it is certainly now over. His outlandish recommendations on Thursday, March 17th suggest that he is not capable of exercising that power wisely.

<u>Radiation Accident Management</u>

At Fukushima, there was no proper management of radiation doses immediately after the realization that the reactor situation was out of control (immediately after the tsunami). There seems to have been no realization in Japan, and probably no realization anywhere else, of the fact that radiation management after an accident should, even must, differ from radiation management immediately before the accident.

Before TMI (before 1980), it was generally accepted that there were certain radiation levels that should not be exceeded. After TMI, and even more after Chernobyl, these were reduced. While it makes some sense to keep, for example, to 5 Rems/yr (0.05 Sv/yr) for a nuclear power worker in ordinary operation it is, I believe desirable to return to the higher figures as soon as an accident goes beyond normal operation. Thus it should be allowed for a worker to plan for 20 Rems (0.2 Sv) for the whole accident, and indeed at Chernobyl 100,000 or more workers got this dose of 20 Rems as "liquidators" (clean up workers). A one time dose of 80 Rems (0.8 Sv) was allowed for an astronaut and for a rare individual "to save lives", 80 Rems was allowed. It is reported that at Fukushima workers were pulled off the job on Sunday and Monday in the Fukushima accident before the Japanese belatedly restored the pre-1980 levels. This probably delayed a proper technical response to the accident. Although at all ages it is important to keep the radiation dose below that, given Acute Radiation Sickness, full use of older workers, particularly volunteers, should be taken. A person over 70 years old with a high accumulated radiation dose will develop cancer only after 20 years and then it is the least of his worries.

It is unclear whether my recommendation of an immediate reversion to the pre-1980 radiation levels would have enabled TEPCO to control the reactors any better. I think they would.

A Lesser Issue: Responsibility of the Media.

News media have been the principal method of communicating with the public, and even with experts. At Three Mile Island, at Chernobyl and at the Tokai incident, the U.S. media failed miserably and forgot their duty. Neither one newspaper, nor the Associate Press quoted the precise NRC press releases. None of the major newspapers even got the units correct. The internet has improved this. Experts can find information directly. But there is still a responsibility to inform the public and, in particular, to explain what the radiation dose levels mean, in terms of public health and to discuss the harrowing decisions those on the spot must make.

I next turn to accident prevention.

The aim is to prevent the undesirable fission products ever coming into contact with the public. There are several barriers:

1. The fuel is in zirconium pellets which are in a zirconium tube. Although design criteria allowed 0.1% of these tubes to leak, probably none did.
2. If the first barrier fails, there is a pressure vessel which should hold them.
3. If the pressure vessel fails there is a containment vessel.

BUT, one must emphasize the importance of keeping barrier 1 intact if possible. Of course it is always important to stop an untoward event as early in the chain as possible. But it is especially important because failure here makes it harder to control. In a BWR one can be close to the reactor in operation as I personally have been—one is shielded by the water in the pressure vessel. Once barrier 1 fails, doses are higher outside the pressure vessel increasing radiation doses for a worker and making subsequent fixes harder.

Man Rems (Person-Sievert) or Rems/man (Sv/person)?

In the preceding paragraphs I have emphasized the dose per person (Rems/man or Sv per person) because that matches the decisions that I was discussing. But in radiation protection, it is common to calculate the collective dose in Man-Rems or Person Sieverts) because when using a linear dose response relationship and multiplying by the appropriate slope (coefficient), this gives the total societal impact. I do this for the next section on comparing disasters and also take a more optimistic slope allowing for a dose rate reduction factor.

Comparison to Other World Disasters

The effects of evacuating or not evacuating should be compared to 15,000 dead and 15,000 missing direct, measurable and definite "dead bodies" from other earthquake and tsunami problems.

Fatal cancers calculated from Natural Background (including medical) exposures (world wide) about 10,000,000 per year. In an average ½ lifetime 300,000,000.

Estimated Deaths from arsenic in Bangladesh assuming everyone has pure water henceforth: 500,000.

Earthquake in Haiti: 200,000.

Earthquake and Tsunami in Japan (prompt deaths: 30,000).

Fatal cancers from Chernobyl in next 60 years (calculated excluding effects of stress): 7,500 in Belarus, Russia and Ukraine. 20,000–30,000 worldwide.

Cancer fatalities from Three Mile Island: 0.7 calculated for the Kemeny Commission.

My rough guess is 500 calculated cancers from radiation from Fukushima. (Originally my incorrect prediction was close to zero. But it could be zero or even slightly negative.)

0-5,000 adverse health effects of evacuation.

I emphasize that the calculated cancers are within the fluctuations of the natural cancer numbers and cannot ever be directly measured.

My Recommendations for Study of Radiation Emergencies.

I have argued since 1980 that there should be detailed study of a number of fundamental issues. (These should be for other pollutant substances and actions also.)

1. Are cancers caused by radiation truly indistinguishable from naturally occurring cancers? Or is it just that a pathologist cannot distinguish? (Use DNA analysis.)
2. What is the effect of dose rate? (Look carefully at such data as the Techa River and the Mayak workers.)
3. What is the effect of disaster stress on cancer? In people? In animals?

Before 1980, the U.S. Nuclear Regulatory Commission asked for an "Emergency Planning Zone" (not an evacuation zone) of 10 miles diameter. After TMI, this became an "evacuation zone" without the detailed discussion such a decision requires. I, personally, was opposed to this implication for automatic evacuation, and testified to an Ontario Royal Commission and others that it was a mistake. I strongly urge the International

Community to reexamine this requirement.

My Recommendations for Continued Nuclear Power

There have been already several discussions on what to do about existing nuclear power plants, and proposals for new ones throughout the world. These include recent fine articles in the *Bulletin of Atomic Scientists*. Many of these discussions echo previous views expressed by the authors asking that certain issues already decided be reconsidered. I here make my comments on several of them: I do not give the detailed discussion that each topic deserves.

1. We should obviously reexamine every existing nuclear power plant to see whether and how whether any specific problem raised by Fukushima can be easily addressed. We must remember that when the first event tree analysis was done in 1976, it was found that the reactor (Surry) could be made 5 times safer at small expense by modification of a control system. I agree with the Japanese report to IAEA that even extreme events such as a combined earthquake, tsunami and bureaucratic muddle should be analyzed by event tree analysis.
2. Since it is claimed, and probably true, that newer designs of reactors would be safer and would have survived the terrible events of earthquake and tsunami, should we not *immediately* retrofit all older reactors? And, in particular, all 23 older GE reactors in the USA? This question has been repeatedly raised over the last 30 years. I believe the question is posed too narrowly. It is too restricted to nuclear plants. Should we not retrofit or get rid of all older energy plants? The cost of retrofits can be high and, in many cases, would close the plants down. Once the question is broadened, we might decide to retrofit, and perhaps close, plants in order of the estimated risk to public health. For example, start by closing every coal fired power plant and only then consider the older nuclear ones.
3. One of the largest risks we face in the world is the risk of starting a catastrophic all out nuclear war. We must remember that bombs with 100 pounds of TNT killed many people in World War II. Yet Sakharov's test explosion at Nuovo Zembla over 40 years ago was of a bomb ONE BILLION times more powerful. Should we not consider this carefully in all applications of nuclear energy or failures to use nuclear energy? In particular, should that not be a fundamental question asked by the Nuclear Regulatory Commission? While it is evident to me that this issue MUST be in the front of all our minds from now to eternity, it is far less clear that the Nuclear Regulatory Commission is the proper place. Already, as noted above, they have internal conflicts between discussion of "ordinary" operation, and reaction to an accident mode which has resulted in the silly statements mentioned earlier. It goes far beyond Fukushima or indeed beyond any other public safety or public safety issue. Where, when and how proliferation questions should be discussed is, perhaps, the most important long-term issue. Maybe I should invert the question. When are the few occasions when non-proliferation need NOT be discussed? It has been urgent for at least the last 60 years.

NUCLEAR AFTER FUKUSHIMA

LADY BARBARA THOMAS JUDGE
Emeritus Chairman, UK Atomic Energy Authority
London, UK

It is common knowledge that the world is set to go from six to nine billion people in the next 20-30 years. In order to cope with the energy challenges encompassed in this daunting thought, there are three questions which many nations must face.

1. Energy security—do we have enough energy?
2. Energy independence—where is the source of that energy located?
3. Climate change—are we changing our climate to the detriment of our children and our children's children?

The only type of energy that answers all of these three questions is nuclear.

1. With respect to energy security, if a country builds a nuclear power plant, it can build it according to its own current and future specifications and needs, and it will have a source of base-load generation which will provide energy 24 hours a day.
2. Energy independence—if a country builds a nuclear power plant within its own borders, it will not need to deal with any other country with respect to the price, quantity or transmission of energy across any national boundaries.
3. With respect to climate change, it is fairly well-known that nuclear energy emits very little carbon, and is one of the best sources of clean energy currently available. This is so particular because renewable energy, worthy as it may be, is only "top-up" energy. That is, if it is solar, it works well when the sun shines, and if it is wind, it works well when the wind blows. Notwithstanding, however, there is a possibility of a cold, dark, still night when, if you were relying upon a renewable source of energy, when you flick the switch, the lights will ***not*** go on.

Accordingly, it ought to be self-evident that nuclear energy should be part of most any country's energy mix.

The accident at Fukushima has, however, made many nations rethink their tentative steps towards building nuclear power plants. Before Fukushima, both Three-Mile Island and Chernobyl were nuclear accidents that had to be considered when rethinking the return to nuclear power, but informed people understood that Chernobyl was an accident waiting to happen. It was the result of old technology, badly maintained with operators who were "fooling around" and not responsible, and the Russians determined not to "make an issue of it".

Then there was Three-Mile Island which was not a failure, but a success. When there was a problem in the reactor, everything turned off, no one died, and indeed no one was hurt.

Fukushima itself was a case of 40-year old technology, badly maintained where there was a cosy relationship between the regulator and the operator, so that much needed repairs were deferred and operating standards were not as high as they ought to have been.

Even without the advent of Fukushima, however, there are certain policy issues which any country must consider in order to determine whether to build nuclear power plants at this time.

These issues are as follows:

1. Politics—nuclear is political.
2. Planning—correct siting of nuclear power plants is of paramount importance as Fukushima has graphically shown us.
3. Skills shortage—there is a dearth of nuclear scientists, physicists and operators which needs to be addressed.
4. Financing—nuclear power plants are expensive and financing is difficult because of the long-term nature of the finance.
5. Parts—if all of the nuclear plants that are being discussed, are actually built, there will be a queue for certain heavy forgings and other important parts.
6. Proliferation—operating a nuclear power plant is not proliferating in itself but there are issues with respect to the front-end (enrichment) and the back-end (reprocessing).
7. Waste—the common belief is that the best way to deal with nuclear waste is by deep geological storage. Until sites are chosen however, and construction is completed, dry cask storage on site will be often used.
8. The Press—reporting on a good story about nuclear does not sell newspapers.
9. Radiation—fear of radiation is difficult to educate against, as it is irrational and unscientific.

Notwithstanding the difficulties posed by some of the issues set forth above, many countries in the world are choosing to continue with their nuclear plans. Some countries, however, that were hesitant to begin consideration before Fukushima (especially those with strong green parties) are using the opportunity to withdraw, so that, whereas China, India, Turkey, Abu Dhabi, Finland, the Netherlands and the UK, among others, are continuing with their nuclear plans, countries like Germany, Italy, Switzerland and Sweden are withdrawing.

It appears that the developing world understands that it needs nuclear energy and will not be deterred, whereas the developed world is more cautious.

From my own point of view, I believe that nuclear is not ***the*** answer but it is an important ***part*** of the answer. The world needs a bouquet of energy sources. We need oil, gas, coal, renewables and we need nuclear, and without all of them, we will not be able to deal with the difficult energy questions we will all sooner or later be forced to face.

FORESTS AS A CHANCE FOR SUSTAINABLE DEVELOPMENT OF RURAL AREAS IN THE LIGHT OF UN CLIMATE CHANGE CONVENTION AND UN CONVENTION ON BIODIVERSITY

JAN SZYSZKO
Forestry Science, University of Warsaw
Warsaw, Poland

CO-AUTHORS: AXEL SCHWERK, IZABELA DYMITRYSZYN, KATARZYNA SZYSZKO, MAGDALENA BODZENTA, AGATA JOJCZYK, JAROSLAW MALCZYK

SUSTAINED DEVELOPMENT

Sustained development is prompt economic development combined with rational use of environmental resources and respect for human rights.

Some of the indicators of sustainable development of rural areas

1. Increase rural income.
2. Unemployment drop.
3. Reduced emission of greenhouse gases according to the UN Framework Climate Change Convention and the Kyoto Protocol.
4. No disappearance (return) of native species according to UN Biodiversity Convention.

Facts

- Increase of GHG concentration.
- Loss of biodiversity.
- Deforestation.
- Forest degradation.
- Increase of unenployment in rural areas.
- Decrease the number of workers in rural areas.
- Degradation of soil—decreasing of biomass production.
- These problems are connected with each other. Carbon as a key factor.

Why?

Industry
Land use
Intensification of agriculture
Forest degradation

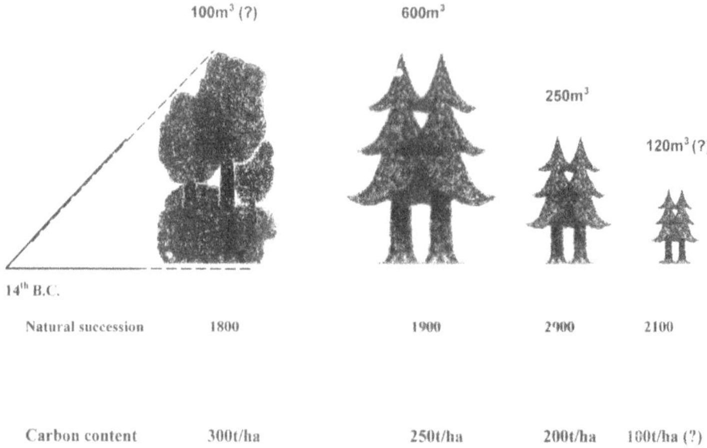

Carbon distribution in native forests in cool temperate forests

Vegetation:	160 t/ha
Soil:	140 t/ha
Litter:	25 t/ha
Total:	325 t/ha

Our main purpose:

Reducing the concentration of CO_2 in the atmosphere.
Protection and return of native species.
Increase of place of work in rural areas.

Some Proposals:

Renewable energy
CCS and absorption of CO_2 based on Land Use Land Use Change an Forestry.

Biodiversity and carbon distribution in young plantations

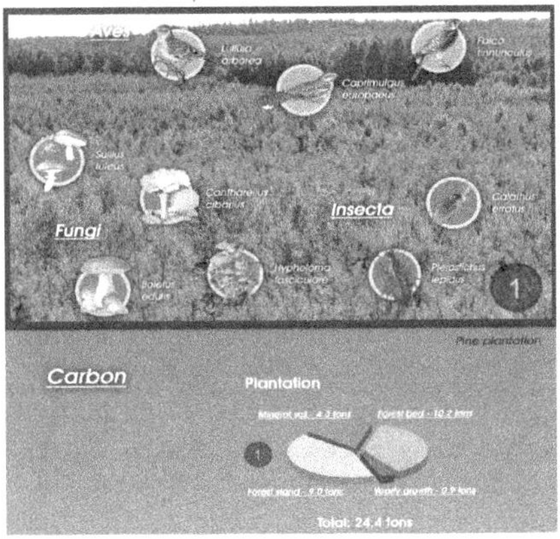

Carbon/ha: 24.4 t
Carbon dioxide/ha: 89.47 t

Value*/ha: 2071.2 €
Value-incr./year: 76.3 €

** Price of 1 t of carbon dioxide on 15.08.2008: 23.15 Euro (www.pointcarbon.com)*

Biodiversity and carbon distribution in old pine forests

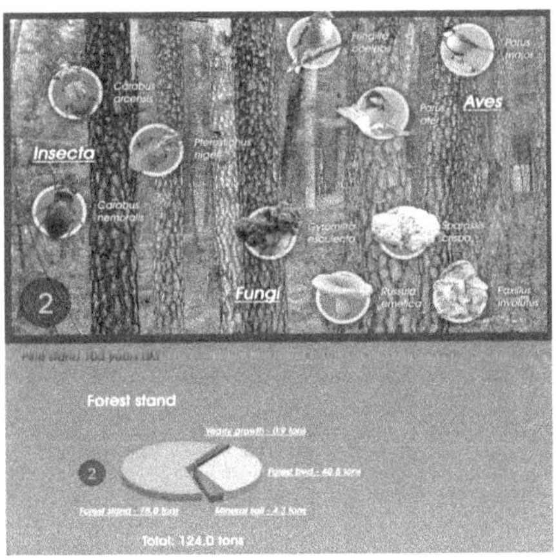

Carbon/ha: 124.0 t
Carbon dioxide/ha: 454.67 t

Value*/ha: 10525.6 €
Value-incr./year: 176.3 €

** Price of 1 t of carbon dioxide on 15.08.2008: 23.15 Euro (www.pointcarbon.com)*

Biodiversity and carbon distribution in beech forests

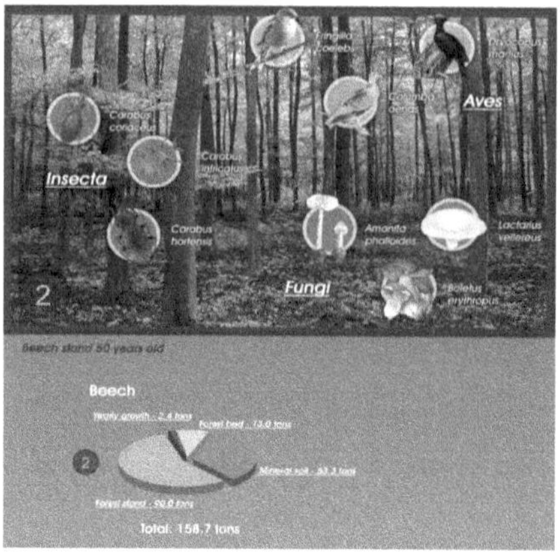

Carbon/ha: 158.7 t
Carbon dioxide/ha: 581.90 t

Value*/ha: 13470.0 €
Value-incr./year: 203.8 €

* *Price of 1 t of carbon dioxide on 15.08.2008: 23.15 Euro (www.pointcarbon.com)*

Carbon contents in natural living systems are responsible for occurrence of native species.

Forest is a chance for sustained development of rural areas if we include them to the European Trade Emision System

WHY?

<u>Price of one ton of carbon dioxide</u>

1. ETES—12.3 euro (www.pointcarbon.com) price on 16 August 2011

<u>The cost of absorbing one ton of carbon dioxide</u>

1. CCS—60-100 euro
2. Forest (purchase of land, afforestation)—3 euro

<u>Because we have in EU</u>

- Many millions hectares of degradated land for forestation.
- One hectare absorbs 16 tonnes of carbon dioxide per year.
- It is 16 x 12.3 euro = 196.8 euro/year www.pointcarbon.com (price on 16 August 2011)

<u>Because we have in the EU</u>

- Many millions of hectares of degradated forests for regeneration.
- One hectare is about 1.5 tonnes absorption of carbon dioxide per year in 100 years.
- It is 1.5 x 12.3 euro = 18.45 euro/year (price on 16 August 2011)

These amounts can be used to create jobs in rural areas only.

<u>Because we have in the EU</u>

- Several millions hectares of degradated agricultural soils for regeneration.
- One hectare is about 1 tonne absorption of carbon dioxide/year in 100 years.
- It is 1 x 12.3 euro = 12.3 euro/year (price on 16 August 2011).

We must develop plans for sustainable development on a regional level, which are connected to climate, culture and biodiversity for creation of places of work in the rural landscape.

CLIMATE PHYSICS, FEEDBACKS, AND REDUCTIONISM—WHEN DOES REDUCTIONISM GO TOO FAR?

RICHARD S. LINDZEN
Earth, Atmospheric and Planetary Sciences, Massachusetts Institute of Technology, Cambridge, USA

INTRODUCTION

The public perception of the climate problem is somewhat schizophrenic. On the one hand, the problem is perceived to be so complex that it cannot be approached without massive computer programs. On the other hand, the physics is claimed to be so basic that the dire conclusions commonly presented are considered to be self-evident. Consistent with this situation, climate has become a field where there is a distinct separation of theory and modeling. Commonly, in traditional areas like fluid mechanics, theory provides useful constraints and tests when applied to modeling results. This has been notably absent in current work on climate. In principle, climate modeling should be closely associated with basic physical theory. In practice, it has come to consist in the almost blind use of obviously inadequate models. In this paper, I would like to sketch some examples of potentially useful interaction with specific reference to the issue of climate sensitivity.

It should be noted that the above situation is not strictly the fault of modelers. Theory, itself, has been increasingly idealized and esoteric with little attempt at real interaction. Also, theory in atmospheric and oceanic dynamics consists in conceptual frameworks that are generally not mathematically rigorous. Perhaps, we should refer to it as physical or conceptual reasoning instead. As we shall see, when reductionism goes beyond the constraints imposed by these frameworks, it is probably going too far though reductionism remains an essential tool of analysis.

THE GREENHOUSE EFFECT

Any discussion of anthropogenic climate change must begin with the greenhouse effect. To be sure, the greenhouse effect, itself, does not lead to concern. Concern depends on the alleged presence of strong positive feedbacks in the climate system. The crucial question is whether the high sensitivity implied by such feedbacks is correct.

However, all attempts to estimate climate sensitivity depend on how the climate greenhouse actually works. Despite the concerns with the greenhouse effect that have dominated environmental thinking for almost a quarter of a century, the understanding of the effect is far from widespread. Part of the reason is that the popular depiction of the effect as resulting from an infrared 'blanket' can be seriously misleading, and, as a result, much of the opposition that focuses purely on the radiation is similarly incorrect. The following description is, itself, somewhat oversimplified; however, it is probably adequate for understanding the underlying physics.

First, one must recognize that the troposphere, the layer of the atmosphere in contact with the surface, is a dynamically mixed layer (as schematically illustrated in Figure 1). For a gaseous atmosphere, mixing requires that the resulting atmosphere is characterized by temperature decreasing with altitude. The rate of decrease is approximately 6.5K/km which is sometimes taken as an approximation to the moist adiabatic lapse rate, but the real situation is more complicated. To be sure, in the tropics, the mixing is effected by moist convection, but outside the tropics, the mixing is accomplished mostly by baroclinic eddies. Moreover, the moist adiabat in the tropics does not have a uniform lapse rate with altitude (viz the 'hot spot' discussed in Section 3a). For our immediate purposes, the important facts are that the lapse rate is positive (not zero or negative), and relatively uniform over most of the globe.

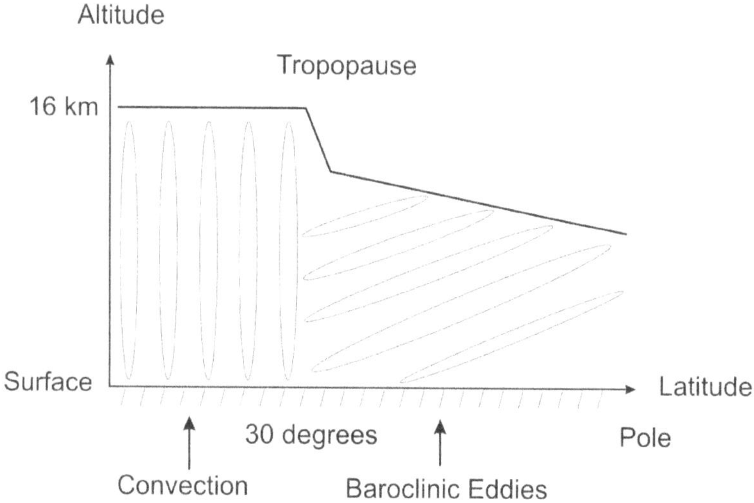

Fig. 1: Schematic of the dynamically mixed troposphere. Vertical convective exchange dominates the tropics, while baroclinic eddies dominate the extratropics.

Second, one must recognize that gases within the atmosphere that have significant absorption and emission in the infrared (i.e., greenhouse gases) radiate to space with a flux characteristic of the temperature of the atmosphere at about one optical depth (measured from space downward). To be sure, this level varies with wavelength, but the average emission level is about 5-6 km above the surface and well within the troposphere.

Third, adding greenhouse gases to the atmosphere must elevate the average emission level, and because of the first point, the new emission level is colder than the original emission level. This reduces the outgoing infrared radiative flux, which no longer balances the net incoming solar radiation. Thus, the troposphere, which is a dynamically mixed layer, must warm as a whole (including the surface) while preserving its lapse rate (Figure 2).

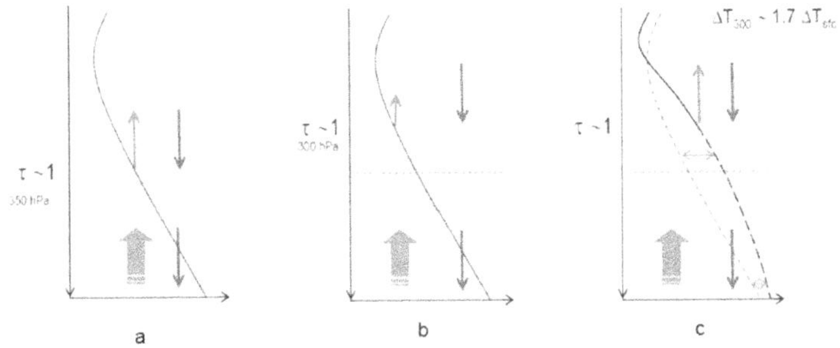

Fig. 2: a) Situation with atmosphere in equilibrium with space. b) Situation when added greenhouse gas elevates emission level to a cooler level, leaving radiative imbalance that constitutes the radiative forcing. c) Re-equilibration with moist adiabat.

Note that this mechanism can be shown to lead to the simple result that doubling CO_2 gives rise to warming of about 1°C. This would not lead to significant concern. Larger warming calls for positive feedbacks, and it is the determination of these feedbacks that is a central question in climate.

<u>Some implications of the above picture of the greenhouse.</u>

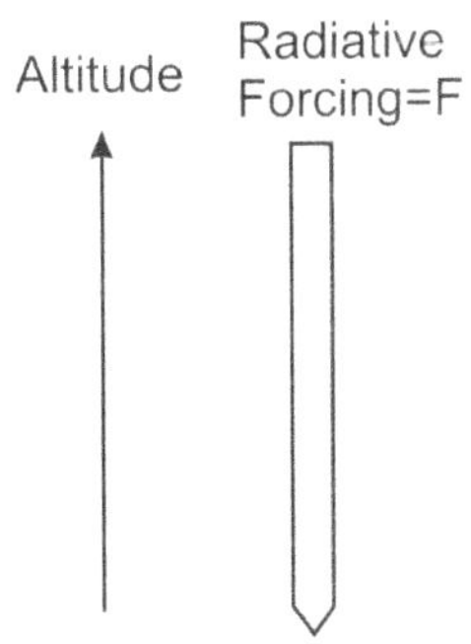

Fig. 3: Schematic of non-divergent TOTAL flux.

These points also lead to the almost non-divergence of the total flux with altitude as illustrated in Figure 3. In order for the dynamically mixed troposphere to warm as a whole, flux imbalance at the top of the atmosphere must approximately equal flux imbalance at the surface. The total flux consists in radiative flux, sensible heat flux, and latent heat flux. At the top of the atmosphere, the flux is exclusively radiative, while at the surface, the flux is primarily in the form of latent heat flux (i.e., evaporation). That evaporation at the surface must approximately follow radiative imbalance imposed at the top of the atmosphere may, at first, seem counter-intuitive. However, as noted in Lindzen, Hou and Farrell (1981), this is achieved by internal changes in the jump in relative humidity and temperature across the near surface turbulent boundary layer.

Later in this paper, we will consider using space-based observations to measure radiative forcing and climate sensitivity. From the above, we see that an alternative to observing outgoing radiation from space may be to measure fluxes at the surface. This too will be discussed later.

Implications for response time.

The approximate non-divergence of flux is also the rationale for assuming that radiative forcing is acting at the surface in simple energy balance models.

$$C_{\text{land}} \frac{\partial \Delta T_{\text{land}}(t)}{\partial t} - \frac{\nu}{A_{\text{land}}}(\Delta T_{ml}(t) - \Delta T_{\text{land}}(t)) + \frac{B}{\text{gain}} \Delta T_{\text{land}}(t) = \Delta Q(t)$$

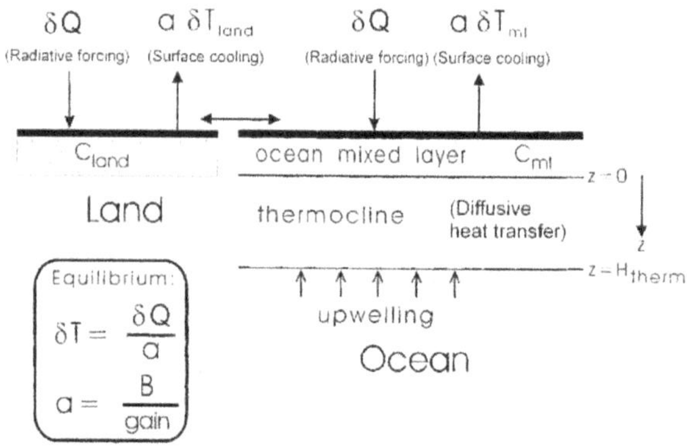

Fig. 4: Geometry and equation for a simple energy balance box model of the climate system response to radiative forcing (from Lindzen and Giannitsis, 1998).

Note that high gain (sensitivity) implies weak thermal coupling between the atmosphere and ocean. Such coupling is obviously important for air-sea interactions. This immediately leads to an important question that we will not have time to go into. Current coupled general circulation models all do poorly in simulating important phenomena involving air-sea coupling such El Nino, the Pacific Decadal Oscillation, and the Atlantic Multi-decadal Oscillation. They also suffer from secular drift. Would reducing sensitivity (even artificially) improve simulations of ENSO, PDO, etc., and eliminate problems of drift? Modelers have generally avoided this question.

Here, we will focus on the crucial point that climate sensitivity (essentially $\Delta T/\Delta\Phi$) translates immediately into the time scale for response of the system, with high sensitivity associated with long response times. Lindzen and Giannitsis (1998) looked at the long term response to sequences of volcanic eruptions. (The short term response (as we see from Figure 5a)—one to two years—to single volcanoes did not distinguish among sensitivities. To be sure, higher sensitivities are associated with slower recoveries, but, in all cases the 'tail' is too small compared to measurement uncertainties. However, as we see from Figure 5b, the response to a sequence of volcanoes) allows subsequent volcanoes to build on the tails of predecessors, and thus, provides discernible distinctions among different sensitivities.

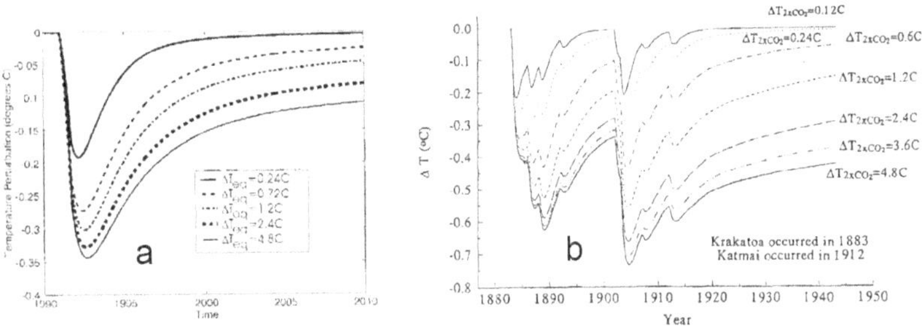

Fig. 5: a) The response to a single volcano for different sensitivities. b) The response for a sequence of volcanoes between Krakatoa in 1883 and Katmai in 1912.

Figure 6 compares calculated results for various sensitivities plus increased forcing from CO_2 with observed time series. Visually, the data shows no evidence of volcanoes except for short blips.

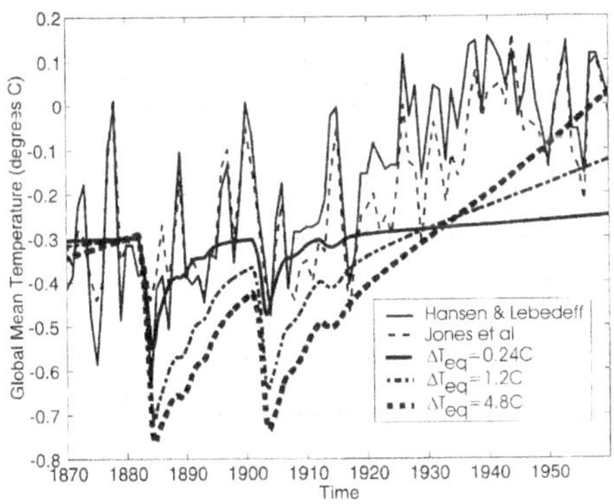

Fig. 6: Comparison of calculated responses v. observed temperature records.

Another possible approach to estimating time scales emerges from Roe (2009). He shows that power spectrum for Pacific Decadal Oscillation can be simulated by an AR(1) process with a response time of 1.6 +/- 0.8 years. Moreover, it should be possible to examine the PDO index in models to see if the time series corresponds to a much longer response time. This would be strong evidence that model sensitivity was excessive. Note that current models do not simulate the PDO. We are currently beginning such a study.

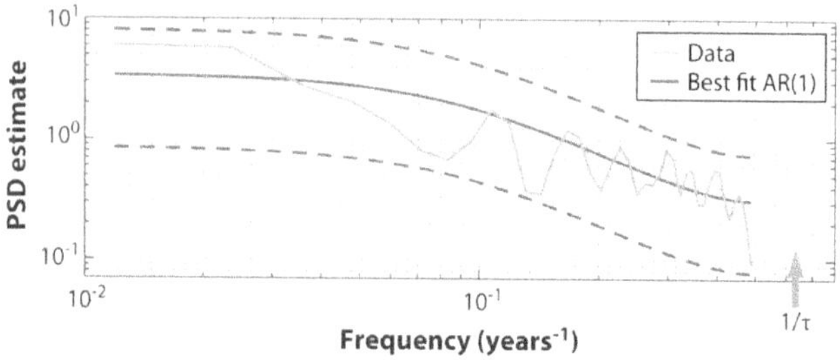

Fig. 7: Power spectral density of Pacific Decadal Oscillation Index (grey line) and best fit to AR(1) which is achieved with a choice of t=1.6 years (solid green line). Dashed green lines show confidence bounds. After Roe (2009).

If models display a longer τ, it would confirm that model sensitivity is excessive.

Additional comments.

It should be emphasized that if dynamical mixing were to have led to an isothermal atmosphere, then there would be no warming due to added greenhouse gases. In the *counterfactual* case that mixing were to have lead to increasing temperature with altitude, then added greenhouse gases would actually cool the atmosphere. In brief, greenhouse warming depends crucially on the existence and properties of dynamic mixing within the troposphere, and not simply on the radiative picture.

The structure imposed by the dynamics determines how the warming at the characteristic emission level is manifested at the ground.

SOME OTHER IMPORTANT PHYSICAL CONCEPTS: THE MOIST ADIABAT AND THE ROSSBY RADIUS.

Any picture of the thermal structure of the atmosphere clearly displays two important features: 1) Temperatures are nearly horizontally homogeneous in the tropics, and 2) the vertical profile of temperature in the tropics closely follows the moist adiabat.

The Moist Adiabat.

The moist-adiabatic lapse rate—(or saturated-adiabatic lapse rate.) is the rate of decrease of temperature with height along a moist adiabat. It is given approximately by Γ_m in the following:

$$\Gamma_m = g\frac{1+\frac{L_v r_v}{RT}}{c_{pd}+\frac{L_v^2 r_v \varepsilon}{RT^2}}, \qquad (1)$$

where g is gravitational acceleration, c_{pd} is the specific heat at constant pressure of dry air, r_v is the mixing ratio of water vapor, L_v is the latent heat of vaporization, R is the gas constant for dry air, ε is the ratio of the gas constants for dry air and water vapor, and T is temperature. This expression is an approximation to both the reversible moist adiabatic lapse rate and the pseudoadiabatic lapse rate. When most of the condensed water is frozen, this may be replaced by a similar expression but with L_v replaced by the latent heat of sublimation. A more detailed description can be found in Wikipedia. In Figure 8 we see the meridional distribution of the temperature response to a doubling of CO_2 from four typical models.

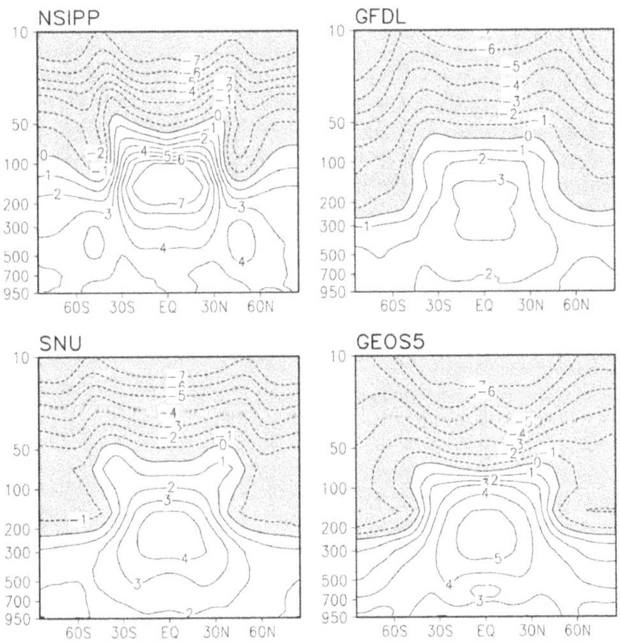

Fig. 8: Zonal-mean meridional distributions of temperature change associated with a doubling of CO_2. Units are degrees Kelvin. From Lee, et al. 2008.

The response is characterized by the so-called hot spot (i.e., the response in the tropical upper troposphere is from 2-3 times larger than the surface response). The models are likely correct in this respect since the hot spot is simply a consequence of the fact that tropical temperatures approximately follow the moist adiabat. This is essentially a consequence of the dominant role of moist convection in the tropics.

However, we see in Figure 9 that the temperature trends obtained from observations fail to show the hot spot.

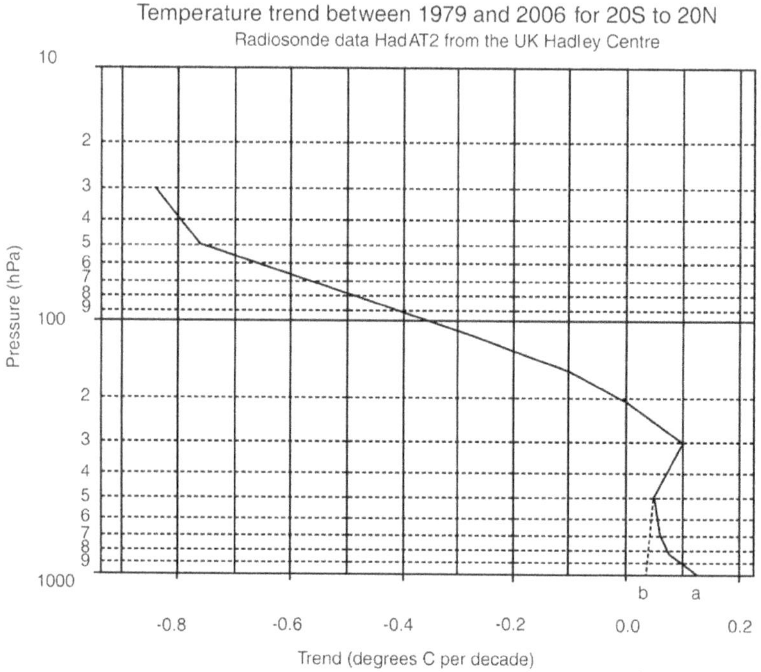

Fig. 9: Temperature trend as a function of pressure level (in hectopascals) for period 1979-2006 in the tropics (20S-20N) based on balloon data analyzed by the Hadley Centre. 'a' shows the observed trend at the surface. 'b' is explained in the text.

In point of fact, it seems likely that some of the recent temperature data must be wrong!

The resolution of the discrepancy demands that either the upper troposphere measurements are wrong, the surface measurements are wrong or both. If it is the surface measurements, then the surface trend must be reduced from 'a' to 'b'. Although, it is generally ill-advised to estimate climate sensitivity from observed changes in temperature (simply because of ignorance of all the relevant processes including especially natural internal variability on time scales of centuries or less), it would be very difficult to simulate the trend at 'b' with models having their current sensitivity.

Given how small the trends are, and how large the uncertainties in the analysis, such errors are hardly out of the question. In fact there are excellent reasons to suppose that the error resides in the surface measurements. To understand this requires an awareness of the Rossby Radius of Deformation.

Rossby Radius of Deformation

In dynamic meteorology, there is something called the Rossby Radius. It is the distance over which variables like temperature are smoothed out. This distance is inversely proportional to the Coriolis Parameter (twice the vertical component of the earth's rotation), and this parameter approaches zero as one approaches the tropics so that temperature is smoothed over thousands of kilometers (Detailed formulas can again be found on *Wikipedia*).

However, this smoothing is only effective where turbulent diffusion is small. Below about 2 km, we have the turbulent trade wind boundary layer, where such smoothing is much less effective so that there is appreciable local variability of temperature. In practice, this means that for the sparsely sampled tropics, sampling problems above 2 km are much less important than at the surface (Lindzen and Nigam, 1987). Thus, errors are more likely at the surface.

An important philosophical point to this little exercise is that neither ambiguous data nor numerical model outputs should automatically be assumed to be right or wrong. Both should be judged by basic, relatively fundamental theory—where such theory is available.

A FUNDAMENTAL PROBLEM WITH ONE DIMENSIONAL CLIMATE THINKING

Although the tropics may be important for global warming, focusing on the tropics misses the fact that climate change has generally involved the temperature difference between the tropics and higher latitudes.

This means that we must consider the meridional transport of heat. However, outside the tropics, such transport takes place mostly along isentropic surfaces. Figure 10 (taken from Sun and Lindzen, 1994) shows the distribution of zonally averaged contours of potential vorticity and potential temperature (essentially a measure of entropy) with respect to pressure level and latitude. We see a clear delineation between the tropics where potential vorticity contours are nearly vertical and the extratropics where potential vorticity tends to follow isentropes.

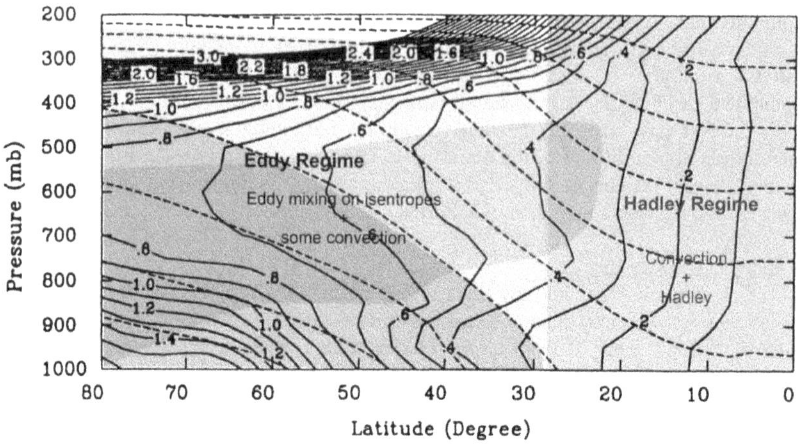

Fig. 10. Contours ofzonally averaged potential temperature (dashes) and potential vorticity (solid contours) as functions of pressure level and latitude. See Sun and Lindzen (1994) for details.

Baroclinic neutrality (when the instability generating cyclonic disturbances is neutral) is basically a condition on the slope of isentropes wherein the last isentrope leaving the tropical surface becomes the polar tropopause. An attempt at a more basic theoretical foundation for this is in T. Schneider, 2004.

Note that heat from the tropics is essentially transported to the upper troposphere of the extratropics. How is it communicated to the surface? According to Kirk-Davidoff and Lindzen (1998), radiative exchange between the surface and high clouds plays a major role (since at the cold temperatures involved, there is little water vapor). We will look at this a little more later in this paper.

In general, the whole one dimensional approach that we have been describing probably represents an oversimplification when forcing is regional. For example, in the Milankovich theory of glaciation cycles, the important quantity is insolation at high latitudes in summer. Indeed, spatially varying forcing is probably the rule rather than the exception. In the case of Milankovich Theory, summer insolation varies by the order of 50 watts per square meter. Its magnitude determines whether snow that always accumulates in winter, will or won't melt away in the summer. If it does not, then snow will accumulate over many years leading to the major glaciations of the ice ages. The changes in summer arctic insolation are huge compared to the small changes in radiative balance associated with the change in CO_2 from 180 ppmv to 280 ppmv found in Antarctic ice cores. It is implausible to think that the ultimate radiative balance is significantly affected by this (much less driven by this). Moreover, as shown by Roe (2007), for example, the Milankovich Theory excellently accounts for the cycles of glaciation. The atmosphere has many degrees of freedom that would allow it to adjust to balance in response to Milankovich forcing. Only in the case of relatively global forcing (as is associated with CO_2 or solar variability) is it likely that the one dimensional reasoning that we focus on in this paper will be appropriate. Otherwise, the one dimensional approach that is commonly taken to climate already represents a very important example of reductionism being taken too far. Before, returning to this, a brief consideration of what determines the meridional temperature distribution may prove of interest.

As already noted, heat is transported from the tropical surface to the polar tropopause region. As already pointed out by Lindzen and Farrell (1980), this leads to a much smaller equator-to-pole temperature difference than is currently observed. Indeed the temperature difference is closer to what is estimated to have been the difference during the Eocene. They noted that getting the currently observed difference required taking into account the arctic inversion (the fact that arctic temperatures increase rather than decrease with height through much of the arctic troposphere). Indeed, if one looks at observed distributions of current temperature as a function of latitude for various altitudes (viz Figure 11) we see that above about 500 hPa, the equator-to-pole temperature difference is, indeed, much smaller than at the surface. Understanding the arctic inversion remains an important but largely ignored problem. Interestingly, when current General Circulation Models attempt to simulate Eocene conditions by increasing CO_2, they get almost uniform warming at all latitudes while preserving the present day equator-to-pole surface temperature difference. This is seen in Figure 12.

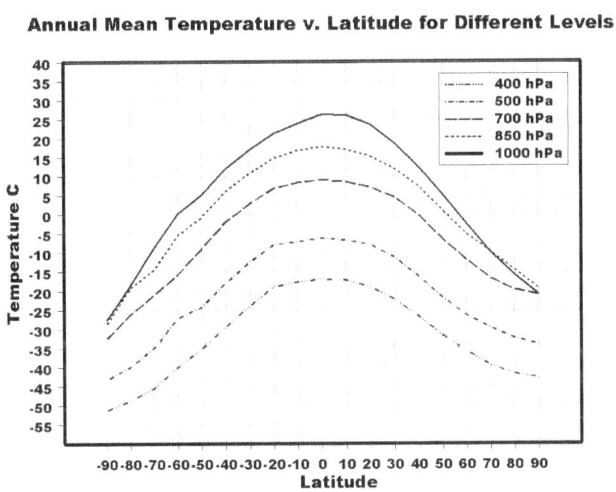

Fig. 11: Latitude variations in annual mean temperature for various pressure levels.

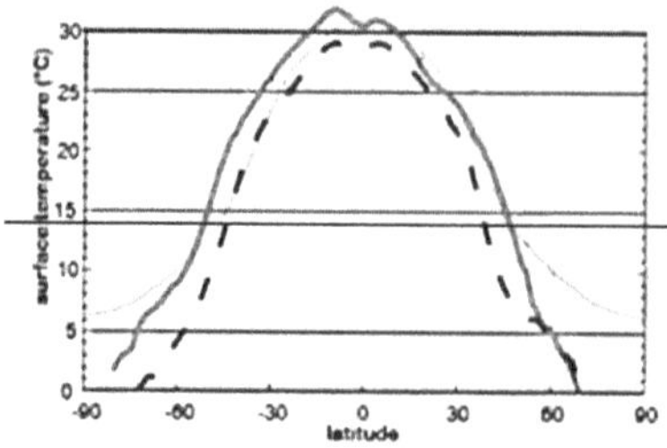

Fig. 12: Surface temperature distributions. The black dashed line shows the current distribution. The red line shows the model simulation for elevated CO_2. The blue shading shows the estimated Eocene distribution from proxy data.

ARCTIC TEMPERATURES: A PROBLEM WITH ANNUAL AVERAGES.

Another area where we may reasonable suspect that our reductionism has gone too far is in the use of annual averages. Figure 13 shows monthly temperatures for the arctic for 4 typical years.

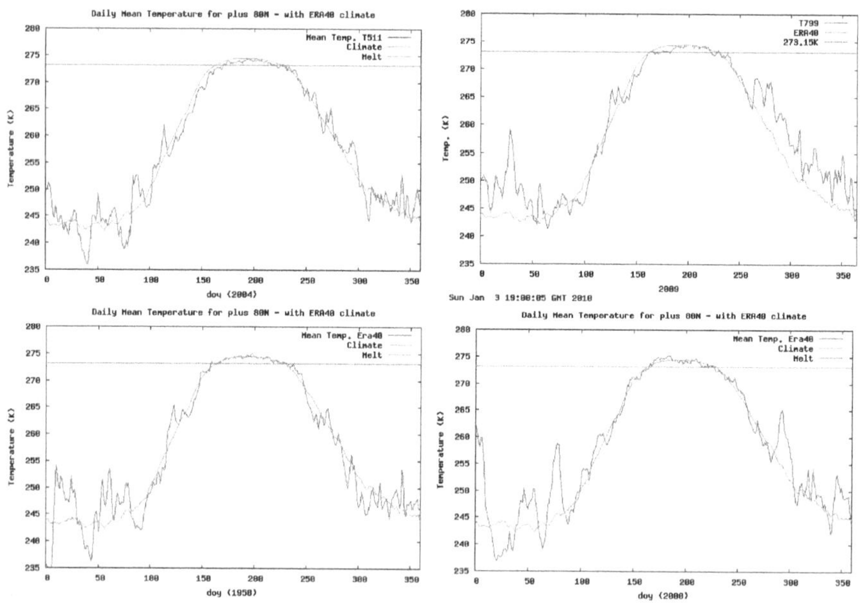

Fig. 13: Arctic temperatures as a function of date for four arbitrarily chosen years.

What we see in Figure 13 is that summer temperatures are surprisingly stable. In winter we see immense fluctuations in temperature—often as large as 20°C. Focusing on the small residues of these large changes misses some crucial aspects of the physics. What Figure 13 illustrates is that during summers, when there is sunlight, temperatures are largely determined by local radiative balance and this does not seem to be changing much. However, during the winter night, temperatures would be even colder than they are but for the transport of heat from lower latitudes. This transport involves turbulent eddies or storms. Understanding arctic temperatures must involve understanding why these storms erratically penetrate to the arctic, and how heat transported to the arctic tropopause region is communicated to the surface. Annual means would appear to obscure what is actually going on.

RETURN TO ONE-DIMENSIONAL THINKING: FEEDBACKS.

Despite the limitations of one-dimensional thinking, it does appear to be appropriate to situations involving global forcing as might be produced by well mixed greenhouse gases or by variations in solar output. Here feedbacks are essential. The study of climate feedbacks also involves some very basic theory. Recall that feedbacks are essential to the concern over greenhouse warming. Without positive feedbacks, the climate sensitivity is generally found to be small. Figure 14 illustrates how feedbacks operate. Basically, for a positive feedback, increased surface temperature acts to reduce outgoing radiation beyond what would otherwise be emitted. For an arbitrary temperature fluctuation, this would mean that the outgoing radiation would be less than one would expect from simple black body radiation from the emission level.

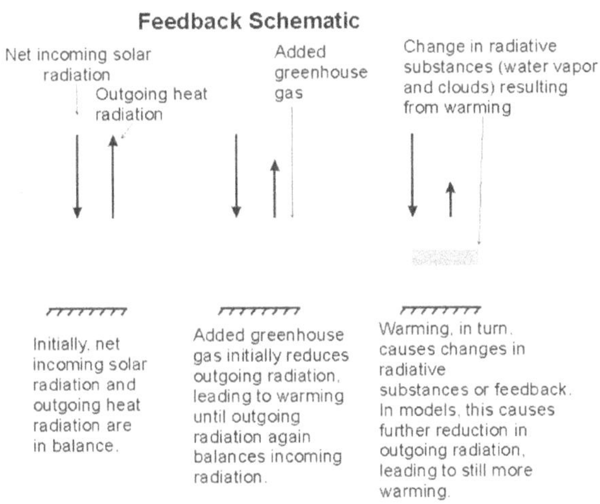

Fig. 14: Schematic depiction of the implications of added greenhouse gases and feedbacks for the balance of radiation as seen from space.

This suggests that one might determine feedbacks by seeing whether observed surface temperature fluctuations were accompanied by decreases in outgoing radiation (relative to black body radiation) indicating positive feedback or by increased outgoing radiation indicating negative feedback. One is able, in principle, to use satellite data from ERBE and CERES (that measure net outgoing radiation in both the visible and infrared portions of the spectrum) to test the preceding situation, and to quantitatively evaluate climate feedback factors. As we have already noted, these are related to climate sensitivity by the following equation:

$$\Delta T = \frac{\Delta T_0}{1-f}, \qquad (2)$$

ΔT_0 is the zero feedback response to a doubling of CO_2. It is about 1°C. It turns out moreover, that the model intercomparison program has the models used by the IPCC, forced by actual SST, calculate outgoing radiation. So one can use the same approach with models, while being sure that the models are subject to the same surface temperature fluctuations that applied to the observations.

While this might appear to be a straightforward task, in practice, it is rather difficult. The two main difficulties involve basic physical considerations.

First, not all time scales are appropriate for such studies. Greenhouse warming continues until equilibrium is reestablished. At equilibrium, there is no longer any radiative imbalance. If one considers time intervals that are long compared to equilibration times, then one will observe changes in temperature without changes in radiative forcing. The inclusion of such long time scales thus biases results inappropriately toward high sensitivity. Equilibration times depend on climate sensitivity. For sensitivity on the order of 0.5°C for a doubling of CO_2, it is on the order of years, and for higher sensitivities it is on the order of decades. In order to avoid biasing sensitivity estimates, one should restrict oneself to time intervals less than a year. There is also the need to consider time intervals long enough for the relevant feedback processes to operate. For water vapor and cloud feedbacks, these time scales are typically on the order of days. For practical time resolution, this is generally not a problem. Time scales on the order of 1-3 months are, thus, certainly appropriate for sensitivity studies. Longer time scales also involve 'pollution' from seasonal effects, etc.

The second problem is more difficult. Outgoing radiation varies (especially in the visible) for reasons other than changing surface temperature (volcanoes, non-feedback cloud fluctuations). Such changes primarily cause surface temperature fluctuations rather than respond to surface temperature fluctuations.

Lindzen and Choi (2011) dealt with the first problem by restricting consideration to short [O(3 mos)] fluctuations in surface temperature. They dealt with the second problem by considering lagged regressions. The details are to be found in that paper, and will not be repeated here—though we will review some results. Interestingly, there are quite a few papers that ignore both the above problems and use simple regressions (Trenberth, et al. 2010; Dessler, 2010; Foster and Gregory, 2006; Murphy, 2010). The results, as we will show, can be very misleading. When the feedback is unambiguous, then there is a single peak in ΔFlux/ΔSST occurring for a positive lag, whose sign indicates the sign of the feedback. Similarly, the correlation, R, peaks at roughly the same lag. This, surprisingly, is the case for the longwave feedback which is unambiguously negative (corresponding to a positive increase in flux) as may be seen from Figure 15. This result has recently been confirmed by Trenberth et al. (2009), and is suggestive of the proposed infrared iris effect (Lindzen, Chou and Hou, 2000). However, problem 2 is evident for the shortwave feedback where we see a characteristic S-shaped curve for both ΔFlux/ΔSST and correlation. The results are characterized by two peaks in R: one associated with a lead and therefore corresponding to non-feedback fluctuations in flux; the other associated with a lag, and therefore corresponding to a feedback. The feedback is again negative, but the results for zero lag are dominated by the non-feedback changes and spuriously suggest a positive feedback. The approach in Lindzen and Choi (2011) was to use the value of the negative shortwave feedback corresponding to the local maximum in R^2. The resulting sensitivities obtained from models are shown in Table 1. Also shown are the estimated sensitivities claimed for the models (Table 2). The two appear to differ, but, as can be seen, the 90% confidence intervals for the models confidence level, it could be slightly positive). The reason for this follows immediately from Equation 2. As can be seen in Figure 16, once the feedback factor exceeds 0.25, small changes lead to large changes in sensitivity. For feedback factors less than 0.25, the sensitivity is only weakly dependent on the feedback factor.

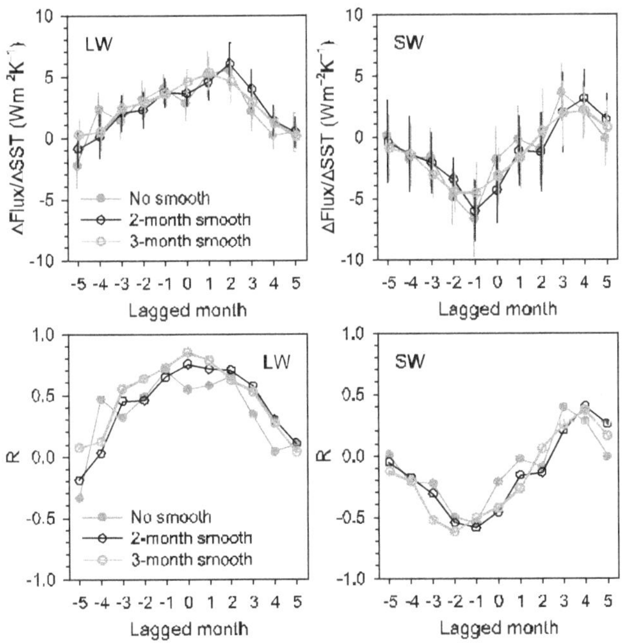

Fig. 15: The impact of smoothing and leads and lags on the determination of the slope (top) as well as on the correlation coefficient, R, of the linear regression (bottom).

Table 1. Comparison of model equilibrium climate sensitivities (in K) for a doubling of CO_2 defined from IPCC AR4 and estimated from feedback factors in Lindzen and Choi (2011). The estimated climate sensitivities for models as well as their confidence intervals are given for 90%, 95%, and 99% confidence levels.

Models	IPCC AR4	Estimate in this study			
	Sensitivity	Sensitivity	Confidence interval of sensitivity		
			90%	95%	99%
CCSM3	2.7	8.1	1.6 – Infinity	1.4 – Infinity	1.1 – Infinity
ECHAM5/MPI-OM	3.4	1.7	0.9 – 8.0	0.9 – 28.2	0.8 – Infinity
FGOALS-g1.0	2.3	7.9	2.2 – Infinity	2.0 – Infinity	1.6 – Infinity
GFDL-CM2.1	3.4	2.2	1.1 – 351.4	1.0 – Infinity	0.8 – Infinity
GISS-ER	2.7	2.5	1.5 – 8.7	1.4 – 16.4	1.2 – Infinity
INM-CM3.0	2.1	2.7	1.3 – Infinity	1.2 – Infinity	1.0 – Infinity
IPSL-CM4	4.4	10.4	2.1 – Infinity	1.8 – Infinity	1.4 – Infinity
MRI-CGCM2.3.2	3.2	Infinity	2.5 – Infinity	2.0 – Infinity	1.4 – Infinity
MIROC3.2(hires)	4.3	2.2	1.3 – 6.4	1.2 – 10.0	1.1 – Infinity
MIROC3.2(medres)	4	2.4	1.3 – 14.7	1.2 – Infinity	1.0 – Infinity
UKMO-HadGEM1	4.4	1.7	1.0 – 8.8	0.9 – 38.9	0.8 – Infinity

Table 2. Results from observations from Lindzen and Choi (2011)

Sensitivity, mean	0.7
Sensitivity, 90%	0.61.0
Sensitivity, 95%	0.51.1
Sensitivity, 99%	0.51.3

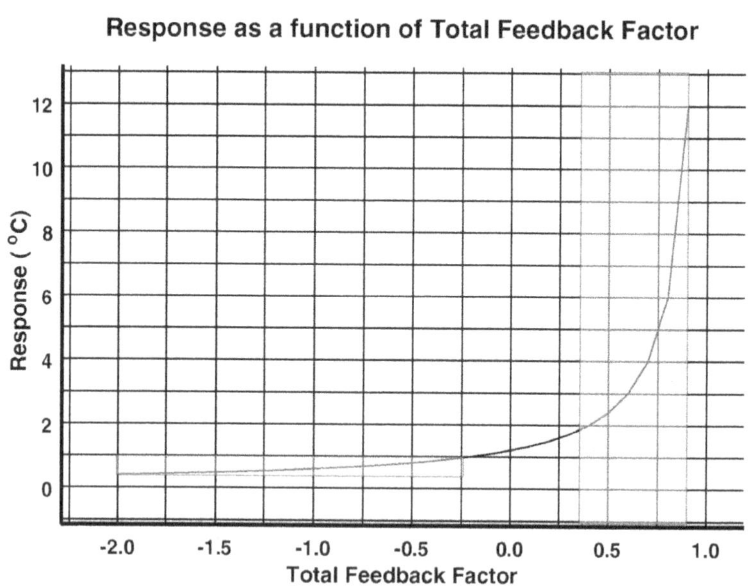

Fig. 16: Sensitivity vs. feedback factor.

ESTIMATING CLIMATE SENSITIVITY WITH SURFACE FLUXES.

Recall that the approximate non-divergence of flux illustrated in Figure 3 allows us to relate the top of the atmosphere fluxes in the figure explaining feedbacks to surface fluxes (i.e., Figure 14). It turns out that heat fluxes at the surface are dominated by latent heat fluxes (i.e., evaporation). The radiative imbalance in the third panel of Figure 14 determines the climate sensitivity and hence the change in temperature once the system re-equilibrates. At the surface, the imbalance will be approximately between the net incoming solar radiation and the evaporation.

For simplicity, let's ignore shortwave feedbacks. Then the net incoming solar flux remains constant. Initially, evaporation balances net incoming solar flux, and after equilibration, evaporation once again must equal the net incoming solar flux, but the temperature will be different. Thus, relative humidity must change so as to prevent evaporation from increasing. This is a surprising result, since the heart of the positive water vapor feedback is the assumed constancy of relative humidity. In point of fact, rh does not have to change much to hold evaporation constant and even in models, rh varies with warming.

At short time scales, the change in evaporation will have to match the climate sensitivity. To be sure, this depends on the absence of short wave feedbacks and on ignoring sensible heat flux—both of which may be serious (and are being looked at in a more serious study). However, as an exercise in numerology, it is interesting to look at some results.

Wentz, F.J. et al. (2007) used bulk aerodynamic formulas and space based observations to measure how evaporation changed with temperature and compared their results with GCM results.

In GCMs, E (evaporation) increased from 1-3% for each degree increase in temperature. Observationally, E increased 5.7%. Now a 1% change in E corresponds to about 0.8 watts m-2. Climate sensitivity is essentially $\Delta T/\Delta F$. More specifically,

EC=ΔEvaporation/ΔT (in units of percent change per degree)
CF=Radiative Forcing due to doubling of CO_2=3.6 Watts m^{-2}
FL=Heat Flux associated with EC=0.8 Watts m^{-2} x EC
Climate sensitivity=CF/FL

Table 3. Dependence of evaporation on temperature and derived sensitivity for both observations and models.

Source	EC (percentage change in E per degree)	Climate Sensitivity (Degrees C)
Model Range	1-3	1.5-4.5
Observe	5.7	0.8

The value obtained is surprisingly close to what was gotten using top of the atmosphere fluxes in Section 7. In a recent paper, Stephens and Hu (2010) argued that the above value for the observed EC is excessive. However, the argument was somewhat circular since it depended on model outputs with positive feedback.

VARIABILITY OF CLIMATE SENSITIVITY.

My last point is that feedback factors are not constant. In point of fact, feedback factors can vary because cloud radiative properties are modified by aerosols and possibly by cosmic rays. Figure 17 from Choi et al. (2010), for example, shows a significant correlation between dust frequency at a particular level and fraction of clouds that are supercooled. The relevance of this is that clouds consisting in supercooled water droplets have a greater albedo than do ice clouds. Thus, the presence or absence of dust impacts the optical properties of clouds. Given this, it seems likely that if the feedback factor is around 0.8, then sometime during the earth's 4.5 billion years, it exceeded one. I would suggest that if the sensitivity were really around 5C for a doubling of CO_2, then we would not be here discussing it.

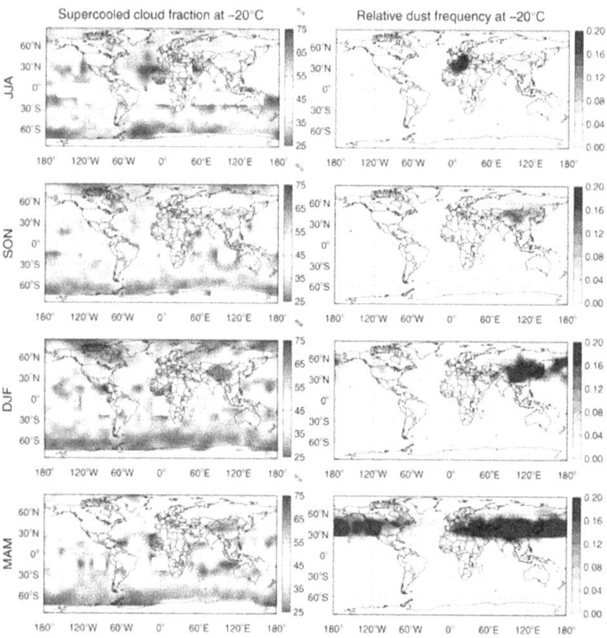

Fig. 17: Seasonal variation in SCF and relative dust frequency (to the highest dust 478 frequency), both at the –20°C isotherm. Pattern correlation coefficients between SCF 479 and dust frequency for the Northern Hemisphere are negative for all seasons: –0.1, –0.5, 480 –0.5, and –0.6 for JJA, September-October-November (SON), December-January-481 February (DJF), and March-April-May (MAM), respectively.

CONCLUDING REMARKS.

This paper considers approaches to estimating climate sensitivity involving the basic physics of the feedback processes rather than attempting to estimate climate sensitivity from time series of temperature. The latter have to assume a perfect knowledge of all sources of climate variability – something generally absent. The results of a variety of independent approaches all point to relatively low sensitivities. We also note that when climate change is due to regional and seasonal forcing, the concept of one dimensional climate sensitivity may, in fact, be inappropriate.

REFERENCES

1. Choi, Y.-S, R.S. Lindzen, C.H. Ho, and J. Kim (2010) "Space observations of cold-cloud phase change." *PNAS*, **107**:11211–11216.
2. Dessler, A.E. (2010) "A determination of the cloud feedback from climate variations over the past decade." *Science*, **330**:1523-1527.
3. Forster, P.M., and J.M. Gregory (2006) "The climate sensitivity and its components diagnosed from Earth Radiation Budget data." *J. Climate.*, **19**, 39-52.
4. Kirk-Davidoff, D.B. and R.S. Lindzen (2000) "An energy balance model based on potential vorticity homogenization." *J. Climate*, **13**:431-448.
5. Lee, M.-I., M.J. Suarez, I.-S. Kang, I.M. Held, and D. Kim (2009) "A Moist Benchmark Calculation for Atmospheric General Circulation Models." *J. Atmos. Sci.*, **21**:4934-4954.
6. Lindzen, R.S. and B. Farrell (1980) "The role of polar regions in global climate, and the parameterization of global heat transport." *Mon. Wea. Rev.*, **108**:2064-2079.
7. Lindzen, R.S. and S. Nigam (1987) "On the role of sea surface temperature gradients in forcing low level winds and convergence in the tropics." *J. Atmos. Sci.*, **44**:2418-2436.
8. Lindzen, R.S. and C. Giannitsis (1998) "On the climatic implications of volcanic cooling." *J. Geophys. Res.*, **103**:5929-5941.
9. Lindzen, R.S., M.D. Chou, and A.Y. Hou (2001) "Does the Earth have an adaptive infrared iris?" *Bull. Amer. Met. Soc.* **82**: 417-432.
10. Lindzen, R.S. and Y.S. Choi (2011) "On the Observational Determination of Climate Sensitivity and Its Implications." *Asia-Pacific J. Atmos. Sci.*, **47**: 377-390.
11. Murphy, D.M, and Coauthors (2009) "An observationally based energy balance for the Earth since 1950." *J. Geophys. Res.*, **114**:D17107.
12. Roe, G.R. (2006) "In defense of Milankovitch." *Geophys. Res. Ltrs.*, **33**:L24703, doi:10.1029/2006GL027817.
13. Roe, G.R. (2009) "Feedbacks, Timescales, and Seeing Red." *Annu. Rev. Earth Planet. Sci.* **37**:93-115.
14. Schneider, T. (2004) "The tropopause and thermal stratification in the extratropics of a dry atmosphere." *J. Atmos. Sci.*, **61**:1317–1340.

15. Stephens, G.I. and Yongxiang Hu (2010) "Are climate-related changes to the character of global-mean precipitation predictable?" *Environ. Res. Lett.*, **5,** doi:10.1088/1748-9326/5/2/025209.
16. Sun, D.Z. and R.S. Lindzen (1994) "A PV view of the zonal mean distribution of temperature and wind in the extra-tropical troposphere." *J. Atmos. Sci.*, **51**:757-772.
17. Trenberth, K.E. and J.T. Fasullo (2009) "Global warming due to increasing absorbed solar radiation." *Geophys. Res. Ltrs.*, **36**:L07706, doi:10.1029/2009GL037527.
18. Trenberth, K.E., J.T. Fasullo, C. O'Dell, and T. Wong (2010) "Relationships between tropical sea surface temperature and top-of-atmosphere radiation." *Geophys. Res. Lett.*, **37**:L03702.
19. Wentz, F.J. et al. (2007) "How much more rain will global warming bring?" *ScienceExpress*, **31.**

SESSION 2

ENERGY

FOCUS: GLOBAL NUCLEAR ENERGY ISSUES AFTER FUKUSHIMA

ENERGY PERSPECTIVE AND NUCLEAR ROLE AFTER FUKUSHIMA DAIICHI ACCIDENT

TETSUO YUHARA
University of Tokyo, the Canon Institute for Global Studies
Tokyo, Japan

HIROSHI UJITA
Tokyo Institute of Technology, Department of Nuclear Engineering, International Nuclear Human Resource Development (Hitachi), Tokyo, Japan

INTRODUCTION

Here, the Fukushima Daiichi Severe Accident is described and the direction for nuclear power after the accident is discussed. Energy mixture formation against global warming is further examined. Taking the effort for energy-saving as major premise, carbon-sequestration for fossil fuel, renewable energy and nuclear energy should be altogether developed, which means best energy mix is achieved, under the CO_2 constraint around 450 ppm atmosphere.

SEVERE ACCIDENT OF FUKUSHIMA DAIICHI NUCLEAR POWER PLANT

What happened in the Fukushima Daiichi Nuclear Power Plant

East-Japan Great Earthquake is described, as follows:

Magnitude of the Earthquake (Mw) is 9.0, which is ranked 4th largest in the world. All 15 plants located east coast of north Japan suffered Earthquake, while every plant was successfully shut down as same as Kashiwazaki plants suffered large Earthquake in 2007.

The first and highest tsunami wave was observed at 15:51 (65 min. after the earthquake) of March 11th at a point 50 km north of the Fukushima Daiichi Nuclear Power Station. All plants were attacked by Tsunami, while observed Tsunami height only in Fukushima Daiichi Nuclear Power Plant is over actual ground level (or design height), and therefore Loss of Power phenomena have occurred in 3 plants in Daiichi site as summarized in Table 1.

Status of the Nuclear Power Stations hit by the Earthquake and the Tsunami, and accident process are summarized as below:

1. On March 11th, there was a huge magnitude 9.0 earthquake that occurred off the Pacific Coast of Japan. All Nuclear Power Plants (NPPs) on the east coast were shut down safely (See Figure 1(a)). Every emergency cooling function succesfully operated and was cooling the reactor core.
2. About one hour later, a huge tsunami hit Japan. However, all NPPs continued cooling down operations, except for Fukushima Daiichi NPP (1F), where

Residual Core Cooling system still operated (See Figure 1(b)).

3. Hydrogen explosions occurred at the units in which cooling function was lost, which may have been caused by a Zirconium-water reaction. In addition, the Reactor Building was damaged (1F1, 3 and 4) (See Figure 1(d)).
4. For the units that lost the cooling system, water is continuously being injected from outside (1F1-4) (See Figure 1(d)).
5. Quite a large amount of radioactive material has been released into the atmosphere and sea, and continuous efforts are being made to bring the situation under control. Radiation level decreases gradually for three month to steady level (See Figure 2).
6. Extensive recovery efforts are being continued. Stable cooling is scheduled to take three months and cold shutdown is scheduled to take another 3 to 6 months (See Figure 3).

Table 1 Status of the Nuclear Power Plants (NPPs) hit by the quake and the tsunami

NPS	Unit	Type	MWe	Before Quake	After Quake	After Tsunami	Accel[1] (gal)	Tsunami Height
Higashi Dori	1	BWR-5	1,100	Outage	Cold Shutdown	Cold Shutdown	---	---
Onagawa	1	BWR-4	524	Operating	Automatic Scram	Cold Shutdown	587 [529]	Design: 9.1m Ground level: 13.8m (Observed: 13m)
	2	BWR-5	825	Reactor Start	Automatic Scram	Cold Shutdown	607 [594]	
	3	BWR-5	825	Operating	Automatic Scram	Cold Shutdown	573 [512]	
Fukushima Daiichi (1F)	1	BWR-3	460	Operating	Automatic Scram	*Loss of cooling*	447 [489]	Design: 5.7m Ground level: 10m (1F1-4) 13m (1F5&6) (Observed: 14-15m)
	2	BWR-4	784	Operating	Automatic Scram	*Loss of cooling*	550 [438]	
	3	BWR-4	784	Operating	Automatic Scram	*Loss of cooling*	507 [441]	
	4	BWR-4	784	Outage	Cold Shutdown	*Cold Shutdown*	319 [445]	
	5	BWR-4	784	Outage	Cold Shutdown	Cold Shutdown	548 [452]	
	6	BWR-5	1,100	Outage	Cold Shutdown	Cold Shutdown	444 [448]	
Fukushima Daini (2F)	1	BWR-5	1,100	Operating	Automatic Scram	Cold Shutdown	254 [434]	Design: 5.2m Ground level: 12m (Observed: 6.5-7m, Locally >14m)
	2	BWR-5	1,100	Operating	Automatic Scram	Cold Shutdown	243 [428]	
	3	BWR-5	1,100	Operating	Automatic Scram	Cold Shutdown	277 [428]	
	4	BWR-5	1,100	Operating	Automatic Scram	Cold Shutdown	210 [415]	
Tokai Daini	—	BWR-5	1,100	Operating	Automatic Scram	Cold Shutdown	225 [400]	Ground level: 8m (Observed: 5.4m)

[1]Observed maximum response acceleration (horizontal) on the basement of Reactor Building.

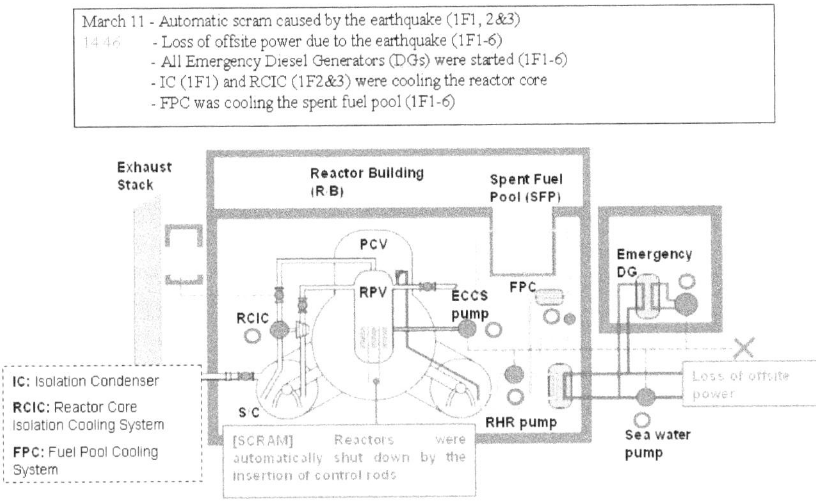

Fig. 1(a): Accident Progression—just after the earthquake.

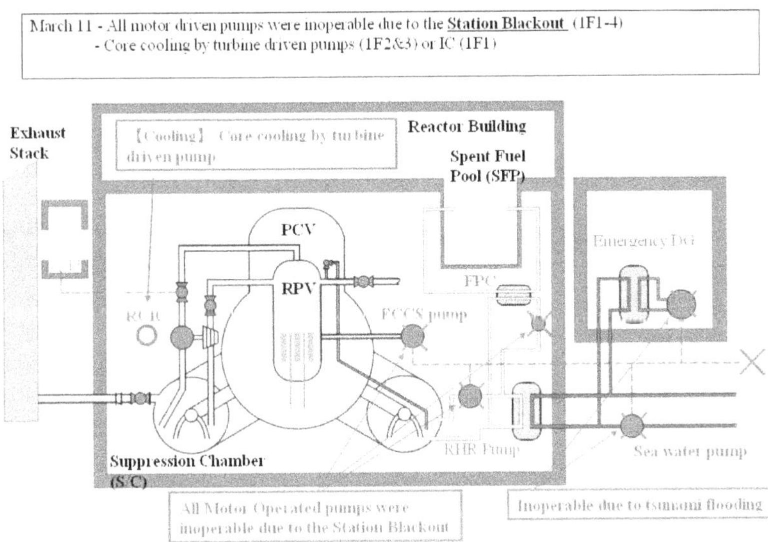

Fig. 1(b): Accident Progression—just after the tsunami hit.

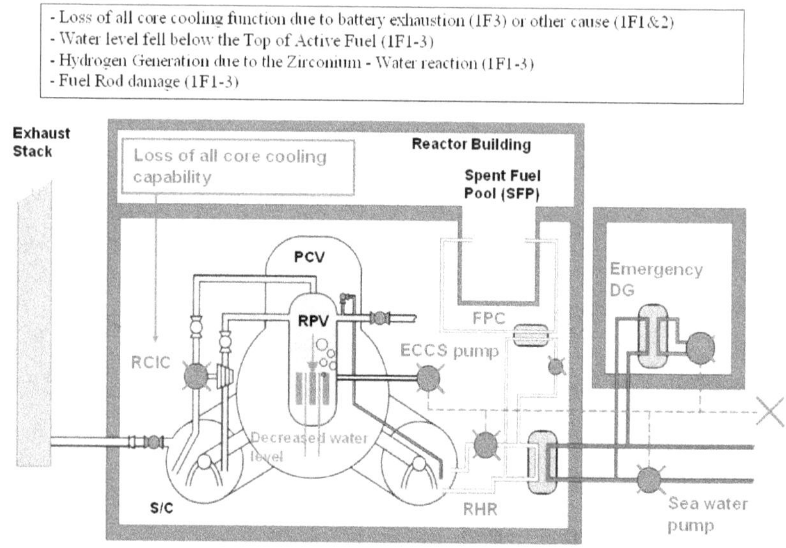

Fig. 1(c): Accident Progression—Loss of core cooling.

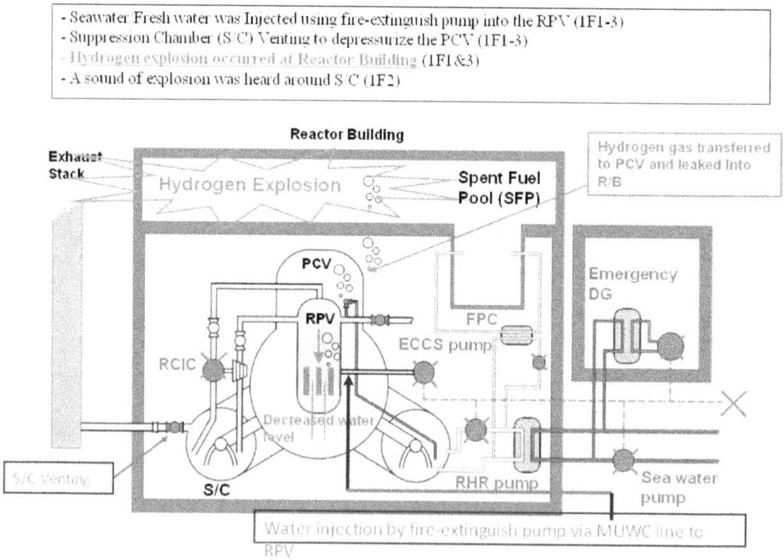

Fig. 1(d): Accident Progression—S/C vent and Water injection.

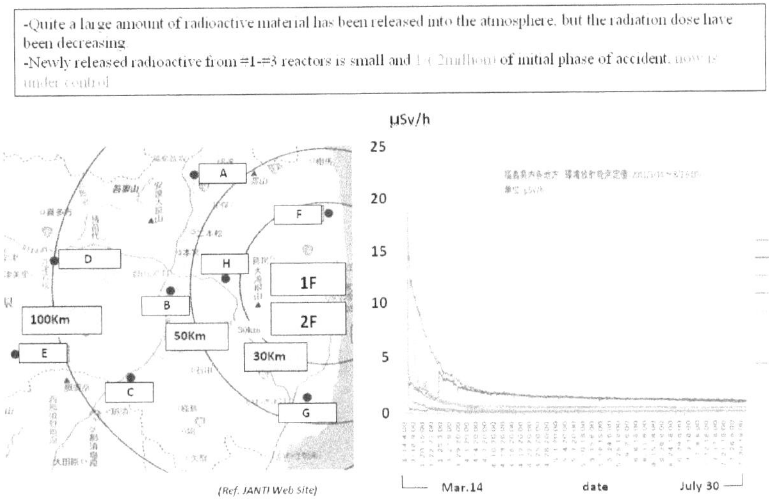

Fig. 2: Radiation dose around the NPP.

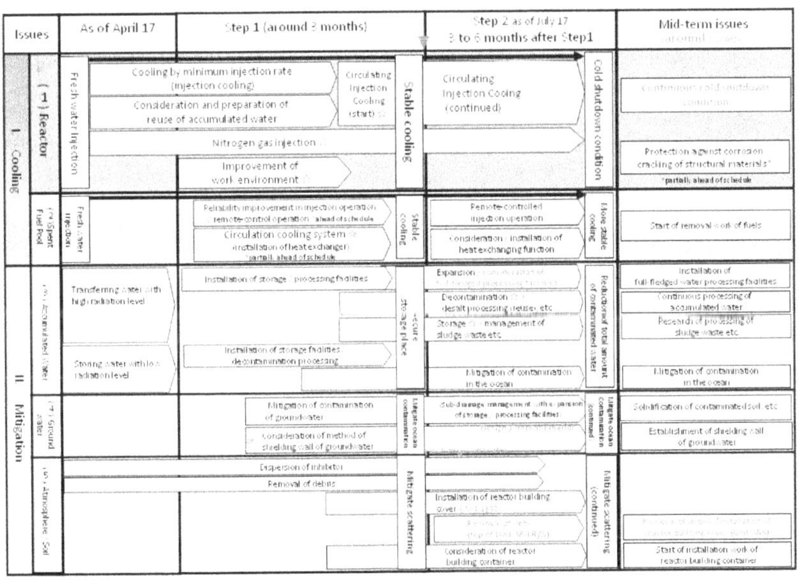

Fig. 3: Future Schedule—Roadmap towards Restoration.

As the total safety function status is summarized in Figure 4, shutdown was successfully done, while cooling and containment functions were damaged.

- "Shutdown" was secured by automatic simultaneous insertion of all control rods at the earthquake
- Transmission line was damaged by the quake; diesel generators started but subsequently were lost due to the Tsunami, leading Station Black Out.
- Most of the "Cooling" function of reactor and spent fuel pool were lost by the loss of power supply.
- Highly contaminated water has been found in turbine buildings. "Containment" function is presumed to be impaired.

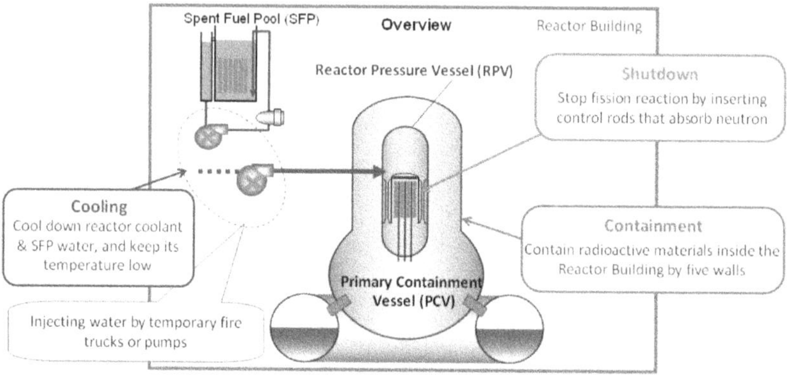

Fig. 4: Station Black Out with Loss of Heat Sink.

International Nuclear and Radiation Scale rating of three accidents are compared in Figure 5. Fukushima is assessed as Level 7 but release amount is one order less than Chernobyl even with three units damaged. The reason why the release from Fukushima and TMI is less than Chernobyl is due to reactivity control capability and fission product containment potential in the water.

Figure 6 shows Frequency-Consequence Curves (only for early fatalities) for Severe Accidents in various energy chains, OECD 1969-2001. Even the Chernobyl accident, the risk is very small than other energy industries, and we cannot plot Fukushima or TMI accidents because there were no fatalities (people death). The issue to follow in the near future would be psychological problem such as Post-traumatic stress disorder for evacuating people, as appeared in Chernobyl case.

<u>Countermeasures</u>

Countermeasures, which are lessons learned from accident, are summarized here:

- Against Station Black-Out
  - Ensuring the water tightness of essential equipment facilities, such as Metal-clad switchgears and Emergency Diesel Generators.
  - Secure power supply system through diversification of Power Supply sources, Fukushima Daiichi Nos. 5 and 6 were safe due to diversification.
  - Secure robust cooling functions of reactor and containment vessel.
- Against Severe Accidents

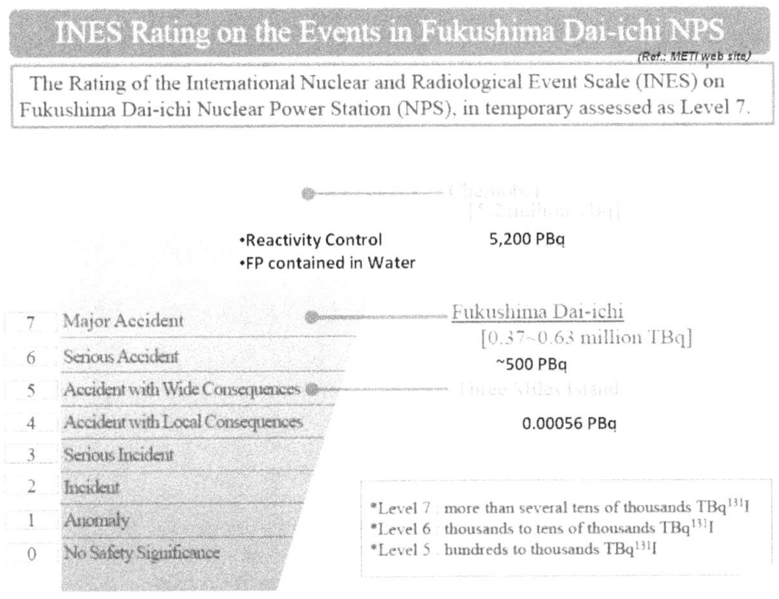

Fig. 5: INES Rating of three accidents.

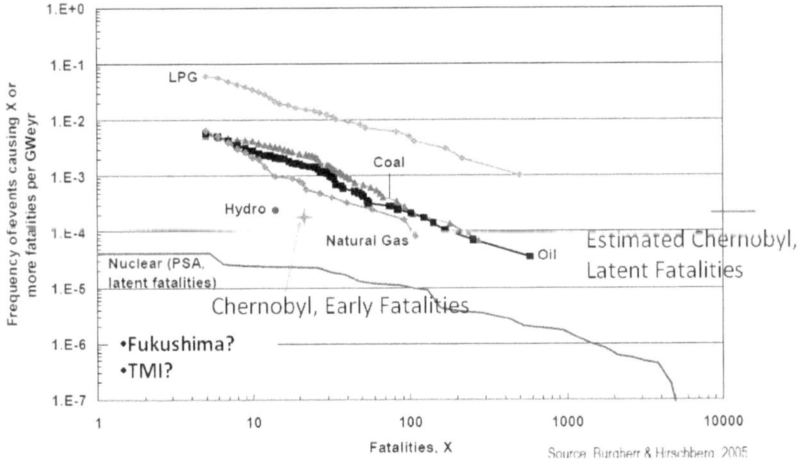

Fig. 6: Frequency-Consequence Curves for Severe Accidents in Various Energy Chains OECD 1969–2001.

Severe Accident means core melting and damage of Containment Vessel. So far Severe Accident is considered as Beyond Design Basis Accident (B-DBA), those measures should be extensively evaluated, thorough Accident Management (AM) measures.

1. Enhancement of prevention measures of hydrogen explosion
2. Enhancement of Containment Venting system

The causes of accident and countermeasures of safety on the NPP are considered on the aspects of regulations, designs and operations. To protect the Severe Accident under the station black out accident, safety design of NPP is further examined once more.

RETHINKING NUCLEAR POWER PLANT SAFETY ISSUES

Accident root cause analysis

First of all, Fukushima Daiichi accident issues will be presented. "Assuming while not considering such event!" That is the phrase to represent the Fukushima Daiichi Accident problem. First priority for safety related personnel is that Human Factor and Common Mode Failure are always worth keeping in mind.

- Station Blackout: Loss of offsite Power and Emergency DG failure
- Site with multiple units, Earthquake and Tsunami multiple events
- Everything would be done completely, while by formalism
- Imagination lack on risk or severe condition by responsible personnel

There are two items to be discussed, which are considered to be root cause of the Accident. One is "Rare Event" treatment; the other is "Crisis Management" problem, while both are due to lack of imagination by responsible personnel. Rare Event is high consequence with low frequency. Low consequence with high frequency event is easy to treat by commercial reason, while it is very difficult to handle the rare event even the risk is just the same. Unexpected event has been used frequently, but it is the risk-benefit issues to assume or not. Tsunami Probabilistic Risk Analysis has been carried out, and safety related personnel knew the severity of the effect well. Regardless of the initiating event, lack of measures to "Complete Loss of Power" is to be asked.

There are many Crisis Management problems as follows:

- Delay in initial response
- Delay in decision making
- Delay in external support request
- Poor collaboration among government (Prime Minister Kan), bureaucrats (NISA, JNES), and interested party (TEPCO)
- Poor information disclosure in emergency situation

After all, it is a matter of organizational culture. Anyway, rare event occurred on one occasion, measures had to be taken. Fukushima Daiichi Nuclear power plants as "National Privatization" destroyed by large-scale disasters should be treated same as infrastructure systems as a national policy.

What should be done to improve the safety?

Safety design principle is "Defense in Depth" concept as shown below, which should be therefore, reconsidered reflecting the accident causes.

- Preventing damage
- Failure expansion mitigation: autonomous characteristic, inherent safety (intrinsic safety)
- Incident prevention: a fail-safe, fool-proof, redundancy, diversity
- Incident expansion mitigation: confinement, control release
- Environmental effects mitigation: evacuation

Usual systems focus on the forefront function, such as preventing damage, expansion mitigation, or incident prevention, while safety critical systems increases attention to back-up function, if it has a large enough impact on the environment. Common Mode Failure, such as External Initiating Event, which is usually Rare Event, or auxiliary systems failure is difficult to install to Defense in Depth design.

- Earthquake and Tsunami, Off-site Power and EDG and Battery
- Sea Water Cooling

The application of the Swiss Cheese Model as proposed by James Reason of the University of Manchester, to the Fukushima accident indicates operational problems. Fallacy of the defense in depth has frequently occurred recently. The plant system was apparently safe enough as operators not consider system safety. Consequently the safety culture degraded, and any future incidents would more easily become organizational accidents. Such a situation requires a final barrier that is crisis management.

A concept of "Soft Barrier" has been proposed. There are two types of safety barriers, one is Hard Barrier that is simply Defense in Depth. The other is Soft Barrier, which maintains the hard barrier as expected condition, makes it perform as expected function. Even when it does not perform its function, human activity to prevent hazardous effect and its support functions, such as manuals, rules, lows, organization, social system, etc. Soft Barrier can be further divided to two measures, one is "Software for design": Common mode failure treatment, Safety logic, Usability, etc. The other is "Humanware for operation", such as operator, maintenance personnel, Emergency Procedure, organization, management, Safety Culture, etc.

MITIGATION OF GLOBAL WARMING BY OPTIMUM ENERGY MIXTURE

Setting and meeting the common target of CO_2 emission and energy mix are proposed on the basis of climate science and engineering model simulations. The feasible target for emission scenario named "over-shoot and zero emission" is proposed. International cooperation would enable the practical mitigation against global warming, based on cost minimum optimization over the world. Innovative and useful technologies should be introduced with well balances among renewable energy, nuclear energy and fossil energy.

Target of emission scenario and long term energy vision against global warming is discussed here. Feasible CO_2-emission scenario of the world, based on scientific concept is set as the target. To meet the target, world energy mix fitting to this CO_2 emission scenario was optimized well-balanced between investment and merit of fuel savings. International system on simple and transparent corporations, excluding speculation for carbon offset has been proposed. Alternative and feasible global emission pathway by scientific analysis based on target of global mean temperature rise, to limit the global surface temperature rise to approximate 2 degree C compared to pre-industrial levels, to decrease the CO_2 concentration by zero emission after a peak over the target concentration by Global Emission Pathway: Z650 as shown in Figure 7(a), which means that total accumulation CO_2 in this century is 650GtC. Compared with the IPCC 450 ppm equilibrium stabilization scenario, the proposed global emission pathway, so called Z650 scenario, allows a relatively large emissions during the near short term as shown in Figure 7(b). Therefore, the Z650 scenario will generate a higher CO_2 concentration in the near future. The concentration will exceed 450 ppm in a short period, but it will decrease and stabilize at a lower level. As the result, the Z650 scenario will cause a similar temperature rise with the IPCC 450 ppm scenario as shown in Figure 7(c). Multi gas GHG concentrations: max 2.3 degree C in 2100 and saturating 2 degree C towards 2200.

Result of simulation is shown below, in which system cost minimum optimization under Z650 restriction all over the world.

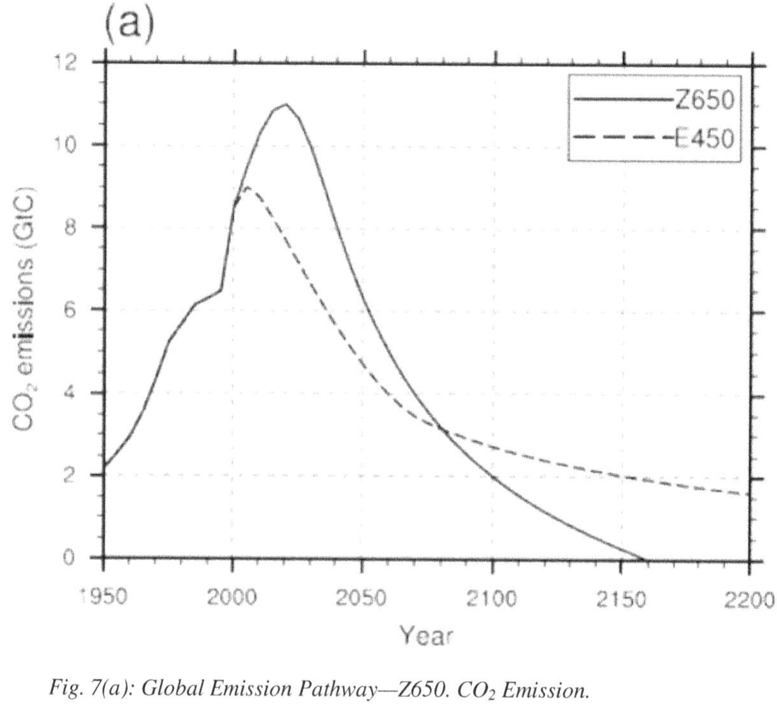

Fig. 7(a): Global Emission Pathway—Z650. CO_2 Emission.

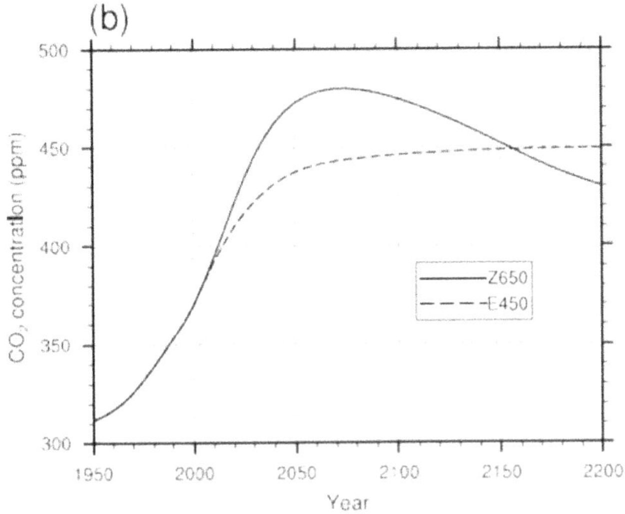

Fig. 7(b): Global Emission Pathway—Z650. CO_2 concentration,

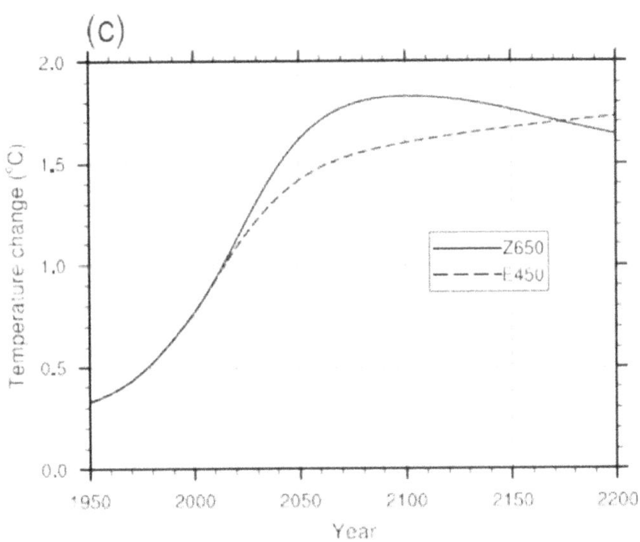

Fig. 7(c): Global Emission Pathway—Z650. Rise of mean global temperature.

Total Primary Energy for Z650 shown in Figure 8(a) continuously increases up to

2100, where both nuclear and renewable energy increase. Peak of fossil fuel consumption is around 2030. As the results, portion of Fossil: Nuclear: Renewable is 5: 2: 3 in 2050, while 3: 2: 5, in 2100. Region Total Primary Energy for Z650 is also examined as shown in Figure 8(b). In industrialized countries, total primary energy is almost constant up to 2100, where share of fossil fuel gradually decreases and share of renewable energy mainly increases alternatively. In developing countries, total primary energy continuously increases up to 2100, where peak of fossil fuel consumption is around 2040, and both nuclear and renewable energy increase remarkably. As CO_2 emission is summarized in Figure 8(c), developing countries would be able to release CO_2 much until 2030. As investment and fuel saving costs are summarized in Figure 8(d), in industrialized countries (2005 level = 1) investment cost is \$5Tri, while earn fuel saving \$4Tri, and in developing countries (2005 level = 1) investment cost \$8Tri, while earn fuel saving \$9Tri.

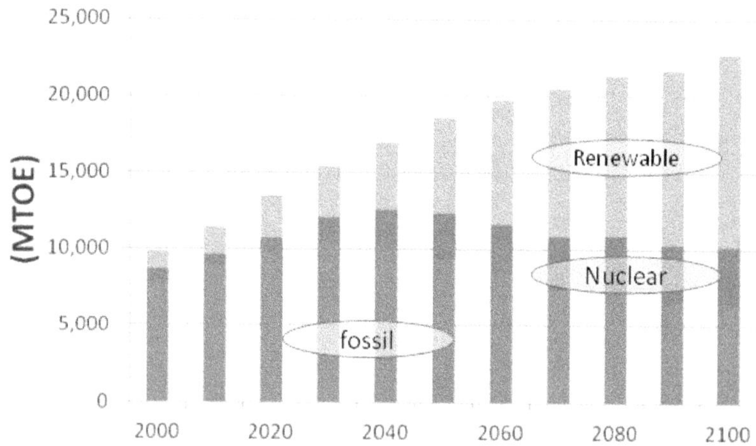

Fig. 8(a): Total Primary Energy for Z650.

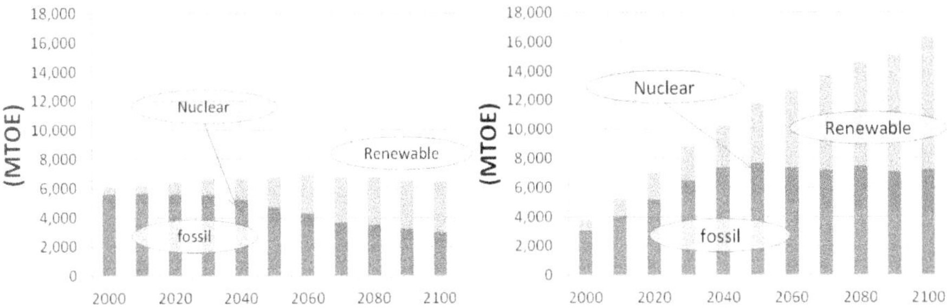

Fig. 8(b): Region Total Primary Energy for Z650.

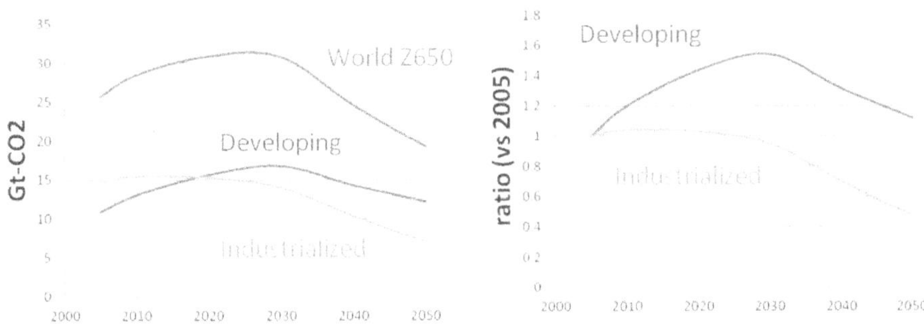

Amount of CO2 emission Ratio vs. 2005

		2005	2030	2050
WORLD	Z650 Emission	1.0	1.2	0.75
Industrialized Countries			0.9	0.5
	USA		0.9	0.4
	EU-15		0.85	0.5
	JAPAN		0.8	0.4
Developing Countries			1.6	1.1
	CHINA		1.6	0.9
	INDIA		1.9	1.7

Fig. 8(c): CO_2 emissions of Z650 scenario.

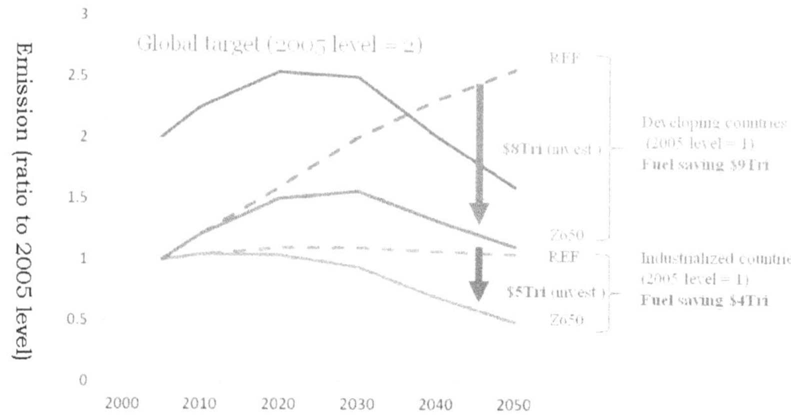

Fig. 8(d): Result of simulation Investment and fuel saving cost.

INNOVATIVE TECHNOLOGY AND DEPLOYMENT

Innovative technology and deployment to achieve world under CO_2 constraint are proposed as follows:

- Innovative technology
  1. Replacements of clean and highly efficient fossil-power plants (Combined cycles with natural gases and gasified coals) efficiency: $\eta = 30\%$ to 50%~60%
  2. Introductions of advanced nuclear power plants (3rd + and 4th generation reactor systems) and advanced spent fuel recycle systems
  3. Stabilization of unstable renewable energy with large-scaled battery and stable supply with grid networks
  4. Fuel alternative in transport sector (Fuel cell vehicle and Electric vehicle)
  5. Energy saving in industrial sector, especially steel , cement, paper and pulp and chemical industries

- Deployment
  1. New systems instead of the current CDM's short coming (i.e., additionality)
  2. Speedy and fair deployment systems of advanced technology
  3. Excluding speculations in carbon offset trading mechanism

The feasible target for new emission scenario (Overshoot and Zero-Emission) instead of traditional concept and energy mix against global warming has been proposed. It demonstrated that international cooperation would enable the practical mitigation against global warming, based on cost minimum optimization over the world. It identified that innovative and useful technologies should be introduced over the world. International cooperation should be advanced with new system instead of additionality of CDM in Kyoto Protocol and without speculation of carbon trading.

ENERGY ISSUE AND ROLE OF NUCLEAR ENERGY AFTER THE FUKUSHIMA DAIICHI ACCIDENT

Premise here is that "Global warming is an invariant problem!", "Energy security is also the invariant problem!". The long-term energy demand and supply simulation to minimize the total energy system cost was conducted for energy prediction during the 21st Century in the world. Taking the effort for energy-saving as major premise, carbon-sequestration for fossil fuel, renewable energy and nuclear energy should be altogether developed, which means energy best mix is achieved, under the CO_2 constraint around 450 ppm atmosphere.

Nuclear phase-out scenario, in which new nuclear plant construction is prohibited, is possible even considering the issue of global warming, from simulation as shown in Figure 9. Until around 2050, when nuclear is mostly replaced by natural gas. Since 2080, the contribution of the fossil power declined, large-scale introduction of the Fuel Cell. Nuclear phase-out scenario has two problems; increasing energy costs, little room for countermeasure and large uncertainties of technology. Therefore, rational use of nuclear power is requested, that is each country should make decision, Japan and several European countries will be also phase out, while China, India and ASEAN countries will continue to be introduced. If the accident happens again anywhere in, it will become the global phase-out. Therefore, rational unified safety standards (organizational structure, design and operation, regulations) should be reviewed based on the Fukushima Daiichi Problem world-wide analysis and established in the world.

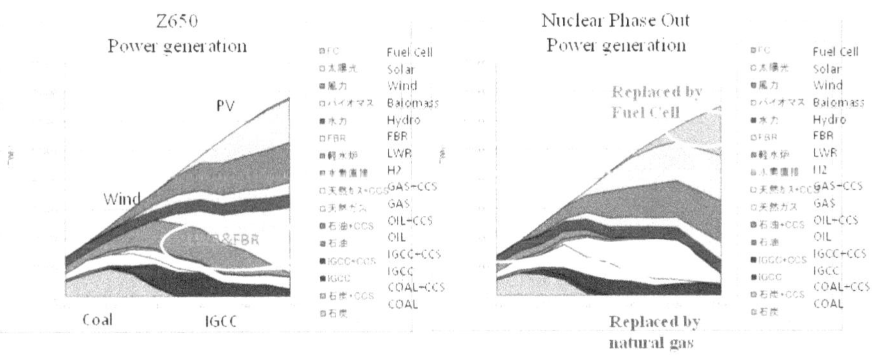

Fig. 9: Z650 and Nuclear power phase out (no newly established since 2020).

Table 2 summarizes the results. In Z650 case, idealized energy mix is obtained, well-balanced between investment and fuel saving. In Nuclear Phase Out (NuPO) case, unbalance between investment and fuels saving, and the burden on developing countries increases largely

Table 2 Summary of results/CO_2Emission, Additional investment and fuel savings in REF, Z650, and NuPO.

	CO_2 Emission/2005 = 1.0		**Acc. Emission**	**Acc. Cost ($Tri)**		
	2030	2050	**($GtCO_2$)~2050**	**(2010~2050)**		
REF (No restriction of CO_2)						
World	1.5	1.7	1483	326		
Industrialized C	1.1	1.0	635	155		
Developing C	2.0	2.6	848	171		
Z650 (Total CO_2 Emission 650GtC in this century)						
	CO_2 Emission ratio		**Acc. Reductions**	**Additional**	**Fuel**	**Balance**
	2030	**2050**		**Cost**	**Saving**	
World	1.2	0.75	372	13	-13	0
Industrialized C	0.9	0.5	128	5	-4	+1
Developing C	1.6	1.1	245	8	-9	-1
Z650 and NuPO (Nuclear phase out)						
World	1.2	0.75	373	20	-7	+13
Industrialized C	0.9	0.4	159	7	-4	+3
Developing C	1.7	1.2	214	13	-3	+10

CONCLUSION

Cause of Fukushima Daiichi Nuclear Accident is simple and primitive, and can be overcome. The main cause was lack of water tightness of switchgear, and/or lack of a water proof emergency diesel generator. Evolutions of safety technologies of nuclear power plants, such as emergency residual heat removal system have been in progress for this half a century. The future energy outlook should not be designed by technologies in 1960's. Now we are in the age of "Generation 3+" Reactor system, which have passive safety characteristics.

Nuclear power and renewable energy should be two wheels towards low carbon societies against global warming with economical growth. Under the new concept on emission scenario and with feasible advanced technologies, industrialized and developing countries can reach to low carbon societies with the optimized energy mix by all-over-the-world-cost minimum simulation. An international cooperation system against global warming should be re-established on the basis of Science, Feasibility, and Fairness.

REFERENCES

1. H. Ito, JANTI Report to American Society for Mechanical Engineers, May 12, 2011.
2. Burgherr and Hirschberg, 2005, Stefan Hirschberg, 22.06.2006, NEA/Paris 13.
3. T. Matsuno et al. "Stabilization of the CO_2 concentration via zero-emission in the next century", presented at the 1st CIGS Symposium on Oct. 27, 2009.
4. RCP Database (version1.0), IIASA Homepage. http://www.iiasa.ac.at/web-apps/tnt/RcpDb.
5. T. Yuhara et al., "Towards the harmony—Principles for the new climate regime", presented at the 2nd CIGS Symposium on Sep. 16, 2011.

GLOBAL IMPLICATIONS OF THE FUKUSHIMA DISASTER FOR NUCLEAR POWER

THOMAS B. COCHRAN, PH.D.
Natural Resources Defense Council, Inc.
Washington DC, USA

ABSTRACT

This paper reviews implications for the global use of nuclear power stemming from the nuclear accident at the Fukushima Dai-ichi Nuclear Power Station following the earthquake and tsunami on March 11, 2011. It includes: a) a review of numerous lessons learned from Fukushima related to accident precursors, design and operation of plant safety systems, emergency planning and nuclear plant regulation and safety culture: b) an examination of the historical frequency of severe fuel damage accidents at nuclear power plants and comparison of that frequency against the safety goals of the U.S. Nuclear Regulatory Commission; c) an examination of the population densities in the vicinity of nuclear power stations worldwide to assess the potential impact of radioactivity releases of Fukushima-scale or larger at other stations; and d) a review of reactor sites that are in areas of high risk to earthquakes and tsunamis.

INTRODUCTION

This paper follows closely and updates U.S. Senate testimony by Thomas Cochran, April 12th, which was updated in a statement by Thomas Cochran and Matthew McKinzie before the U.S. National Academies' Nuclear and Radiation Studies Board on May 26th, and again at an International Workshop on Nuclear Energy Safety in Beijing on June 29th. We also draw on several excellent reports and presentations, including:

1. Staff of Congressman Edward J. Markey, "Fukushima Fallout, Regulatory Loopholes at U.S. Nuclear Plants," May 12, 2011;[1]
2. Naoto Sekimura, "Overview of the Accident in Fukushima Daiichi Nuclear Power Plants," U.S. National Academies' Nuclear and Radiation Studies Board, May 26, 2011;[2]
3. Japanese Government, "Report of Japanese Government to the IAEA Ministerial Conference on Nuclear Safety: The Accident at TEPCO's Fukushima Nuclear Power Stations," June 2011;[3]
4. International Atomic Energy Agency (IAEA), "IAEA International Fact Finding Expert Mission of the Fukushima Dai-ichi NPP Accident Following the Great East Japan Earthquake and Tsunami," 24 May–2 June, 2011, Report to the IAEA Member States, 16 June 2011;[4]

---

[1] http://markey.house.gov/docs/05-12-11reportfinalsmall.pdf
[2] http://dels.nas.edu/resources/static-assets/nrsb/miscellaneous/SekimuraPresentation.pdf
[3] http://www.kantei.go.jp/foreign/kan/topics/201106/iaea_houkokusho_e.html
[4] http://www-pub.iaea.org/MTCD/meetings/PDFplus/2011/cn200/documentation/cn200_Final-Fukushima-Mission_Report.pdf

5. Charles Miller, et al., "Recommendations for Enhancing Reactor Safety in the 21st Century: The Near Term Task Force Review of Insights from the Fukushima Dai-ichi Accident," U.S. Nuclear Regulatory Commission (U.S. NRC), SECY-11-0093, July 12, 2011;[5]
6. Union of Concerned Scientists (UCS), "U.S. Nuclear Power after Fukushima: Common Sense Recommendations for Safety and Security, July 2011;[6] and
7. Arjun Makhijani, "Preliminary review of Recommendations for Enhancing Reactor Safety in the 21st Century: The Near-Term Task Force Review of Insights from the Fukushima Dai-ichi Accident, U.S. Nuclear Regulatory Commission," Institute for Energy and Environmental Research (IEER), July 12, 2011.[7]

ISSUES RAISED BY THE FUKUSHIMA DISASTER

There is a long list of issues and lessons learned from Fukushima that can be found in the reports identified above:

Category A. Accident initiators:[8]

1. Need to strengthen measures to protect against:
    a. natural external hazards:[9]
        i. earthquakes[10]

---

[5] http://pbadupws.nrc.gov/docs/ML1118/ML111861807.pdf

[6] http://www.ucsusa.org/assets/documents/nuclear_power/ucs-rpt-nuclear-safety-recs.pdf

[7] http://a4nr.org/wp-content/uploads/2011/07/July-19-IEER-review-NRC-report-on-Fukushima.pdf

[8] See IAEA, "IAEA International Fact Finding Expert Mission of the Fukushima Dai-ichi NPP Accident Following the Great East Japan Earthquake and Tsunami," 24 May – 2 June, 2011, Report to the IAEA Member States, 16 June 2011, p. 16, "Lesson 1: There is a need to ensure that in considering external natural hazards:

- the siting and design of nuclear plants should include sufficient protection against infrequent and complex combinations of external events and these should be considered in the plant safety analysis – specifically those that can cause site flooding and which may have longer term impacts;
- plant layout should be based on maintaining a 'dry site concept', where practicable, as a defense-in-depth measure against site flooding as well as physical separation and diversity of critical safety systems;
- common cause failure should be particularly considered for multiple unit sites and multiple sites, and for independent unit recovery options, utilizing all on-site resources should be provided;
- any changes in external hazards or understanding of them should be periodically reviewed for their impact on the current plant configuration; and
- an active tsunami warning system should be established with the provision for immediate operator action."

[9] Japan Government, "Report of Japanese Government to the IAEA Ministerial Conference on Nuclear Safety: The Accident at TEPCO's Fukushima Nuclear Power Stations," June 2011, p. 28, Lesson 1. Also, Charles Miller, et al., "Recommendations for Enhancing Reactor Safety in the 21st Century: The Near Term Task Force Review of Insights from the Fukushima Dai-ichi Accident," U.S. Nuclear Regulatory Commission (U.S. NRC), SECY-11-0093, July 12, 201i, p. 30, Recommendation 2, "The Task Force recommends that the NRC require licensees to reevaluate and upgrade as necessary the design-basis seismic and flooding protection of SSCs for each operating reactor."

ii. tsunamis
iii. floods
iv. hurricanes and typhoons
v. tornadoes[11]
vi. the implications of predicted sea-level rise and increased storm surges due to climate change.[12]

b. on-site hazards:
   i. fires, including seismically induced fires[13]

c. malevolent acts:
   i. external assault; terrorist attacks[14]
   ii. insider threats

2. Need to periodically review the magnitude and potential consequences of accident initiators.[15]
3. Need to improve warning systems for tsunamis and tornadoes.

Category B. Compliance with safety goals:

4. On a worldwide basis, power reactors are not adequately safe.[16]
5. The U.S. NRC should create an enhanced-design-basis framework to replace the design-basis plus beyond-design-basis framework now in place;[17] extend regulations to cover severe accidents;[18] which additional accident scenarios should be treated as part of the design-basis?[19]
6. Determine whether old reactor designs, e.g., GE BWRs with poorly designed Mark 1 and Mark 2 containments, are sufficiently safe to continue operation or have their licenses extended, even with subsequent safety upgrades.[20]
7. Apply the same type of timeliness goals to nuclear plant safety that it [the U.S. NRC] does for business-related requests from reactor owners.[21]
8. Require new reactors to be safer than existing reactors.[22]
9. Assign an appropriate value to human life in cost-benefit analyses.[23]

---

[10] Arjun Makhijani, "Preliminary review of Recommendations for Enhancing Reactor Safety in the 21st Century: The Near-Term Task Force Review of Insights from the Fukushima Dai-ichi Accident, U.S. Nuclear Regulatory Commission," Institute for Energy and Environmental Research (IEER), July 12, 2011, p. 14, Recommendation 5: The U.S. NRC should have its own independent seismology department.
[11] IEER, pp. 10-12, Recommendation 4.
[12] NRDC, see discussion in this paper.
[13] USNRC, Near-Term Task Force Review, p. 32, Recommendation 3; UCS, "U.S. Nuclear Power after Fukushima: Common Sense Recommendations for Safety and Security, July 2011, p. 3, Recommendation: Enforce Fire Protection.
[14] UCS, p. 3, Recommendation: Improve Protection against Terrorist Attacks.
[15] U.S. NRC, Near-Term Task Force Review, p. 30, Recommendation 2.2; IEER, p. 14, Recommendation 6.
[16] NRDC analysis.
[17] U.S. NRC, Near-Term Task Force Review, p. 22, Recommendation 1; IEER, p. 3.
[18] USC, p. 2, Recommendation: The U.S. NRC should extend the scope of its regulations to include the prevention and mitigation of severe accidents.
[19] NRDC question.
[20] IEER, p. 16, Recommendation 8.
[21] UCS, p. 3. Recommendation: Set Timeliness Goals for Safety Issues.
[22] UCS, p. 3, Recommendation: Strengthen Safety Standards for New Reactors; IEER, p. 18, Recommendation 12.

10. Probabilistic risk assessments (PRAs) should be updated;[24] U.S. NRC should perform its own independent PRAs as part of the license review process.[25]
11. Suspend relicensing and granting new licenses. [26]
12. Consider alternatives to nuclear power.[27]

Category C. Reactor systems to cope with severe accidents:[28]

13. Secure AC power—offsite and emergency diesel generators and generator fuel.[29]
14. DC power—longevity of batteries.[30]
15. Ensuring the water tightness of essential equipment facilities.[31]
16. "For severe situations, such as total loss of off-site power or loss of all heat sinks or the engineering safety systems, simple alternative sources for these functions including any necessary equipment (such as mobile power, compressed air and water supplies) should be provided for severe accident management."[32]
17. "Such provisions as are identified in [Category C, Lesson 10] should be located at a safe place and the plant operators should be trained to use them. This may involve centralized stores and means to rapidly transfer them to the affected site(s)."[33]
18. "Nuclear sites should have adequate on-site seismically robust, suitably shielded, ventilated and well equipped buildings to house the Emergency Response Centres, with similar capabilities to those provided at Fukushima Dai-ni and Dai-ichi, which are also secure against other external hazards such as flooding. They will require sufficient provisions and must be sized to maintain the welfare and radiological protection of workers needed to manage the accident."[34]
19. "Emergency Response Centres should have available as far as practicable essential safety related parameters based on hardened instrumentation and lines such as coolant levels, containment status, pressure, etc., and have sufficient secure communication lines to control rooms and other places on-site and off-site."[35]

---

[23] UCS, p. 4, Recommendation: The U.S. NRC should increase the value of human life in its analyses so it is consistent with other government agencies.
[24] Japan Government, p. 40, Lesson 27.
[25] IEER, p. 17, Recommendation 10.
[26] *Ibid.*, pp. 19-20, Recommendation 14.
[27] *Ibid.*, p. 20, Recommendation 15.
[28] U.S. NRC, Near-Term Task Force Review, p. 37, Recommendation 4.
[29] Japan Government, p. 29, Lesson 2.
[30] *Ibid.*
[31] Japan Government, p. 32, Lesson 8.
[32] IAEA, p. 16, Lesson 2.
[33] IAEA, p. 16, Lesson 3.
[34] IAEA, p. 16, Lesson 4
[35] IAEA, p. 17, Lesson 5.

20. Secure robust emergency core cooling systems for the reactor and the pressurized containment vessel (PCV).[36]
21. Insure that systems are capable of coping with large volumes of contaminated water, e.g., in the event of a breach of the reactor pressure vessel (RPV).[37]
22. Improve instrumentation for measuring status of reactors and spent fuel pools.[38] "Greater consideration should be given to providing hardened systems, communications and sources of monitoring equipment for providing essential information for on-site and off-site responses, especially for severe accidents.[39]
23. Common cause failure should be particularly considered for multiple unit sites and multiple sites, and for independent unit recovery options, the utilization of all on-site resources should be provided. "External events have a potential of affecting several plants and several units at the plants at the same time. This requires a sufficiently large resource in terms of trained experienced people, equipment, supplies and external support. An adequate pool of experienced personnel who can deal with each type of unit and can be called upon to support the affected sites should be ensured."[40]
24. "The international nuclear community should take advantage of the data and information generated from the Fukushima accident to improve and refine the existing methods and models to determine the source term involved in a nuclear accident and refine emergency planning arrangements."[41]

Category D. Spent fuel pool issues:

25. Secure robust cooling functions of spent fuel pools.[42]
26. See Lesson 22 above.
27. Placement of spent fuel pools in new reactor and NPS designs.[43]
28. Move spent fuel from wet pools to dry casks as soon as practicable.[44]

Category E. Adequacy of the responses to severe accidents:

29. Hydrogen production due to steam cladding interactions following uncovering of the core; enhancement of prevention measures.[45] "The risk and implications of hydrogen explosions should be revisited and necessary mitigating systems

---

[36] Japan Government, p. 30, Lesson 3.
[37] NRDC Recommendation.
[38] Japan Government, pp. 34-35, Lesson 14 and U.S. NRC, Near-Term Task Force Review, p. 46, Recommendation 7.
[39] IAEA, p. 17, Lesson 10.
[40] IAEA, p. 17, Lesson 7 and Japan Government, p. 31, Lesson 6.
[41] IAEA, p. 18, Lesson 13.
[42] Japan Government, p. 30, Lesson 4; U.S. NRC, Near-Term Task Force Review, p. 46, Recommendation 7; and IEER, p. 16, Recommendation 9.
[43] Japan Government, p. 32, Lesson 7.
[44] UCS, p. 3, Recommendation: Move Spent Fuel to Dry Casks; IEER, p. 17, Recommendation 11; NRDC recommendation.
[45] Japan Government, p. 32, Lesson 9.

should be implemented."[46] Consider alternatives to zircaloy fuel cladding to reduce the risk of hydrogen explosions.[47]

30. Enhancement of the containment venting systems.[48]
31. Training programs for responding to severe accidents.[49]
32. Central control of emergency supplies and equipment and setting up rescue team.[50]
33. Thorough accident management measures.[51]
34. Improvement of accident response environment.[52] Severe Accident Management Guidelines and associated procedures should take account of the potential unavailability of instruments, lighting, power and abnormal conditions including plant state and high radiation fields.[53]

Category F. Adequacy of radiation monitoring:

35. Enhance radiation exposure measurement systems for routine operations and during accidents.[54]
36. Adequate identification and forecast of the effect of released radioactive materials.[55]
37. Enhance radiation exposure management for workers. "Large scale radiation protection for workers on sites under severe accident conditions can be effective if appropriately organized and with well led and suitable trained staff."[56]
38. "Exercises and drills for on-site workers and external responders in order to establish effective on-site radiological protection in severe accident conditions would benefit from taking account of the experiences at Fukushima."[57]
39. Enhance radiation exposure management of members of the public.[58]

Category G. Adequacy of the emergency responses:[59]

40. Response to combined emergency of both large-scale natural disaster and prolonged nuclear accidents.[60]
41. Establishment of clear division of labor between relevant central and local organizations.[61]

---

[46] IAEA, p. 17, Lesson 8.
[47] IEER, p. 19, Recommendation 13.
[48] Japan Government, p. 33, Lesson 10; U.S. NRC, Near-Term Task Force Review, p. 41, Recommendation 5; and p. 43, Recommendation 6 and IEER, p. 16, Recommendations 7 and 8.
[49] Japan Government, p. 34, Lesson 13.
[50] Japan Government, p. 35, Lesson 15.
[51] Japan Government, p. 31, Lesson 5.
[52] Japan Government, p. 33, Lesson 11.
[53] IAEA, p. 17, Lesson 6.
[54] Japan Government, p. 34, Lesson 12 and p. 36, Lesson 17.
[55] Japan Government, pp. 37-38, Lesson 21.
[56] IAEA, p. 18, Lesson 14.
[57] IAEA, p. 18, Lesson 15.
[58] NRDC recommendation.
[59] U.S. NRC, Near-Term Task Force Review, p. 49, Recommendation 8, p. 56, Recommendation 9, and p. 57, Recommendation 10; UCS, p. 3, Recommendation: Strengthening Emergency Planning.
[60] Japan Government, p. 35, Lesson 16.

42. Enhancement of communication relevant to accidents.[62]
43. Enhancement of response to assistance by other countries and communication to the international community.[63]
44. Clear definition of widespread evacuation area and radiological protection guideline in nuclear emergency.[64]
45. Determine which reactor sites are located in areas that cannot be adequately evacuated, with consideration given to the need to evacuate in the immediate aftermath of a natural disaster.[65]
46. U.S. NRC should require a complete revaluation of emergency planning zones, evacuation, and population protection, especially in densely populated areas where evacuation in zones extending out 30 or 40 miles will be essentially impossible and asking ten million or more people to stay indoors for extended periods of months is also equally impossible.[66]
47. The U.S. NRC needs to revamp its entire regulatory structure for emergency planning to take into account (i) the protection of children, required by Executive Order 13045, (ii) the fact that large numbers of people may not be able to return to contaminated homes and jobs for prolonged periods, if ever, and that this needs special attention in densely populated areas, (iii) the irregular nature of the high radiation areas that develop, (iv) cleanup costs in the aftermath of accidents, (v) the need for real-time monitoring equipment to be in place far beyond the 10-mile planning zone, (vi) the need to take meteorological factors into account, often in real time, when informing the population (as, for instance, about potential rainfall and milk contamination), and (vii) the need to thoroughly reevaluate the viability of nuclear power plants in densely populated zones where evacuation in a 30-, 40-, or even 50-mile radius (depending on meteorological factors) may not be feasible.[67]
48. The NRC needs to publish scientific guidelines for communicating radiation risks that respect the established science, notably the EPA's Federal Guidance Report 13 and the National Academies BEIR VII report.[68]
49. Improve the adequacy of models used to calculate offsite environmental consequences and economic damage associated with a severe accident lasting for weeks.
50. Determine which reactor stations impose an undue economic risk to the local, regional or state economy in the event of a partial core melt accident.[69]
51. "The use of IAEA Safety Requirements (such as GS-R-2) and related guides on threat categorization, event classification and countermeasures, as well as

[61] Japan Government, p. 36, Lesson 18.
[62] Japan Government, pp. 36-37, Lesson 19.
[63] Japan Government, p. 37, Lesson 20.
[64] Japan Government, p. 38, Lesson 22.
[65] NRDC question.
[66] IEER, p. 9, Recommendation 1.
[67] *Ibid.*, pp. 9-10, Recommendation 2.
[68] *Ibid.*, p. 10, Recommendation 3.
[69] NRDC question.

Operational Intervention Levels, could make the off-site emergency preparedness and response even more effective in particular circumstances."[70]

52. "The use of long term sheltering is not an effective approach and has been abandoned and concepts of 'deliberate evacuation' and 'evacuation-prepared area' were introduced for effective long term countermeasures using guidelines of the ICRP and IAEA."[71]

Category H. Robustness of the safety infrastructure established at the nuclear power station:

53. Reinforcement of safety regulatory bodies.[72]
54. Establishment and reinforcement of legal structure, criteria and guidelines.[73]
55. Human resources for nuclear safety and nuclear emergency preparedness and response.[74]
56. Securing independency and diversity of safety system.[75] "Particularly in relation to preventing loss of safety functionality, the robustness of defense-in-depth against common cause failure should be based on providing adequate diversity (as well as redundancy and physical separation) for essential safety functions."[76]

Category I. Safety culture:

57. Raise awareness of safety culture at the nuclear reactor station.[77]
58. Need for improved safety culture at the regulatory authority.
59. "Nuclear regulatory systems should ensure that regulatory independence and clarity of roles are preserved in all circumstances in line with IAEA Safety Standards."[78]
60. Need for an industry sponsored safety program such as the U.S. Institute for Nuclear Power Operations (INPO).

SOME IMPLICATIONS FOR NUCLEAR POWER REACTORS WORLDWIDE

As noted by this lengthy list of issues and lessons learned, there are numerous concerns raised by the Fukushima nuclear disaster that bear on the safe operation and regulation of nuclear power reactors worldwide. Here we will offer brief observations regarding four issues:

- Are operational nuclear power reactors safe?

---

[70] IAEA, p. 17, Lesson 11.
[71] IAEA, p. 17, Lesson 12.
[72] Japan Government, pp. 38-39, Lesson 23.
[73] Japan Government, p. 39, Lesson 24.
[74] Japan Government, p. 39, Lesson 25.
[75] Japan Government, pp. 39-40, Lesson 26.
[76] IAEA, p. 17, Lesson 9.
[77] Japan Government, p. 40, Lesson 28.
[78] IAEA, p. 18, Lesson 16.

- Which reactors are located in areas that represent the greatest challenge if evacuation of the surrounding population is required or impose an undue economic risk to the local, regional or national economies in the event of a partial core melt accident?
- Which reactors are located in areas that are in areas of high risk to earthquakes and tsunamis?
- What improvements should be made to address the issue of hydrogen production caused by fuel cladding interactions with steam?

Are operational nuclear power reactors safe?

To address this issue we first address what constitutes an acceptable level of safety and then compare the history of power reactor operations against this benchmark.

What constitutes an acceptable safety risk?

The U.S. NRC Inspection Manual defines safety and compliance as follows:

> As commonly understood, safety means freedom from exposure to danger, or protection from harm. In a practical sense, an activity is deemed to be safe if the perceived risks are judged to be acceptable. The Atomic Energy Act of 1954, as amended [AEA], establishes "adequate protection" as the standard of safety on which NRC regulation is based. In the context of NRC regulation, safety means avoiding undue risk or, stated another way, providing reasonable assurance of adequate protection for the public in connection with the use of source, byproduct and special nuclear materials.
>
> The definition of compliance is much simpler. Compliance simply means meeting applicable regulatory requirements.[79]

Martin G. Malsch, former U.S. NRC Acting General Counsel, notes that the AEA was initially silent on 1) how undue risk should be defined, and 2) when non-safety

[79] U.S. NRC, "NRC Inspection Manual," Part 9900 Technical Guidance, Operation – Safety and Compliance, September 11, 2007. Martin G. Malsch, former U.S. NRC Acting General Counsel, similarly notes:

> The basic statute providing for the licensing and regulation of nuclear power plants in the United States is the 1954 Atomic Energy Act (AEA), which has been amended periodically.
>
> The AEA provides that licenses will be granted only if it is shown that there is adequate protection of the public health and safety. Historically, this has always been understood to mean that there must be reasonable assurance of no undue radiological risk to public health and safety. A license may be suspended or revoked if new information indicates this reasonable assurance no longer exists.

considerations (such as economic costs) may be taken into account in making undue risk decisions.[80]

Following the TMI accident, the U.S. NRC further defined "undue risk" in terms of safety goals. In response to industry concerns in the 1980's about the NRC's failure to account for the economic costs in safety decisions, the NRC developed a framework whereby economic costs would be considered in some decisions but not in others.

Prior to the U.S. NRC formulating its safety goals, in July 1985, the U.S. NRC Advisory Committee on Reactor Safeguards (ACRS) stated:

> We believe that the Commission should state that a mean core melt frequency of not more than 10^{-4} per reactor year (one in 10,000 reactor-years) is an NRC objective for all but a few, small, existing nuclear power plants, and that, keeping in mind the considerable uncertainties, prudence and judgment will tend to take priority over benefit-cost analysis in working toward this goal.[81]

The following year, on August 4, 1986, the U.S. NRC published a final policy statement on safety goals, which states in part:

> Severe core damage accidents can lead to more serious accidents with the potential for life-threatening offsite release of radiation, for evacuation of members of the public, and for contamination of public property. Apart from their health and safety consequences, severe core damage accidents can erode public confidence in the safety of nuclear power and can lead to further instability and unpredictability for the industry. In order to avoid these adverse consequences, *the Commission intends to continue to pursue a regulatory program that has as its objective providing reasonable assurance, while giving appropriate consideration to the uncertainties involved, that a severe core damage accident will not occur at a U.S. nuclear power plant.*[82] (emphasis added)

The policy statement also included two qualitative and two quantitative goals. With regard to the later the policy statement says:

---

[80] Martin G. Malsch, "U.S. Nuclear Power Reactor Safety Goals and Protection from Natural Phenomena," International Workshop on Nuclear Safety: Improving Safety in the Aftermath of the Fukushima Crisis, Beijing, China, June 29, 2011.

[81] ACRS letter from D. A. Ward to N. J. Palladino, Subject: ACRS comments on proposed NRC safety goal evaluation report (17 July 1985); cited in David Okrent, "The Safety Goals of the Nuclear Regulatory Commission, *Science*, **236**, 296-300 (17 April 1987).

[82] U.S. NRC, "Safety Goals for the Operation of Nuclear Power Plants; Policy Statement," *Federal Register* 30028, August 21, 1986. See also, U.S. NRC, *Federal Register* **51**, 28044 (4 August 1986); cited in David Okrent, "The Safety Goals of the Nuclear Regulatory Commission, *Science*, **236**, 296-300 (17 April 1987). See Also: Richard R. Meserve, Chairman, NRC, "The Evolution of Safety Goals and Their Connection to Safety Culture," Speech before the Atomic Energy Society of Japan/American Nuclear Society Meeting on Safety Goals and Safety Culture, Milwaukee, Wisconsin, June 18, 2001 http://www.nrc.gov/reading-rm/doc-collections/commission/speeches/2001/s01-013.html

The Commission has decided to adopt the following two health effects as quantitative objectives concerning mortality risks to be used in determining achievement of the qualitative safety goals—

- The risk to an average individual in the vicinity of a nuclear power plant of prompt fatalities that might result from reactor accidents should not exceed one-tenth of one percent (0.1 percent) of the prompt fatality risks resulting from other accidents to which members of the U.S. population are generally exposed.
- The risk to the population in the area near a nuclear power plant of cancer fatalities that might result from nuclear power plant operation should not exceed one-tenth of one percent (0.1 percent) of the sum of cancer fatality risks resulting from all other causes.

Richard A Meserve, then Chairman of the U.S. NRC, noted in 2001:

In 1990, the Commission provided additional guidance to the staff regarding the Safety Goals, endorsing surrogate objectives concerning the frequency of core damage accidents and large releases of radioactivity.[2] The numerical value of one-in-ten-thousand for core damage frequency (CDF) was cited as a "very useful subsidiary benchmark...." In addition, a conditional containment failure probability of one-tenth was approved for application to evolutionary light water reactor designs. This resulted in a large release frequency of one in one-hundred-thousand, since containment failure is necessary for a large release to occur. These values have evolved into the "benchmark" values of 10^{-4} for CDF and 10^{-5} for large early release frequency (LERF), as discussed in Regulatory Guide 1.174 for use in risk informed regulatory decision-making.[3]

[2.] U.S. NRC, Staff Requirements Memorandum on SECY-89-102, "Implementation of the Safety Goals," June 15, 1990.
[3] U.S. NRC, Regulatory Guide 1.174, "An Approach for Using Probabilistic Risk Assessment in Risk-Informed Decisions on Plant-Specific Changes to the Licensing Basis, July 1998.

U.S. NRC Regulatory Guide 1.174, to which Commissioner Meserve referred, states:

While there is no requirement to calculate the total CDF [core damage frequency], if there is an indication that the CDF may be considerably higher than 10^{-4} per reactor year, the focus should be on finding ways to decrease rather than increase it. Such an indication would result, for example, if (1) the contribution to CDF calculated from a limited scope analysis, such as the individual plant examination (IPE) or the individual plant examination of external events (IPEEE), significantly exceeds 10^{-4}, (2) a potential vulnerability has been identified from a margins-type analysis, or (3) historical experience at the plant in question has indicated a potential safety concern.[83]

In sum, the U.S. NRC now assumes the following quantitative safety goals for accident risk:

- The risk to an average individual in the vicinity of a nuclear plant of prompt fatality should not exceed 0.1% of the prompt fatality risk from other accidents (or 10^{-7}/yr), and related to the U.S. NRC's desire for defense in depth, two subsidiary quantitative safety goals:
  - a core damage frequency (CDF) goal of 10^{-4}/yr, and
  - a large early release frequency (LERF) goal of 10^{-5}/yr.
- And the risk to the population in the area near a nuclear power plant of cancer fatalities that might result from nuclear power plant operation should not exceed one-tenth of one percent (0.1 percent) of the sum of cancer fatality risks resulting from all other causes.

Neither "core damage" nor "core damage frequency' was precisely defined when these safety goals were enunciated. We thus turned to the U.S. NRC's web-based glossary and a report by the Institute of Nuclear Power Operations (INPO) for further insight. In its web-site glossary, "Core damage frequency (CDF)" is defined by the U.S. NRC to mean "An expression of the likelihood that, given the way a reactor is designed and operated, an accident could cause the fuel in the reactor to be damaged;"[84] and "Large early release frequency (LERF)" is defined as "the frequency of those accidents leading to significant, unmitigated releases from containment in a time frame prior to effective evacuation of the close-in population such that there is a potential for early health effects."[85] As will be seen from the discussion below, INPO distinguishes major fuel damage events caused, for example, by improperly managed reactivity additions, thermal-hydraulic transients, obstructions in coolant channels and overheating, from minor fuel damage events resulting, for example, from fabrication flaws.[86]

Initially the U.S. NRC's quantitative goals were not used in individual reactor licensing decisions because of reservations about probabilistic risk assessment. Moreover, as Martin Malsch notes, in developing the goals, no explicit consideration was given to potential environmental damage from severe accidents, the severe accident safety risk to populations (and opposed to individual safety risk), the increased severe accident risk from multi-unit sites, or the potential for large socioeconomic costs of severe accidents. Also, as Malsch notes:

---

[83] NRC, Regulatory Guide 1.174, "An Approach for Using Probabilistic Risk Assessment in Risk-Informed Decisions on Plant-Specific Changes to the Licensing Basis," November 2002. http://www.nrc.gov/reading-rm/doc-collections/reg-guides/power-reactors/rg/01-174/

[84] http://www.nrc.gov/reading-rm/basic-ref/glossary/core-damage-frequency.html

[85] U.S. NRC Regulatory Guide 1.174, footnote 3.

[86] See Institute of Nuclear Power Operations, "In-reactor Fuel-damaging Events: A Chronology 1945-1990," INPO 91-008, July 1991. http://research.greenpeaceusa.org/?a=download&d=434

> Reservations about PRA [probabilistic risk assessment] are now largely resolved; for example NRC's reactor inspection and oversight program now includes quantitative risk metrics based on the 10^{-4}/yr CDF goal; an existing reactor with a CDF significantly greater than 10^{-4}/yr would not be allowed to continue operation without further risk reduction.
>
> Other countries' CDF goals range from 10^{-4}/yr to 10^{-5}/yr.
>
> Other countries' LERF goals range from 10^{-5}/yr to 10^{-7}/yr.
>
> But comparisons are difficult because of different definitions of core damage and large early release.[87]

Reassessing the frequency of reactor core damage events.

There have been enough major fuel damage events, i.e., core damage, that we can ask whether the operational nuclear power plants throughout the world are safe enough as a group. For this purpose we will use as our primary benchmark the CDF goal of 10^{-4}/yr. As noted above, experts differ regarding the level of core or fuel damage that needs to occur before the event should be counted in assessing CDF. With this caveat in Table 1 we record 25 instances of fuel damage that have occurred in 24 nuclear power plants.[88] The 24 reactors are: [89]

1. Experimental Breeder Reactor-I (EBR-I)
2. Sodium Reactor Experiment (SRE)
3. Stationary Low-Power Reactor No. 1 (SL-1)
4. Enrico Fermi Reactor-1
5. Ågesta
6. Chapelcross-2
7. St. Laurent A-1
8. Lucens Experimental Power Reactor
9. St. Laurent A-2
10. Dresden-3
11. Three Mile Island-2
12. Hatch-1
13. Surry-1

[87] Martin G. Malsch, "U.S. Nuclear Power Reactor Safety Goals and Protection from Natural Phenomena," International Workshop on Nuclear Safety: Improving Safety in the Aftermath of the Fukushima Crisis, Beijing, China, June 29, 2011.

[88] Oyster Creek Generating Station experienced two fuel failure events.

[89] Absent from this list are events at reactors that were not designed to produce electricity, including events with major fuel damage at research and test reactors—National Research Experiment (NRX) (December 12, 1952) and Heat Transfer Reactor Experiment (HTRE-3) (November 8, 1958), Westinghouse Testing Reactor (WTR) (April 3, 1960), Oak Ridge Research Reactor (July 1, 1963)—and at military production reactors that were not dual-use and connected to the grid—Hanford (October 3, 1954), Hanford (January 5, 1955), Windscale (October 7, 1957). Windscale Pile 1, an air-cooled, graphite-moderated plutonium production reactor, was destroyed by a fire in October 1957. Windscale Piles 1 and 2 were both permanently shut down following the fire in Pile 1.

14. Arkansas Nuclear One-1
15. Oyster Creek (two core damage events)
16. Atucha-1
17. Chernobyl-4
18. Limerick-1
19. Pickering A-1
20. Hadden Neck (Connecticut Yankee)
21. Greifswald-5
22. Fukushima Dai-ichi-1
23. Fukushima Dai-ichi-2
24. Fukushima Dai-ichi-3

Following INPO's lead, in Table 1, events involving minor fuel damage resulting from fabrication flaws or damage that occurred during fuel handling are excluded.[90] Also excluded are fuel damage events resulting from vibration or erosion from coolant flow (baffle jet impingement) because such damage could have occurred without the reactor ever being critical.[91]

All 24 reactors produced electricity. Twenty-two of the reactors—all except EBR-I and SL-1—were connected to the grid during some period of their operation. Seven reactors—Lucens, TMI-2, Chernobyl-4, Greifswald-5 and Fukushima Dai-ichi-1, -2, and 3—were permanently shut down as a consequence of reactor accidents resulting in partial core melt or fuel damage. After 18 accidents at 17 reactors, the fuel melt or damage did not result in immediate closure of the plant; rather the damage was repaired and the reactor was restarted. However, in three cases—SER, SL-1 and Fermi-1—the reactors only operated for a short period of time following the fuel damage incident. Sixteen of the 24 power reactors are now permanently shut down. Eight reactors—Dresden-3, Hatch-1, Surry-1, Arkansas Nuclear One-1, Oyster Creek, Atuche-1, Limerick-1 and Pickering A-1—are still operational.

In assessing the historical core melt frequency among nuclear power reactors, the number counted depends on how the issue is framed. EBR-I, SL-1 and Lucens can be excluded on the basis that they were experimental reactors. EBR-I produced electricity but was not connected to the grid. The SL-1design was abandoned after the accident. Although it was the first U.S. reactor to supply electricity to the grid, the SRE could be excluded because it was primarily a research reactor. Ågesta was an early small design. Chapelcross-2 and St. Laurent A1 and A2 were dual use military reactors, producing plutonium for weapons and electricity for civilian use. Ågesta was also intended to be a production reactor. Thus, reasoned arguments can be advanced for excluding up to eight reactors/events of the 25 events at the 24 reactors. It is difficult to make a case that any of the remaining 17 events at 16 reactors should be excluded.

Also notable is how the frequency of fuel damage events varies over time. There were two events in the 1950s, six in the 1960s, three in the 1970s, 11 in the 1980s, and three to date in the 2010s. The absence of events between 1990 and the Fukushima events

[90] See Institute of Nuclear Power Operations, "In-reactor Fuel-damaging Events: A Chronology 1945-1990," INPO 91-008, July 1991.
[91] *Ibid.*

may reflect a lack of documentation rather than a lack of events. We will assume there were no, or only a few, events during this period. Thus, before Fukushima the historical CDF data suggested that power reactor operations were becoming significantly safer over time—a case that has become more difficult to make after Fukushima.

Worldwide, there have been 141 nuclear power plants that have been shut down after becoming operational with a total generating capacity of about 43,811 mega-watts electric (MWe) and about 3,077 reactor-years of cumulative operation.[92] Excluding EBR-I and SL-1, 14 experienced fuel damage.[93] Thus, one in ten (141/14 = 10.1) shut down power reactors experienced some form of fuel damage during their operation. Even excluding experimental and dual-purpose production reactors (EBR-I, SRE, SL-1, Ågesta, Chapelcross-2, Lucens and St. Laurent A-1 and A-2, for example) one in 15 (108/7=15.4) shut down power reactors experienced some form of fuel damage during their operation.

Of the power reactors that have been shut down, seven—Lucens, TMI-2, Chernobyl-4, Greifswald-5, Fukushima Dai-ichi-1, -2, -3—were shut down as a direct consequence of partial or full core melt accidents—one in 20 (141/7 = 20.1) reactors, or one for every 440 reactor-years (3,077/7 = 440) of operation. Excluding experimental (e.g., Lucens) and dual-purpose production reactors, one in 18 (108/6 = 18.0) were shut down as a direct consequence of partial core melt accidents. With the same exclusions only about nine out of ten giga-watts (GW) (40,545-4,068)/40,545 = 0.9) of nuclear power plant capacity have been closed without experiencing a fuel damage accident. One out of 10 GW (40,545/4,068 = 10) of nuclear power plant capacity have been closed as a direct result of a fuel melting accident.

Worldwide, there have been 593 nuclear power reactors that have operated approximately 14,400 reactor-years.[94] Thus, to date, the historical frequency of core damage is about one in 600 reactor-years (14,400/23 = 626). Even if one excluded nine of the fuel damage events—thus discounting the least severe core damage events—the CDF is still one in 1,000 reactor-years, or an order of magnitude greater that what the U.S. NRC deemed acceptable.

Worldwide, there have been 433 Light Water Reactors (LWRs) that have operated, of which 74 have been shut down and 359 remain operational. Among LWRs there have been 13 core damage events in 12 LWRs. Five of these were serious partial core melt downs—TMI-2, Greifsweld-5, Fukushima 1, -2, -3. More than one percent of the LWRs have experienced a serious core melt down. Four of the five generated a hydrogen explosion. One of the four hydrogen explosions (at TMI-2) was contained, but then we are not counting the hydrogen explosion at Fukushima Dai-ichi-4 which appears to have resulted from the core melt event at Fukushima Dai-ichi-3.

[92] This sum excludes production reactors that were not dual-use and connected to the grid and the U.S. reactors, EBR-I, SL-1, ML-1, PM-1, PM-2A, PM-3A, SM-1, SM-1A and Sturgis. Although it produced electricity, EBR-I and SL-1 are excluded because they were not connected to the grid. The German KNK-I and KNK-II reactors are treated a one reactor.

[93] The 14 shut down reactors that experienced fuel damage counted here are SER, Fermi-1, Ågesta, Chapelcross-2, Lucens, St. Laurent A-1 and A-2, TMI-2, Chernobyl-4, Hadden Neck, Greifswald-5, and Fukushima Dai-ichi-1, -2 and -3.

[94] See footnote 92.

Worldwide, there have been 117 Boiling Water Reactors (BWRs) that have operated approximately 3,100 reactor-years. Thus, to date, the historical frequency of core-melt accidents in BWRs is about one in 1,000 reactor-years (3,100/3 = 1,033).

Worldwide, there have been 49 BWRs with Mark 1 containments (the type at Fukushima) and 12 with Mark 2 containments. Five with Mark 1 containment (Millstone Unit 1 and Fukushima Dai-ichi Units 1-4) have been permanently shut down. These 61 BWRs have operated for 1,900 reactor-years to date. Thus, to date, the historical frequency of core-melt accidents in BWRs with Mark 1 and 2 containments is about one in 630 reactor-years (1,900/3 = 633).

Clearly, the historical frequency of core damage events worldwide does not measure up to the safety objectives of the U.S. NRC. On the whole, the operational reactors worldwide are not sufficiently safe. If one believes U.S. reactors on the whole are safer than those in most other countries, the logical conclusion is that the reactors in these other countries are somewhat less safe than indicated by the worldwide historical frequency of core damage.

If nuclear power is to have a long-term future, greater attention must be given to the safety of current operational reactors worldwide. Older obsolete designs should be phased out rather than having their licenses extended.[95] Reactors currently under construction worldwide and those on the drawing board must be made safer than operational reactors.

We offer two additional observations regarding the use of probabilistic risk assessments (PRAs) and the NRC safety goals. First, the NRC increasingly relies upon core-melt frequency estimates from PRA studies. While PRA analyses provide a useful *relative* measure of safety, given the large uncertainties of many input assumptions we do not believe quantitative risk calculations reliably measure the *absolute* frequency of low-probability catastrophic accidents. As but one example of why the absolute frequency values are unreliable, prior to Fukushima PRAs did not include scenarios where major events such as a hurricane, tornado or other severe weather conditions result in the loss of offsite power for more than 24 hours. Rather, the PRA analyses assumed offsite power would be restored within a few hours, and thus the PRAs were typically not run for more than 24 hours after the accident is initiated.

For the second observation we return to the two U.S. NRC's quantitative safety goals:

- The risk to an average individual in the vicinity of a nuclear power plant of prompt fatalities that might result from reactor accidents should not exceed one-tenth of one percent (0.1 percent) of the prompt fatality risks resulting from other accidents to which members of the U.S. population are generally exposed.
- The risk to the population in the area near a nuclear power plant of cancer fatalities that might result from nuclear power plant operation

[95] In the United States 71 of the 104 operational reactors have been granted 20-year license renewals by the NRC. See U.S. NRC, "Status of License Renewal Applications and Industry Activities," updated July 25, 2011. http://www.nrc.gov/reactors/operating/licensing/renewal/applications.html

should not exceed one-tenth of one percent (0.1 percent) of the sum of cancer fatality risks resulting from all other causes.

While these risk limits are typically compared against PRA estimates of frequency of accident scenarios multiplied by their consequences, we have taken a different approach. We compare the Fukushima consequences to the safety goals. In this comparison the operation of the Fukushima Dai-ichi nuclear plants, even accounting for the radioactive releases from the disaster, meet these two NRC safety goals. There were no prompt radiation fatalities due to the Fukushima accident. There were about 60,000 persons living within 20 km (12.6 mi) of Fukushima Dai-ichi. Over their lifetimes one would expect about 12,000 (20 percent) cancer fatalities from other causes. One-tenth of one percent of this amount is 12 cancer fatalities. Since the population was evacuated, less than 12 cancer fatalities are expected in this population from Fukushima.

Thus, on their face, we believe these two individual-risk safety goals are inadequate. At the very least, explicit consideration needs to be given to potential environmental damage from severe accidents, the severe accident safety risk to populations (as opposed to individual safety risk), and the potential for large socioeconomic costs of severe accidents.

Which reactors are located in areas that represent the greatest challenge if evacuation of the surrounding population is required or impose an undue economic risk to the local, regional or national economies in the event of a partial core melt accident?

Immediately following the earthquake and tsunami residents within 20 kilometers (km) (12. 4 miles) of Fukushima Dai-ichi were evacuated and those between 20 and 30 km (12.4 to 18.6 miles) were advised to remain in their homes as shelter or voluntarily evacuate. Subsequently, the Japanese government considered extending the evacuation zone to 30 km but ended up establishing a 20 milli-Sievert per year (mSv/y) or 2 rem/y dose limit for establishing which areas would be evacuated. Also notably, shortly after the Fukushima accident began to unfold the U.S. NRC was so concerned regarding how the accident might progress that it recommended that U.S. citizens stay at least 50 miles away. Some criticized the U.S. NRC Chairman for this action. Given the uncertainties at the time the decision was made, and the fact that the Japanese mandatory evacuation standard, 20 mSv/y, ultimately impacted people out to about 75 km (46.5 miles), we believe the U.S. NRC action was appropriate.

people within 20 km (12.4 miles), 160,000 within 30 km (18.6 miles), and 2 million within 80 km (50 miles) of the Fukushima Dai-ichi reactor station.

Worldwide there are 135 reactor sites that have a greater number of people residing within 30 km of the reactor station than were residing within 30 km of Fukushima Dai-ichi; including 21 reactor stations with more than one million people within 30 km.[96] Topping the list is the 125 MWe Karachi Nuclear Power Plant (KANUPP) in Pakistan, which has 8.3 million people within 30 km of the reactor. Two reactor stations on Taiwan—Kuosheng (2 operational reactors) and Chin Shan (2 operational reactors) have 5.5 million and 4.7 million people, respectively, living within 30 km; two adjacent stations in South Korea—Kori and Shin-Kori (5 operational

[96] Calculations by Declan Butler, Senior Reporter, *Nature*, France, private communication.

reactors, 3 under construction and 2 planned) and two on the Chinese mainland—Guangdong/Daya Bay (2 operational reactors) and Ling'ao (4 operational reactors) have more than 3 million people within 30 km. A severe nuclear accident at any one of these stations could have devastating consequences for the entire country. Figure 1 displays a histogram of populations within 30 km of a reactor, worldwide.

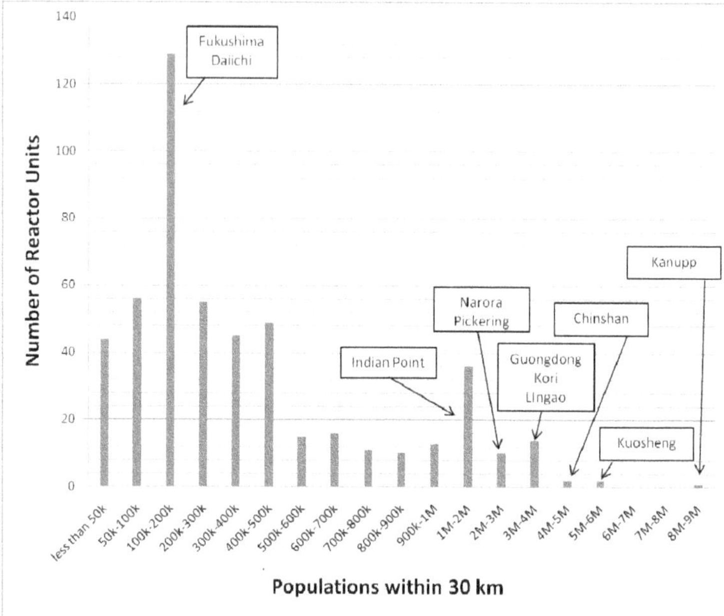

Fig. 1: Histogram of populations within 30 km of nuclear power reactors, worldwide.

Which reactors are located in areas that are in areas of high risk to earthquakes and tsunamis?

In a database maintained at NRDC, we have recorded the coordinates for 442 operational reactors in 31 countries, the majority of which are in North America and Western Europe. The geographic distribution of reactors worldwide may be analyzed with respect to two natural hazards that contributed to the accident at Japan's Fukushima Dai-ichi nuclear power plant: seismic and tsunami risks.

Seismic hazard data was obtained from the Global Seismic Hazard Assessment Program (GSHAP), a demonstration program launched in 1992 by the International Lithosphere Program with the support of the International Council of Scientific Unions, and endorsed as a demonstration program in the framework of the United Nations International Decade for Natural Disaster Reduction. The GSHAP data is available for download at http://www.seismo.ethz.ch/static/GSHAP/global/, and consists of gridded seismicity hazard values in 0.1 decimal degree intervals in latitude and longitude. These seismicity hazard values are given in units of Peak Acceleration (m/s^2) with 10% Probability of Exceedance in 50 Years. In these units, 0.0 to 0.8 are classified as a "Low Hazard;" 0.8 to 2.4 as a "Moderate Hazard;" 2.4 to 4.0 as a "High Hazard;" and greater than 4.0 as a "Very High Hazard." Figure 2 charts the number of operational reactors located in the given seismic hazard zones.

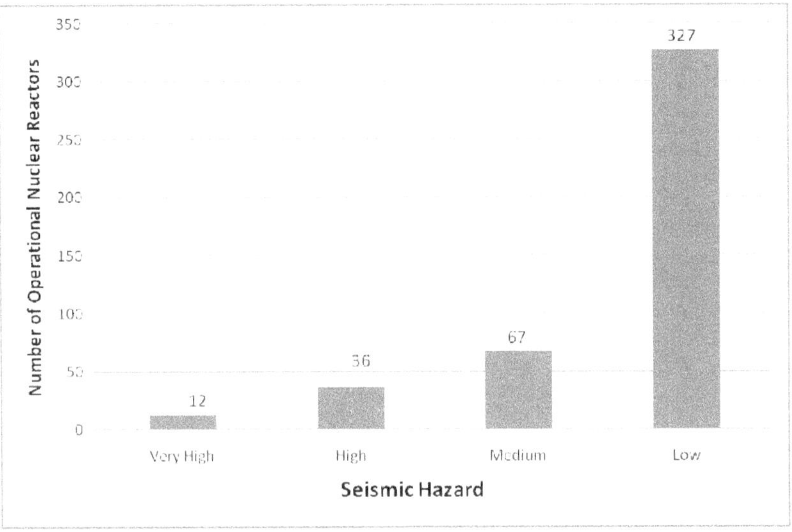

Fig. 2: Numbers of operational reactors within very high, high, medium and low seismic hazard areas, as determined by the GSHAP seismic hazard data.

The 12 operational nuclear reactors within very high seismic hazard areas are located in Japan and Taiwan, including all of Taiwan's six operating reactors.

In May of 2011 Japanese Prime Minister Naoto Kan requested that the Hamaoka Nuclear Power Plant be shut down, due to predictions that an earthquake of magnitude 8.0 or higher has an 87% likelihood of occurring in the area of the plant within the next 30 years. The plant remains shut and the reactors may begin decommissioning, pending the outcome of lawsuits. The three nuclear reactors at the Onagawa Nuclear Power Plant remain in cold shutdown following the March 11th earthquake and tsunami.

In addition to the six operating nuclear reactors in Taiwan located in a very high seismic hazard area, two additional reactors are under construction at Lungmen: Lungmen Unit 1 is expected to begin commercial operation at the end of 2011.

Of the 36 operating nuclear reactors in the high seismic risk category, 29 are in Japan, 4 in the United States and one each in Armenia, Iran (Bushehr) and Slovenia (Krško). The 67 operating nuclear reactors in the medium seismic risk area include 15 in Japan, 10 in France, 5 in South Korea, and 5 in the United States. Figure 3 shows a map of seismic hazard areas in Asia and the locations of operating nuclear reactors.

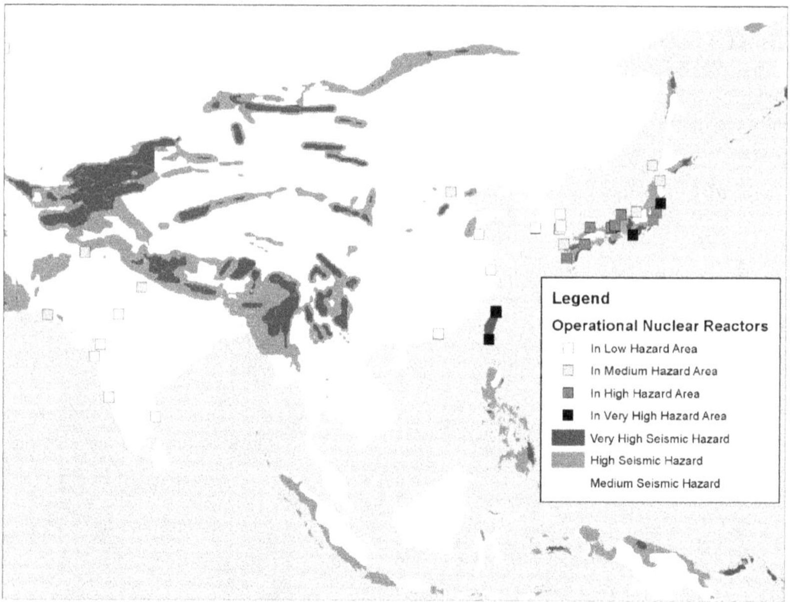

Fig. 3: GSHAP seismic hazard data and operational nuclear reactors in Asia.

The degree to which an earthquake will damage a nuclear reactor depends not only on the magnitude of the earthquake, but also on the reactor's seismic design basis. We note that the reactors at Fukushima Dai-ichi appear to have withstood the March 11th earthquake, but the damage from the earthquake to the electric grid contributed to causing the station blackout conditions.

A world database of tsunami events is maintained by the U.S. National Oceanic and Atmospheric Administration (NOAA) National Geophysical Data Center (NGDC), at http://www.ngdc.noaa.gov/hazard/tsu_db.shtml. Tsunami events are mapped in Figure 4, with increasing symbol size for increasing wave height. The majority of tsunami events for wave heights greater than 10 meters have occurred in: Indonesia, the United States, Japan, Tonga (an archipelago in the South Pacific Ocean) and in the Russian Far East regions. The locations of operating nuclear reactors are also shown in Figure 4.

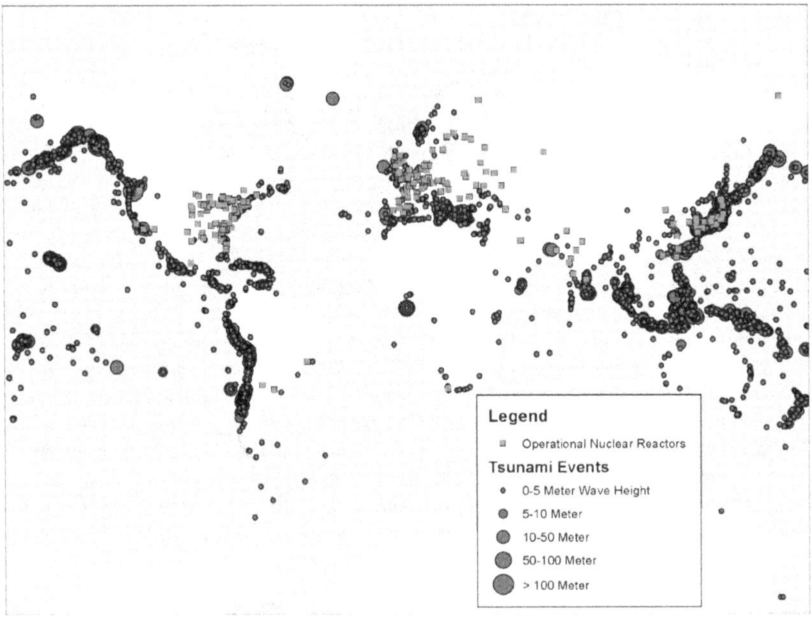

Fig. 4: NOAA historical tsunami event locations and operational nuclear reactors. With respect to earthquake and tsunami hazards, and large nearby populations, Taiwan's six reactors represent outliers in terms of high risks and consequences from a nuclear reactor accident. In order to explore these implications further, we used the U.S. Department of Defense computer model HPAC (Hazard Prediction and Assessment Capability, version 4.0.4, April 2004, Defense Threat Reduction Agency), which can calculate a radioactive source term for incidents at commercial nuclear reactors.[97]

For a given reactor HPAC permits the user to compute two pre-packaged nuclear facility scenarios:

> A moderate incident involving core damage from the result of a reactor accident, assuming that the core containment is not breached. The source term mainly involve the noble gas components of the reactor inventory.

---

[97] The HPAC documentation describes the code as: "...a counter proliferation, counterforce tool that predicts the effects of hazardous material releases into the atmosphere and its collateral effects on civilian and military populations. HPAC assists warfighters in destroying targets containing weapons of mass destruction (WMD) and responding to hazardous agent releases. It employs integrated source terms, high-resolution weather forecasts and particulate transport algorithms to rapidly model hazard areas and human collateral effects."

A severe incident involving very serious damage to the reactor core and an associated breach in the reactor containment. This results in a large release of radionuclides to the environment.

The HPAC code contains information on specific commercial nuclear reactors used in the radioactive source term calculation, including: geographic coordinates, thermal power, and a radionuclide inventory in the core. For the severe incident calculation, HPAC computes 5 releases of radioactivity with a total duration of two days.

The Chinshan and Kuosheng nuclear power stations lie on the northern tip of Taiwan, less than 50 kilometers from downtown Taipei. The results of HPAC calculations for Chinshan-1 and Kuosheng-1 are shown in Figures 5 and 6, respectively. In each case a severe radioactivity release was assumed; historical Taiwan weather data were used—for a release beginning at noon in August—assuming no precipitation and broken cloud cover; and terrain effects were included. As can be seen from Figures 5 and 6, under these conditions the plumes would travel south from the reactors over areas of high population density in the direction of Taipei. The dose rates contours are displayed in these figures, and may be compared with the evacuation threshold used by the Japanese government with respect to the Fukushima Dai-ichi accident of 0.22 mrem per hour, or about mid-way between the orange and yellow contours in Figures 5 and 6. Population density is also shown in these figures, along with the computed expected populations within the given contour.

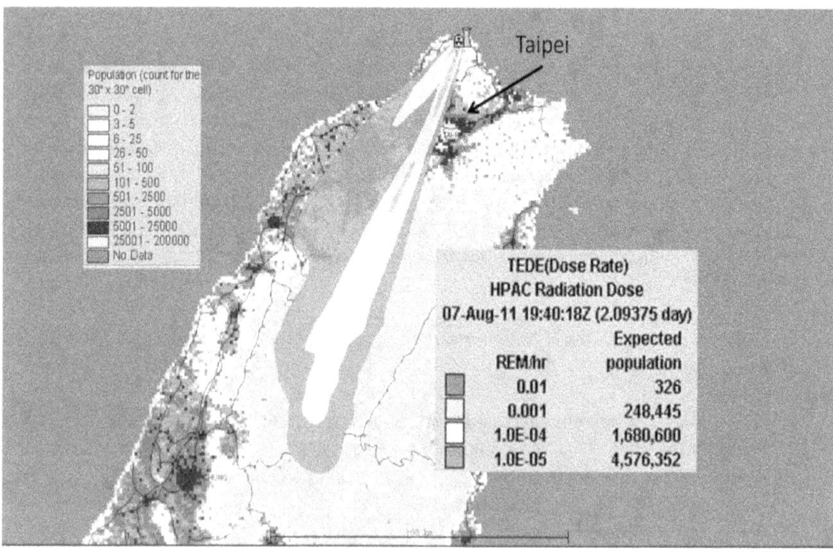

Fig. 5: HPAC calculation of dose rates from a severe incident at Taiwan's Chinshan-1.

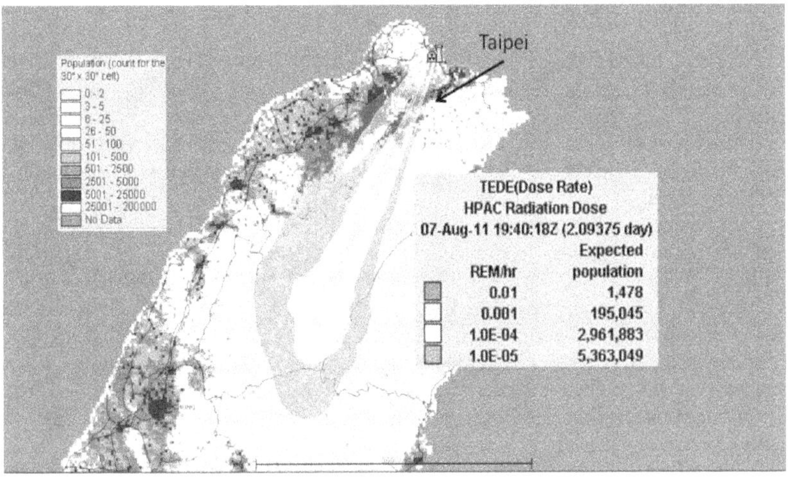

Fig. 6: HPAC calculation of dose rates from a severe incident at Taiwan's Kuosheng-1.

An HPAC severe incident calculation for Maanshan-1 (not shown) predicts that the plume would travel southwards out to sea for similar weather conditions. However, the coastal residents in the vicinity of the reactor are predicted to experience a high dose rate, greater than 10 mrem per hour, a population estimated by HPAC to be 3,500 persons.

<u>What additional improvements should be made to cope with hydrogen production in the event of zircaloy fuel clad interactions with steam?</u>

In 2001, the OECD Nuclear Energy Agency reported that computer analyses using "[t]he available Zircaloy-steam oxidation correlations were not suitable to determine the increased hydrogen production in the [CORA and LOFT LP-FP-2] experiments."[98] The LOFT LP-FP-2 experiment, conducted in 1985, is considered "particularly important in that it was a large-scale integral experiment that provides a valuable link between the smaller-scale severe fuel damage experiments and the TMI-2 accident."[99] Furthermore, an Oak Ridge National Laboratory report from 1990, discussing the CORA-16 experiment, states that "[t]he predicted and observed cladding thermal response are in excellent agreement until application of the available Zircaloy

[98]Nuclear Energy Agency (NEA) Groups of Experts, OECD Nuclear Energy Agency, "In-Vessel and Ex-Vessel Hydrogen Sources," NEA/CSNI/R(2001)15, October 1, 2001, Part I, B. Clément (IPSN), K. Trambauer (GRS), W. Scholtyssek (FZK), Working Group on the Analysis and Management of Accidents, "GAMA Perspective Statement on In-Vessel Hydrogen Sources," p. 9. http://www.oecd-nea.org/nsd/docs/2001/csni-r2001-15.pdf.

[99]NEA. OECD, "In-Vessel Core Degradation in LWR Severe Accidents: A State of the Art Report to CSNI," NEA/CSNI/R(91)12, January 1991, S.R. Kinnersly, *et. al.* p. 3.23. http://www.oecd-nea.org/nsd/docs/1991/csni-r91-12.pdf

oxidation kinetics models causes the low-temperature [1,652-2,192°F] oxidation to be under-predicted."[100]

The U.S. NRC is currently reviewing a petition for rulemaking PRM-50-93,[101] that among other things, requests that the U.S. NRC require that the rates of energy release, hydrogen generation, and cladding oxidation from the Zircaloy-steam reaction considered in computer analyses of postulated loss-of-coolant accidents (LOCA) be based on data from multi-rod bundle severe fuel damage experiments (e.g., the LOFT LP-FP-2 experiment).

If indeed, as experimental data indicates, computer analyses under-predict the total amount of hydrogen that would be generated in an accident, the design basis accidents of light water reactors need to be reassessed to determine the extent to which their power ratings should be reduced.

We have now experienced five serious partial core meltdown accidents in light water reactors—at TMI-2, Greifswald-5 and Fukushima Dai-ichi-1, -2, and -3—more than one percent of the total that have operated. Four of these (Greifswald-5 being the exception) were accompanied by a hydrogen explosion, with the three at Fukushima causing serious damage.

The United States Department of Energy is funding research and development of advanced fuel cladding designs based on silicon carbide (SiC). SiC is a ceramic material that has high strength and very low chemical reactivity at high temperatures when compared to zirconium alloys. SiC decomposes at 4900°F compared to a melting point of 3370°F for zirconium. The use of such fuel cladding materials would substantially diminish the likelihood of hydrogen production in core melt accidents. Consequently the development, qualification and required use of these advanced fuel cladding materials should be accelerated.

PRINCIPAL CONCLUSIONS

The historical frequency of core melt accidents worldwide does not measure up to the safety objectives of the U.S. NRC. On the whole the operational reactors worldwide are not sufficiently safe. If nuclear power is to have a long term future greater attention should be given to current operational reactors. Older obsolete designs should be phased out rather than having their licenses extended. We should also revisit whether the newer reactors currently under construction worldwide and those on the drawing board are safe enough. Because of differences in the number of reactors, the reactor safety cultures and the quality of regulatory oversight the next nuclear power plant disaster is more likely to occur somewhere outside of the United States.

In light of an improved scientific understanding of the full range of natural and man-made "beyond design basis" events that could strike nuclear plants, the risk of core melt followed by failure of containment should be stringently reevaluated for existing and new reactors located in areas of high population density and high risk from natural

---

[100] L. J. Ott, Oak Ridge National Laboratory, "Report of Foreign Travel of L. J. Ott, Engineering Analysis Section, Engineering Technology Division," ORNL/FTR-3780, October 16, 1990, p.3. http://www.osti.gov/bridge/servlets/purl/6434331-64LUUI/6434331.pdf

[101] Mark Leyse, PRM-50-93, November 17, 2009, located at www.nrc.gov, ADAMS Public Documents (Accession No.: ML093290250).

disasters. Under these revised conditions, the feasibility of an adequately protective evacuation and the potential for severe economic damage extending 30 km and beyond should be reassessed. The severity of the resulting radiological and other risks to life, property, and natural resources should inform regulatory decisions regarding which units should have their operating licenses shortened or construction permits revoked.

Countries should make comprehensive reassessments of the risks due to all the principal severe accident precursors, including the risks associated with earthquakes (beginning with those in areas of known high seismic risk), flooding, fires and malevolent acts at all reactor sites, and tsunamis, hurricanes and typhoon of coastal reactors. These studies and assessments should be conducted before further reactor license extensions are granted.

Table 1: Nuclear Power Reactors that have Experienced Fuel Damage.[102]

1. **Experimental Breeder Reactor I (EBR-I)**
   Location: Idaho National Laboratory, Idaho, USA
   Reactor type: Liquid Metal Fast Breeder Reactor (LMFBR)
   Power: 1.2 MWt; 0.2 MWe
   History: initial criticality: December 20, 1951; world's first electricity generating nuclear power plant; on November 29, 1955 a slow scram instead of a required fast scram was activated manually to terminate a planned transient. (For these experiments, the automatic fast scram capability, which was added to the reactor to compensate for a design weakness, was bypassed in the automatic mode.) About 50 percent of the core melted. The sodium-potassium coolant boiled, forcing molten fuel into the blanket-reflector region. Reactor shut down December 30, 1963.

2. **Sodium Reactor Experiment (SER)**
   Location: Santa Susana Field Laboratory, California, USA
   Reactor type: sodium-cooled graphite-moderated thermal power reactor
   Power: 20 MWt; 6.5 MWe
   History: initial criticality: April 25, 1957; first produced electricity July 1957; operated 2 years, partial core melt accident between 12 and 26 July 1959, resulting in melting of as much as one-third of the fuel; shutdown 26 July 1959 (It appears to have been operated for several days with its core partially melted.); converted to HEU-Th fuel; second core operations began September 1960; permanently shutdown February 1964.

3. **Stationary Low-Power Reactor No. 1 (SL-1)**
   Location: National Reactor Testing Station (now Idaho National Laboratory)
   Reactor type: experimental, gas-cooled, water-moderated
   Power: 3.3 MWt; 300 kWe

---

[102] Much of the history data are from the Institute of Nuclear Power Operations, "In-reactor Fuel-damaging Events: A Chronology 1945-1990," INPO 91-008, July 1991.

History: initial criticality March 1961; on January 3, 1961, in the process of working on control rod mechanisms, the center control rod was lifted manually by an excessive amount. A violent power excursion (prompt criticality) resulted, impaling the operator on the ceiling of the building and causing three fatalities, the only three people present. The facility was permanently shut down May 1964

4. **Enrico Fermi Unit 1 Reactor**
   Location: Newport, Lagoona Beach, Frenchtown Township, Monroe County, Michigan, USA
   Reactor Type: Liquid Metal Fast Breeder Reactor (LMFBR)
   Power: 200 MWt; 65 MWe (gross); 61 MWe (net)
   History: initial criticality 23 August 1963; commercial operations began August 1966; partial fuel melt accident 5 October 1966, two of the 105 fuel assemblies melted, but no contamination was recorded outside the containment vessel; closed November 1972

5. **Ågesta Nuclear Power Plant**
   Location: near Farsta, Sweden
   Reactor Type: Pressurized Heavy Water Reactor (PHWR)
   Power: 80 MWt; supplied 55 MWt heat; 12 MWe (gross); 10 MWe (net)
   History: grid connection and commercial operation May 1964; in 1968 15 fuel assemblies failed due to spacer grid spring relaxation and flow vibration. The reactor was returned to service with modified fuel and permanently shutdown June 2, 1974.

6. **Chapelcross Unit 2 Nuclear Power Plant**
   Location: Annan, Dumfreshire, Scotland, United Kingdom
   Reactor Type: gas-cooled, graphite moderated; Magnox
   Power: originally 180 MWt, up-rated progressively to 265 MWt, originally 23 MWe (gross) progressively up-rated to 60 MWe (gross); 50 MWe (net)
   History: startup May 1959; while under evaluation for the commercial reactor program experienced a partial blockage in a single fuel channel May 1967, contamination was limited to one region of the core; shut down 29 June 2004

7. **Lucens Experimental Power Reactor**
   Location: underground at Lucens, Moudon, Switzerland
   Reactor Type: heavy water moderated gas-cooled reactor
   Power: 30 MWt, 7 MWe
   History: start up 1962; generated electricity for the first time in January 1968; on January 21, 1969, the reactor was damaged irreparably during a startup because fuel channels were obstructed by corrosion products from moisture leakage into the carbon-dioxide reactor coolant. Fuel cladding melted, pressure tubes ruptured, and a violent reaction of molten metal with the carbon-dioxide coolant or heavy-water moderator occurred. Two-

thirds of the core fuel inventory was released into the moderator. The reactor vessel was damaged severely, and over five tons of contaminated heavy water poured into the fuel-handling room beneath the reactor. The reactor is permanently shut down.

8. **Saint-Laurent A-1 Nuclear Power Plant**
   Location: St. Laurent-Nouan, Loir-et-Cher, Centre, France
   Reactor Type: gas-cooled, graphite moderated
   Power: 1570 MWt; 405 MWe (gross), 390 MWe (net)
   History: grid connection 14 March 1969; commercial operation June 1969; on October 17, 1969, a coolant flow restrictor was loaded into the fuel channel by manually overriding an automatic stop on the on-line refueling machine. One hundred and eighty kilograms of fuel were melted during subsequent operation. While off-site radioactivity releases were minimal and no one was injured, the reactor was shut down for one year to complete cleanup operations end repair. The reactor was permanently shut down 27 May 1992.

9. **Dresden Unit 3 Nuclear Power Plant**
   Location: Morris, Grundy County, Illinois, USA
   Reactor Type: Boiling Water Reactor (BWR), BWR-3 (GE2), Mark 1 containment
   Power: 2,725 MWt; 867 MWe (gross)
   History: initial criticality January 1, 1971; on October 31, 1974, the control rods were operated in a manner intended to achieve better burnup of the fuel in the bottom of the core. The control rod movements did not adhere to guidelines provided by the fuel vendor. The technical review completed prior to positioning the control rods was inadequate. A xenon transient was started, causing power peaking near the bottom of the core. The clad was perforated in localized areas, resulting in a minor release of radioactivity. Replacement of a number of fuel elements was required. The nuclear power plant is currently operational.

10. **Saint-Laurent A-2 Nuclear Power Plant**
    Location: St. Laurent-Nouan, Loir-et-Cher, Centre, France
    Reactor Type: gas-cooled, graphite moderated
    Power: 1690 MWt; 465 MWe (gross) (uprated to 530 MWe (gross)), 450 MWe (net)
    History: started November 1970; grid connection 9 August 1971; commercial operation November 1971; heat excursion causing some fuel melting 13 March 1980; permanently shut down 27 May 1992

11. **Three Mile Island Unit 2 Nuclear Power Plant**
    Location: Londonderry Township; Dauphine County, Pennsylvania, USA
    Reactor Type: Pressurized Water Reactor (PWR)
    Power: 2,568 MWt, 808 MWe (gross); 776 MWe (net)

History: initial criticality December 1978; partial core melt accident March 28, 1979; decommissioned 1979; plant conditions, including reactor coolant system water inventory, were misinterpreted during a transient triggered by a minor fault in a non-safety system. The high-pressure safety injection system was prematurely shut off, reactor coolant pumps were turned off, and a stuck-open pressurizer relief valve was not diagnosed and isolated. As a result, the reactor core was severely damaged, and highly contaminated water spilled into the containment and auxiliary buildings.

12. Edwin I. Hatch Unit 1 Nuclear Power Plant

Location: Baxley, Appling County, Georgia, USA

Reactor Type: Boiling Water Reactor (BWR), BWR-4 (GE3), Mark 1 containment

Power: 2,436 MWt; 810 MWe (gross)

History: initial criticality September 12, 1974; on September 20, 1981, while operating at 100 percent power, large quantities of cold feedwater were added to the reactor vessel in a short period of time causing a power transient that was controlled by the insertion of control rods. The control rod geometry that was utilized resulted in increased local power peaking. The resultant fuel failures caused air ejector activity to rise to 0.3 Ci per second. (The reactor had been operated before the transient with a high concentration of copper in the coolant for an extended period, possibly weakening the fuel rod cladding.) The nuclear power plant is currently operational.

13. Surry Unit 1 Nuclear Power Plant

Location: Gravel Neck, Surry County, Virginia, USA

Reactor Type: Pressurized Water Reactor (PWR), 3-loop

Power: 2,441 MWt, 822.6 MWe (gross); 781 MWe (net)

History: initial criticality July 1, 1972; on March 9, 1983, fuel rods were discovered damaged due to metal chips that had been left in the reactor coolant system. The steam generators had been replaced previously, and necessary standards of reactor system cleanliness had not been maintained. Five years of operation were required to remove the resultant contamination from the reactor coolant system. The nuclear power plant is currently operational.

14. Arkansas Nuclear One Unit 1 Nuclear Power Plant

Location: Russellville, Pope County, Arkansas, USA

Reactor Type: Pressurized Water Reactor (PWR), 2-loop

Power: 2,568 MWt; 883 MWe (gross)

History: initial criticality August 6, 1974; on September 26, 1983, a control rod, which had been misaligned approximately 90 inches for 12 days, was realigned while the reactor operated at 100 percent power. The resulting xenon oscillations further increased the already high local power peaking, damaging the fuel and increasing gaseous activity. The nuclear power plant is currently operational.

15. Oyster Creek Generating Station

Location: Lacey and Ocean Townships, Ocean County, New Jersey, USA

Reactor Type: Boiling Water Reactor (BWR), GE BWR/2, Mark 1 containment

Power: 1,930 MWt; 641 MWe (gross)

History: initial criticality May 3, 1969; on December 19, 1984, control rods were withdrawn to increase power. Local power exceeded prescribed limits when the on-line core monitoring system, not properly programmed and tested, underestimated peaking power. Gaseous activity subsequently increased. Post-shutdown inspection found fuel damaged in 44 assemblies. The nuclear power plant is currently operational.

16. Atucha Unit 1 Nuclear Power Plant

Location: Lima, Buenos Aires, Argentina

Reactor Type: Pressurized Heavy-Water Cooled and Moderated Reactor (PHWR)

Power: 1,100 MWt; 357 MWe (gross); 335 MWe (net)

History: initial criticality 1974; on June 14, 1985, with the plant operating in a load-following mode (normal operating mode was base-loaded), operators were asked to increase power. Control rod motion to compensate for xenon burnout resulted in abnormally high local power peaking. Indications of this excessive power generation were misinterpreted, and reactor operation continued. The excessive local power in the core caused 46 fuel rods to fail, releasing fuel pellets into the reactor coolant system. The nuclear power plant is currently operational.

17. Oyster Creek Generating Station

Location: Lacey and Ocean Townships, Ocean County, New Jersey, USA

Reactor Type: Boiling Water Reactor (BWR), GE BWR/2, Mark 1 containment

Power: 1,930 MWt; 641 MWe (gross)

History: initial criticality May 3, 1969; on December 18, 1985, power was increased too rapidly causing both excessive flux peaks and excessive local power increase rates. Computer program data, relied upon to guide power increases, inappropriately was based on previous fuel cycle operating data. Fuel rod failures resulted in increased levels of gaseous activity. The nuclear power plant is currently operational.

18. Chernobyl Unit 4 Nuclear Power Plant

Location: Pripyat, Ukraine SSR (now Ukraine)

Reactor Type: RBMK-1000 (graphite-moderated water-cooled)

Power: 3,200 MWt; 1,000 MWe (gross); 925 MWe (net)

History: destroyed in full-core melt accident 26 April 1986; during the performance of a turbine-generator coastdown test, a severe reactivity excursion was experienced that, with the accompanying power and pressure surge and fire, destroyed the reactor and the surrounding building. The test procedure had not been reviewed adequately from a safety standpoint. Management control of the evolution was not maintained: the test procedure was not followed, control rods were withdrawn to an unauthorized configuration, and several safety functions were bypassed. These and other errors, coupled with serious design problems, resulted in a large insertion of positive reactivity and a rapid, uncontrolled power excursion.

19. Limerick Unit 1 Nuclear Power Plant
Location: Pickering, Ontario, Canada
Reactor Type: Boiling Water Reactor (BWR), BWR-4 (GE3)
Power: 3,293 MWt; 1,138 MWe (gross)
History: initial criticality December 22, 1984; on March 25, 1988, the reactor was operated over several months with copper concentrations in the reactor coolant at twice the recommended limit, weakening the fuel rod barrier metal. Subsequent fuel rod failures caused reactor power to be limited to 50 percent for the remaining six months of the fuel cycle. The nuclear power plant is currently operational.

20. Pickering A Unit 1 Nuclear Power Plant
Location: Pickering, Ontario, Canada
Reactor Type: Pressurized Heavy-Water Cooled and Moderated Reactor (PHWR)
Power: 1,744 MWt; 540 MWe (gross)
History: initial criticality 1971; on November 22, 1988, power was raised rapidly to avoid a shutdown due to xenon buildup while operating with an abnormal control rod configuration. This abnormal reactor control rod configuration compensated for reactivity loss incurred due to the unavailability of the on-line refueling equipment. Some damage to 191 fuel rods occurred. The nuclear power plant is currently operational.

21. Hadden Neck (Connecticut Yankee) Nuclear Power Plant
Location: Hadden Neck, Connecticut, USA
Reactor Type: Pressurized Water Reactor (PWR), 4-loop
Power: 1,825 MWt, 590 MWe (gross)
History: grid connection August 7, 1967; on October 9, 1989, 480 fuel rods were damaged during the operating cycle by metal chips present in the coolant system. During the 16-month operating cycle, coolant activity indicated only minor fuel damage was occurring. The thermal shield had been repaired during the preceding outage without maintaining the necessary standards of cleanliness and imposing restrictions on forced circulation of reactor water before cleanup was completed. Some of the debris became lodged in locations in the reactor system that were difficult

to inspect and clean. This debris led to the fuel damage through fretting of the cladding. The reactor was permanently shut down December 5, 1996.

22. Greifswald Unit 5 (KGR-5) Nuclear Power Plant

Location: Lubmin, GDR (now Germany)

Reactor Type: VVER-440, Model V-230, Pressurized Water Reactor (PWR)

Power: 1,375 MWt; 440 MWe (gross); 408 MWe (net)

History: grid connection 24 April 1989; commercial operation November 1, 1989; on November 24, 1989,three out of six cooling water pumps were switched off for a test; a fourth pump broke down and control of the reactor was lost; 10 fuel elements were damaged—partial core melting. The accident was reportedly attributed to sticky relay contacts. The reactor was permanently shutdown November 24, 1989.

23. Fukushima Dai-ichi Unit 1 Nuclear Power Plant

Location: Ohkuma, Fukushima Prefecture, Japan

Reactor Type: Boiling Water Reactor (BWR), GE BWR/2, Mark 1 Containment

Power: 1,380 MWt; 450 MWe (gross); 439 MWe (net)

History: initial criticality 10 October 1970; grid connection 17 November 1970; commercial operation 26 March 1971; partial core meltdown following earthquake on 11 March 2011

24. Fukushima Dai-ichi Unit 2 Nuclear Power Plant

Location: Ohkuma, Fukushima Prefecture, Japan

Reactor Type: Boiling Water Reactor (BWR), TOS1 (GE BWR/4), Mark 1 Containment

Power: 2,381 MWt; 794 MWe (gross); 760 MWe (net)

History: initial criticality 10 May 1973; grid connection 24 December 1973; commercial operation 18 July 1974; partial core meltdown following earthquake on 11 March 2011

25. Fukushima Dai-ichi Unit 3 Nuclear Power Plant

Location: Ohkuma, Fukushima Prefecture, Japan

Reactor Type: Boiling Water Reactor (BWR), TOS1 (GE BWR/4), Mark 1 Containment

Power: 2,381 MWt; 794 MWe (gross); 760 MWe (net)

History: initial criticality 28 January 1978; grid connection 24 February 1978; commercial operation 12 October 1978; partial core meltdown following earthquake on 11 March 2011

Excluded from this list because there is no report that the fuel failed is:

1. Shika Unit 1 Nuclear Power Plant

Location: Shika-machi, Ishikawa, Japan

Reactor Type: Boiling Water Reactor (BWR)

Power: 11,593 MWt; 540 MWe (gross)

History: grid connection January 12, 1993; commercial operation July 30, 1993; on June 18, 1999, a criticality event that occurred while the reactor was shut down for a planned inspection. The event occurred because of mismanagement of the routine testing of the hydraulic control rod system, which caused a pressure increase in the control rod withdrawal pipes. This resulted in the unwanted drop of three control rods, which lead to a local criticality event. The chain reaction was terminated by operators' intervention 15 minutes later. The reactor is currently operational.

NUCLEAR POWER IN THE MIDDLE EAST FOLLOWING FUKUSHIMA

ADNAN SHIHAB-ELDIN
Kuwait Foundation for the Advancement of Sciences
Safat, Kuwait

INTRODUCTION

This paper sets out the economic and environmental case for nuclear power in the MENA region and the challenges of deployment. It reviews the effect of the accident at Fukushima in Japan on nuclear programs worldwide and suggests the remaining outlook for the region.

THE CASE FOR NUCLEAR POWER IN MENA

Oil continues to contribute a significant share to the economy of most MENA countries as is shown in Figure 1. Most remain critically dependent on oil exports. However they also have some of the world's highest rates of growth of population. This, and growing prosperity, is driving up to a 9% annual increase in hydrocarbon fuel consumption for domestic power production. There is, therefore, a clear imperative to reduce consumption of oil for domestic energy production to increase exports and national income.

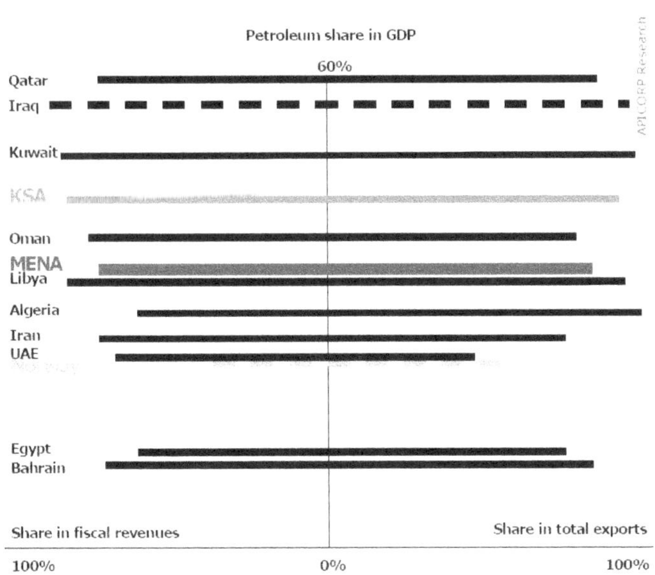

Fig. 1: Contribution of Petroleum to MENA Economies. (Source: APICORP Research[1])

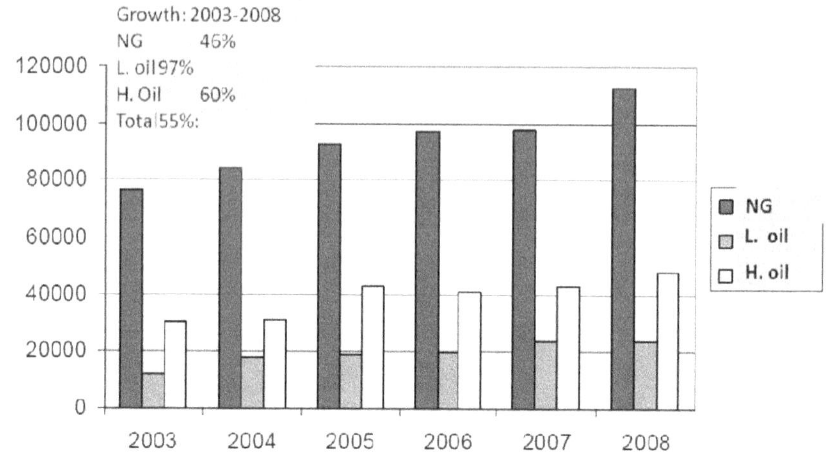

Source: Arab Union of Electricity Generators.[2]

Fig. 2: Fuel Consumption for Domestic Power Production in MENA Countries.

The potential is huge for improved energy efficiency and conservation in buildings in all sectors of the economy. The least cost candidate is the removal of energy subsidy. However, energy efficiency requires much greater regulation of new building design and construction and the more difficult task of improvement by retrofitting the existing stock of mainly residential buildings which is the predominant present source of demand. Conservation requires education and outreach to the entire population and without very severe drivers has very long timescales. It is unlikely that these measures will significantly reduce domestic fuel energy demand before 2020.

Of alternatives, even in advanced countries which have actively pursued them for decades, the contribution of renewable energy to generation is predominantly from wind and hydro and remains relatively small. In MENA, wind resource is less and hydro negligible. Despite relatively high solar resource and continual research interest, present adoption of renewables is among the lowest in the world because of the historical perception of ready availability of hydrocarbon fuels and an environment not friendly to new technologies. Even with an acceleration of effort, a significant renewable generation contribution before 2025 is unlikely.

Natural gas is the optimum hydrocarbon for power production (Figure 2) but as shown in Table 1, its availability in most GCC states is limited, due to its association with oil production, which would result in reliance on LNG imports raising concerns of cost and security of supply. However, without nuclear generation most MENA countries will inevitably become increasingly dependent on imports and their price will then become the primary determinant of future generation fuel mix and plant choice.

Table 1. Opec Proven Gas Reserves.

	(billion cu m)	Associated Reserves	Non-Associated Reserves
		% share	% share
Algeria	4,504	..	..
Angola	270	..	..
Ecuador	9	..	..
Iran, I.R. 2/	28,080	21%	79%
Iraq 3/	3,170	70%	30%
Kuwait	1,780	(~90%)	(~10%)
Libya, S.P.A.J. 4/	1,540	35%	65%
Nigeria 1/	5,215	52%	48%
Qatar 1/	25,257	2%	98%
Saudi Arabia 3/	7,305	60%	40%
UAE	6,072	(70%?)	(~30%)
Venezuela 1/	4,838	85%	15%

Source: OPEC—ASB[3]

Although because of relatively small populations, total carbon dioxide emissions are small, per capita values and intensity relative to economic activity are among the world's highest (Figure 3). As world attention to global warming increases, these specifics are an increasing concern regarding national and regional responsibility and world image, reinforcing the economic drivers for reduction of domestic oil consumption discussed above.

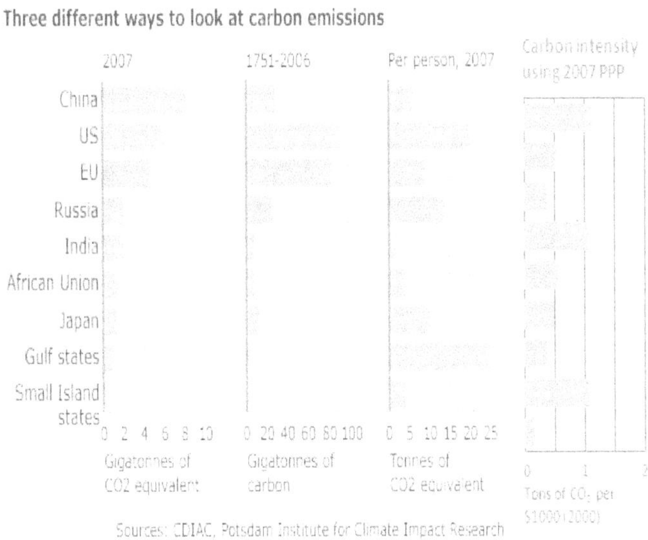

Fig. 3: CO_2 emissions and intensities. (Source: APICORP Research)

The economic competitiveness of nuclear power relative to coal and gas fired generation and the most competitive renewable technology, wind has been extensively studied. Most recent results from NEA/IEA study are summarized for major world regions in Figure 4 and for major industrial nations in Figures 5 and 6.

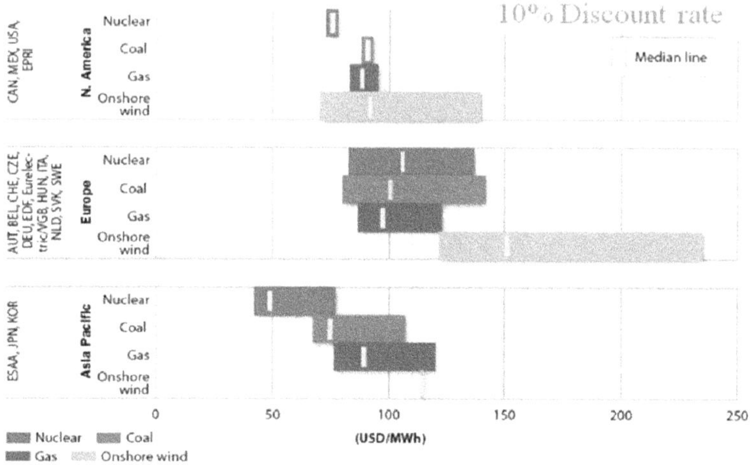

Fig. 4: Comparison of Projected Costs of Electricity Generation for World Regions. (Source: NEA/IEA 2010 Study[4])

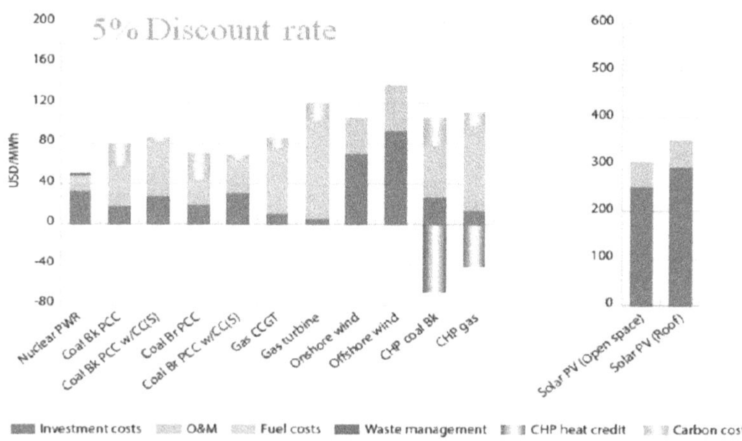

Fig. 5: Comparison of Projected Costs of Electricity Generation in Germany. (Source: NEA/IEA 2010 Study)

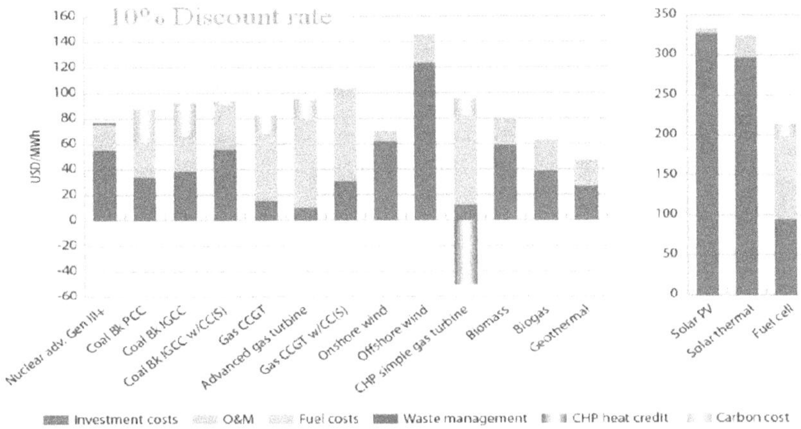

Fig. 6: Comparison of Projected costs of electricity generation in USA. (Source: NEA/IEA 2010 Study)

The competitiveness of nuclear in MENA with fossil fuels and renewable has been demonstrated by in-depth studies in recent years for several MENA countries. Detailed studies have been conducted in 2010 by several international consultants for the Kuwait National Nuclear Energy Committee (KNNEC). Figure 7 shows comparisons of nuclear, gas and oil fired generation with and without expected CO_2 tax for a range of discount rates. The effects of discount rates on each of these technologies are presented in Figure 8.

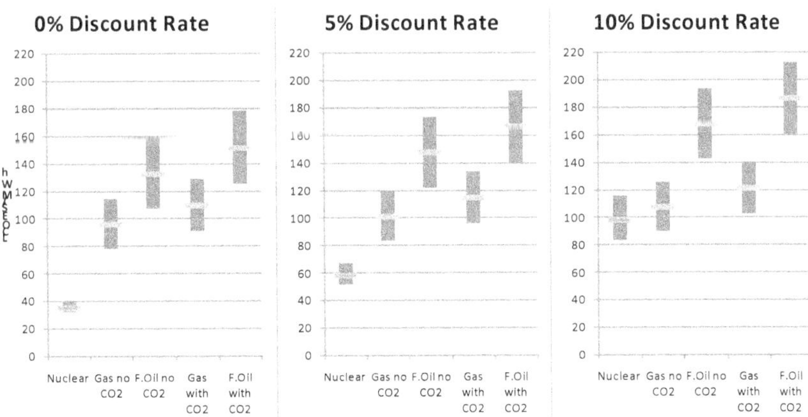

Fig. 7: Levelized Cost of Electricity in GCC. (Source: KNNEC (Study 2010)[5])

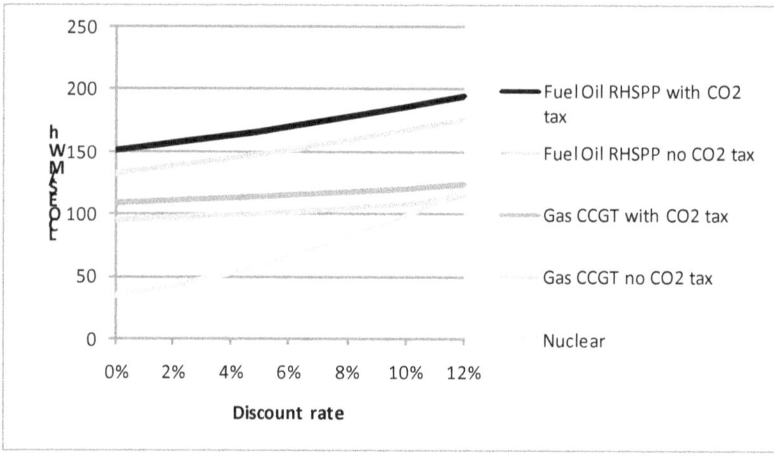

Fig. 8: Economic Viability of Nuclear—Significance of Discount Rate. (Source: KNNEC (Study 2010)

The critical sensitivities of those studies clearly indicate the nuclear plant to discount rate and gas fired plant to fuel price as shown in Figure 9. Nuclear is competitive below 6.4% discount rate @ 8$/Mbtu gas price (equivalent to an oil price of 52$/Bbl) or at 8.1% with 30$/ton CO_2 tax.

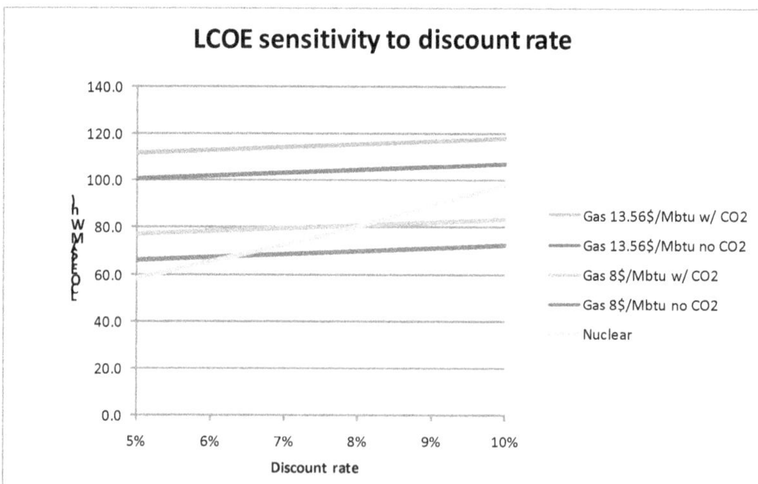

Fig. 9: Nuclear vs. Gas Firing (Effect of discount rate and fuel price). (Source: KNNEC (Study 2010)

The Kuwait study was extended to assess overall economic savings (netback) from a maximum nuclear deployment limited because of nuclear operating characteristics to system base load and considering a wide range of scenarios. In all probable cases, significant savings result, even with a low-probability adverse case nuclear breaks even. Results are summarized in Figure 10.

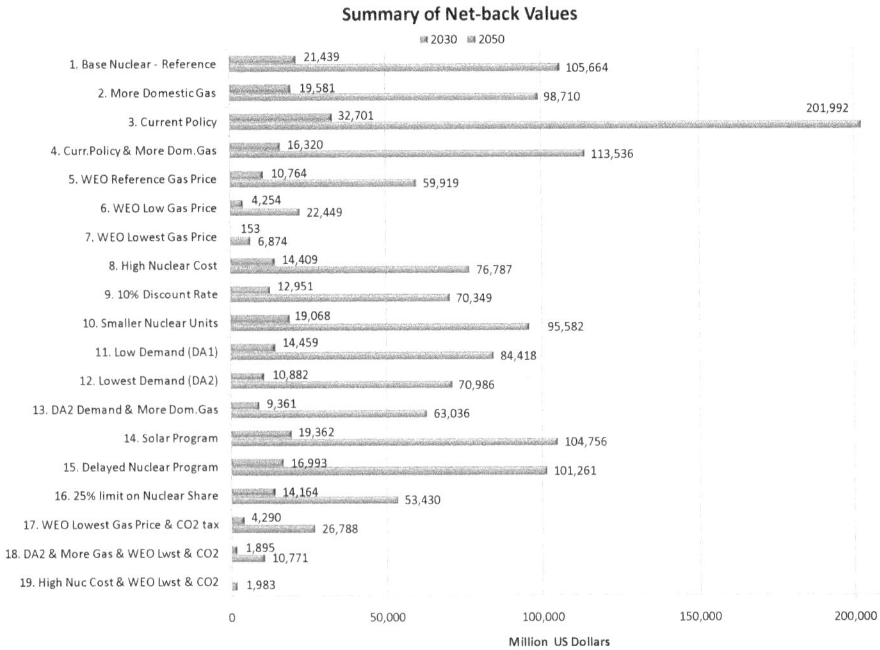

Fig. 10: Summary of Economic Assessment Results. (Source: KNNEC (Study 2010))

The studies demonstrate that the limits for economic competitiveness of nuclear power in oil exporting countries in the MENA are:

1. If imported gas price < 7$/MBTU
2. If oil <$46/Bbl
3. If overnight cost > $8000/MW
4. If electricity load remains at 12 GW
5. If discount rate > 12%
6. Natural gas prices is the most important determinant of economic feasibility of nuclear power

The economic case is, therefore, extremely robust.

CHALLENGES FOR INTRODUCTION OF NUCLEAR IN MENA

A number of important challenges face the successful deployment of safe and secure nuclear power in MENA region. These are:

1. Achieving informed political decisions and national consensus.
2. Development of National Workforce for NPP Education and Training Programs.
3. Safety and security of nuclear installations including meeting emergencies beyond design basis.
4. Establishment of safety culture within the operating organization.
5. Establishment of an independent regulator working to international standards and closely linked to international safety authorities.
6. Establishment of a framework of nuclear law and applicable safeguards.
7. Ensuring nuclear fuel supply through redundant international sources.
8. Establishing a reliable strategy for nuclear waste and management and storage with multiple redundant options.
9. Choice of an appropriate Generation 3 + Nuclear Plant technology.
10. Choice of a business model and public/private sharing of costs and risks.
11. Selection of the organization/partner to operate nuclear plant business model.

To launch new nuclear build, the IAEA recommends a roadmap to address 19 issues in 3 phases each with a major corresponding Milestone, as shown in (Figure 12). The expected timescale for the introduction of nuclear power in a new country, like in MENA region, is estimated to take from 10–15 years. Typical activities for the first two phases are as follows:

Phase 1

Pre- feasibility study
Identification of sites
Reviewing business model and financing
Setting up necessary organizations
Initial human resource development
Public information and consensus building

Phase 2

Detailed feasibility study & site investigations
Establish legal framework
Plan fuel supply and waste disposal
Continued 1 human resource development
Continued public information and consensus building

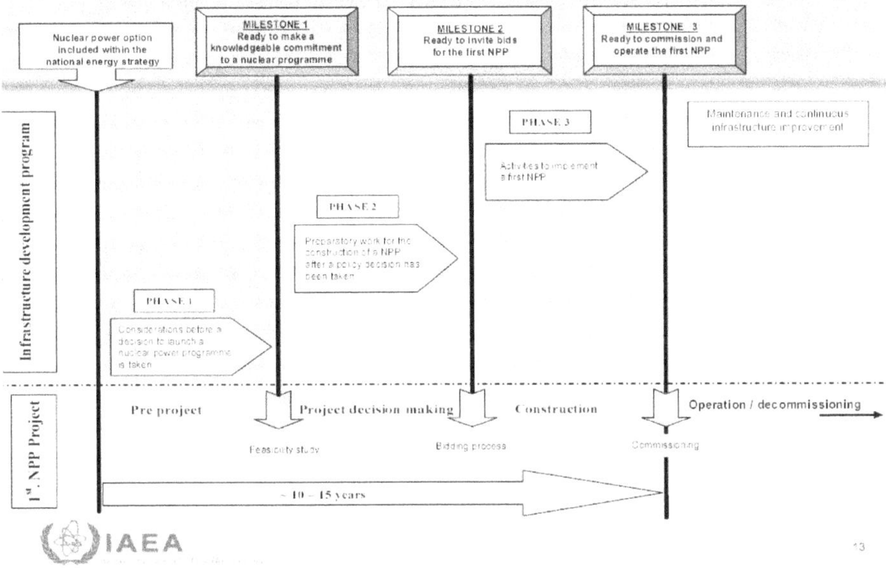

Fig. 12: Roadmap from concept to Operation—Milestones.[6]

IMPACT OF FUKUSHIMA

In the wider world, Germany, Switzerland and Italy have decisively rejected nuclear power plants. They have committed to decommission present plants as urgently as possible and abandoned all future nuclear plans. Japan is seeking strategies to reduce or eliminate nuclear power. Other countries including China, India, France, USA and South Korea have reacted in a more measured manner, retaining their programs whilst carrying out safety reviews of present and future plants. Countries considering or commencing nuclear programs have a similar spread of response.

The status of nuclear programs in MENA countries are summarized in Table 2. The UAE program is well advanced. With additional safety reviews, it is proceeding with construction on schedule. Of the countries considering nuclear programs in MENA region, only Kuwait has decisively rejected nuclear power. Response of all the other countries is more measured and their programs appear to be continuing. Jordan and Egypt remain committed, but face a number of challenges including financing and siting with rising public opposition in Jordan's case. That of Saudi Arabia, considering several options including building 16 reactors, is likely to be the next to mature.

Table 2. Nuclear Power Programs in MENA, pre and post Fukushima.

Country	Pre-Fukushima	Post-Fukushima
UAE	• Announced Plans to build fleet of 14 Reactors • Completed IAEA Phase I & II Milestones • Establishment of ENEC and FANR (regulator) • Completed site characterization, one selected • Selected Korean technology and contracted KIPCO for 4 nuclear reactors, 1st NPP 2017 • Participation in GCC joint NPP studies	• No changes, but launched comprehensive safety review of the program, including design • Program moving along at scheduled pace so far , (possibly few months ahead of schedule)
Kuwait	• Established in 2009 KNNEC to explore NP • Completed Phase I of IAEA milestones • Participation in GCC joint NPP studies	• KNNEC dissolved, ➔ KISR • Kuwait no longer actively pursuing construction of NPP
Saudi Arabia	• Government announced plans to consider implementation of NP program in 2009 • Royal Decree in 2010 established King Abdullah City for Atomic and Renewable Energy (KACARE) • Participation in GCC joint NPP studies	• Completed comprehensive planning and feasibility studies • Announcement of plans to build 16 reactors throughout KSA, • No details, but other options studied, waiting for decision
Turkey	• Research infrastructure in place • Announced plans to build 2 NPP, BOO agreement with Russia and Japan	• No change in plans • Go ahead with plans to build 1st NPP with Russia, talks with U.S.
Country	• Pre-Fukushima	• Post-Fukushima
Iran	• Bushehr NPP construction delayed, start 2011; part of ambitious plan started in 70s • Continuation of NFC program, enrichment up to 20%; despite UN sanctions	• The Bushehr NPP was connected to grid at low power (extra safety tests ordered) • Confirmed plan for 20 new reactors
Egypt	• Program in existence since 1955; Plans in 70s to build 10 NPP by 1999 put on hold (delay to ratify NPT until 81) • resurrected in 2007 calling for 4 NPP by 2025 • 5 potential sites currently exist, one selected	• No clear signs of a shift in policy • Announced safety reviews • Delay expected due to political restructuring, waiting confirmation by to-be-elected assembly
Jordan	• Compelling economic case led to plans and establishment of JAEC and JNRA (regulator) • Phase II of IAEA Milestones document completed; Tendering for first NPP ➔ 3 technologies chosen, was about to decide! • Site identification and characterization done • Contract for research reactor signed in 2009; • Contract to exploit U reserves & finance NPP	• Mounting public opinion backlash, opposition to desert site; other sites are earthquake zones; • Program continuing, so far, but Financing is major obstacle • Re-evaluation of strategy moving forward (slowing down expected)
Others	• Bahrain, Qatar, Oman, Morocco… All made various announcements not matched by concrete plans, time table and actions	• No new announcements so far, • Morocco moving ahead—hosting IFNEC Ministerial meeting in 2012

Whilst it is too early for full assessment, some observations may be made regarding the effect on future nuclear plant in MENA. There will be significant effects on design to improve safety and security affecting emergency power supplies, alternative AC power sources and design of spent fuel pools. These could apply to the forthcoming UAE nuclear plants. In general, there will be additional emphasis on:

1. Emergency planning at a very early stage of plant planning
2. Spent fuel storage remote from the reactors
3. Plant siting away from earthquake zones, tsunamis and other possible severe natural disasters

It is evident that the effect of Fukushima is critically dependent on informed decision making and consensus based on long-term considerations. The Arab spring may delay nuclear power program in MENA over the short term, but it opened the door for the possibility of assuring at national consensus a pre-requisite for sustainability.

CONCLUSIONS

The principle regional driver for nuclear power to release oil production, thus increasing export revenues remains, provided natural gas prices remain above $8/MBtu.

Before Fukushima, it was expected that about 20 reactors would be operating in 5 Arab countries before 2030 and that 20–30% of electricity generation could be nuclear by 2050.

Following Fukushima, it is probable that build up will be much slower with, at most, 8 reactors by 2030 in 2 or 3 countries, contributing 10–20% generation by 2050. The balance of generation must be met by gas fired plants, LNG imports and a greater emphasis on renewables.

The challenges of nuclear implementation in MENA are daunting. They include in particular safety and security, fuel supply and waste disposal and human resource development remain; but studies conducted so far show that they can be overcome. The most difficult issue now is to develop an informed national consensus. Paradoxically, in MENA, in this issue, the unexpected events of Fukushima, will in the longer term, be aided by the even more unexpected and seminal events of the “Arab Spring”.

REFERENCES

1. “Economic Commentary”, Vol.3, No. 5-6, (June 2008). APICORP Research,
2. “Energy Consumption in Electricity Sector in Arab Counties”, Arab Union of Electricity Generators, 9th Arab Energy Conference, 9-12 May, 2010, Doha—Qatar.
3. OPEC Annual Statistical Bulletin, 2010
4. Projected Cost of Generating Electricity 2010 Edition”, NEA/IEA 2010
5. “Preliminary Economic Feasibility Study a Civil Nuclear Power Program in Kuwait”, report by Lightbridge to Kuwait National Nuclear Energy Committee (KNNEC), July 2010
6. “Milestones in the Development of a National Infrastructure for Nuclear Power, NE Series Guide NG-G- 3.1, September 2007. IAEA.

NUCLEAR POWER DEVELOPMENT IN CHINA

DADI ZHOU
Beijing Energy Efficiency Center (BECon)
Beijing, China

RAPID GROWTH OF ENERGY CONSUMPTION AND GHG EMISSION

- Energy demand increase rapidly, due to high economy growth.
- Total primary energy consumption over 3.2 billion tons of coal equivalents.
- Rank the first in terms of primary energy consumption and GHG emission.
- Primary energy will keep on growth for next decades.
- Annual increase over 150 million TCE likely.
- Based on coal, 70% of the total.

GDP GROWTH REACHED 11.18% IN ANNUAL AVERAGE IN LAST FIVE YEARS

INCREASE IN EMPLOYMENT IN URBAN ANNUALLY

CAPITAL INVESTMENT INCREASE RAPIDLY, OVER 4.2 TRILLION U.S. DOLLARS IN 2010

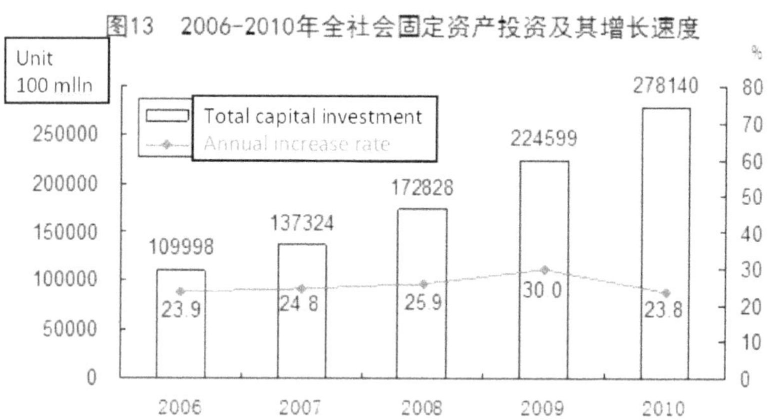

NEW INFRASTRUCTURE CAPACITY CONSTRUCTED IN 2010

- Power generation capacity: 91.18 gw.
- Electric transform capacity over 220 volts: 258.16 TVA.
- Railway: 4986 km.
  - Within which: high speed railway: 1,554 km.
- Double line railway: 3,747 km.
- Railway electrified: 5,948 km.
- Motor road: 104,457 km.
  - Within which: express way: 8,258 km.
- Port capacity: 272.02 million tons.
- Optical fiber cable for communication: 1.66 million km.
- Cell phone exchange capacity: 64.33 million users.
- Building under construction: 4.05 billion square meters.
- Building constructed: 759 million square meters.

TOTAL PRIMARY ENERGY PRODUCTION FROM 1978 TO 2010

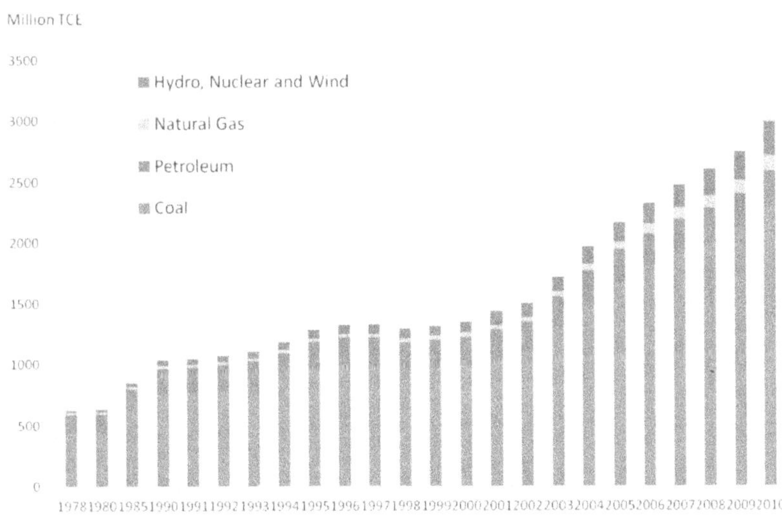

TOTAL PRIMARY ENERGY CONSUMPTION FROM 1978 TO 2010

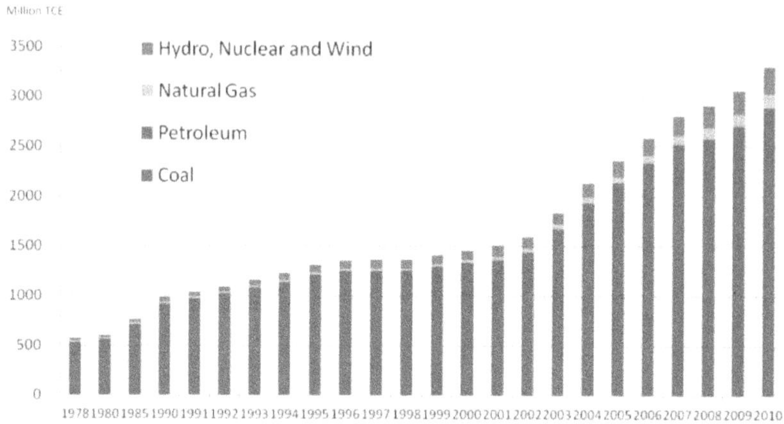

NATURAL GAS CONSUMPTION FROM 1978 TO 2010

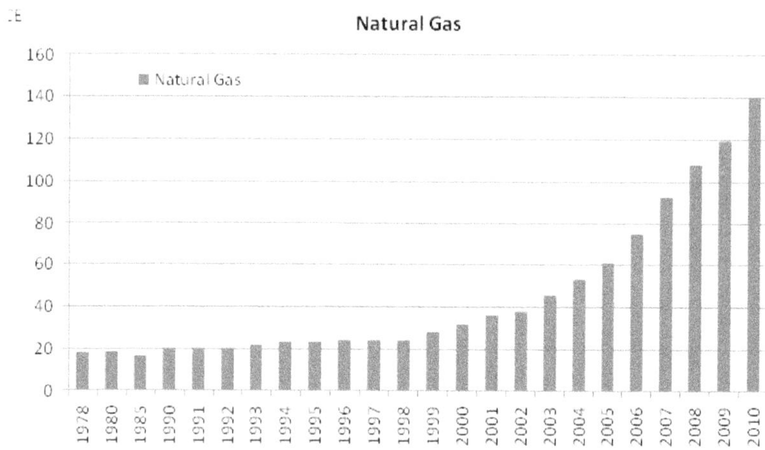

WIND POWER GENERATION CAPACITY FROM 2000 TO 2010

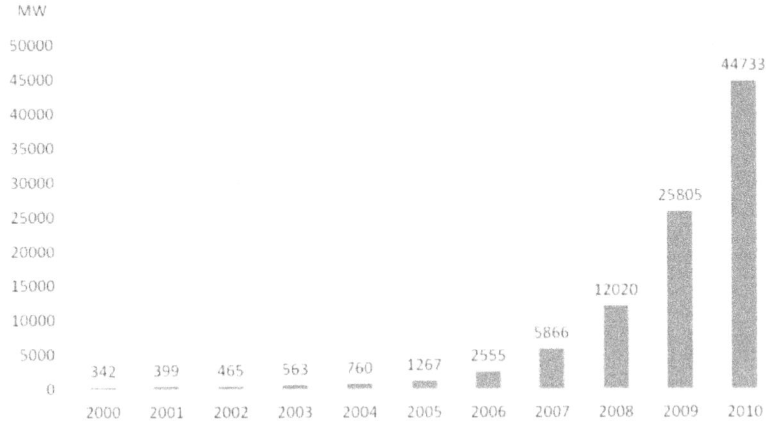

Data source: CWEA

INVESTMENT IN ENERGY INDUSTRY FROM 2003 TO 2010

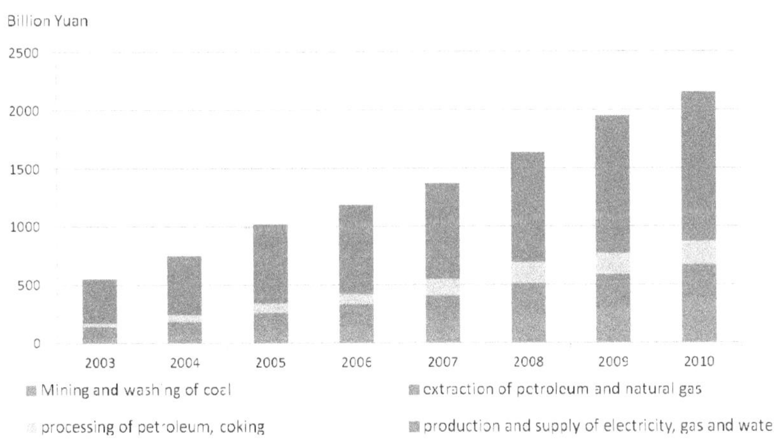

PRIMARY ENERGY DEMAND FUTURE: BAU SCENARIO, CHINA

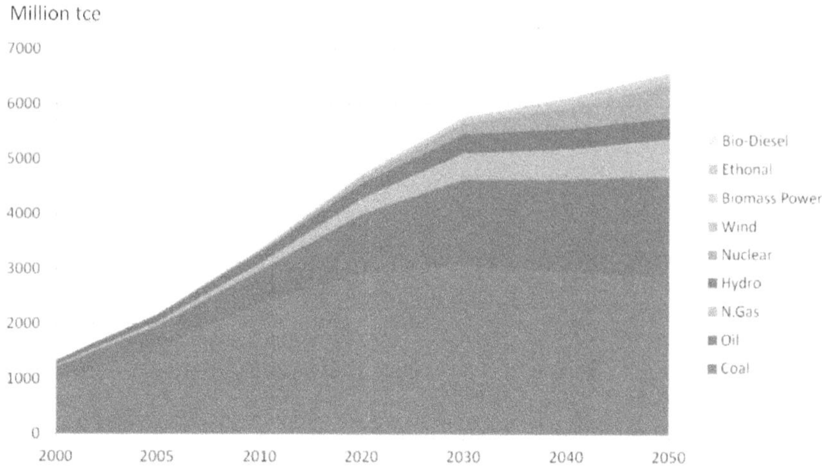

PRIMARY ENERGY DEMAND FUTURE: ENHANCED LOW CARBON SCENARIO

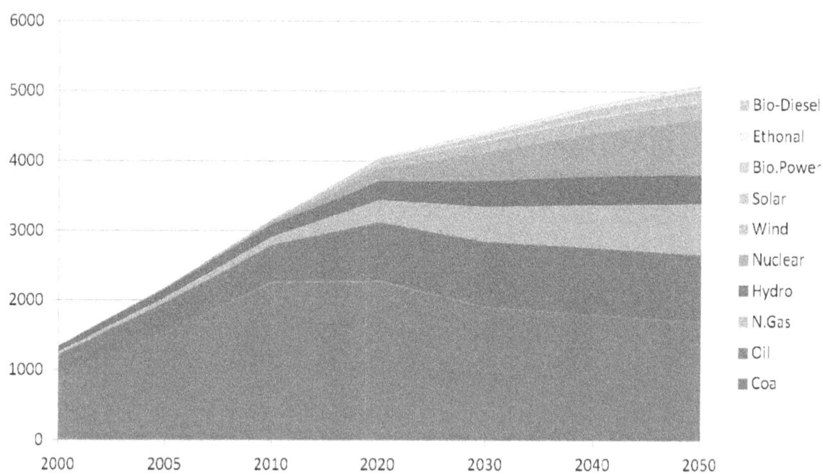

FUTURE PRIMARY ENERGY AVAILABILITY

- Coal is resource available for long, but faces externality constraint.
- Domestic crude oil output will keep stable, increase mainly from import, facing oil price soaring.
- Natural gas will contribute more but need import too.
- Wind power resource about 1000 Gw, may contribute around 600 Mtce at most.
- Biomass resource available less than 300 Mtce as total, with land, water limitation, leading to food security concern.
- Solar, abundant, while only less than 1 Gw by 2010, 20 Gw planned for 2020, but most expensive alternative.
- Hydro may up to 450 Gw, while to overcome reallocation people, and environmental concerns.
- Nuclear is important choice, may share 15% of the TPE by 2050, or more coal.

NUCLEAR POWER DEVELOPMENT IN CHINA

- Started from early 1960s, took long learning time.
- First commercial nuclear power until 1992, speed up only recently, as strategy choice.
- For ensuring energy supply, energy security, low carbon.
- 14 power units commissioned up to now, 11.5Gwe.
- 28 units under construction currently, 31Gwe, with 4 of AP1000, and 2 of EPR1600.
- 49 units in pipe of approval, 57 Gwe, with G-3 share increased.
- 200 Gwe suggested for 2030, 400 Gwe for 2050.
- Strategy of three phase: PWR— FBR—Fusion.

IN OPERATION

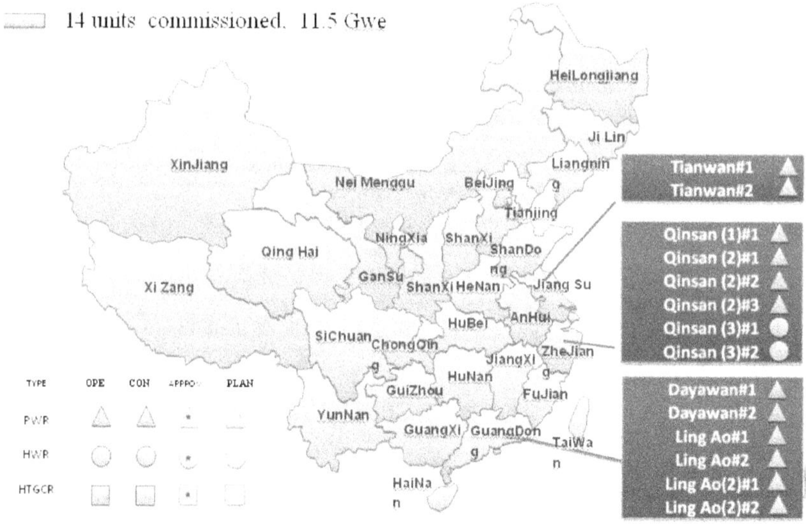

UNDER CONSTRUCTION

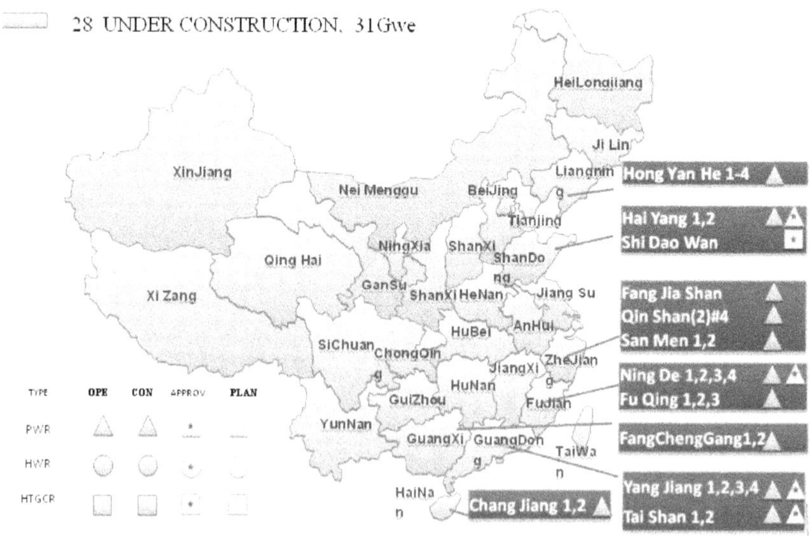

PLANNED, many in land

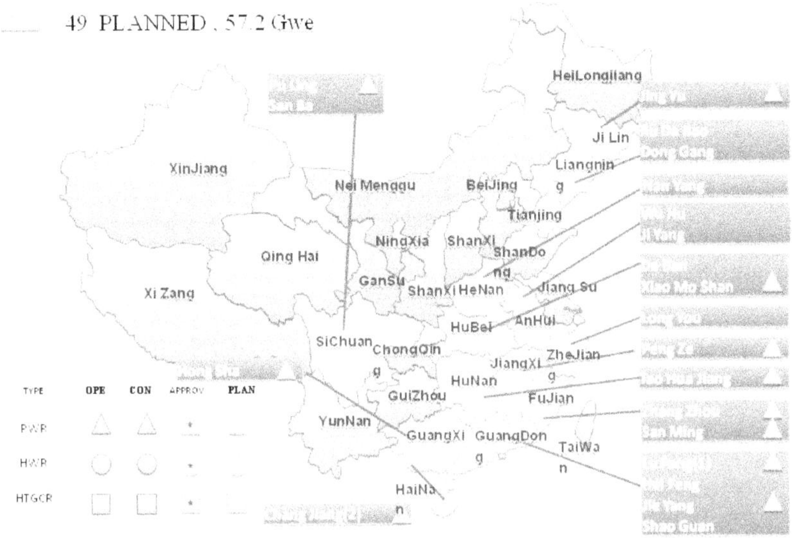

IMPACT OF FUKUSHIMA

- Significant impact of Fukushima on nuclear development of China.
- May slow down the pace of development in short term.
- Raise sufficient emphasis on safety and security of nuclear utilization.
- Promote technology improvement and management, including regulation and supervision.

CURRENT RESPONSE FROM CHINA

- Strict investigation and review of the safety of nuclear power in operation and under construction.
- Suspending new project approval temporarily.
- Safety plan requested by State Council, is being developed.
- Safety report under preparation, to be completed soon.
- Key elements under study.

MAJOR LESSONS AND CONCERNS

- Extreme external events (tsunami) become major course in Fukushima.
- Do we have enough knowledge on possibility of extreme external events?
- Raise question like "is Nuclear power still safe?"
- Do we have sufficient measures to deal with external event, and avoid crises?

- Is a nuclear crisis like Fukushima acceptable in China?
- Scientists and engineers should answer first.

IMPROVEMENTS

- Studies and enhancement on response measures to combat extreme event, both in plant and out of plant.
- Strengthen facility and design for possible external threats, earthquake, flood, tsunami, etc.
- Technical measures on the way, urgent external power, plug in system improvement, hydrogen explosion release, etc.
- Speed up new reactors application
- Strict regulation, management and institutional capacity building.
- Public education and transparency.

MAJOR CHALLENGES

- Safety concern on in-land water system.
- Radioactive pollution of river and basin, if Fukushima-like happened.
- Long time land pollution and evacuation of high density population.
- Political responsibility of crisis.
- Public acceptability and democratic process.
- Man-made external threads, war, terror.

OPTIMISTIC FOR FUTURE

- Nuclear power remains a strategy choice.
- Safety and security improvement, both on technology and management.
- Improvement on crisis management.
- Reliable energy supply, competitive price.
- Low life lost record for nuclear events, even in serious crises.
- Low carbon property.
- Substitute of coal, both on carbon and life lost.
- Improvement of knowledge on radioactive impact and damage.

COOPERATION NEEDED

- Ultimately, is a social issue, not only technical issue
- Improvement on scientific knowledge of nuclear safety, international forum and special reports for scientific consensus
- Agreement by major nuclear nations, on better information sharing on management and risk control, serving as important signal and substantial message
- Better exchange on social and political issues related to nuclear policy and public policy, especially success experiences

- Improve IAEA function, focus from anti-proliferation to safety assurance, to resume and enhance confidence
- International convention on nuclear power facility protection, including during the military conflict

NUCLEAR POWER PLANTS IN SOUTH KOREA

DR. JUNG-UCK SEO
Former Minister of Science and Technology, President Emeritus of IEEK
Seoul, Korea

In 1957, the Republic of Korea (South Korea) became a member of the International Atomic Energy Agency (IAEA). In 1958, the first president of the Republic, Dr. Syngman Rhee joined the Atoms for Peace Program, and legislated the Atomic Energy Law. In 1959, the Office of Atomic Energy was established, and a nuclear reactor for research was installed in 1962. In 1972, construction of the first nuclear power reactor began, and started up in 1977. The first nuclear power reactor in South Korea commenced commercial operation in 1978.

Eight nuclear power reactors were under construction in the early 1980s. Currently twenty-one nuclear power reactors are in operation, seven under construction, and many more in the planning stage. The expansion of nuclear power plants has been an integral part of the national energy policies. It will get thirty-five nuclear power reactors up and running by 2024. Eventually, nuclear electric power will make up about 50% of the nation's total electric power demand, up from the current 30%.

The Korean peninsula is not earthquake-insured at all. South Korea needs a comprehensive check-up on the safety of nuclear power reactors and on the design and construction practices to withstand human errors and natural disasters.

After Fukushima, the world nuclear power industry is faced with a grand challenge for the safety, security and code of conduct. Keeping pace with world-wide efforts to facilitate the public acceptance of nuclear power plant construction, South Korea is reengineering its nuclear power industry based on the lessons learned from Fukushima, and recommendations and suggestions of the Integrated Regulatory Review Service (IRRS) of the IAEA.

Despite the decisions made by some countries to shutdown nuclear power reactors or put off a decision to extend the operation of existing reactors, South Korea recognizes the fact that the benefits from nuclear energy are greater than any other alternatives at the moment.

So far, South Korea has demonstrated the best safety practices in the world, and its nuclear power industry has had only minor incidents that would lead to automatic shutdown. Up until now, not a single case has been classified as an accident.

The ratio of unplanned outages of electric power generation during a given period of time is a measure of the performance of a nuclear power reactor. According to the IAEA, South Korea's ratio is 0.5% compared to the world-wide average of 5.5% at the moment.

The energy policy in South Korea has been driven by the energy security and the need to minimize energy imports. South Korea is determined to be a global leader in the atoms for peace program world-wide.

South Korea imports more than 96% of its energy demand. It is challenging for renewable energy and fuel-cell technologies. The government is a key driver in reducing dependence on energy imports. Since energy prices have been high and volatile in recent years, energy imports are eroding the country's balance of payments. Moreover, South

Korea's exports are energy-intensive heavy and chemical industries, which make the country's exports less competitive in the face of expensive energy.

South Korea has already come a long way with progressive energy efficiency measures to reduce the country's demand for energy as well as its reliance on imported energy. Although the country has successfully reduced its demand of primary energy, the country's imports of energy are staggering.

As an indication of South Korea's reliance on energy, the World Bank's Development Indicators include measures of energy use per capita. This chart compares South Korea to major high-income countries as well as the world. There are a few critical facts for South Korea to reflect.

South Korea's energy use per capita (4693 kg) is much higher than the world average (1835 kg). Whereas energy use per capita in the United States, France, Germany, and the United Kingdom have all followed a downward trend as of 2004, South Korea's energy use per capita has continued to increase.

Although the United States exceeds South Korea in terms of energy use per capita, it would seem from data comparing the same countries' gross domestic product (GDP) per capita, that for every unit of energy used in the United States (and more so in France, Germany, and the United Kingdom), goes further in supplying GDP per capita.

South Korea has an ambitious national energy plan for 2008-2030 to reduce energy intensity by 46% according to an ABB South Korea Energy Efficiency Report. This intense, policy-driven pressure on energy in South Korea will widen and open both existing and new opportunities for clean energy technology development in South Korea. South Korea's national energy plan for 2008-2030 calls for a significant reduction in its energy intensity, which is typically measured as energy (kgoe) per unit of gross domestic product. South Korea aims to reduce its energy intensity by 46% between 2007 and 2030. At its most basic, energy intensity measures the benefit, in monetary terms, for a unit of energy.

South Korea's newest energy plan is very ambitious and the estimated energy savings goal for 2030 is almost 38 million tons of oil equivalent (Mtoe), 44% of which should be from industry (17 Mtoe), 32% from the households and services sector (12 Mtoe), 19% from the transport sector (7 Mtoe), and 5% from the public sector (1.9 Mtoe) according to an ABB South Korea Energy Efficiency Report. Part of this latest plan is a program to integrate what the country refers to as "new and renewable energy" (NRE). This program focuses on renewable and efficient technologies, namely, fuel cells.

As I mentioned earlier, South Korea imports 96.6% of its energy demand. What is South Korea relying on for its electricity? It is a large share of coal, nuclear, gas, and oil. Electricity production is a more precise measure of energy consumption in the stationary sector like homes, industry and public sector but does not include the transport sector which largely relies on oil.

The energy for electricity production in South Korea consists of Coal (42.95%), Nuclear (33.81%), Gas (18.22%), Oil (3.44%), Hydro (1.25%), Waste (0.11%), Biomass (0.11%), Wind (0.10%), Tide (0.00%), Geothermal (0.00%), Solar Photovoltaic: (0.06%), Solar thermal (0.00%) and Others (0.02%).

South Korea's residential program, the Million Green-Houses Distribution Policy, aims to integrate new and renewable energy into one million homes by 2020. Fuel cells qualify for this project as of 2009. Local governments have been encouraged to fund new and renewable energy projects with national subsidies since the mid-1990s.

Subsidies for regional deployment were as high as $146 million in 2008. The national government has mandated the integration of NRE into public buildings such as schools, city halls, hospitals, and the like. Typically, buildings are more than 3,000 meters squared, making them suitable for commercial-size installations.

South Korea is also offering long-term, low interest loans for consumers or manufacturers of new and renewable energy. In addition, it is offering tax deductions and reduced tariffs on imported NRE technology.

Although it is unclear whether it is possible for South Korean households to benefit from feed-in tariffs (FIT), FITs do seem to exist for businesses. In addition, the South Korean government subsidizes FITs to varying degrees based on the technology and size of the installation.

In July 2011, a team of nuclear safety experts, so-called the IRRS Missions, headed by the IAEA deputy director general and also head of the department of nuclear safety and security, confirmed that South Korea has strong safety regulations on operation of atomic energy facilities and prevention of accidents.

This was the first IRRS mission that included a review of the regulatory implications from the Fukushima accident. It is designed only to provide an objective view, recommendation, suggestion or recognition of good practice on South Korea's nuclear safety regulations.

The team, consisting of 16 nuclear safety experts from 14 countries, stated that South Korea already had in place a technically capable and effective nuclear safety regulatory program. It also noted that South Korea's response to the Fukushima accident had been timely, prompt and effective. Communications with the public, development of actions for improvement and coordination with international stakeholders were of high quality.

Mr. William Borchardt, the IRRS Missions team leader and also executive director of operations at the U.S. Nuclear Regulatory Commission, especially recognized the South Korean authorities for their openness and commitment to sharing their experience with the nuclear safety community worldwide.

The IRRS Missions are so crucial to the enhancement of nuclear safety worldwide. One of the key recommendations of the team for South Korea was to set up regulations on how to manage spent fuel and decommission nuclear plants. South Korea willingly supports the global nuclear safety regime and provides training at national and international levels.

Key agenda of the IRRS Missions to South Korea:

1. Responsibilities and Functions of the Government
2. Global Nuclear Safety Regime
3. Responsibilities and Functions of the Regulatory Body
4. Management System of the Regulatory Body
5. Authorization
6. Review and Assessment
7. Inspection
8. Enforcement

9. Regulations and Guides
10. Emergency Preparedness and Response
11. Regulatory Implications of the Fukushima Accident

Recommendations of the IRRS Missions:

1: The regulatory framework should require decommissioning plans for nuclear installations to be constructed and operated. These plans should be updated periodically

2: The Government of Korea should continue the process of clearly defining the responsibilities within the new regulatory body and avoid overlaps between the Secretariat, Korea Institute of Nuclear Safety (KINS) and the Advisory Committee. Resources and staff should be allocated commensurate with those responsibilities.

3: The Korean government should provide for development of a management system, to cover all activities of the new Nuclear Safety Commission and its Secretariat. This management system should be in place, at least the main parts of it, as soon as the new organization takes over responsibility.

4: KINS should develop a process on Resource Management in order to achieve a fully integrated Management System.

5: The Regulatory body should initiate the extension of the legal basis of the licensing process in order to ensure that:

- In case of the submittal of a report on a license amendment in minor matters, whenever the reported change has safety significance the licensee is required to submit a safety assessment on the possible consequences of the modifications, and
- In case of the submittal of a report on a license amendment in minor matters, whenever the reported change has safety significance the licensee shall not commence to realize the modification prior to the answer to its notification from the regulatory body.

6: The Regulatory body should initiate the process to modify the Atomic Energy Act in order to eliminate the option of replacing a suspension of the licensed activity by financial penalty when the safety violation would rightly call for suspension of the activity.

7: The Regulatory body should initiate a change in the regulations in order to:

- Require a Quality Assurance Plan to be submitted when licensing a research reactor of any size. The requirements on the plan shall reflect the safety importance of the facility to be constructed in line with the graded approach, and
- Require emergency preparedness organization and emergency preparedness plans for research reactors of any size. The organization and plan shall follow graded approach and shall be commensurate with the threat posed by the facility

8: The new Nuclear Safety Commission should recognize the current potential for conflict between technical and enforcing inspectors and ensure measures are introduced to reconcile these differences in an effective manner.
9: Daily inspection programs should be reviewed by KINS to ensure they are founded on the safety significance of the structures, systems and components such that they are inspected in a graded and systematic manner.
10: Emergency planning zones should be defined in accordance with the IAEA Requirements.

Suggestions of the IRRS Missions:
1: In order to ensure the effectiveness of the new Nuclear Safety Commission, the selection criteria for the members should reflect the need of independence and nuclear safety competence within the Commission.
2: The new Nuclear Safety Commission should prepare human resources plans for the Secretariat that provides appropriate staff to enable the accomplishment of its administrative function in support of the Commission without undue burden.
3: The new Nuclear Safety Commission should establish an advisory committee, with similar capabilities as the existing Nuclear Safety Committee to support the decision-making process.
4: KINS should revise its Management System Manual to include an overall description of the basic processes, how they relate to each other as well as a description of all types of documentation used within the Management System.
5: The new Nuclear Safety Commission and KINS should describe in their Management System Manuals what means they plan to use in order to ensure a common understanding of safety culture, to support individual and groups to carry out work in a safe way, to reinforce a learning and questioning attitude, and to continually develop and improve the safety culture.
6: The new Nuclear Safety Commission and KINS should supplement their Management Systems with a process or procedure for managing organizational change in order to ensure that regulatory efficiency and effectiveness are not compromised.
7: The regulatory body should initiate updating the Enforcement Regulation in order to extend the scope of the Safety Analysis report so that design extension conditions and probabilistic safety assessments (PSA) are adequately covered.
8: The regulatory body should consider harmonization of approaches used for determination of radiological consequences in the Safety Analysis Report and Radiological Environmental Report.
9: In connection with the licensing of new reactors, the regulatory body should consider harmonization of licensing acceptance criteria and off-site intervention levels for design basis accidents and establish criteria for design extension conditions.
10: The regulatory body should ensure it implements a formal process to observe and assess the inspection methods and techniques of all of its inspectors to ensure they are being conducted in a suitable, consistent and effective manner.

11: The regulatory body should complete a review of its decision making processes that have been applied to significant events to determine whether the appropriate enforcement actions were taken and whether improvements are required to its decision making processes and associated enforcement strategies.
12: Although stakeholder involvement is encompassed in the drafting process for regulations and guides, general public involvement should be enhanced, especially by making them aware of the drafts being developed well before they are submitted to KINS Technical standard committee.

Good Practices recognized by the IAEA-IRRS Missions:
1: Korea has a clear and structured national approach to set out in its policies and the corresponding implementation plans with well-defined priorities for the further development of the legal framework.
2: The establishment of the International Nuclear Safety School which provides a large variety of training and education opportunities for members of the regulatory body as well as for students and other interested parties
3: Korea strongly promotes a global nuclear safety regime through multi or bilateral actions, with countries having established nuclear programs and those embarking on nuclear power, in technical, regulatory and research domains.
4: The regulatory body has developed and implemented a comprehensive Operating Experience Feedback (OEF)-system that is based around a number of effective initiatives including: collection and trending of data from a wide range of sources; well-populated and presented information databases; numerous key communication events; and close working with the licensees and other stakeholders on a national and international front.
5: KINS has an effective practice to recruit successors 3 years before the actual retirement of experienced staff, in order to preserve knowledge and provide continuity.
6: The "safety experience" training course of KINS dedicated to students/parents, teachers, opinion leaders has been shown to positively influence public understanding and acceptance of the regulatory body's activities.
7: KINS has an excellent comprehensive integrated computerized information and data management system for establishing, maintaining and retrieving adequate records relating to the safety of facilities and activities (MIDAS and 19 additional information systems). This contributes considerably to improve effectiveness and efficiency of the regulatory performance.
8: In addition to enhancing efficiency of the review and assessment process by means of continuous monitoring, evaluation and feedback the KINS publishes every year a book of "Good Practices (Best Regulation)" which contains selection of exceptional results of the regulatory works and findings.
9: KINS maintains and utilizes its internal capability for performing independent audit calculations by means of a number of deterministic and probabilistic computer codes including development and validation of such codes, and sharing the computer codes and relevant experience with other regulatory bodies.

10: The annual Quality Assurance (QA) inspection plan produce by KINS ensures that all major vendors to the Korean nuclear program are subjected to a KINS QA inspection at least once in the 12-month period, including all overseas suppliers.

11: KINS Technical Standard Committee plays a key role in reviewing draft comments by giving expert opinions. A significant number of external experts allow for broader stakeholder input. As per regulations, the nuclear safety committee also includes a wide range of experts outside the Regulatory body.

12: The drafting process for regulations and guides explicitly includes identification then comparison to domestic and international standards, including IAEA safety standards. Since nearly a decade, there have been several comparison exercises between Korean regulations and guides and IAEA safety standards, in an effort to improve harmonization with those standards

13: The operation and continuous enhancement of Atomic Computerized technical Advisory system for a Radiological Emergencies (AtomCARE) and the development of a Nationwide Integrated Management System for Environmental Radiation/Radioactivity Monitoring is a good example of integrating information and data gathering systems into an effective and efficient national emergency response organization.

14: The Regulatory body has implemented an exemplary outreach program, including the integration of volunteering experts and first responders (Unsupervised Record Extraction SysTem or U-REST), as well as raising the awareness and understanding of the population regarding the issues of nuclear and radiological emergencies.

15: The Korean national response to the Fukushima accident was well-coordinated and addressed key areas in a short timeframe. In particular, national environment radiation monitoring was reinforced, contamination of goods and people was monitored at airports and harbours, public concerns were addressed by significant communication involvement, and cooperation with Japan was conducted through staff support and technical meetings.

TOWARD MANIFESTLY SAFE NUCLEAR FISSION-BASED BULK ELECTRICITY SUPPLY

LOWELL WOOD AND ROBERT PETROSKI
Hoover Institution, Stanford University, University of California, Lawrence Livermore National Laboratory, Stanford, CA, USA

INTRODUCTION

Global consumption of electricity is forecasted to increase significantly in the next few decades, as economic growth allows billions of persons in developing countries to raise their standards of living. Availability of electricity is also important in driving economic growth, since electricity permits modernization and is tightly coupled with per capita GDP, as well as quality of life and life expectancy. For supplying this needed electricity, nuclear energy holds several major advantages over other forms of generation. It is cost competitive and has a manifest ability to scale (>70% of electrical energy in France, 30% in Japan, 20% in the USA). The low-volume localized waste from nuclear energy means that it produces no air pollution and has no climate change impact. The amount of nuclear energy available on Earth can be effectively limitless, through breeding of nuclear fuel in fast reactors or economical extraction of uranium from seawater.

Despite these advantages, nuclear energy faces two major impediments to its widespread adoption. First is the risk of a large release (>1 megaCurie) of radioactive materials to the biosphere, such as the accidents at Chernobyl and Fukushima, which can lead to very severe social, economic, and human consequences. Second is the possibility that materials or facilities intended for nuclear energy can be diverted toward the production of nuclear weapons.

For nuclear energy to overcome these impediments, it must obviate these risks by becoming safe in both a *basal* sense as well as a *manifest* sense. In this context, basal safety refers to minimizing the risks of radiological release or material diversion, through improved technology, operating practices, and public policy. Manifest safety refers to making the basal safety of nuclear energy plain for all to see, and is achieved by making these safety features of an obvious character, and implementing them in a comprehensively transparent manner. This paper examines the technical aspects of making manifest safety for nuclear energy a reality.

PRESENT SAFETY OF NUCLEAR ENERGY

Nuclear energy has a documented history as being one of the safest forms of electrical generation available. Comprehensive end-to-end studies of different forms of generation show that on a deaths per TWh basis, nuclear is comparable in safety to hydro and wind power, 10 times safer than gas, 40 times safer than coal, and 50 times safer than oil (Starfelt and Wikdahl, 2001; Burgherr and Hirschberg, 2008). Unlike electricity derived from fossil fuels, nuclear energy does not pose a health risk due to production of airborne pollutants. Mining requirements and deaths are significantly lower for nuclear than for coal energy. The safety culture that permeates the nuclear energy industry means that it experiences a rate of workplace injuries a factor of ten lower than the average for electric utilities (World Association of Nuclear Operators, 2004; Bureau of Labor Statistics, 2011).

Despite nuclear energy's excellent safety record, public perception of nuclear energy has never been uniformly positive. Part of this is due to the unfamiliar, eerie nature of radiological hazards: ionizing radiation is invisible, can have effects across long distances, and can cause genetic damage. Another important reason is that nuclear energy has experienced three major accidents, at Three Mile Island, Chernobyl, and Fukushima. Even though there was no significant release of radioactive material at Three Mile Island, the accident turned public opinion against nuclear energy and effectively halted the growth of nuclear energy in the United States. Meanwhile, Chernobyl represents a worst case nuclear accident, leading to a significant loss of life (30 prompt fatalities, 19 delayed fatalities from radiation sickness, and 15 fatal thyroid cancers), contaminating ~100,000 km^2 of land, and exposing a large population to low dosings of radiation (UNSCEAR, 2008).

The health impacts of widely distributed low radiation doses due to accidents like Chernobyl and Fukushima are too small on a fractional basis to be measured. In referring to the Chernobyl accident, the United Nations Scientific Committee on the Effects of Atomic Radiation (UNSCEAR) states: "there is no scientific evidence of increases in overall cancer incidence or mortality rates or in rates of non-malignant disorders that could be related to radiation exposure." (UNSCEAR, 2011) However, the large size of the exposed populations means that even tiny fractional impacts on life expectancy or cancer incidence could aggregate to a large total effect. Also important for the goal of manifest safety is the perception of health impacts due to radiation. UNSCEAR also states that "there is a tendency to attribute increases in the rates of all cancers over time to the Chernobyl accident," so irrespective of what the actual health effects may be, the *perceived* effects and the consequences for nuclear energy are significant. In order to achieve manifest safety, the probability and magnitude of radiological release need to be minimized to such an extent, and in such a readily apparent manner, that large releases are *perceived* as not credible events.

The risk of civilian nuclear materials or facilities being diverted toward creating nuclear weapons has to date been readily suppressed by national and international policies that prevent such actions from occurring. Such diversions have only occurred as deliberate acts by sovereign governments wishing to acquire nuclear weapons, and even then, just seven of the thirty countries with nuclear power plants possess nuclear weapons. A technical issue pertinent to manifest safety is the question of what technologies best allow this state of affairs to be sustained if there is a widespread adoption of nuclear energy around the globe. Another central issue is determining the best ways to manage existing fissile material stores, most of which currently reside in spent fuel pools at operating nuclear power plants.

MANIFEST SAFETY OF NUCLEAR REACTORS

The goal of manifest safety is achieved by buying-down the risk of a large scale radioactive release to as low as a level as desired, and doing so in a manner that is fully visible to everyone. In modern nuclear reactors which employ defense in depth, such releases can only occur if the core is disrupted, the reactor vessel or pressure boundary fails, and both primary and secondary containments fail. None of these events can occur in a modern reactor so long as the afterheat from decay of fission products in the core can be successfully removed as quickly as it is generated. In addition to afterheat removal, other topics of significance include additional safety measures beyond assuring afterheat removal, safeguarding against external events such as earthquakes, tsunamis, and storms, and ways in which modern computing and communications technology can be used to bring about manifest safety.

Afterheat removal

Afterheat removal is the single most important issue for ensuring the safety of a nuclear reactor. Failure to remove afterheat resulted in the meltdowns at Three Mile Island and Fukushima. (The Chernobyl disaster was caused by a combination of operator error and the presence of severe design deficiencies that would not be permitted to exist in modern nuclear reactors.) The reason afterheat removal is so important is because the energy available in decaying fission products is sufficient to melt and vaporize a nuclear reactor core, a condition that persists for many months after complete cessation of full-power operation.

The importance of afterheat removal is highlighted by probability risk assessment (PRA) based analyses of core damage frequencies. The majority of event sequences that can result in core damage involve the failure of afterheat removal systems. As a result, the modern sodium-cooled PRISM reactor, which has passive decay heat removal capability, has a PRA-based core-damage frequency of less than 10^{-7}/year, roughly two orders of magnitude smaller than currently-operating light water reactors (General Electric, 1987; U.S. Nuclear Regulatory Commission, 1990).

The most robust afterheat removal systems are those that require no power or active systems to function, and no operator signals to activate. An example of this would be a heat exchanger that continuously removes ~5% of reactor power from the primary coolant loop, so the availability of indefinite afterheat removal would be constantly assured. This removed heat could be used for a low grade heat application, such as thermocouple banks that can serve as emergency power supplies. Flow in the primary coolant and in the heat removal system would be driven by natural circulation. Either the ambient atmosphere or a local body of water could be used as an effectively infinite ultimate heat sink.

Similar to a continuously operating afterheat removal system, but bearing less of a power production penalty, would be passively operating systems that would actuate if needed for afterheat removal. An example of such a system would be the isolation condenser in some boiling water reactor designs, in which naturally circulating steam in the primary coolant loop passes through a heat exchanger and rejects its heat into a large open pool of water. Such a system must be designed with sufficient heat removal capability (either indefinite or at a minimum several days worth) to allow decay heat to decline after the reactor is shut down. Furthermore, such a system must be capable of being actuated and operated without external power.

Finally, while passive and automatic afterheat removal systems are the best way to manifestly guarantee afterheat removal, an additional reduction in radiological release risk can be achieved by implementing robust and redundant backup power supplies. Such systems would take the form of multiple generators, batteries, and fuel cells, capable of supplying several percent of the station's generating capacity. They would be emplaced in several locations, with all systems (including power transmission pathways) hardened against earthquakes, tsunamis, and other external hazards. Having such systems in place would guarantee that a station blackout would never need to be endured, even in the event of a sustained loss of offsite power. In this way, active afterheat removal systems would be assured to function and plant operability could be maintained in all conditions.

Safety beyond afterheat removal

While ensuring afterheat removal accounts for reducing the majority of the risk of core damage or radiological release, additional protections can be implemented that can further reduce risk to any level desired. If all such systems fail and a core meltdown occurs, its consequences can be minimized by employing one or more "core catchers", which dilute and spread out molten core material, allowing it to cool and solidify, thereby manifestly obviating "China Syndrome" fantasies in which the core melts its way into the earth. Likewise, releases of radioactive materials in an accident can be minimized through robust, layered "vault-within-a-vault" containment boundaries with automatic closures and cooling systems. Another useful option would be the ability to convert the containment into a confinement, in which air and steam are allowed to escape to prevent damaging high pressures, but are routed through a sequence of getters and cold traps so that airborne fission products would react and/or condense out and in all cases remain localized within the plant.

Similarly, a proactive approach can be employed in which volatile fission products are removed from the core and sequestered, thereby separating them from the afterheat energy that would cause them to be released. Such a system could function by using a porous fuel form that allows volatile fission products to escape into a fission gas plenum, then by periodically venting the plenum into a gas collection-&-storage sub-system. Doing this would greatly reduce the potential magnitude of any possible radiological release from the core. By layering a series of such defenses and boundaries between the core and the biosphere, each capable of preventing release of radioactive materials, it is possible to drive risk to as low a level as may be locally desired-to-be-purchased.

The definitive option for manifest safety of nuclear reactors is the Teller-Sakharov specification, i.e., underground emplacement of the entire nuclear island at 30–100 m depths and isolated from the local water-table. Alternatively, the entire power plant can be placed in an underground cavern, e.g., as was done in the USSR. This approach would ideally take advantage of long-cycle core designs with automated operations and minimal need for refueling. Power from the core would be transmitted to generators on the surface through coolant pipes extending upward to underground heat exchangers, which would transfer heat to an intermediate coolant or the power cycle working fluid. These coolant pipes and all other connections with the surface can be safeguarded using redundantly actuated one-time pipe closures, which if needed could be used to completely isolate the nuclear island from the biosphere. Such a reactor would manifestly and definitively address both the risk of radiological release and material diversion from the reactor. Even if all other engineered safeguards and systems fail, no radiation would be released to the environment, and no human lives would be put at risk. The primary cost and technical challenges of such a reactor system would lie in digging and filling the hole, and designing a reactor core suitable for significantly-autonomous operation.

Safeguards against external events

In modern nuclear power plants, the PRA-calculated risk of core damage due to external events such as earthquakes and fires is greater than that due to internally initiated events such as equipment failure. External events are the more challenging because they can occur in concatenated fashions, e.g., an earthquake followed by a tsunami and fires, and can simultaneously challenge a large number of reactor systems. These qualities alone emphasize the need for diverse and redundant systems, each with sufficient capacity to perform an entire critical function, such as shut down the reactor, remove afterheat, and prevent radiological releases. They also highlight the importance of robust passive systems with few failure modes, which thus can be counted on to operate during these challenging events.

For earthquakes, given that nuclear reactors are designed to operate for at least forty years, the challenge is to engineer them to withstand Richter 8 earthquakes which occur fairly regularly and Richter 9 earthquakes that occur once every several decades, such as the 2011 Tōhoku earthquake. Assuring structural integrity during such events is a well-understood engineering problem, with the basic 'recipe' being sufficient usage of reinforced concrete and rebar, and design of structural frequencies far from seismic ones to preclude resonance phenomena. Likewise, internal equipment and plumbing can be protected using these principles, while taking advantage of seismic isolation, anelastic dampening, and inertial buffering wherever feasible. Ensuring a safe shutdown in the face of an incipient earthquake can be accomplished using rapid-response scram systems, preferably activated in advance by functional rings of seismographs around the reactor site.

Reactor structures must similarly be able to withstand the dynamic side-loadings caused by tsunamis and storms, and to endure intervals of complete submergence, due to 10-100 m tsunamis or in extended storm surges that can last a day or more. With tsunamis there is the additional challenge of being potentially left briefly without cooling water, during tsunami early stage sea-retreat. Again, the recipe for structural integrity is use of sufficient quantities of high-strength concrete. Vital systems such as backup generators must be placed in watertight enclosures with sufficiently high air intakes, or at a high enough elevation to avoid being submerged. In addition, submarine-style provisions must be made for plant submergence, keeping in mind that such events would likely be accompanied by loss of grid power and ground transport. These would include internal life support and communications, and on-site availability of remotely operable heavy equipment such as fire trucks and cranes.

One additional advantage of having robust structural integrity and plant self-sufficiency is a much stronger ability to withstand 21st century-grade, competently-engineered vandalism. Such vandalism might be more akin to multiple anti-tank rounds fired from an overhead helicopter instead of merely a truck bomb at the site fence. Again, protection is provided by an adequacy of high-durability mass interposed between the external threat and reactor systems. In addressing these engineered threats, active site defenses can also be employed to whatever levels may be socio-politically acceptable.

The engineered safeguards described above, with their simple principles of operation, can manifestly drive down the risk posed by both natural and man-made external threats.

Applying modern technology to manifest safety

Outside of the simple physical defenses considered so far, modern information processing and communications technology can also be used to improve nuclear reactor safety, and be instrumental in making it manifest for all to see. First, use of pervasive and redundant sensors, diagnostics, and robotic monitoring allows plant operators to have constant "situational awareness" of all plant operating conditions, and avoid the possibility of an unknown failure during operation. The data from these sensor systems can be tied to detailed reactor simulations, which would be able to supply real-time feedback to reactor operators, as well as provide canned moment-by-moment responses based on pre-modeled simulations of all conceivable plant emergencies. This makes it far easier for operators to perform appropriately in situations that might otherwise be panicked and far-worse-than-ideal. Such capabilities are enabled by the past half-century of nuclear engineering design and operational experience, and by supercomputing clusters already having ~10 billion times the computational power that was available when Generation I and II reactors were designed.

The ability to perform advanced reactor simulations would not even need to be present on site, as modern rapid, pervasive, and highly redundant telecommunications allow the necessary data to be shuttled to and from any reactor site to any computational system. The presence of these high-performance communication systems means that no reactor operating staff ever needs to "go it alone" in the face of a crisis, and can teleconference with the best of planetary nuclear engineering wisdom, brought in at a moment's notice by ubiquitous telecomm devices—even personal ones.

Modern technology will also play a central role in making nuclear safety *manifest*, by making reactor safety performance clear to see for anyone wanting to see it. Advanced simulations would allow all postulated plant occurrences to be modeled in detail and exhibited to the public and decision makers. The ever decreasing cost of imaging, data storage, and transmission means that a continuous record of a plant's existence can be created and disseminated: for example, 3 megabyte/second (HDTV-level) 24/7/365 documentation of a plant's history from conception to decommissioning would have a present-day aggregate HDD-based storage cost of just ~$100,000. Ultimately, not only will it become possible for "everyone to watch everyone", but it will also become "universal best practice" as a form of peer review in continuous, world-wide operation.

MANIFEST SAFETY OF THE NUCLEAR FUEL CYCLE

In addition to ensuring the safety of nuclear reactors, it is equally important to make manifest the safety of all aspects of the nuclear fuel cycle, from the creation of new nuclear fuel to its storage and disposal after use. Not only does the risk of diversion come into play when nuclear fuel is outside a reactor core, but spent fuel also poses its own radiological risk, as demonstrated by releases from the spent fuel pool at Fukushima Daiichi unit 4. These issues, particularly those concerning material diversion, have significant local and international policy components to them.

This section will examine the technical options that can play a difference in improving manifest safety. Because of this, the focus will be on *diversion risk*, rather than *proliferation risk*. "Diversion risk" refers specifically to the risk that a subnational actor steals nuclear material and is able to create a nuclear weapon (or steals an entire weapon), while "proliferation risk" refers to the risk of a State acquiring nuclear weapons, in particular from covertly co-opting a civilian nuclear program.

Front end of the nuclear fuel cycle

The front end of the current LWR-based nuclear fuel cycle involves enriching natural uranium to produce usable fuel. From a material diversion standpoint, commercial enrichment facilities producing low-enriched uranium pose a minimal risk. This is because if any low-enriched uranium is diverted from the facility, it could not be turned into highly enriched uranium without some separate enrichment capability, and if such a capability existed, then the highly enriched uranium could also be produced from natural uranium (albeit at a significantly slower rate). Diverting a portion of the enrichment facility itself for producing highly enriched uranium could not be done without the knowledge and consent of the enrichment plant owner-operator. An important safeguard to have is inspection capabilities that are able to discern whether a complicit operator is allowing highly enriched uranium to be produced.

Despite the low risk posed by commercial enrichment facilities as source of weapons usable material, it is still desirable to have a smaller number of enrichment facilities around the world, located in countries with adequate stability and infrastructure to keep such facilities highly safeguarded. First, this would prevent the widespread diffusion of enrichment technology, especially as the technology matures and has lower capital, power, and space requirements. Second, it improves the proliferation risk of nuclear energy not to have numerous countries with the ability to convert their peaceful enrichment capability into weapons programs.

A robust international fuel supplier market would be instrumental in keeping global enrichment capability concentrated. It would give an economic incentive for countries without enrichment capability to buy enriched fuel from others, since that would be cheaper than developing a homegrown infrastructure. A market with multiple fuel vendors would also obviate potential monopoly-supplier issues.

Reactor systems that require less enriched fuel would improve these incentives and also help keep total global enrichment capacity low. Reactor designs with reduced enrichment requirements include heavy-water- and graphite-moderated reactors, light water reactors designed for higher fuel utilization, as well as high-burnup fast reactors. A future possibility for high-burnup fast reactors is a breed-and-burn reactor, which would be able to directly burn unenriched uranium by converting it to fissile plutonium and then burning this in a single pass. Such a reactor would allow a nuclear infrastructure to sustain itself indefinitely with no fissile input whatsoever, which also achieving very high uranium utilization.

The other alternative to relying on enriched uranium as fissile input is to substitute it with plutonium, such as in MOX fuel. The plutonium can be sourced from spent nuclear fuel or existing weapons stockpiles. Going this route carries with it potentially much larger diversion risks, since fuel products containing plutonium can readily be converted into weapons materials by chemically separating the plutonium from the uranium, a much easier process than enriching uranium. Therefore, any process involving plutonium needs to be extensively safeguarded, a topic discussed in greater depth below. Meanwhile, one of the upsides of using plutonium to make new fuel is that it allows one to actually decrease the global inventory of plutonium, while putting plutonium where it is hardest to divert: in the core of an operating reactor.

Finally, it should be noted that some of the diversion disadvantages of plutonium based fuels are obviated by a thorium-uranium fuel cycle, in which uranium-233 bred from as-mined thorium-232 can be mixed with natural uranium to prevent it from being isolated by chemical means.

Back end of the nuclear fuel cycle

The back end of the nuclear fuel cycle potentially poses a much greater risk for materials diversion because spent fuel contains fissile plutonium (or fissile uranium-233, in a thorium breeder) that can be chemically separated from the rest of the fuel and potentially used as weapons material. It is therefore very important to vigorously guard all forms of used fuel, from when it's still in a reactor to when it is either burned as recycled fuel or permanently disposed of.

When fuel is either inside a core or cooling in a spent fuel pool, diversion is challenging because the fuel is hot, both thermally and radiologically, in large discrete assemblies, and subject to the same protections from intrusion as the rest of the nuclear power plant. Similarly to in an enrichment plant, it would be important to have robust inspection and monitoring capabilities that can detect an operator-complicit diversion.

The spent fuel in a spent fuel pool also represents a radiological risk, because it requires cooling and can melt if cooling is lost for a sufficiently-prolonged interval. The fact that it is aged relative to the freshly burned fuel in the core means that cooling problems would arise over much longer timescales: days to weeks instead of minutes to hours. Again, the same principles employed for ensuring core afterheat removal should be employed: simple, passive cooling systems with effectively limitless heat sinks, such as cooling towers that reject heat to the atmosphere and gravity-driven water flows from a local cooling source. Spent fuel pools should be engineered to avoid major leaks, such as through below-grade emplacement, and have high-reliability means to be refilled, if leakage or boiling-off nonetheless occurs.

After cooling in a spent fuel pool, fuel can be transferred to dry storage, in which the fuel is sealed in an inerted canister and placed into a massive concrete storage cask. Natural convective-draft airflow through the cask provides necessary cooling, and such casks can be licensed to operate for a century or longer, while most all (>90%) of the initially-present fission-product inventory decays. These casks also offer protection from diversion, since they are extremely durable and heavy (~100 tons being a representative weight). Basic safeguards from diversion still need to be present, and can include intelligent below-ground emplacement, remote monitoring, and an adequate guard force.

Transportation of spent fuel from a reactor to an interim storage, reprocessing, or disposal facility needs to occur under similar protections. Spent fuel is already transported in massive, extremely durable steel and concrete casks, primarily to avoid any possibility of radioactive material escaping from the cask in case of an accident or sabotage. The size of the casks makes them a difficult target for diversion or the materials inside them, but a guard force and monitoring are needed to prevent attempts at sabotage or hijacking.

Whenever chemical separations types of reprocessing are employed, a much higher level of safeguarding needs to be employed, above and beyond that found at a nuclear reactor or enrichment plant. This is because the possibility of diversion becomes a much larger threat: only a small fraction of material would need to be diverted from a commercial-scale reprocessing plant to constitute a weapon, and it would be extremely challenging to design processes and safeguards that could account for every kilogram of fissile material for a plant processing tonnes of such material each year. This is in contrast to at an enrichment plant, in which an entire enrichment cascade would be need to be diverted, or at a reactor or spent fuel storage facility, in which an entire fuel assembly would need to be diverted or taken apart.

The best approach for achieving manifest safety in reprocessing is to first minimize the amount of reprocessing and number of reprocessing facilities, then subject all such facilities to an extremely tight, 'special military'-level of control and security. Reprocessing options that produce actinide mixtures instead of separated plutonium only strengthen diversion resistance slightly, because of the relative ease with which plutonium can be extracted from such mixtures. Greater resistance to diversion could be achieved by improving one's ability to successfully account for all fissile materials. For example, recycling entire fuel assemblies without reprocessing (e.g., taking discharged fuel elements from a sodium-cooled reactor and burning them further in a lower-power lead-cooled reactor) offers a potential way to extract more energy from fuel without making it any easier to divert.

Fuel cycle options which minimize the amount of reprocessing include the present once-through LWR fuel cycle in the USA, as well as once-through fuel cycles using any reactor type. Here reactors that can support higher fuel burnups have innate advantages arising from better fuel utilization and lower mass flows. It may be possible to defer the need for significant reprocessing for a century or more by using reactors with such enhanced uranium utilization, allowing more expensive albeit much more abundant sources of uranium to be economically utilized.

Permanent geologic disposal nuclear waste is the most straightforward way to greatly reduce or eliminate the diversion risk posed by spent nuclear fuel. Therefore, demonstrating spent fuel disposal is an important facet of manifest safety. Spent fuel disposed deep underground would be extremely difficult to retrieve and therefore comparatively easy to safeguard against diversion. If deep borehole emplacement is used, then the spent fuel becomes effectively impossible to recover, except at astronomical cost. In addition to reducing diversion risk and aiding with manifest safety, demonstration of disposal technology would also allow nuclear energy to achieve *manifest cleanliness*, which would be an enormous advantage in comparison to chemical fuel-burning forms of electricity generation.

CONCLUSION

Contemporary fission-based electricity supply is far safer-in-retrospect than any scalable alternative, is cost-competitive with these alternatives, and possesses a comparatively tiny carbon footprint. Even with these advantages, the radiological nature of nuclear energy and the potentially large impact of a nuclear accident make it essential to pursue the goal of manifest safety.

Manifest safety is achieved by further suppressing nuclear energy risks and providing comprehensive peer and public visibility into the safety measures employed. Reducing the risks of nuclear energy by large factors can be readily achieved via any-or-all of many basal safety measures. These measures operate in a series of layered defenses: one set guards against internal and external accident initiators, another guarantees the ability to remove afterheat, and a still another undergirds the other two by preventing radiological releases in the event of a core disruption. At each stage, redundant systems and passive mechanisms with few modes of failure are employed to reduce risk to any level desired.

The safety advantages of these measures can be made manifest by making their design, implementation, and operation fully transparent to the greater nuclear community and beyond that to all of the interested public. In this manner, nuclear energy can be relied on to supply plentiful electricity to power rapid development and increasing prosperity around the world.

REFERENCES

1. Bureau of Labor Statistics, *Injuries, Illnesses, and Fatalities Data Tables,* 2011, available at http://www.bls.gov/iif/#tables.
2. Burgherr, P.; Hirschberg, S., "A comparative analysis of accident risks in fossil, hydro and nuclear energy chains," *Human and Ecological Risk Assessment*, **14**(5), 947 – 973, 2008.
3. General Electric, *PRISM Preliminary Safety Information Document*, GEFR-00793, Appendix A, December 1987.
4. Starfelt, N.; Wikdahl, C., "Economic Analysis of Various Options of Electricity Generation—Taking into Account Health and Environmental Effects," *Proceedings of Ecological Aspects of Electric Power Generation,* Warsaw, Poland, November 14-16, 2001.
5. UNSCEAR, "Sources and Effects of Ionizing Radiation," *UNSCEAR 2008 Report to the General Assembly with Scientific Annexes*, Annex D, 2008.
6. UNSCEAR, "The Chernobyl Accident: UNSCEAR's assessments of the radiation effects," 2011, available at: http://www.unscear.org/unscear/en/chernobyl.html #Health.
7. U.S. Nuclear Regulatory Commission, *Severe Accident Risks: An Assessment for Five U.S. Nuclear Power Plants, Final Summary Report*, NUREG-1150, 1990
8. World Association of Nuclear Operators, *WANO 2004 Performance Indicators*, 2004, available at: "http://www.wano.org.uk/PerformanceIndicators/PI_Trifold/ PI_2004_TriFold.pdf".

SESSION 3

WATER & POLLUTION

FOCUS: WATER SCARCITY AND POLLUTION

WATER NEEDED FOR INCREASING DEMANDS FOR FOOD, ENERGY, ECOSYSTEMS, AND HEALTH

FRANK L. PARKER
Environmental Engineering
Vanderbilt University, Nashville, USA

CAN MATT DAMON BRING CLEAN WATER TO AFRICA?
(Ellen McGirt July 22, 2011: http://www.fastcompany.com/magazine/157/can-this-man-save-this-girl)

When water makes headlines with a Hollywood star, you know the problem's time with the public has come.

Student involvement—21st Birthday gift: she sent money to Africa for clean water. She then spent spring break working there on clean water projects.

<u>Goals of United Nations' Millennium Declaration: September 2002 to Reduce Extreme Poverty by 2015</u>

1. Eradicate extreme poverty & hunger*
2. Achieve universal primary education
3. Promote gender equality and empower women
4. Reduce child mortality*
5. Improve maternal health*
6. Combat HIV/AIDS, malaria, and other diseases

7. Ensure environmental sustainability*
8. Develop a global partnership for development

4 of 8 goals are *dependent upon clean water and adequate sanitation http://www.un.org/millenniumgoals/bkgd.shtml

There is no Substitute for Water—the Critical Resource

("Making Sense of the Water Resources That Will Be Available for Future Use" Hossain Faisal, Dev Niyogi, Jimmy Adegoke, George Kallos and Roger Pielke, Sr., *Eos*, Vol. 90, No. 17, 26 April 2011.)

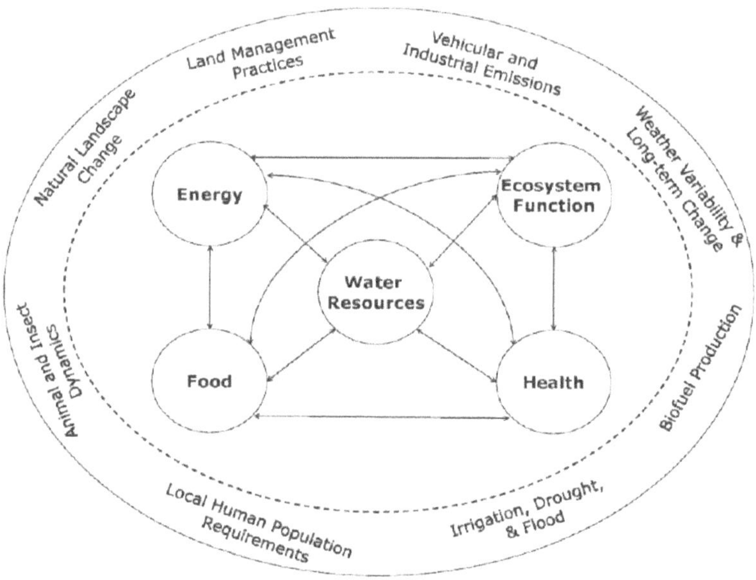

Complicated Problem: Discuss only **Water Resources** and their relationship to **Energy**, **Food**, **Ecosystem, Functions** and **Health**

Objectives of the Water Talks

- The Competing Uses for Water
- The Health Dimensions of the Problem
- The Human Dimensions of the Problem
  - Peace through Science and Technology
  - Cooperative Israeli and Palestinian Solutions

    Our colleague, Alfred Abed Rabbo, co-editor of *Water Wisdom*, preparing the groundwork for a cooperative and sustainable water management in the Middle East, who was to present today is in the hospital and unable to be here. However, his paper is available in the back of the room. We wish him a speedy recovery and good health.

- Getting the WFS Planetary Emergencies Project more involved in the solution of world wide water problems

WATER

The Energy Challenge: Global energy consumption to grow by 50% by 2030.
(Mike Hightower, Suzanne A. Pierce, *Nature* Vol:452, N:7185, pp 253-386.)

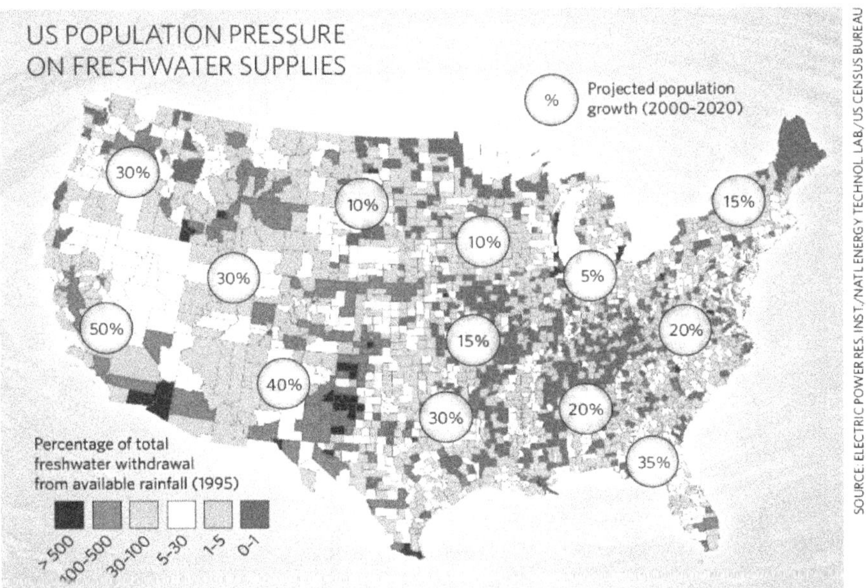

Water Conflict Chronology—Over 400 Incidents
(November, 2008. Peter H. Gleick, Pacific Institute for Studies in Development, Environment and Security)

Date	Parties Involved	Basis of Conflict	Violent Conflict or in the Context of Violence	Description
3000BC	Ea, Noah	Religious Account	Yes	Sumerian Legend, Bible
2008	Pakistan	Terrorism	Yes	Taliban Threat to Blow up Warsak Dam

Earliest and latest entries: http://worldwater.org/conflictchronology.pdf.

State of Georgia (Atlanta) wants to change the State of Tennessee Boundary line so that the Tennessee River would be in Georgia.

<u>The Changing Water Paradigm-a Look at Twenty-first Century Water Resources Development</u>

(Peter H. Gleick IWRA Water International, March 2000)

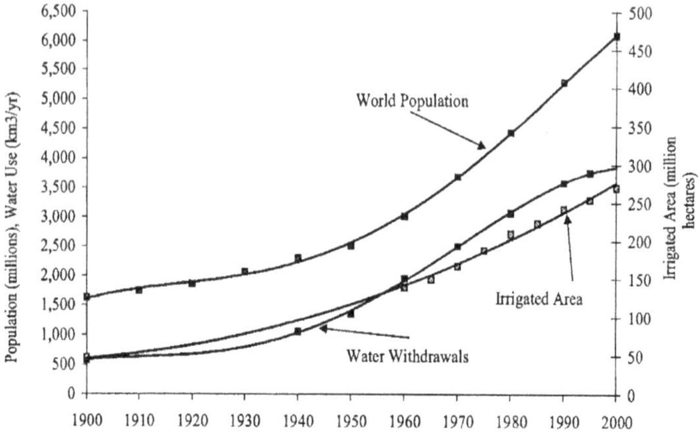

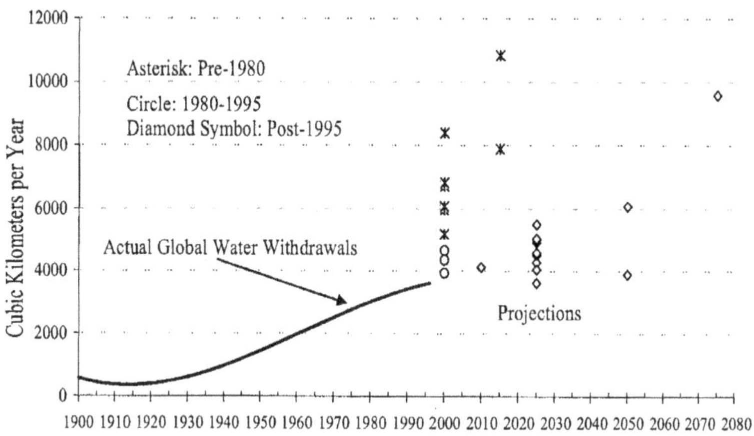

Water Scenarios: Projected and Actual Global Water Withdrawals.

There is a shift away from reliance of finding new sources of water to a growing emphasis on incorporating ecological values into water policy, a reemphasis on meeting basic human needs for water services and a conscious breaking of the ties between economic growth and water use.

Desalination-Increased Availability of Water

(3rd UN World Water Development Report: "Water in a Changing World 2009" *Earthscan*)

- There are 14,451 desalination plants on-line with a total capacity of 59.9 million m^3. It is projected that by 2014, more than the equivalent of a new river Thames will be added each year to the world's renewable freshwater resources and by 2020, twice that amount will be added.
  "700 Desalination Plants Commissioned in 2008". Nov 11, 2009. *Water and Waste Water News.*
- Feedstock for desalination plants is derived from brackish water, 48 percent and seawater, 52 percent.
- Ashkelon Reverse Osmosis Desalination Plant, 100 million m^3/yr at $0.52/m3. 15% of domestic demand. 2010 (http://www.water-technology.net/projects/israel/)

Engineering Solutions to Drought

(Great Green Wall of Africa Generic .PDF, Tunis, 2008 introductory note No. 3.)

The Great Green Wall Initiative, has not been conceived as a wall made up of trees planted across the Sahara, but rather as a set of cross-sectoral actions and interventions aimed at the conservation and protection of natural resources with a view to achieving development and, particularly, alleviating poverty.

The Republic of Benin, Burkina Faso, The Central African Republic, The Republic of Chad, The Republic of Côte d'Ivoire, The Republic of Djibouti, The Arab Republic of Egypt, Eritrea, The Gambia, The Republic of Ghana, The Republic of Guinea Bissau, The Republic of Liberia, The Great Socialist People's Libyan Jamahiriya, The Republic of Mali, The Kingdom Morocco, The Republic of The Niger, The Federal Republic of Nigeria, The Republic of Senegal, The Republic of Sierra Leone, The Somali Democratic Republic, The Republic of The Sudan, The Togolese Republic, The Republic of Tunisia.

Terrorism and Water Infrastructure

- At risk, 75,000 dams, thousands of miles of pipes, aqueducts, water distributions, sewer lines, 160,000 public drinking water facilities and 16,000 publicly owned waste water treatment facilities.
- Worries are physical destruction, bio-terrorism and cyber attacks. (Claudia Copeland and Betsy Cody, "Terrorism and Security Issues Facing the Water Infrastructure Sector," *CRS*, May 21, 2003, RS21026; "Biological Attack—Human Pathogens, Biotoxins, and Agricultural Threats," NAS and DHS, 2004)
- Risks: "raw" water systems, low water intakes, "medium" water treatment, "high" water distribution, and "very high" interlocking dependencies. ("How Safe is Our Water? The Threat of Terrorism," http://www.ionizers.org/water-terrorism.html)
- Updating and hardening of infrastructure, better monitoring, interconnectedness of water supply systems, protection against cyber and vandalism attacks, and multiple barriers including storage capacity, enhanced treatment systems and mutual aid. ("Safety of Our Nation's Water," Richard G. Luthy, testimony on *H.R. 3178 and the Development of Anti-Terrorism Tools for Water Infrastructure*, November 14, 2001, http://www7.nationalacademies.org/ocga/testimony/HR_3178.asp)

Kepner Tregoe Risk Analysis of Various Attack/Threat Scenarios for Water

(*Securing our Water Supply: Protecting a Vulnerable Resource.* Dan J. Kroll, 2006, PennWell Books.)

Most material is classified. Qualitative analysis by an employee of a water treatment company.

Site of Attack/Attack Type	Cost	Mass Casualties >1000	Skill Level Required	Cleanup and Repair Costs	Difficulty gaining access and mounting attack	Total Risk Score >Score = >Risk
Source Water/Physical	2	2	1	2	2	8
Untreated Water Transport/Physical	8	1	7	4	8	1792
Distribution System/Physical	9	1	10	2	10	1800
Finished Water Storage/Physical	9	1	9	6	9	4374
Untreated Water Storage/Physical	5	4	6	7	6	5040
Source Water/Contamination	4	6	8	6	6	6912
Treatment Plant/Physical	8	4	6	7	7	9408
Untreated Water Storage/Contaminant	7	6	8	5	8	13440
Untreated Water Transport/Contaminant	8	6	8	5	8	15360
Treatment Plant/Contaminant	9	9	8	7	8	36288
Finished Water Storage/Contaminant	9	9	10	10	9	72900
Distribution System/Contaminant	10	10	10	10	9	90000

Response to Water Scarcity and Competition

(3rd UN World Water Development Report: "Water in a Changing World 2009" http://www.unesco.org/water/wwap/wwdr/wwdr3/pdf/13_WWDR3_ch3.pdf)

- Seventy percent of fresh water withdrawals are for agriculture and can reach as high as 90 percent in some developing countries.
- Irrigated agriculture accounts for 40 percent of global food production. Industrial use is 23% and domestic use (household water, drinking water and sanitation) account for 8%.
- Producing enough food for one day requires 3,000 liters of water, 2-5 liters of water/day for drinking. (Global Water Security, Global Water Partnership, 2009)
- Solving water problems requires realistic plans, political buy-in, available funds, how much they contribute to development priorities, poverty alleviation, and ecosystem health.

ENERGY AND WATER

Competition between Energy and Water

(Energy-Water Nexus Overview Brochure, Sandia National Laboratory)

- Water needed for electricity generation, coal and nuclear, accounts for 39% of all freshwater withdrawals (but not consumed) in the USA.
- Each kW/h from coal requires 25 gallons (94.7 liters) of water. In the USA, the largest population increases are in the most water deficient areas.
- Fuel from oil shale requires 2-5 gallons (7.7-37.4) of water/gallon of fuel.
- Biofuels can require as much as 1,000 times as much water as ordinary oil refining.
- Drought in France in 2003 resulted in a 15% loss of power from its nuclear power plants and 20% of what it normally receives from its hydroelectric plants. Energy production is reduced when cooling water is too warm.

<u>Water Use Efficiency for Different Generation Types</u>

(Energy and Water, Mike Hightower, EW_ROADMAP_UCOWR-06.PDF)

OL = Open loop cooling, CL = Closed Loop Cooling, CC = Combined Cycle
*IGCC = Integrated Gasification Combined-Cycle, includes gasification process water

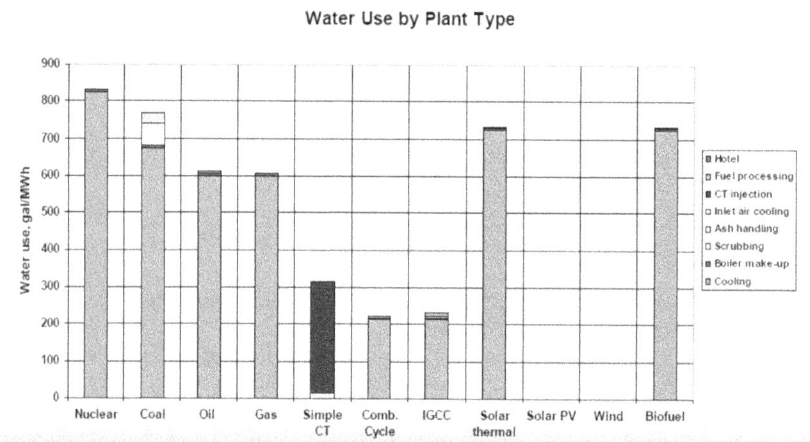

Emerging Water Demands for Future Water Development
(Energy and Water, Mike Hightower, EW_ROADMAP_UCOWR-06.PDF)

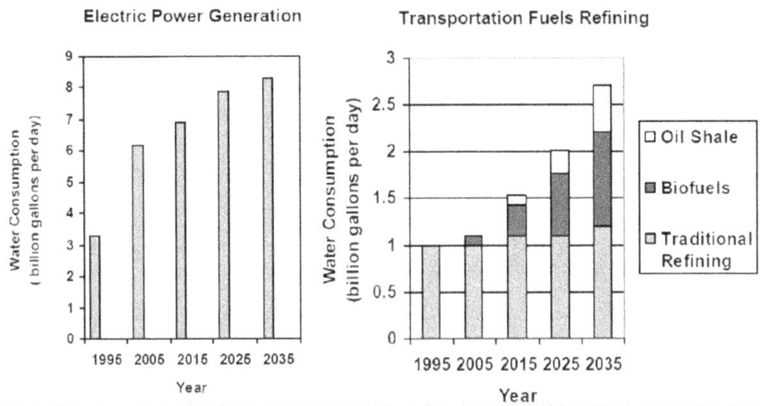

FOOD AND WATER

Percent of Agricultural Water Threatened

Under the Dry (CSIRO) Climate Change Scenario, Given 2050 M&I Withdrawals and EFRS.

("Competition for water for the food system". Kenneth Strzepek and Brent Boehlert *Phil. Trans. R. Soc.* B (2010) 365:2927–2940.)

Municipal and industrial (M&I) water demands, environmental flow requirements (EFRs)

Primary Factors Affecting Crop Prices

("Why Have Food Commodity Prices Risen Again?"/WRS-1103, Ronald Trostle, Daniel Marti, Stacey Rosen, Paul Westcott, June 2011, Economic Research Service/ USDA.)

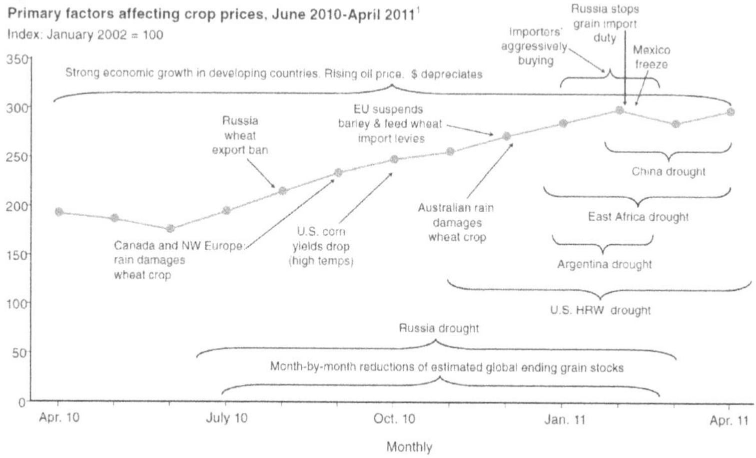

Conclusions

("Why Have Food Commodity Prices Risen Again?"/WRS-1103, Ronald Trostle, Daniel Marti, Stacey Rosen, Paul Westcott, June 2011, Economic Research Service/ USDA.)

Prospects for the future path of prices depend on many developments. As before, high prices will provide incentives to increase global agricultural production. Farmers around the world will make production decisions about allocating available land to various competing crops and about how much additional production inputs to use.

The relationship between corn used for fuel and the price of food is more complicated than originally thought.

ECOSYSTEMS AND WATER

Importance of Functioning Aquatic Ecosystems

(Heather Mackay, July 2001, The Water Page—Water for Ecosystems www.africanwater.org/aq_eco_july_01.htm)

- Bacteria process organic waste products and maintain good water quality;
- Riparian vegetation mitigates floods and provides more stable river and spring flows;
- More reliable flow regimes can be utilized for food production, transport, water supply or to support terrestrial ecosystems and wildlife;
- Health ecosystems ensure maintenance of biodiversity and, hence, resilience to the pressures of utilization.

Platte River major bird flyaway—need sufficient water at specific time to prevent plant growth so that nesting can occur. River also used for water supply, irrigation, recreation and power generation. How to allocate the water?

HEALTH AND WATER

World Health Organization

(25 June 2008. http://www.who.int/features/qa/70/en/index.html)

Annually, safer water could prevent:

- 1.4 million child deaths from diarrhea;
- 500 000 deaths from malaria;
- 860 000 child deaths from malnutrition; and
- 280 000 deaths from drowning.

In addition, 5 million people can be protected from being seriously incapacitated from lymphatic filariasis and another 5 million from trachoma.

Women spend long hours hauling water to the detriment of their health, education and other employment:

Gujarat India	3-4 hours per day
in summer	5-6 hours per day
Nepal	80 liters of water 4-5 times a day
if close to home	10-15 times per day-200 liters per day

There is no Substitute for Water—the Critical Resource

("Making Sense of the Water Resources that will be Available for Future Use," Hossain, Faisal, Dev Niyogi, Jimmy Adegoke, George Kallos, and Roger Pielke Sr. *Eos*, Vol. 90, No. 17, 26 April 2011.)

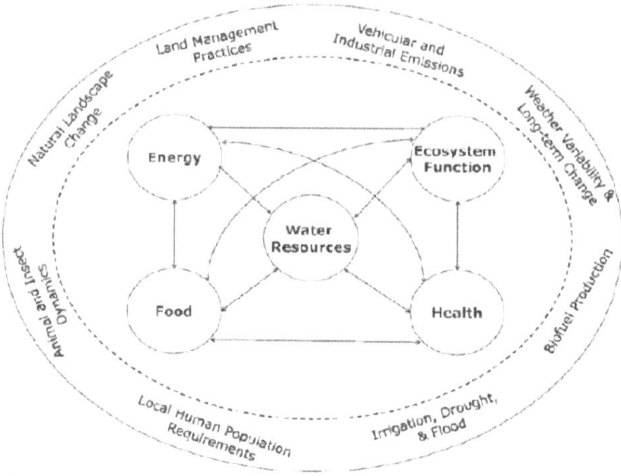

CONCLUSIONS

Harvard Water Resources Study is, most likely, the most influential study in the entire field. However, Fiering and Rogers concluded that it was a failure as most of its recommendations were ignored in the U.S. and infrequently used in the rest of the world. For such complicated problems, multi-attribute utility theory was thought to be the solution. Also a failure: Detlof von Winterfeldt wrote about its failure in the solution of the U.S. Department of Energy's Radioactive Waste problems.

Muddling Through seems to be the dominant solution.

AN OVERVIEW OF WATER ISSUES RELATED TO PUBLIC HEALTH

BERNARD D. GOLDSTEIN AND BARBARA PAVLIAKOVA
Dept of Environmental and Occupational Health, Graduate School of Public Health, University of Pittsburgh, Pittsburgh, USA

INTRODUCTION

Clean water is essential to health. Provision of clean water for drinking and for hygiene was central to the 19th century sanitary revolution. It has played a critical role in the advance in life expectancy and improved health indices that characterized the past century. In the developed world, clean water for drinking and for sanitary purposes is taken for granted, with few citizens aware of the complex processes that deliver potable tap water to their homes and remove personal wastes. Yet much of the rest of the world does not have these benefits. As a result, diarrheal disease is an endemic problem in developing countries.

The lack of access to water and sanitation systems is estimated to cause 9.1% of the global burden of disease, of which a disproportionate portion affects children under 14.[1,2] However, while diarrheal disease is by far the most prevalent of risks associated with poor sanitation and water, it is only one of many risks to human health related to water; others are posed by water-borne toxins, illnesses transmitted by vectors that breed in water; and global climate change that is altering water distribution and availability. World population growth also continues to stress existing clean water resource availability. Few citizens in the developed world are aware of the importance of water to their health until things go wrong, but in the developing world there is a high awareness of the value of water to wellness and productivity in daily life. Water in these regions is important due to the direct health risks and health benefits that it may pose to humans, but also in its role in agriculture, economic development and poverty relief.[3,4,5]

WATER AND INFECTIOUS DISEASE

Poor sanitation, hygiene, and water quality results in water and food contamination with a variety of parasites, bacteria, and viruses that can cause diarrheal and other diseases. These issues are particularly prominent and are responsible for high mortality burdens in developing countries.[6] Overall, it is estimated that there are 2 billion cases of diarrheal disease in children under age 5 each year, more than half of which occur in Sub-Saharan Africa or South Asia. For example, in India, diarrheal diseases account for 32 percent of mortality in children under the age of five.[7] Controlling diarrheal diseases is considered to be crucial to achieving the Millennium Development Goal of reducing child mortality.
There are broader implications of childhood diarrheal disease. Children affected by persistent diarrheal disease in early childhood are more likely to experience chronic malnutrition or nutritional deficiencies, such as anemia, which is associated with developmental problems, cognitive delay, and stunting.[8] A study of Peruvian children living in a Lima slum found that approximately one-third of children studied were stunted by malnutrition and diarrhea caused by *giardia lamblia* by the time they were nine years old.[9] This group of children was significantly more likely to score lower on intelligence

tests than non-stunted counterparts. Children in developing countries are also at risk of stunting through consumption of aflatoxin on maize and groundnuts.[10] Lower cognitive function affects the education and employment potential of affected children and adolescents, future household income, and also has repercussions on national economic growth.

The global nature of infectious disease challenges related to clean water and food is dramatically demonstrated by recent events in Haiti and in Europe. A major cholera epidemic was brought to Haiti by Nepalese soldiers on a humanitarian mission aimed at helping recovery from a devastating earthquake. It is estimated that as of July 2011 there have been 344,623 cases and 5,397 deaths.[11] Due to the lack of sanitary conditions, particularly access to clean water and poor sanitation, this epidemic continues. The recent epidemic of renal disease in Germany and, to a lesser extent in France, after a number of false attributions, is now being traced to an Egyptian seed producer which sold fenugreek seeds contaminated with E. coli O104 to a British seed distributor which then sold the contaminated seeds to farms in Germany and France. The sprouts were sold to consumers in Germany by an organic farm presumably adhering to appropriate organic farm practices. This once again demonstrates that organic farming practices are not necessarily free of health risk. The relation to water and public health sanitation of the fenugreek issue is two-fold. First, one can speculate about the availability of sanitary latrines at the source of the seed. Further, as this particular bacteria not only produces high levels of shiga toxin but also has a "sticky" coat making it difficult to remove from food,[12] questions can be raised about the suitability of usual water cleaning practices for agricultural products. Would food irradiation have prevented the diseases and deaths?
Developed countries are not immune from water borne infectious disease. A thoroughly investigated outbreak of diarrheal illness in a Canadian community found an unusually high level of rain led to manure contamination of drinking water wells that should have been treated with chlorine but were not because of human mistakes, failure of oversight, lack of training and absence of back-up systems. Over 1,700 cases of diarrhea and 7 deaths were identified in a community of 4,800 people.[13]

While chlorination is advocated as one of many ways to prevent water-borne disease outbreaks, it is not effective against all infectious agents. Cryptosporidium, a protozoan parasite, in its cyst stage is relatively resistant to chlorine but can be filtered from drinking water.[14] Notable outbreaks include Milwaukee in 1993 where in a two week period over 400,000 cases of diarrheal illness and more than 100 deaths were attributed to a malfunctioning water treatment plant.[15] Recent outbreaks have occurred in two locations in Sweden and one in Ireland. While usually a self-limited illness, cryptosporidiosis can be fatal to those with compromised immune systems as well as to the very young and old. Rates of infection and of secretion of cryptosporidium cysts are higher in developing countries. Chlorination of drinking water also leads to chlorinated hydrocarbons, some of which are believed to be carcinogenic. However, the risk of cancer is far less than the risk of a serious infectious illness

TOXIC CONTAMINATION OF WATER—NATURAL AND INTRODUCED

Concerns about water contaminants are not limited to microbes and parasites. In West Bengal, India and in Bangladesh upwards of 10 million residents have experienced the risk of unsafe water supplies due to high arsenic levels from natural deposits contaminating shallow tube wells used for daily water needs. Hair, skin swabs, and nail clippings were taken from a sample of individuals living in arsenic-affected villages. An average of 93% and 73% of Bangladesh and West Bengal samples, respectively, tested above normal levels.[16] Due to the lack of alternative water sources and high arsenic contamination levels of wells, nearly 25% of Bangladeshis are suffering from skin lesions due to arsenic poisoning.[17] Arsenic contamination also puts the population at risk for other diseases including skin, bladder, kidney, and lung cancer and diseases of blood vessels of the legs.[18,19] In this case, the desire of international organizations to decrease diarrheal diseases caused by surface water contamination led to the impetus to dig the tube wells, which ultimately led to a high burden of preventable arsenic-related disease.[20] Despite the fact that arsenic testing is relatively inexpensive, and arsenic contamination of natural aquifers had previously been observed in Asia, water was never tested for arsenic until after arsenic poisoning was evident in the population. The expected death toll has been said to be over one million from the various vascular and carcinogenic effects of arsenic.

Human activity can introduce toxins into water sources. These toxins can have detrimental effects on health and increase the risk for serious disease. For example, Minamata disease caused by severe mercury poisoning was recognized as a problem in southern Japan in the 1950s. Wastewater containing high levels of mercury was dumped into Minamata Bay where the heavy metals bioaccumulated in fish and shellfish eaten by the local population.[21] Health repercussions resulting from these actions are still experienced by residents.

More recently, gold mining in many countries in Latin America, Africa and Asia poses health challenges to workers and communities. Mercury is used in the extraction process, diluted with water, and has contaminated water resources, increasing prevalence of heavy metal poisoning.[22] Two pathways to mercury poisoning have been evident. One is the direct exposure of workers to inorganic mercury vapor which causes nervous system effects. The second is caused by organic mercurials, such as methyl mercury, which are even more potent neurotoxins and particularly affect developing fetuses. Methylation of mercury occurs through bacterial processes often occurring many miles downstream from the release of inorganic mercury. The resultant biomagnification through the food chain leads to high mercury levels in fish eaten by humans. Worldwide, higher levels of mercury in seafood, particularly fish such as tuna and swordfish which are high on the food chain, have resulted in recommendations to limit fish intake by pregnant women. Global mercury uptake, including high mercury levels in seafood far from mercury sources, appears related to long range transport of mercury emitted during coal consumption.

In the Nigerian state of Zamfara, over 400 children died of lead poisoning in 2010 due to high exposure during the process of illicitly mining gold.[23,24] The Zamfara case highlights how an intersection between the human poverty and richness of the land can lead to devastating health consequences. In this case, locals had few options for income other than illegal gold extraction. Underlying causes suggested included globalization of agriculture; global climate change leading to desertification; high birth rates; lack of education; and a multinational mining company.

Toxic contamination of water is not unique to developing countries; in the United States one out of six homes receives water from private wells that are particularly susceptible to microbial or chemical contamination.[25] Agricultural practices can lead to fecal contamination or contamination with pesticides or nitrogen—with children at highest risk. A new threat in the United States and elsewhere is drilling for oil or natural gas resources using fracking and other newer technologies. This often occurs in rural areas where well water is more likely to be used.[26,27] Trace elements exist in water, whether added through natural processes or through contamination. However, the health effects from their consumption remains unclear, for example, literature surrounding selenium suggests that trace amounts may contribute to good health while other authors find the reverse.

Modern technology has also given us new pharmaceuticals and chemical agents, including plasticizers, which are increasingly found in water bodies and in drinking water. It has also given us more sensitive analytical techniques allowing the detection of agents not previously recognized to be in water. Consumer products such as perfluorooctanoic acid and a variety of estrogenic pharmaceutical agents used for birth control are among those that have been identified in water and have given rise to concerns about health. The increasing use of grey water in agriculture raises the possibility that ingested hormonal agents appearing in the urine may be taken up in animal feed thereby entering milk or meat products.

VECTOR-BORNE PATHOGENS

Many vectors, such as black flies and mosquitoes, that spread diseases such as onchocerciasis, malaria, and dengue, require water to breed. Areas at risk of attracting mosquitoes include man-made containers used for water storage, discarded containers and tires, large leaves, and shallow ditches that collect water during the rainy season. Malaria is the most prevalent of vector borne diseases, causing nearly one million deaths, mainly in children in sub-Saharan Africa.[28] Dengue has a smaller disease burden, infecting 50 million world-wide, however unlike malaria, dengue does not have any effective treatment options available. Environmental modifications, by reducing stagnant water sources, and larvicides are currently the major approaches to reduce the breeding grounds of the day-time biting *Aedes* mosquito and the prevalence of disease.[29,30]

In addition agricultural practices are inter-linked with the prevalence of disease-spreading vectors. For example, in Sri Lanka, rice field development has increased the available breeding-grounds for mosquitoes carrying Japanese encephalitis and resulted in widespread epidemics.[31] Increases in vector-borne diseases have also been linked with decreased productivity, poor yields, and reduced incomes for workers falling ill.

Developed countries are not exempt from the spread of vector-borne diseases. In the past two decade, West Nile virus (WNv) has appeared in the United States and increased in prevalence. In Kern County, California, neglected swimming pools provided breeding grounds for the mosquito vector responsible for the spread of WNv.[32] Economic conditions have been linked to increases in WNv in the United States, because of the association of home and pool maintenance with per capita income and rates of home foreclosures.[33]

In an increasingly globalized society it is not surprising that diseases are emerging in new areas, such as in the case of WNv spreading to North America from Africa. In addition models have also predicted that climate change will increase the risk of contracting vector-borne diseases such as malaria, onchocerciasis, yellow fever, and dengue in areas where disease is already endemic. Climate change will potentially lead to higher prevalence and have higher vector survival rates in warmer locations.[34] Climate change will affect most vectors and the pathogens in their geographic distribution, development, reproduction rates, behavior, and survival. Increases in temperature increase the longevity and infectiousness of mosquitoes that could spread malaria, dengue, WNv, or yellow fever.[35,36] However, there is some skepticism about global climate change as a cause of an increase in vector-borne diseases because the complexity of interactions between environmental, human, vector, and parasite factors which ultimately determine the potential extent of the disease. These factors either have not been taken into consideration or are difficult to consider and incorporate in current models.[37] It is also anticipated that other diseases, such as Lyme disease, will decrease in prevalence due to shifts in their viable geographic range in both Europe and the United States.[38]

DISTRIBUTION AND AVAILABILITY OF WATER

The growing global population puts an increased strain on limited potable water resources. Currently only 0.01 percent of water is potable and accessible. It is inevitable that alterations in the weather-related distribution of rainfall will put an additional stress on water resources in water scarce places. This potentially has meaningful effects on food security, famine, and malnutrition in areas such as sub-Saharan Africa. In Kenya, key agricultural areas have experienced a reduction in rainfall ascribed to warming of the Indian Ocean.[39] In west-central Kenya rain declines of over 100mm per year have already occurred, which could have significant effects on the rain-fed agriculture predominant in the area. Kenya is also the location of a large refugee camp for Somalis who are fleeing both drought and conflict in their country. In regions that experience high poverty and malnutrition rates, the impact of climate change on agricultural production and food security is by no means certain, however, it creates long term policy problems exacerbated by the complexity of the issue and population growth.

Increased urbanization puts strain on water resources.[40] The high growth rates of cities, especially those in the developing world, increase the risk of contamination of existing inadequate water resources and sanitation systems. Slums and illegal squatting settlements, such as the favelas of Brazil, are known as areas of poverty, high population densities, lack of sanitation services, poor water quality, and as a result areas with high risks of disease to the population.

A shortage of clean water in hospital or other emergency circumstances may increase the likelihood of death, even when lifesaving interventions have been performed. Without clean reliable water sources to use in surgery and hospital out-patient settings, the threat of infection of wounds and surgical scars increases and death becomes much more likely. In Iraq in 2003, when conflict was at a high point, Baghdad hospital staff treating patients with bullet wounds and burns from blasts was in a constant struggle to maintain a sterile hospital environment due to frequent shortages in clean water and electricity.[41] The absence of clean water is particularly important in refugee circumstances where overcrowding, lack of sanitation, poor hygiene, and malnutrition contributes to high diarrheal rates and a high risk of epidemics.

WATER RESOURCES, PUBLIC HEALTH AND SUSTAINABLE DEVELOPMENT

Management of water resources for public health is central to sustainable development. But the health component has not always been at the forefront of sustainability considerations. The sustainability paradigm is usually defined to contain three components: environmental, economic and social, with health being subsumed under the social leg. This downgrading of health can be problematic—for example, it is difficult to conceive of sustainable development in much of the African continent without addressing malaria and HIV. But the concept of sustainability, which attempts to maximize benefits across the three legs of environmental, economic and social (health), is highly pertinent to water issues which affect all three components in the paradigm. Major consideration of water issues in relation to sustainability can be anticipated at the Rio plus 20 meeting in Brazil next summer

SUMMARY

Water is a critical resource. While problems exist in western countries in ensuring water quality to their citizens, these problems are compounded in developing countries due to the challenges posed by poverty, water scarcity, poor water and sanitation delivery and disposal systems, population density, poor regulatory systems, and water access, to name a few. Scientists, donors, and policy makers will need to focus on providing innovative solutions to problems surrounding clean water, and recognize that clean water is inextricable from not only vast range of health problems, but also development and environmental issues that are key in the sustainable economic growth and productivity of developing countries.

REFERENCES

1. Choffnes, E.R., Mack, A. (2009) *Global Issues in Water, Sanitation, and Health: Workshop Summary.* Washington, D.C.: The National Acadamies Press.
2. UNICEF, World Health Organization. (2009) "Diarrhea: Why children are still dying and what can be done?"
3. Bartram, J., Cairncross, S. (2010) "Hygiene, sanitation, and water: Forgotten foundations of health." *PLoS Med.* 7:e1000367.
4. Cairncross, S., Bartram, J., Cumming, O., Brocklehurst, C. (2010) "Hygiene, sanitation, and water: What needs to be done?" *PLoS Med.* 7:e1000365.
5. Coussens, C. (2009) *Global environmental health: Research gaps and barriers for providing sustainable water, sanitation, and hygiene services.* Workshop Summary. Washington, D.C.: National Academies Press.
6. World Health Organization. (2003) *Emerging issues in water and infectious disease.* France: World Health Organization.

7 Bassani, D.G., Kumar, R., Awasthi, S., et al. (2010) "Causes of neonatal and child mortality in India: A nationally representative mortality survey." *Lancet.* 376:1853-1860.

8 World Health Organization. Water Sanitation and Health (WSH): "Water-related diseases: Anaemia." Available at: http://www.who.int/water_sanitation_health/diseases/anemia/en/. Accessed 7/13/2011.

9 Berkman, D.S., Lescano, A.G., Gilman, R.H., Lopez, S.L., Black, M.M. (2002) "Effects of stunting, diarrhoeal disease, and parasitic infection during infancy on cognition in late childhood: A follow-up study." *Lancet.* 359:564-571.

10 Khlangwiset, P., Shephard, G.S., Wu, F. (2011) "Aflatoxins and growth impairment: A review." *Crit Rev Toxicol.*

11 World Health Organization. "Fighting the rise in cholera cases in Haiti." Available at: http://www.trust.org/alertnet/news/fighting-the-rise-in-cholera-cases-in-haiti. Accessed 07/06/2011.

12 Bielaszewska, M., Mellmann, A., Zhang, W., et al. (2011) "Characterisation of the Escherichia coli." *Lancet Infect Dis.*

13 Hrudey, S.E., Hrudey, E.J. (2009) "Prevention is painfully easy in hindsight: Fatal E. coli O157:H7 and campylobacter outbreak in Walkerton, Canada." In: Choffnes, E.R., Mack, A., eds. *Global Issues in Water, Sanitation, and Health: Workshop Summary.* Washington, D.C.: The National Acadamies Press:127-148.

14 Chen, X.M., Keithly, J.S., Paya, C.V., LaRusso, N.F. (2002) "Cryptosporidosis." *N Engl J Med.* 346:1723-1731.

15 Davis, J.P., Mac Kenzie. W.R., Addiss, D.G. (1993) "Lessons from the massive waterborne outbreak of cryptosporidium infections, Milwaukee." In: Choffnes ER, Mack A, eds. *Global Issues in Water, Sanitation, and Health.* Washington, D.C.: The National Acadamies Press; 2009:108-126.

16 Chowdhury, U.K., Biswas, B.K., Chowdhury, T.R., et al. (2000) "Groundwater arsenic contamination in Bangladesh and West Bengal, India." *Environ Health Perspect.* 108:393-397.

17 Breslin, K. (2000) "Arsenic in Asia; water at its worst." *Environ Health Perspect.* 108:A224-A224.

18 Argos, M., Kalra, T., Rathouz, P.J., et al. (2010) "Arsenic exposure from drinking water, and all-cause and chronic-disease mortalities in Bangladesh (HEALS): A prospective cohort study." *Lancet.* 376:252-258.

19 World Health Organization. Arsenic. Available at: http://www.who.int/ipcs/assessment/public_health/arsenic/en/. Accessed 7/13/2011.

20 Goldstein, B.D. (2001) "The precautionary principle also applies to public health actions." *Am J Public Health.* 91:1358-1361.

21 Ministry of the Environment, Government of Japan. Minamata Disease: The history and measures. Available at: http://www.env.go.jp/en/chemi/hs/minamata2002/. Accessed 7/13/2011.

22 Twerefou, D.K. (2009) "Mineral exploitation, environmental sustainability and sustainable development in EAC, SADC and ECOWAS regions." African Trade Policy Center; ATPC-79.

[23] Moszynski, P. (2010) "Lead poisoning in Nigeria causes 'unprecedented' emergency." *BMJ*. 341:223.

[24] Saleh, J. "Lead poisoning in Zamfara State, Nigeria: A review." Available at: http://environmentandhealth.wordpress.com/2011/01/06/lead-poisoning-in-zamfara-state-nigeria-a-review/. Accessed 07/07/2011.

[25] Committee on Environmental Health, Committee on Infectious Diseases, Rogan,W.J., Brady, M.T. (2009) "Drinking water from private wells and risks to children." *Pediatrics*. 123:1599-1605.

[26] University of Pittsburgh, Graduate School of Public Health. Center for Health Environments and Communities. Available at: http://www.chec.pitt.edu/. Accessed 7/13/2011.

[27] FracTracker: Tracking the impacts of the natural gas industry-Piloting in the Marcellus Shale region. Available at: http://www.fractracker.org/. Accessed 7/16/2011.

[28] World Health Organization. (2010) *Malaria*. Fact Sheet 94.

[29] Massad, E., Coutinho, F.A. (2011) "The cost of dengue control." *Lancet*. 377:1630-1631.

[30] World Health Organization. (2009) *Dengue*. Fact Sheet 117.

[31] Mutero, C.M., McCartney, M., Boelee, E. (2006) "Understanding the links between agriculture and health: Agriculture, malaria, and water-associated diseases." *IFPRI* Focus 13, Brief 6.

[32] Reisen, W.K., Carroll, B.D., Takahashi, R., et al. (2009) "Repeated west nile virus epidemic transmission in Kern Country, California, 2004-2007." *J Med Entomol*. 46:139-157.

[33] Harrigan, R.J., Thomassen, H.A., Buermann, W., Cummings, R.F., Kahn, M.E., Smith, T.B. (2010) "Economic conditions predict prevalence of West Nile virus." *PLoS One*. 5:e15437.

[34] Ostfeld, R.S. (2009) "Climate change and the distribution and intensity of infectious disease." *Ecology*. 90:903-905.

[35] Khasnis, A.A., Nettleman, M.D. (2005) "Global warming and infectious disease." *Arch Med Res*. 36:689-696.

[36] Physicians for Social Responsibility. Health implications of global warming: Vector-borne and water-borne diseases. PSR; n.d.

[37] Reiter, P. (2008) "Global warming and malaria: Knowing the horse before hitching the cart." *Malar J*. 7:S3.

[38] Shope, R. (1991) "Global climate change and infectious disease." *Environ Health Perspect*. 96:171-174.

[39] United Nations Environmental Programme. (2011) "Environmental Science/Policy report: A dryer, more crowded kenya?" 4-5.

[40] World Health Organization. "What are the health risks related to overcrowding?" Available at: http://www.who.int/water_sanitation_health/emergencies/qa/emergencies_qa9/en/index.html. Accessed 7/15/2011.

[41] United Nations. "WHO Health Briefing on Iraq." Available at: http://www.who.int/features/2003/iraq/briefings/wednesday9/en/. Accessed 07/12, 2011.

WATER IN THE MIDDLE EAST: CRISIS, MANAGEMENT, CHALLENGES AND PEACE

ALFRED ABED RABBO (IN ABSENCE)
Water and Soil Environmental Research Unit, Bethlehem University
Palestine, Israel

There appears to be little substantive differences between Israeli and Palestinian assessments of available water resources and their condition. For the most part, the sides are no longer arguing about facts or data, but rather allocation and policy. There are, of course, clear objective disparities in the hydrological circumstances of the two populations which inform each party's positions.

The average per capita consumption of 50 liter per day in the West Bank and the 13 liter per day per capita (suitable for drinking purposes) in Gaza place constant pressure on the stability and socio-economic conditions of a future Palestinian state. Without a sufficient and safe supply of water, Palestinians believe that it will be difficult to ensure a sustainable future, especially during a period of consecutive drought years.

The Palestinians perceive the quantities of water allocated to them as a core political issue in the final status negotiations. Because of past sensitivities and the historic dynamics of occupation, they prefer to view the issue in terms of "rights". Indeed, Palestinians have always held that their water rights should extend to their "indigenous and shared ground water aquifers" as well as to surface waters that run through their jurisdiction, in particular the Jordan River. It should, however, be emphasized that never have the Palestinians demanded "equal" quantities of water—but rather only what they argue is their fair share. While Palestinians would like water quantity issues resolved before moving on to the problem of sustainable management, development and planning to protect water quality, it is likely that they will agree that the two be done in tandem.

Palestinians insist that water rights be resolved according to principles of international law which they believe will guarantee them sufficient quantities and the sovereignty to utilize and control their water resources. The basic axiom of international water law though is rather vague, requiring only that all sides receive *a reasonable and equitable share.* This should not be deemed problematic as the principle is sufficiently amorphous and given to sufficiently flexible interpretations to allow Israel to agree to have it as a basis for resolving the existing disputes. To be sure, the two sides do not agree about the legality of past water developing for Israeli settlements, nor to the legitimacy of Israel's "historic rights" to water resources that originate in the West Bank. Yet, resolving these differences ultimately belong to the "tit for tat" bickering of the past. They should not stand in the way of a pragmatic approach that can reach accommodation about water distribution, management and common efforts to improve water quality.

There are some differences in the data that the two sides will bring to the table, particularly in the area of potential aquifer recharge and present and future needs. Such disparities can be clarified and resolved either through joint scientific commissions or the utilization of third party arbiters. Ultimately, these differences are not excessive, and any remaining gaps will probably be no greater than production in a large desalination plant.

While in the past, Palestinians preferred to focus on water rights rather than expanding available sources, the steady rise in population over the past decade appears to have changed this position. Indeed, talking about water "needs" appears to be a more constructive way to get past disagreements which sabotaged progress in the past. Both sides agree that the growing gap between supply and needs of Palestinian communities makes additional conventional and non-conventional water resources essential. Moreover, putting in place water policies based on a sustainability assessment of the water resource taking into consideration socio-economic, governance and environmental issues, are considered to be an important stage in the move toward sustainability.

There is also broad acknowledgement of the significance of significant upgrades in waste treatment and environmental infrastructure. This is not a concession on the part of Palestinians but a clearer recognition of the consequences of continued neglect in this field. Poor sanitation services, poor management of sewage and solid waste and over application of fertilizers and pesticides in the agricultural sector and over-extraction and reduction in storage volumes have caused pollution to the Palestinian aquifers and harmed their quality life far more acutely than Israel's. Contamination of water will also minimize the already limited quantities of water resources in Palestine, exacerbating Palestinian shortages, even after future Israeli concessions.

While Palestinians correctly point out that occupying Israeli forces never built a wastewater treatment plant in the West Bank during all years of occupation, they also realize that relieving themselves of the "historic blame" will do little to improve their quality of life. The peace process brings with it rare opportunities to receive considerable funds to "jump start" a Palestinian waste treatment system that is based on tertiary treatment that can supply high quality to Palestinian agriculture. There is a growing recognition that most Palestinian farmers will be hard pressed to receive any water as the Palestinian population continues to grow and the quality of life improves.

In short, the present quantity and quality of the water resources on both sides are well understood. It is clear that regardless of the ultimate allocation, present supply is insufficient for the population that will be living in the region in the near future. The steady contamination of these resources exacerbates the situation significantly. Yet, there are reasons for optimism including:

- The feasibility of expanded water production via desalination;
- Greater Palestinian commitment to expanded water supply from waste water reuse (and desalination);
- The increasing commitment on both sides to environmental protection;
- The proven interest of donor nations in establishing water infrastructure.

All these point to water as a source of future cooperation rather than conflict.

JORDAN RIVER

More pages may have been written about the Jordan River than any other water resource in the world. Such keen interest internationally certainly has little to do with the size of the river which, for most of the year naturally, is a modest stream. Rather, the historic, spiritual and religious significance gives it a place in the world's imagination that has never been utilized from an economic point of view.

Hence, future agreements over the Jordan have to clearly consider what its optimal role might be as a regional resource. Economically, rather than talking about adding modest amounts of tomatoes to local farmers' yields, it makes more sense to pursue an ambitious tourist initiative that can bring the three riparians of the lower Jordan together. The economic potential of a peaceful Jordan River as a spiritual center, a resort center or even a center for water sports is substantial. But to attract investment and visitors, the river will have to be undergoing a considerable makeover.

The first step towards the reclamation of the Jordan River needs to involve water quantity. The Jordan River's present environmental state is the result of de fault policy decisions of Israel, Jordan and Syria who have preferred transfer of water for agricultural or other objectives to preserving the integrity of the stream. This is a policy which can change today with the emergence of alternative water sources. At a time when the international community is seriously considering a multi-billion dollar project to bring water to the Dead Sea, surely the cost-effective advantages of returning the Jordan River's flow to its natural level should not be overlooked.

Even if a reasonable permanent flow in the river returns some of its basic aesthetic properties, development efforts will not be successful without substantial improvement of water quality. If baptisms, swimming or participating water sports endanger visitor health due to water contamination, tourist initiatives will ultimately be unsuccessful.

Pollution into the Jordan comes from numerous sources and countries—and its abatement will require intervention in Jordan, Israel, Palestine and Syria. Future management strategies for the stream should involve a basin-wide commission with all the riparians. Just as the "Nile Initiative" has been a locus for international involvement and support, so should the Jordan create a framework that will readily for international agencies to provide funding for sustainable, peace-building initiatives. While this might not be possible with Syria at present, surely a Jordanian-Palestinian-Israeli agency to oversee flow, water quality and tourism along the stream makes sense.

HARMONIZED DRINKING WATER STANDARDS

There is a significant gap in the quality of drinking water available to Palestinian and Israeli households. While Israel's drinking water quality has largely improved, there are many examples of chronic contamination in Palestinian West Bank communities while drinking much of the water supplied in the Gaza Strip that has, for some time, been defined as unhealthy.

For the foreseeable future, Palestinian and Israeli drinking water systems will remain intertwined. Today—some 40.3 million cubic meters of drinking water is supplied by Israel's Mekorot water utility to houses in the West Bank—well over 60% of present municipal use. (An additional 3.2 million m^3 of water is still delivered by Mekorot to the Gaza Strip.) This is among the highest quality water presently available. To change the associated infrastructure and piping that supply this water will take many years and may not make hydrological or economic sense. This means that drinking water standards between the two parties must at least be harmonized and, in the long-run, should probably be identical.

While economic differences exist, the reality is that if Palestinian municipalities provide drinking water that is of poor quality, the population will simply choose to purchase bottled water. This constitutes a disproportionate economic burden on the poorest populations. Bottled water currently sells at rates of 36 cents/liter in the West Bank. While this is orders of magnitude higher than tap water, large segments of the population are paying for it.

Adopting tougher drinking water standards through harmonization in and of itself will not be enough to improve the Palestinian situation. There are fundamental infrastructure measures which will be required for the present contamination to be reduced. For example, septic tanks will need to be cemented so that wastewater does not escape and percolate directly into the ground and reach drinking water sources. Cisterns, which in many Palestinian villages are so critical for capturing rainwater, are subject to considerable biological contamination from bacterial outbreak, bird excrement and waterborne diseases. Because taps have filters, the population is frequently unaware of the actual water quality and illness is common. Education along with drinking water protection measures and disinfection kits are needed. Establishing and upgrading sewage treatment is of course a critical effort in virtually every Palestinian community. Here, it may make more sense to have differential standards, dependent on the ultimate use of the effluent and its potential to contaminate water resources. In either case, efforts to upgrade Palestinian water infrastructure will be extremely costly for a society whose resources are quite limited

Accordingly, a steady process of ratcheting down drinking water contaminant levels is envisioned by Palestinian experts. They compare their situation to Israel's experience. Initial Israeli drinking water standards were low and gradually became more demanding as the country's economic conditions improved. For instance, Israel understood that a standard of 90 mg/l was desirable for nitrates but couldn't afford it. Today they can make this commitment. The same is true of the BOD standard for sewage treatment. Today plants are expected to drop to 10 mg/l whereas initial standards could only require that sewage sit for five days in an oxidation pond.

Palestinians also envision a steady phase-in of higher quality drinking water standards. It can be argued that the present contamination is so severe with pollutants that cause acute health effects (e.g., bacterial pollutants) that the correct approach should be impatience and intolerance for low standards. But for many drinking water standards, a gradual, ratcheting down strategy based on clear timetables and quantitative targets, is one which should be considered favorably by Israeli negotiators, if common drinking-water standards become a negotiating topic.

Because Palestinians receive considerable water from springs, there must be better monitoring to ensure their integrity. But again, identifying polluted water sources is important for public health, but will not change the need to address the actual pollution sources. The Palestinian Water Authority envisions a decentralization process, by which local municipalities will play a larger role in monitoring drinking water quality. This will require considerable upgrading in human capacity, laboratory and field kits, etc. This is an important area for international assistance which will both improve public health and strengthen human and scientific capacity in a new Palestinian state.

PALESTINIAN SEWAGE INFRASTRUCTURE

There is probably no area in water management where the gap between Palestinian and Israeli environmental performance is greater than sewage. In Israel, despite the general sense of progress, some of the assumptions about future waste water utilization may require reconsideration. For instance, as urbanization expands in some regions, the agricultural demand for waste water will continue to drop. Ultimately, transferring effluents great distances to where demand exists may make less economic sense. As the ecological integrity of streams improves, there will be a need to continually improve waste water treatment, and better fresh water sources will be expected. Many environmentalists also argue that the country has made an adequate commitment to sewage infrastructure and that many municipal governments have found other areas to spend local taxes. But if present trends are any indication, it is fair to assume that Israel will continue to expand its usage of sewage treatment and that the quality of effluents will continue to improve.

While Israel has made slow but steady progress first in connecting homes to sewage systems and then upgrading its level of treatment, sewage treatment remains anomalous for the vast majority of Palestinian communities. Existing treatment facilities suffer from lack of maintenance and are increasingly overwhelmed by the growing organic loadings that have come with the population increase. Palestinians have not made sewage treatment a high enough priority. The long-term impact on soil and water resources is still not well characterized but should be a source of apprehension. The effect of mosquitoes, poor odors and periodic outbreaks of related diseases is more apparent and of immediate concern to the general public. In the context of future negotiations, sewage treatment constitutes environmental priority number one. Any negotiated solution and international assistance should reflect this commitment.

There is also probably no other area of water management (with the possible exception of overpumping) where there is a greater need for coordination between Israelis and Palestinians. The integrity of ground water or the aquatic systems in local streams will never be ensured unless dramatic improvement takes place. Common standards for sewage treatment are important because of the impact that sewage discharges have on shared water resources. It is clear from Israel's experience that recharging aquifers with poorly treated sewage will ultimately degrade groundwater quality. Massive irrigation with poorly treated effluents has also led to groundwater contamination, with industrial solvents appearing in wells throughout Israel's coastline, when they were not removed in wastewater treatment plants. Salinization of soils and wells is often attributed to Israel's alacrity to utilize waste water. But there is a limit to the demands that can be made in an international agreement when the issue of sewage treatment standards arises.

There is a link between environmental infrastructure and economic conditions. Several Palestinian sewage treatment plants have been established that ceased to function effectively due to lack of funding. Donors should ensure that financial mechanisms are in place to prevent these dynamics from occurring in the future. More importantly, the limitations of present capacity have clear implications for the establishment of common treatment standards. Essentially, "perfection" is indeed the enemy of the "good". Better to install reasonable secondary treatment across the board, than to have expensive tertiary treatment facilities that are not functioning.

A critical question that needs to be addressed in negotiations involves a coordinated response to the present "transboundary" dynamics of sewage treatment. In recent years, Israel unilaterally has begun to establish treatment facilities to capture and treat Palestinian sewage as it crosses the border. This was initially described as a pragmatic response, taken to protect Israeli water resources given the lack of progress in Palestinian waste water treatment. But it is not clear whether what was once an exigency has turned into a strategy. It is now clear from hydrological research that much of the sewage flowing from the Palestinian Authority percolates into the ground prior to reaching the Israeli treatment facility. A substantial percentage of this "lost" wastewater will make its way to the groundwater.

Accordingly, a final agreement about water should therefore internalize a fundamental axiom of environmental management: the proximity principle. The proximity principle holds that it is essential that water be treated as close to its source as possible. While future infrastructure and normative frameworks should consider waste treatment and *utilization* of effluents at a *regional* level, the focus on *treatment* must be local. Alternatively, and in the interim, untreated wastes can be transported to regional treatment centers via pipelines in order to prevent water loss/contamination and the risk of human exposure while it is in transit.

Until this principal is applied, there remains the thorny issue of costs. Israel has traditionally charged Palestinians for treating their waste in ad-hoc facilties in Yad Hana and in Nahal Beer Sheva. It has not, however, been willing to pay out similar funds in areas like the Kidron valley where Israeli sewage is a source of pollution in the Palestinian Authority. Palestinians justifiably object to this asymmetry. Moreover, they perceive the interim agreement, which requires high levels of coordination with Israel as one of the primary reasons why there has been so little progress in upgrading their sanitary infrastructure during the past decade.

Surely, it is in Israel's hydrological self-interest to facilitate the construction of Palestinian sewage treatment plants—either via monies donated by international donors or generated by local taxes. Moreover, in the long run, siphoning off Palestinian "tax funds" for a sub-optimal sewage plant in Israel will make no sense hydrologically. Eventually, Palestinian plants and reuse of waste will render these facilities useless in any event. Israel should think about making strategic investments from its own funds in Palestinian infrastructure—over the border and close to pollution sources. A long term "sustainable" allocation of water region will require both Israelis and Palestinians to utilize treated effluents in agriculture and as part of their stream restoration program.

THE ROAD AHEAD

The crux of the historic disagreement between Palestinian and Israeli negotiators in the water realm has involved rights to the mountain aquifer. Both sides have conveniently adopted theoretical positions which support their hydrological interests. Hence Israel argues that it enjoys historic rights to the aquifer, pointing to the storage capacity and established wells inside Israel's 1967 borders. Palestinians maintain their rights as riparians, relying on the location of the aquifer recharge area, where the rainfall actually originated. Resolving this "zero-sum-game" dynamic through other—such as "equitable use" or the "needs" of the parties has only been moderately successful. Ultimately, as was the case in the Israeli-Jordanian peace accord, a political compromise must be made, which should be far easier today given the availability of alternative desalinated sources for both sides. Yet, taking concrete measures to protect the aquifer can not wait for such diplomatic resolution.

There is a sense that while politicians have naturally focused on the "allocation" debate which is easy to grasp, the more complex dynamics of joint management and technical challenges for sustainable management of the mountain aquifer have not received adequate attention. As a result, the sources of contamination have not been abated. In particular, municipal sewage and non point source pollution involving runoff from agriculture and from urban sources have not been addressed. The importance of the many projects and practices for protecting the mountain aquifer that are detailed in this chapter is not disputed by technical experts on either side.

Regardless of ultimate water allocation, there needs to be a coordinated system of management and regulation between the two parties that is overseen by technical experts. These experts need to be freed to the extent possible from political constraints and take the necessary measures to ensure future sustainable, high-quality yields of groundwater. Continued neglect of pollution prevention technologies, infrastructures and oversight could lead to a shared resource which is of little value. The grave situation of the coastal aquifer underlying the Gaza Strip is a sobering reminder that hydrological systems will not wait for political harmony, but are easily compromised given the relentless flow of contaminants and uncontrolled tendencies towards extraction beyond sustainable levels.

Upgrading sewage systems in the recharge area remains the single greatest imperative for protecting the mountain aquifer. Over the past years, a litany of sanitary projects has been discontinued in the West Bank and Gaza due to the political instability. Ironically, funds were available from international aid agencies for treating most of the sewage from Palestinian cities. Already, fecal contamination can be measured in many of the West Bank streams, portending water quality problems in the underlying aquifer. An immediate, "emergency" initiative to collect and treat all untreated sewage in the Palestinian sector and shift the existing cesspool systems to connect to these centralized facilities must be initiated immediately,

Even after allocations find a political resolution, water quantity will continue to be an open problem that must be addressed dispassionately and creatively. The effects of the anticipated climate change on aquifer recharge have been sufficiently considered. If projections of the models endorsed by the United Nations International Panel on Climate Change (IPCC) are correct, recharge will be diminished. Indeed, it has been argued that the allocation schemes that emerged from the interim Oslo agreement in the mid-1990s relied on the rainfall levels of the preceding years, which was a relatively wet period. Hence, many wells have essentially been mining groundwater resources. The present drawdown on the aquifer due to continuous pumping will probably grow worse due to effects of global warming, and the hydrological damage could be irreversible.

In the past, agriculture served in Israel as a buffer for drought periods. But, with the shift to wastewater, the impact of cuts to agriculture will be far more painful—with orchards and mature trees being sacrificed rather than a single year's annual yields, for which compensation can be provided. Preparing for such droughts should be done in a coordinated fashion. Desalination offers protection against these fluctuations.

Water Quality Protection

It is likely that water quality will continue to be the subject of mainly domestic law. Because Palestinians and Israelis share many of the water resources, and the failure of either to protect water quality severely affects both parties, it will be essential to find a way to make the two systems for water quality protection compatible. This will be especially important in cases where the actions of one side affect the water quality of the other side, such as in the following cases:

- Wastewater disposed of in one area that reaches the other, either via the surface or through shared groundwater;
- Land disposal of solid waste or building construction in one area that can pollute shared groundwater sources;
- Joint planning for water pollution prevention projects.

Coordination of water quality standards will be essential. Israel traditionally has taken a position which expects the Palestinians to meet Israeli requirements, even as compliance with them in Israel and in Israeli West Bank settlements is frequently deficient. Environmental expectations must be adjusted to meet available capacity. To some extent, the answer to this quagmire involves phasing in more stringent performance standards through realistic timetables. Just as East German industries were allotted several years to meet the higher West German environmental expectations after unification, Israelis and Palestinians will need to agree that Palestinians need a grace period to allow for the improvement of both the physical infrastructure and human capital. Here the role of donor nations in providing resources directed specifically to this goal will be critical. Another possible model is that provided by the U.S. and Mexico cooperation in restoring their shared water resources. Given compelling American interests in environmental improvement, the U.S. demanded higher levels of performance but was willing to participate in the associated investment within Mexico.

While it will be hard for Israeli environmentalists to swallow, it may also be prudent to sacrifice stringency for short-run in the interest of ensuring high compliance levels. The alternative could be creating a pattern of "lip service" and alienation from shared environmental standards for years to come. For the primary legal challenge at present involves implementation and compliance monitoring. The pervasive use of raw sewage by some Palestinian farmers or the historic noncompliance in pollution discharges among industrial and municipal facilities in Israel are examples of the gap that exists between theory and practice.

One difference between the Palestinian and Israeli system for protecting water quality lies in the licensing arrangements. In theory, any business or operation in Israel that is likely to cause water pollution needs a license of some sort that will include pollution prevention provisions; therefore, the regulatory authorities have prior knowledge of any such operations and the specific conditions that apply to a business are set out specifically. Under the Palestinian system, the law sets out general prohibitions on water pollution but no licenses are required. This can make it more difficult to identify and account for all potential sources of pollution and can leave unclear the precise way the general legal requirements apply to specific sources. Both of these characteristics are problematic in the search for greater transparency between both sides. In practice, the Israeli regime does not work as well as it should, and many businesses operate without business licenses or in violation of them. It is recommended that both sides put in place and operate a comprehensive scheme for licensing sources that may pollute the water, that both vigorously enforce the licenses, and that both licensing provisions and enforcement efforts be transparent to everyone on both sides.

At present, only Israel makes substantial use of treated wastewater and has water quality standards for the treated wastewater. If wastewater treatment is undertaken by the Palestinians, it too will need treatment standards. Coordination of the standards for treated wastewater will be necessary where the treated wastewater can flow as runoff from one area to the other, or where it reaches a shared groundwater resource. In other cases, each entity can set its own wastewater treatment standards.

Enforcement

The law is only good when it is observed, and enforcement is not sufficient in either area. No legal system has the funds to catch all violators. The goal of environmental regulatory system on both sides therefore must ultimately be the creation of incentives (and deterrence) among polluters and potential polluters not to violate environmental laws.

At present, the Palestinians find the political situation and pervasive poverty inhibit strong enforcement. It is not now clear whether a change in the political situation would be sufficient to bring about strong and consistent enforcement of the water laws. The Israelis also have a serious enforcement problem, based on a lack of a tradition of fines large and consistent enough to deter violation of the environmental laws. Both sides must greatly improve their enforcement of legal requirements, whatever the obstacles. Otherwise, the existing water resources will be overused and polluted, and mutual blame will continue in a way that displaces the accountability that cooperation will necessitate.

Transparency

Israel has in place a wide reaching Freedom of Information Law, although agency responses to requests for information are not always timely and complete. The Palestinian Authority is working on a Freedom of Information Law. Such laws on both sides are essential to allow the building of trust needed to deal with the shared water resource. Both the Palestinians and the Israelis have active environmental NGOs. These groups should be encouraged to use their respective Freedom of Information Laws to monitor official action, and to share their information with each other.

SESSION 4

ENERGY & POLLUTION

FOCUS: UNCONVENTIONAL NATURAL GAS: BENEFITS AND RISKS

WORLD-WIDE UNCONVENTIONAL GAS POTENTIAL: ENERGY SECURITY AND GHG EMISSIONS

CARMINE DIFIGLIO[1]
Office of Policy and International Affairs, U.S. Department of Energy
Washington, USA

ABSTRACT

World-wide development of shale gas could substantially increase natural gas supply in North America, Europe and Asia. Such a development would significantly decrease natural gas exports from Russia, the Caspian and the Middle East compared to a future without world-wide shale gas development. This paper provides estimates of how shale gas development could affect energy markets. These estimates employ a detailed assessment of world-wide shale gas resources and the costs to produce gas from them.

SHALE GAS RESOURCES

The study relied on an international shale gas database provided by Intek, Inc. (Intek, 2011). This data was collected from a large number of publicly and proprietary data sources. The data includes the location of the shale, the key geologic properties for each deposit, the estimated original and technically recoverable resource, water requirement and availability data, operator data for each company currently active within the shale play, and historical production and well counts for the active shale plays. In addition, the database contains details on the infrastructure and market access requirement as well as the drilling capabilities available for the development of each shale play.

Figure 1 shows Intek's estimated shale gas resources by country. Note the very large reserves in North America (United States and Canada), Europe (Poland, as well as significantly smaller reserves in several other European countries), China and Australia.

[1] Major contributions to the analysis discussed in this paper were provided by Hitesh Mohan, Thomas Alfstad, Robert Anderson, Audrey Lee, and Kevin Easley.

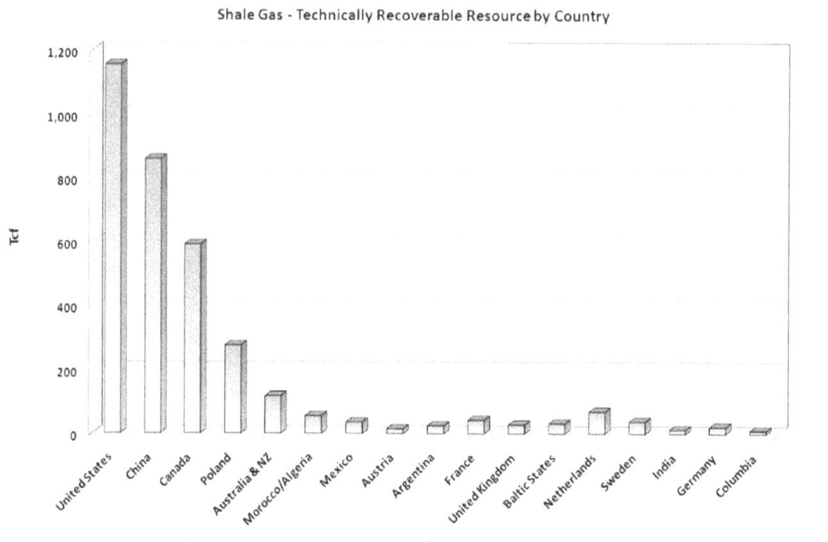

Fig. 1: Shale Gas—Technically Recoverable Resources by Country.

ENERGY TECHNOLOGY PERSPECTIVES MODEL

The Energy Technology Perspectives model (ETP) developed by the International Energy Agency (IEA, 2006; IEA, 2008; IEA 2010a) was modified for this study and the modified version is referred to as the DOE-ETP model. The DOE-ETP model, like most MARKAL-based models, is a partial equilibrium model that incorporates a representation of the physical energy system. The model represents the flow of energy carriers through the physical infrastructure from the resource base through the various energy conversion technologies to the end-user. The model is solved as a cost minimization problem where future states of the energy system are determined by identifying the most cost-effective pattern of resource use and technology deployment over time. This is done while obeying user-defined constraints such as natural resource availability, technology and capital availability, environmental limitations, government policies and government regulations (Fishbone and Abilock, 1981; Hamilton et al. 1992; Loulou et al. 2004).

SCENARIOS

The DOE-ETP model was used to develop two scenarios through 2035. The first (Reference Scenario) was calibrated to match the IEA *2010 World Energy Outlook* Current Policy Scenario (IEA 2010b). The second scenario (Unconstrained Scenario) replaced the shale gas estimates used in the Reference Scenario with the estimates of world-wide shale gas resources and associated production costs described above. It was assumed that only economic factors would limit the development of these estimated shale gas resources. In particular, for the Unconstrained Scenario:

- Costs for pipeline access are included. Assumes open, competitive access to pipelines. No extraordinary legal or regulatory restrictions to pipeline development are considered.
- Water availability is accounted for and fracking fluid disposal costs are included in the production cost estimates.
- No restrictions are included based on political or environmental factors (apart from regulations ensuring well integrity, proper waste water disposal and environmentally conscious management of drilling operations).
- No new tariffs or subsidies.
- No new policies or regulation.
- No "price dumping" or other strategies by exporters of conventional gas to discourage investment in shale gas are employed.

RESULTS

The comparison of the Reference and Unconstrained Scenarios showed large impacts on world-wide natural gas markets. In the Unconstrained Scenario, unconventional gas production increased by 30 trillion cubic feet (TCF). This was partially offset by a 10 TCF decrease in conventional gas production yielding a net increase of worldwide gas production of 20 TCF by 2035 (see Figure 2).

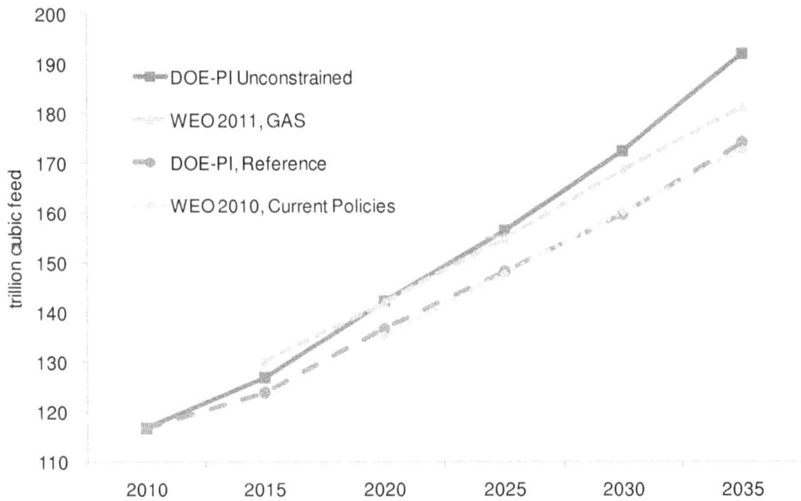

Fig. 2: Increase in World-Wide Natural Gas Production, Unconstrained Scenario vs. Reference Scenario.

The distribution of increased unconventional natural gas production (Unconstrained Scenario compared to the Reference Scenario) is shown in Figure 3 for different countries or regions. Of particular interest is the tripling of natural gas production in China and the doubling of natural gas production in Europe (2035).

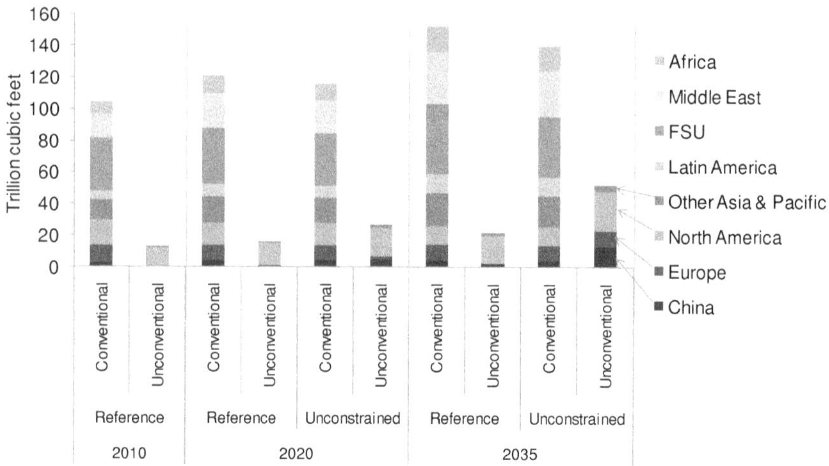

Fig. 3: Changes in Natural Gas Production, Unconstrained vs. Reference Scenarios by Region for Conventional and Unconventional Natural Gas.

Figure 4 shows the impact of the Unconstrained Scenario on regional natural gas trade balances. The figure shows imports of natural gas by major consuming regions (below zero) and exports of natural gas by major gas exporting regions (above zero). Natural gas exports from Russia and countries that comprised the former Soviet Union decreased by 50% and Middle Eastern natural gas exports decreased by 65% (2035).

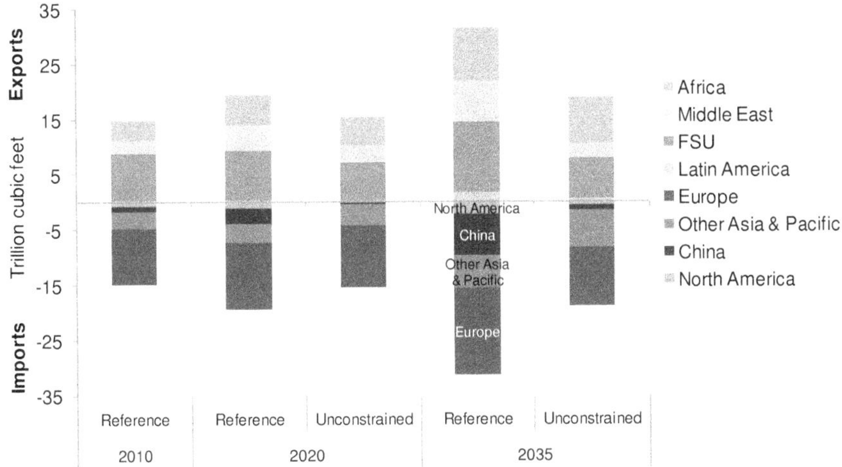

Fig. 4: Changes in the World Natural Gas Trade Balance, Unconstrained vs. Reference Scenarios.

Global CO_2 emissions in 2035 were only reduced by 1.8% in the Unconstrained Scenario. This was due to increased electricity demand and fuel substation among *all* power-sector fuels (coal, renewable, nuclear and oil). While natural gas, in the Unconstrained Scenario, reduces CO_2 emissions by avoiding new coal-fired generation, other displaced power sector options (nuclear and renewable) would have had lower CO_2 emissions than natural gas somewhat offsetting the benefits of using natural gas. Consequently, while increased natural gas supplies can enable reduced CO_2 emissions, they will, by themselves not substitute for explicit policies and measures to reduce CO_2 emissions.

REFERENCES

1. Bain, R.L. (2007) Worldwide Biomass Potential: Technology Characterizations. Report No. NREL/MP-510-42467. National Renewable Energy Laboratory, Golden, CO, USA.
2. Fishbone, L., Abilock, H. (1981) "Markal, a linear-programming model for energy systems analysis: Technical description of the BNL version." *Energy Research*, 5:353-375.
3. Hamilton, L.D., Goldstein, G.A., Lee, J., et al. (1992) MARKAL-MACRO: An overview. Report No. BN48377. Brookhaven National Laboratory, Upton, NY.
4. IEA (International Energy Agency) 2006, *Energy Technology Perspectives 2006*. Organization for Economic Co-operation and Development (OECD)/IEA, Paris, France.
5. IEA 2008, *Energy Technology Perspectives 2008*, OECD/IEA, Paris, France.

6. IEA 2010a, *Energy Technology Perspectives 2010*, OECD/IEA, Paris, France.
7. IEA 2010b, *World Energy Outlook 2010*, OECD/IEA, Paris, France.
8. Intek, Inc., *International Gas Shale Database, Resource, Production and Constraints*, April 2011
9. Loulou, R., Goldstein, G., Noble, K. (2004) Documentation for the MARKAL Family of Models. Available at http://www.etsap.org.

THE EVOLUTION OF THE U.S. NATURAL GAS SUPPLY

DR. BRUCE STRAM
Element Markets
Houston, USA

ABSTRACT

It is well known that energy supply is critical to our modern civilization. Some even call it, with reason, "the master resource." Natural gas (meaning methane) is currently a very significant part of that supply. In use, it offers substantial advantages over other fossil fuels. It is low cost, easy and safe to transport, and gives rise to much less pollution.

Now we have had remarkable developments in hydrocarbon producing technologies which can permit relatively low priced natural gas to be produced from gas-bearing shales. The purpose of this paper is to put this happy occurrence into a historical perspective of natural gas finding over the long term.

The current development is not, in fact, unique. Rather, over the long term, the development of technology has expanded the gas supply far more rapidly than that supply has been exhausted. North America, and in particular, the United States, represents a particularly interesting exemplar of this proposition. Vastly more drilling and exploration for hydrocarbons have occurred in the U.S. than the rest of the world combined. If there were any place on earth wherein the limits to gas supply might be in sight, it is the U.S.

There is no question that the "quality" of U.S. reserves has declined as a consequence of this heavy exploitation. Exploration companies no longer have available huge reserves close to the surface which are easy to find and produce as was the case in the past. But technology evolution has rendered the less attractive deposits more economically accessible. Hydrocarbon finding technology, sensing what lies beneath the earth's crust, has improved dramatically as a consequence of data processing and other technology gains. Thus, more and more resources are accessed at reduced cost.

Further, drilling and production technologies have continually improved with innovations in tools and equipment, also reducing the cost of achieving a given objective.

Finally, new techniques for utilizing these technologies are devised. Fracking, to increase production from tight (low permeability) formations, was innovated in the 1950's. Horizontal drilling was developed in the 1990's in order to exploit reserves with narrow pay zones but with substantial areal extent. New techniques made previously unavailable but known reserves of coal bed methane available.

The current shale gas breakthrough is the result of combining these techniques of fracking and horizontal drilling. The existence of these reserves of gas (and oil) was previously well known. They could be exploited by older recovery methods, but not economically.

The long term story remains the same: there continue to be known vast deposits of natural gas that are currently uneconomic for production. If the future apes the past, technology will develop so as to produce these reserves.

INTRODUCTION

It is well known that energy supply is critical to our modern civilization. Some even call it, with reason, "the master resource." Natural gas (meaning methane) is currently a very significant part of that supply. In use, it offers substantial advantages over other fossil fuels. It is low cost, easily and safely transported, and gives rise to much less pollution.

Now we have had remarkable developments in hydrocarbon producing technologies which can permit relatively low priced natural gas to be produced from gas bearing shales. Just how remarkable is evidenced by an April report from the EIA suggesting that adding identified shale gas resources that are technically recoverable increases the world's supply of recoverable natural gas by 40%.

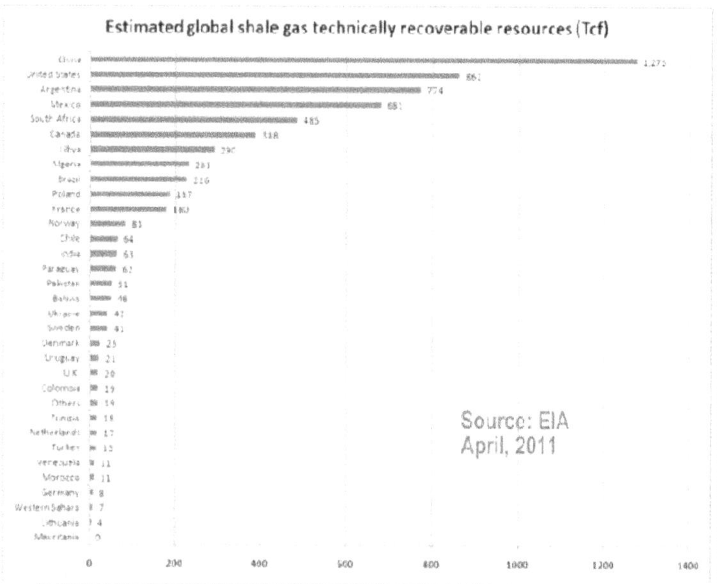

The purpose of this paper is to put this happy occurrence into a historical perspective of natural gas finding over the long term.

The key hypothesis is that this development, while especially dramatic, is not unique. Rather, over the long term, development of technology has expanded the gas supply more rapidly than that supply has been exhausted. The fact that hydrocarbon finding and production is driven by high technology means that more and more resources are accessed. Thus, the long term outlook for natural gas supply is indefinitely large.

North America, and in particular the United States, represents a particularly interesting exemplar of this proposition. Though now neck and neck with Russia in total production, the U.S. cumulative production is much greater than any other nation. (The U.S. produces 22.9% of the world's supply of natural gas while the Russian Federation produces 22.5%.) Further, as might be expected, vastly more drilling and exploration for hydrocarbons has occurred in the U.S than in the rest of the world combined. If there were any place on earth wherein the limits to supply might be in sight, it is the U.S. That this is not the case is a result of technological innovation.

For clarity, it should be understood that this paper is focused on the "resource base" for natural gas in the U.S. The term is less precise and subject to alternative definition than other measures of natural gas resources. For example "proved reserves" are those which are known to exist and are expected to be produced from developed wells or can be accessed with feasible additional development. They may be either "proved developed" (i.e., producing) or "proved undeveloped." Such reserves are the primary assets of oil and gas companies, are reported in financial documents subject to SEC regulations, and are audited by third party engineering firms who specialize in such studies. This is a narrow definition of potential gas. (Even more narrow is proved producing, i.e., the reserves connected to pipelines and that are in production.) The determination is highly related to the costs of producing or developing these reserves. That is, it must be determined that they can be economically produced under current conditions including the price of gas. Thus, if as has occurred recently, the price of gas falls significantly, proved reserves can decline because they won't be produced at the now lower price. (This principally applies to proved undeveloped which require significant additional capital cost to achieve production. The incremental cost associated with flowing gas from already producing wells is very unlikely to fall below even very low prices.)

These concepts obviously have their uses, but are not very appropriate for looking to the future of natural gas. The "resource base" is a more useful concept. It goes to the totality of natural gas known to exist. The Potential Gas Committee, an independent not-for-profit industry volunteer organization, which has been addressing this issue for years, defines the resource base as gas that can be technically recovered. The concept of economic recovery is not considered. This is a more useful concept because, as will be demonstrated, technical capability continually improves so as to make more and more of the resource base economically feasible.

But even this does not fully reflect potential gas resources. There are substantial formations of gas that are not technically feasible today, but might be in the future.

A BRIEF HISTORY OF THE U.S. GAS MARKET

The demand for natural gas that has been developed in the United States has caused more drilling to exploit the resource than any other geographic area. In fact, the whole rest of the world is dwarfed by U.S. drilling statistics.

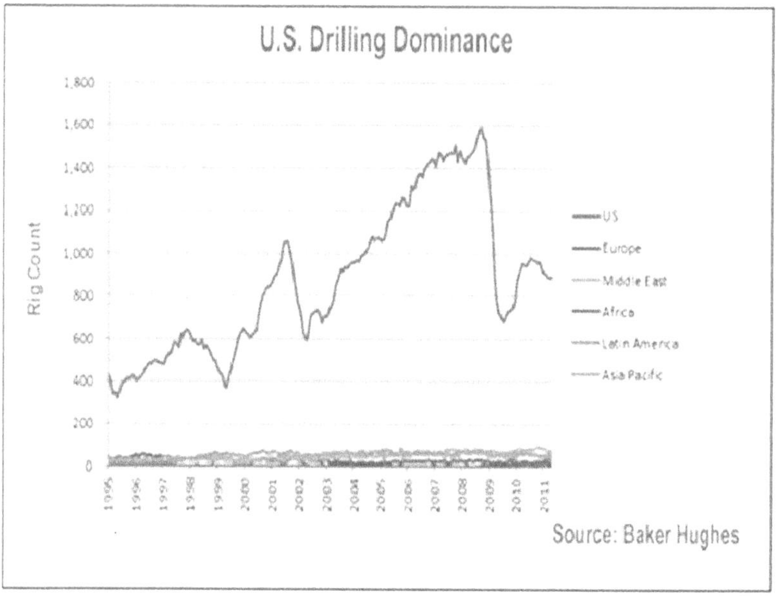

This market developed because the natural gas available in the U.S. Southwest represented a resource whose cost of production was far less than its value.

The early twentieth century development of oil exploration and production in the Southern U.S. (Texas, Louisiana, Oklahoma, and Kansas) is legendary. Vast quantities of this resource were found near the surface. Huge quantities of natural gas were associated with much of this oil. Major natural gas finds also occurred. This natural gas was transported locally by relatively short pipelines but only a fraction of it could be utilized as the long distance pipelines necessary to transport natural gas had not yet been built.

Meanwhile, Northeastern and Middle Western cities built pipelines and distribution networks for much smaller supplies of natural gas and for "town gas" manufactured from coal.

It became increasingly obvious that the value of Southwestern gas supplies was much greater in the North where demand for natural gas for commercial and residential uses was already stronger than in the saturated local markets.

Beginning in the mid-twenties, but with an exponential leap after WWII, improvements in pipeline technology permitted a massive construction of lengthy pipelines from the South to the North to meet demand and reduce this disparity. Southwestern natural gas replaced local supplies of town gas.

The U.S. gas pipeline network is the largest in the world. It is comprised of 305,000 miles of "transmission" lines.[1] Gazprom, by contrast, has 90,000 miles.[2]

---

[1] http://www.eia.gov/pub/oil_gas/natural_gas/analysis_publications/ngpipeline/mileage.html

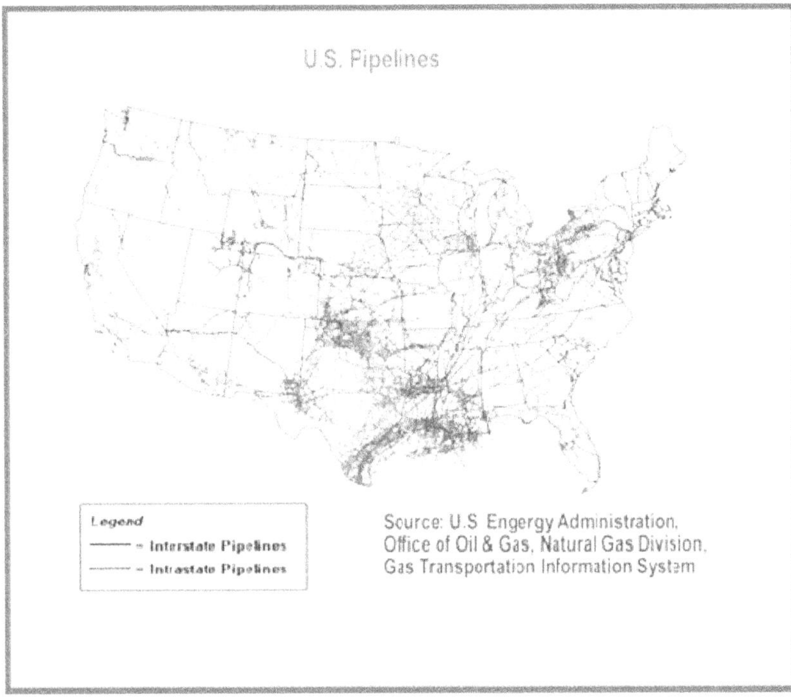

U.S. gas production increased rapidly after WWII, from 2.7 TCF in 1936 to 6 TCF in 1945 and 11 TCF in 1954. Current production is 24 TCF and has leveled off at this level for some time.[3]

[2] http://www.energytribune.com/articles.cfm/379/Russia-A-Critical-Evaluation-of-its-Natural-Gas-Resources
[3] EIA: http://www.eia.gov/dnav/ng/hist/n9070us1A.htm

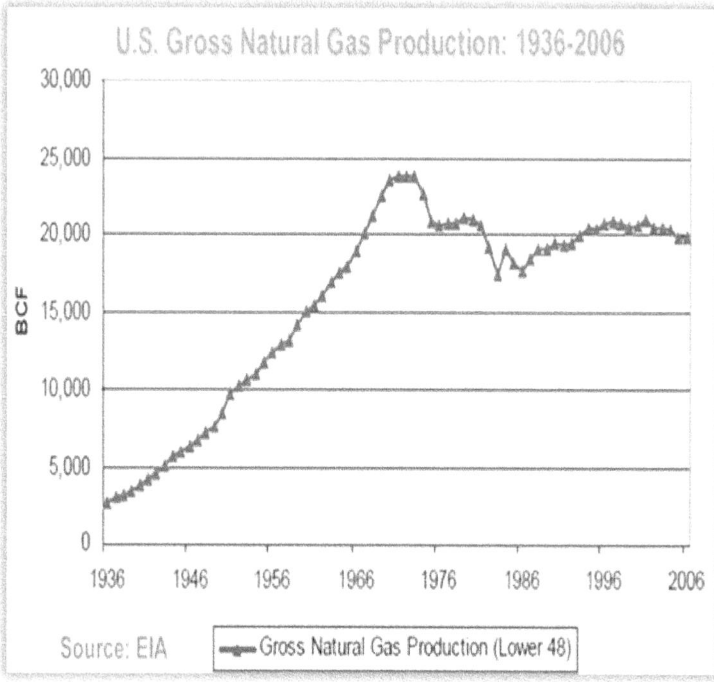

Obviously this resource has been of great value to the U.S. economy. But there is a nagging question: how long can this go on? Given this extended exploitation and drilling, aren't resources close to exhaustion?

The simple answer to these questions is that in a sense, the resource has been substantially depleted. Large "conventional" reserves close to the surface have largely been found. But production has been maintained and increased by the evolution and improvement in technology and techniques which has dramatically reduced the cost of finding a given reserve.

RECOVERY ECONOMICS

The profit incentive drives the massive drilling that has occurred in the U.S. Simply stated, the expected NPV (Net Present Value) of a given drilling project is:

$$\text{Well Cost} \leq \text{NPV} = \text{SF}\sum \text{QP}/(1+\text{ROI})^{\wedge}\text{n}$$

That is, the success factor (SF, the probability of finding natural gas as opposed to a dry hole) times the expected production (Q) for each period, times the expected price for that period, discounted by a hurdle rate of return. (Note that every fundamental element in the equation is probabilistic, suggesting that indeed this is a risky business.) The NPV must be greater than the cost of drilling the well in order to justify the investment.

Technology has reduced well cost with better drilling technology. Improved seismic data evaluation, driven in large part by more powerful computation capability, lets geologists better understand formations below the surface of the earth, reduce the probability of dry holes and better locate wells in order to increase production.

The consequence of these technology developments is that the estimated gas resource base has been rising by more than depletion.

Oil service companies are well aware of the connection between improvements in technologies and techniques and "the master resource." The following are some examples from major oil field service company Schlumberger's informational materials.

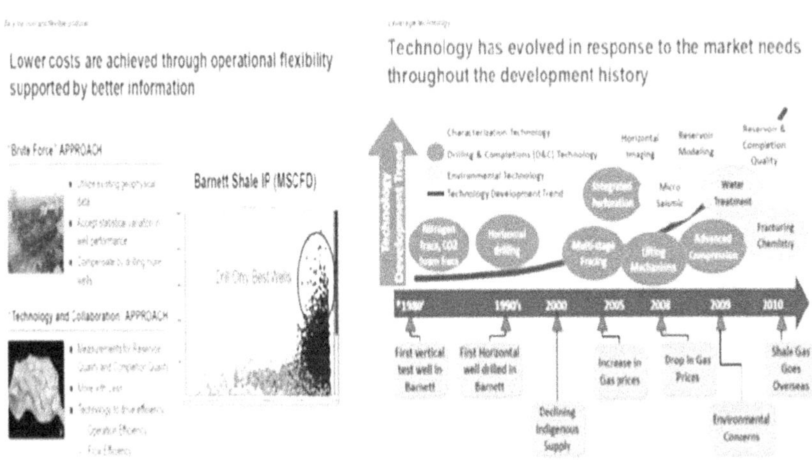

Sample of Schlumberger Informational Materials

PAST GAS SUPPLY SCARES

The U.S. was feared to have run up against the limits of the gas resource base three times in the last 35 years. The first of these occurred in the mid-seventies. Natural gas is primarily used for space heating, residential, commercial, and industrial. As a consequence demand reaches a peak in winter months. In the winter of 76-77, inadequate gas supplies caused a number of school, commercial and industry shutdowns, as well as affecting reserve gas for residential customers. This shortage was not fundamentally due to lack of supply, but to federal regulations which regulated wellhead gas prices so that producers had little incentive to dedicate their gas to interstate markets. Intrastate supplies in producing states were more than adequate.

Ultimately this problem was fixed by deregulation which eventually nationalized the gas market and eliminated the problem of local shortages. However, despite the fact that the shutdowns were not caused by real shortages, policy makers behaved in part as though there were real shortages. Thus, on the road to deregulation, some classes of gas were promised very high prices to induce more supply. Further, natural gas was prohibited from use in power generation in order to husband supplies. This occurred just as combustion turbine technology came to power generation, offering great advantages in terms of capital cost and efficiency.

The combination of interstate pipelines competing for supply and the staged nature of the deregulation meant that producers were contractually promised that almost all of their available gas (well by well) would be taken, and that they'd get a price equal to the highest in a geographic region. These contracts (take available gas or pay for production not taken) were mirrored by contracts between pipelines and northern distribution companies.

Producers, thinking they were promised ever escalating prices and very high takes, got busy. At one point more than 4000 rotary rigs were operating in the U.S., many more than are operating in the world today. But these high prices were unacceptable to consumers, and the take levels were unsustainable as potential production increased dramatically in response to the apparent incentives. The regulatory agency, the Federal Energy Regulatory Commission, cut the Gordian knot by finding that the contracts between the pipelines and distribution companies were null and void. By this time the gas supply had increased dramatically, and with customers freed from contractual obligations to take, prices fell dramatically.

There is a pattern here. Low prices lead to perceived shortages and higher prices (or expectations of higher prices). Higher prices lead to more finding and production and lower prices. But there is a substantial lag in increased production. Much production coming online today is the consequence of decisions made years ago (offshore wells have at least five or six year lags from decision to production.) Further, once in production it makes no economic sense to shut wells in because of low prices.[4] Thus, if prices start falling, they are likely to continue to fall for a period of time, and vice versa.

In the late 1980's, there was another shortage scare. This one was not so readily apparent from available data. For years after the early 80's there had been a natural gas "bubble", i.e. excess supply available for delivery. That was of benefit to consumers but lower prices meant less drilling and obviously the surplus had a limited life. The

[4] Most gas (or oil) wells have very low incremental production costs relative to market prices. Gas shut in today or this year because of low prices cannot be produced tomorrow or next year. Rather the withheld gas is produced over the lifetime of the well and therefore must be discounted by the cost of capital. Prices must rise substantially to justify such a withholding decision. Capital investment decision, however, can and will be deferred. This circumstance also explains why peak winter demand is largely met with gas storage facilities, both at the pipeline source and delivery points. These (almost all underground facilities) are capable of rapid fill and extraction, filled in the warm months and emptied in the winter.

circumstances of the ending raised concern and the question was posed as follows: will gas prices slowly rise to a level which will support enough drilling to maintain supply? Or will there be a "hard landing", a substantial spike in prices, perhaps permanent?

Analysts predicted that prices would rise because they concluded that exploration and production was rapidly becoming less productive: the easy gas had been found and new drilling was bound to be less productive. McKinsey, Cambridge Energy, and EIA came to this same conclusion. In part, this conclusion resulted from a simplistic analysis (McKinsey) that was simply wrong, and a proper EIA analysis which underestimated the potential of technology improvement.

McKinsey based their prediction on a strong observation that gas found per foot drilled had declined dramatically and they took this to be a pure technical phenomenon. But their data, inevitably delayed by several years, reflected a period in which low price expectations were succeeded by high. Drilling had increased dramatically precisely because producers expected to make profits from much smaller target reserves. Finding per foot drilled naturally declined. EIA simply took a very cautious view of the potential for technology improvements.[5]

Producers responded to these price projections by increasing drilling in anticipation of higher prices. Further, as has happened episodically in gas production, a new source of relatively cheap gas was found: coal bed methane. This resource had been known to exist but producers only developed the techniques to recover it in the late 1980's. The prediction of higher prices was to some degree self UN-fulfilling. The hard landing never occurred.

This extended period of low prices led to increased demand, especially in the power generation sector, and prices rose. This led to another supply scare. One response to the higher prices was the building of and planning of more LNG facilities in the last decade. EIA projected in 2008 that the U.S. gas market would require increasing supplies of LNG, 1000 BCF per year in 2010, and steadily increasing after that. However, U.S. supply also responded to those prices, and LNG imports peaked at 100 BCF in July 2007, and had fallen to 36 BCF in July 2010.[6] LNG was simply displaced by cheaper U.S. produced gas.[7]

HOW TO THINK ABOUT NATURAL GAS RESOURCES

A good metaphor for the gas resource base is a pyramid.

The largest and easiest resources to access sit near the top, but are limited in quantity and are least expensive to recover. Substantial lower "quality" resources are known to exist but are too expensive to recover with existing technology at a given time or no known techniques for recovery are available. That there are such substantial

---

[5] Analysts attempting to forecast natural gas resources are confronted with a dilemma. Predicting technology improvements involves speculating on unknown unknowns. If one forecasts based on what is readily known, then one is almost sure to underestimate the impact of technology. Further, innovations in technique are almost by definition impossible to anticipate. If they could be anticipated they would already be in use.

[6] EIA Data from the 2008 *Annual Energy Outlook*

[7] LNG requires very large capital investments in liquefaction and regasification facilities. At present U.S. prices, producers of LNG seek other markets and it is unlikely many planned facilities will be built.

deposits is no great surprise given that methane is one of the simplest molecules and is also among the most common in the universe.

Recovery technology, however, does not remain static, and the lower quality resources provide an incentive for advancing technology. Further technology development has the incentive of recovering more from producing formations, and more conventional resources.

Resource extraction may be thought of as a race between the declining quality or inaccessibility of resources and improved technology and techniques. That's a race that technology has been winning over the last 40 years at least.

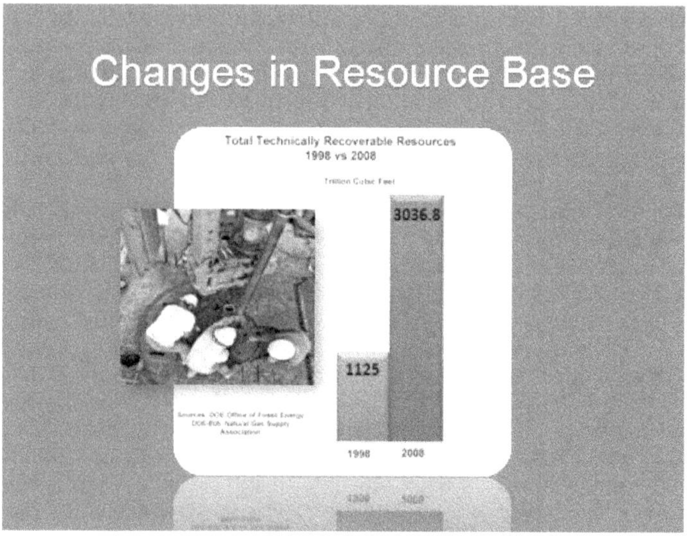

But the existence of the less accessible reserves constitutes an incentive to develop technology to reduce costs, improve recovery, and find innovative techniques to recover what was previously too costly to find and produce. Coal bed methane was known to exist but a method of recovery was wanting. Low permeability (tight) formations were well known to exist, but were only opened to production with the development of fracking. Shale gas formations were well known to exist, but it is the combination of fracking and horizontal drilling that has made that accessible at low prices. Given this view, the constraint on natural gas production is not the resource base. That base is huge in comparison to today's usage. Rather the limit is technological, and the technology is constantly getting better and also advancing through breakthroughs which open large new supplies (tight gas, coal bed methane and shale gas). Technology and resource depletion are in a race, and so far technology has been winning.

The known remaining resource base is huge in comparison to even these large estimated technically recoverable reserves. Those sit atop potential resources of geopressuirzed gas which are thought to range from 5000 to 49000 TCF and methane hydrates from 7000 to 73000 TCF. These are NOT currently technically recoverable, but who is to say that in the unlikely event they were ever needed that technology to recover them could come to the fore.

Technically Recoverable and Potential Resources

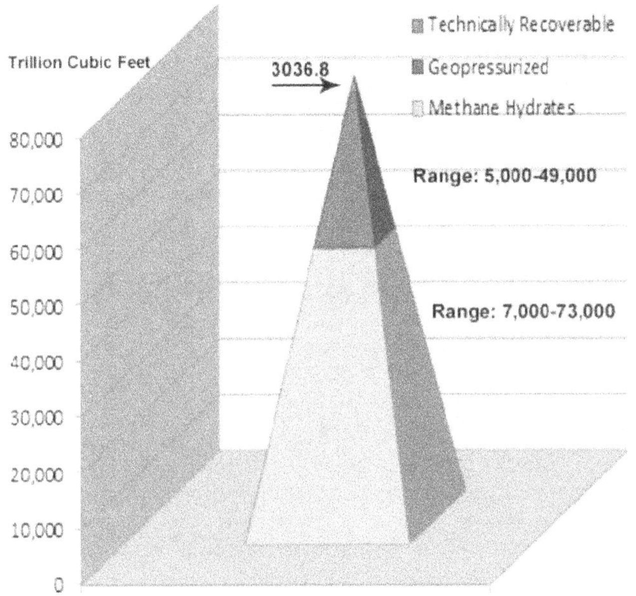

Spurces: DOE-EIA, DOE-Office of Fossil Energy
Natural Gas Supply Association

ENVIRONMENTAL IMPACTS OF UNCONVENTIONAL NATURAL GAS DEVELOPMENT

THOMAS P. BALLESTERO
University of New Hampshire, Department of Civil Engineering
Durham, USA

INTRODUCTION

Technological developments in drilling have resulted in more economic development of deep geologic formations for natural gas. This has led to a dramatic increase in drilling formations worldwide. As with any natural resources extraction practice, there are environmental consequences. The impacts of unconventional natural gas development is explored and discussed, particularly with those facets unique to unconventional natural gas development.

THE WELL SITE

In order to remove natural gas from these tight geologic formations, wells must be placed in the ground that reach the formation deep below the ground surface. The very first step in the drilling process is to create/construct the well pad. This land area can cover 2 hectares or more. The ground at the well pad is often leveled and then fenced off. Gravel parking areas are then established for: onsite offices; parking; trailer, supplies, and equipment storage; tanks, and travel lanes. In ground lagoons may also be constructed to store water or waste liquids.

DRILLING

The drill rig is a conventional machine capable of great depths as well as horizontal drilling. The drilling fluid, although water-based, contains many different additives (Table 1). Some of these additives are benign, some toxic, and some have no toxicity characteristics yet established. However these same additives are used in conventional deep well drilling as well. When the vertical portion of the well is completed, the horizontal portion is then drilled. One vertical wellhead can possess various horizontal legs, much like the spokes emanating from the axle of a bicycle wheel. The vertical and horizontal well portions are lined with metal pipe that is sealed to the formation by a cementitious grout.

HYDRAULIC FRACTURING

When the well construction is complete, the horizontal portion of the metal well casing is perforated along its length. Fluid is then pumped into the well along discrete zones of the perforated horizontal portions. The fluid pressure is increased until it exceeds the strength of the rock just outside of the well. Typically the weakest portion of the rock is at existing fractures, of any scale, and the hydraulic (fluid) pressure emanating from the well causes these fractures to first open, which then may result in the lengthening and widening of the existing fracture. The rock fracturing continues until the pressure is relieved by operators or by fracture extension to more permeable zones. In a perfect elastic response, as soon as the fluid pressure is relieved, the fracture would close. To prevent this, a proppant is used, often sand, to hold the fractures open

Table 1. Drilling Fluids and Additives for Deep Well Drilling

Air
Water
Foam
Polymer (natural and synthetic)
Bentonite clay
Potassium formate
Diesel fuel
Synthetic oils
Calcium carbonate
Ground cellulose
Asphalt
Uintahite (natural asphalt)
Glycols
Corrosion inhibitors

Table 2. Types of Additives Used in Drilling Unconventional Natural Gas Wells.

Additive Type	Principal Compound	Purpose
Dilute acid	Hydrochloric Muriatic	Dissolve minerals Initiate cracks in rock
Biocide	Glutaraldehyde	Minimize microbial growth
Corrosion inhibitor	N,n-dimethyl formamide	Minimize/prevent corrosion
Breaker	Ammonium persulfate	Delay breakdown of gel polymers
Crosslinker	Borate salts	Maintain fluid viscosity at high temperatures
Friction reducer	Polyacrylamide Mineral oil	Minimize friction between fluid and drill stem
Gel	Guar gum Hydroxyethyl cellulose	Thickens fluid to buoy cuttings or proppant
Iron Control	Citric acid	Minimize/prevent deposition/formation of metal oxides
Oxygen scavenger	Ammonium bisulfate	Minimize/prevent corrosion by oxidation
pH adjustment	Potassium carbonate Sodium carbonate	Maintain effectiveness of other compounds/ additives
Proppant	Silica quartz sand	Hold fractures open after hydrofracturing pressure is released
Scale inhibitor	Ethylene glycol	Reduce/minimize deposition on drill stem
Surfactant	Isopropanol	Increase viscosity of drilling fluid

The hydraulic fracturing itself can require 2 to 15 million liters of water. In the vast majority of wells, this water is trucked to sites, and then stored onsite. When the hydraulic fracturing is completed because there is now a direct low pressure connection (the well) to the subsurface, fluids (including the natural gas) flow back up the well.

Although quantities vary widely, anywhere from 20% to 200% of the water used in fracturing the well flows back after fracturing. This backflow water is initially concentrated with the chemicals utilized in the hydraulic fracturing process. Over time, the water quality of the backflow water reflects that of the in situ formation groundwater characteristics.

CONTROL VOLUME AROUND A DRILLING PAD OPERATION

Viewing the drilling and hydraulic fracturing operation in a control volume approach, all elements of the operation that must cross the control volume boundary may be accounted for. Those not uncommon to drilling in general will not get further discussion herein. Those elements specific to hydraulic fracturing will be further discussed as to their potential for environmental concern. Not unique to hydraulic fracturing operations are: the personnel, equipment, the primary well construction, and the drilling fluids. Unique to hydraulic fracturing are: the hydraulic fracturing chemicals, the degree of water trucked to the site, storage of chemicals onsite, the volume of water stored onsite, the quantity of waste fluids trucked off site, and the cumulative impacts of fracturing large regions o a formation.

CHEMICALS

For the chemicals listed in Table 2, some do not yet have laboratory analytical methods yet perfected, some are suspected toxins but have no established toxicity, and some have no established risk assessment. Given the great volumes used, it may be imprudent for widespread use of these chemicals until more is known about them. Even if some are used at low concentrations during hydraulic fracturing, the fact that millions of liters of fracturing fluid is used in the well means that thousands of liters of the original, high-strength chemical must move on roadways and be stored onsite.

FRACTURING ROCK

Due to the nature of in situ stresses in the ground, below 1,000 meters hydraulic fracturing generally creates vertical fractures. Although it is not in the industries best interests to hydraulically fracture beyond the natural gas zone, hydraulic fracturing itself is not an easily controllable process. When viewing the vertical geologic column in total and considering that groundwater for potable uses is located in the uppermost part and the natural gas zones are far below: in the natural state the geology inhibits the vertical movement of natural gas and deep groundwater. Hydraulic fracturing increases the potential communication between these near surface and deeper zones.

NATIVE GROUNDWATER

The native groundwaters at depth naturally often possess nonpotable water quality resulting from the fact that the water has been in contact with the host rock for incredibly long times. Even had no drilling or hydrofracturing chemicals been employed to construct the well, these deep groundwaters would require significant amounts of treatment to make them potable. In addition, they could not be allowed to flow into surface waters without serious consequences to the receiving water ecosystems. Table 3 lists the common compounds naturally occurring in deep groundwaters. Most of these compounds have regulatory limits set for acceptable concentrations in drinking waters, finished (treated) waters, primary contact, and ecosystem health. The native groundwater is mixed in with drilling fluids and hydrofracturing fluids, and all three fluids constitute the wastewater flow from the well site.

Table 3. Natural Compounds in Deep Groundwater Rendering the Water Unpotable. Ranges of concentrations are given for some of the compounds.

Surfactants
Organics
Total Dissolved Solids (30,000 to >230,000 mg/l)
Chloride (15,000 to >110,000 mg/l)
Metals
Barium
Strontium
Selenium
Naturally Occurring Radioactive Materials (NORM)
Radium 226, 228
Gross alpha
Gross beta
Uranium
Radon
High pH
Bromide
Hydrogen sulfide
Ammonium, iron
Manganese
Arsenic
Selenium
Total organic carbon (TOC)

BACKFLOW WASTEWATER

During the drilling and immediately after hydraulic fracturing, the fluid that comes out of the well is termed "production water/fluid". This fluid tends to be highly concentrated with the hydraulic fracturing water and additives. Over time, this fluid is flushed out of the well by the native groundwater and natural gas flowing into the well. Over the long term, the fluid flowing out of the well is termed "backflow water" and starts to mimic the quality of the native formation water. All of these fluid streams constitute the wastewater stream from the site. This stream requires handling, transport, and treatment/disposal.

ONSITE STORAGE INTEGRITY

All fluids/chemicals moved on or off the well site have some form of storage. This includes clean water brought onsite, the chemical additives, and the wastewater. The storage systems vary from chemical containers, to tanker trucks, to lagoons, and each form of storage has its respective potential for leakage. Some storage systems, especially for the more toxic chemicals have redundancy built into their containment, while others do not. Leakage would immediately affect local groundwater and possibly nearby surface waters.

DRILL CUTTINGS AND RESIDUALS

Little is mentioned concerning the solid residuals from the drilling and hydraulic fracturing process. At a minimum, 200 to 400 cubic meters of material is removed to create the well. In addition, any of the solids that settle out of the production water increases this amount. The various rock encountered at each site each have specific chemical and mineral compositions. However a critical environmental condition is that below ground these solids were not exposed to oxygen and typically in an anaerobic environment. Once brought to the ground surface, exposure to the atmosphere can accelerate dissolution of certain minerals that can then lead to water quality problems.

WELL SEAL

When the well is drilled, solid pipe is put into the drill hole in order to hold the hole open. This solid pipe (the well casing) is sealed to the formation in the annular opening between the casing and the rock. Cement, clay, or mixtures of the two are often used to create this seal. The seal prevents the vertical hydraulic communication between formations that would not occur had the well not been drilled. Although well seals are routine, an unknown facet is the longevity of the seal. The well could be abandoned, for natural gas purposes after a few decades; however the seal is expected to last forever.

TRUCKING

The 2 to 15 million liters of water per well brought to the thousands of sites in a region, and adding to that trucking total an equivalent amount or more of production and backflow wastewater, requires thousands of trucks to move these liquids. Truck sizes cannot all be large tractor trailer rigs because many of the roads where well pads are situated cannot handle such loads. These fully loaded vehicles place an enormous burden on local transportation infrastructure in terms of load (weight on bridges, road surfaces, etc.), traffic congestion (due to the slower speeds of the trucks), and visibility (driver sight distance). Added caution exists for the wastewater and chemical trucking given risks of accidents.

WELL ABANDONMENT

Ultimately, when the natural gas supply from a well falls below that economically feasible, the well will be either temporarily capped or permanently abandoned. Temporary capping maintains the direct connection in the well from the surface to the fracture zone. As long as the well casing and well seal remain intact, this should not be problematic. If the well is to be abandoned, it is filled with a cement and/or clay grout. Ultimately infrastructure ages, and although the abandonment procedure is intended to be permanent, steel corrodes and grout cracks. The long term ability of well abandonment to survive does not have sufficient field data to understand its veracity. A poorly abandoned well could create short-circuiting of formations, thereby allowing poorer quality groundwater to reach drinking water aquifers.

TREATING THE HF WASTEWATER

There are various fates for the wastewater generated at hydraulic fracturing sites. Table 4 lists the possibilities. Some options in this list are no longer used in the United States (land application, direct discharge into surface waters) due to the recognition now that such practices were causing environmental degradation.

Table 4. Treatment/Disposal options for Hydraulic Fracture Wastewater

Option
Subsurface injection well
Used to hydraulically fracture other wells
Industrial treatment facility
Brine treatment
Thermal evaporation
Land application
Publicly Operated Treatment Works (POTW)
Direct surface water disposal (dilution)

All of the Table 4 options require the wastewater to be trucked. For deep well injection the looming issue is the capacity for this practice. Increased capacity (additional injection wells) will have to be developed and permitted to accommodate the volumes of water from the hydraulic fracturing industry. The use of the wastewater from one well, especially the flowback water, is a more recent innovation. Ultimately though it becomes a Ponzi scheme in that new gas wells need to be continually developed otherwise there is no market for the wastewater. So although this is a good resolution, it is only an intermediate solution that does require other long term options.

Treatment of the wastewater at industrial wastewater treatment facilities is an expensive option as the primary methods for treating the wastewater is by brine treatment or thermal evaporation. Both processes are energy intensive and ultimately have hazardous waste residuals of which to dispose. New facilities are being constructed and permitted to meet the demand.

Treatment at publicly operated treatment works (POTWs) is slowly being phased out as most POTWs do not remove dissolved compounds: such compounds basically move through the facility and are first diluted by the other incoming wastewater and then further diluted by the receiving water body. It is possible for the hydraulic fracturing wastewater to "upset" the biological processes in a POTW if too much of the flow through the POTW is hydraulic fracturing wastewater.

SAFETY

The potential for accidents on the roads and/or at the well head, especially those that involve the hazardous chemicals and fire, raise risks for emergency personnel and first time responders. Since well pads can often be in rural areas, these safety personnel require proper training in order to understand and effectively respond to site emergencies.

REGIONAL/CUMULATIVE ASPECTS

The geologic formations targeted for hydraulic fracturing cover large geographic regions around the globe. In some regions, thousands of wells per year are now being constructed. Therefore there is a real concern for cumulative impacts of unconventional natural gas development in a region. Some of the cumulative impacts are discussed here.

DRINKING WATER

A few of the disposal options ultimately result with effluent flowing into receiving streams. This can have dramatic consequences to downstream drinking water utilities. These utilities take the water from the streams, improve their quality to drinking water standards, and then supply the water to the population. For utilities serving large populations, a significant amount of testing is required. If hydraulic fracturing wastewater or treated wastewater flows into the receiving water, this may increase the treatment, protocols, and testing burdens on water utilities. For example, if NORM are in the hydraulic fracturing water, then utilities may then be required to test for NORM. If there is high TOC resulting from hydraulic fracturing, this may result with the water having higher trihalomethane formation potential. The bromide in many of the hydraulic fracturing waters can ultimately result in more disinfection by-products (trihalomethanes, haloacetic acids, total organic halogen, and other halogenated organic compounds). For some of the hydraulic fracturing compounds, there is insufficient Maximum Contaminant Level delineation in the Clean Water Act. As stated previously, but especially true for drinking water, for some hydraulic fracturing chemicals there are poorly understood risks.

WATER ACQUISITION

The quantity of water required in regions undergoing natural gas development surpasses the flow of many small streams. At a regional level this is simply a new demand on water resources. However it is not a priority demand. In the summer of 2011, most of the Commonwealth of Pennsylvania was in drought, and hydraulic fracturing water withdrawals were suspended on many streams. Since these withdrawals are permanent removal of water from local aquatic systems, there is a concern that instream flow needs may be impaired at low stream flow times. At the same time, reduced stream flow in and of itself can create other habitat concerns such as aquatic organism passage, lower dissolved oxygen, and higher stream temperatures. These are site specific in nature, but collectively result in regional consequences that can more than offset many of the conservation programs that, over the last few decades, have been diligently working towards restoring native species.

MULTIPLE HYDRAULIC FRACTURE WELLS

One well in a formation may not produce measurable environmental change, however multiple wells fracturing a formation over a regional scale do have such potential. The rock formation in its natural state trapped natural gas and groundwater. It acted as a barrier to the movement of fluids and contaminants. Hydraulically fracturing this same formation creates a lateral highway for fluids and contaminants. The stimulated fractures generally propagate vertically and this increases the hydraulic connection to the near surface aquifers and water bodies. Overlayed on this is the fact that there already exist locations of large fractures in these rocks. Therefore whether it is the lateral formation connection or the vertical fracture propagation, hydraulic fracturing has the potential to intersect some high permeability zones underground. The detailed features of the subsurface environment are largely unknown. Geophysical techniques are used to make estimates of what exists out of sight; however there is a significant degree of interpretation and uncertainty in these estimates. Hydraulic fracturing is not intended to expand fracture zones to high permeability zones, but the probability exists that such subsurface interconnections will occur.

ENVIRONMENTAL JUSTICE

Unconventional natural gas development occurs primarily in rural or existing mining/oil/gas regions. On a regional scale there are many economic benefits to the new developments. At the same time, this places increasing pressure on environmental services. The rural areas do not necessarily possess the political or economic clout to always decide for the public good. It is unfair to allow one sector of the population to bear the brunt of national policy without just compensation and consideration to the ultimate ramifications.

SUMMARY AND CONCLUSION

Unconventional natural gas development has increased dramatically through the use of the hydraulic fracturing technology. As with any other resource extraction, there are environmental consequences. Most of the potential environmental issues raised in this paper can be overcome by: regulatory oversight, industry attention to proper techniques, monitoring, and the development of standards and protocols. There is a much more heightened sense of environmental stewardship today that there was just one century ago, which means today we ask the questions that were not asked previously. If we originally did ask these questions of say coal extraction, we might not be faced today with the hundreds of stream kilometers affected by acid mine drainage. As with any practice, there is no global yes or no, but rather that there are locations where the practice will not result with environmental consequences and other locations that will. The technology exists to address environmental concerns of hydraulic fracturing. Whether we employ the right techniques and effort is a decision that will be made subjected to human prerogatives. As much as we strive to avoid mistakes, errors and accidents will happen. No one planned the Exxon Valdez spill or the Deepwater Horizon spill. Therefore applying the best technology, making prudent decisions, and learning from our mistakes are paramount.

SESSION 5

CLIMATE

FOCUS: COSMIC RAYS AND CLIMATIC PROCESSES

COSMIC RAYS AND CLOUDS: EXPERIMENTS AND OBSERVATIONS

HENRIK SVENSMARK
Center for Sun-Climate Research, Danish National Space Center, Technical University of Denmark, Copenhagen, Denmark

An inscription on a plaque on the wall of the secretariat room at the Centro Ettore Majorana sums up the discovery of cosmic rays:

Here in the Erice maze
Cosmic rays are all the craze
Just because a guy named Hess
When ballooning up found more not less.

More ionization, that is to say. When Victor Hess made his ascents by hydrogen balloon, in 1911-12, going eventually to more than 5000 metres, he expected to escape from the natural radioactivity on and near the ground. But the higher he went, the faster the electric charge dispersed from his gold-leaf electroscope. The important conclusion at the time was that ionizing radiation was coming from outer space. But 100 years later, as I shall explain, the fact that Hess found relatively weak cosmic rays at low altitudes turns out to be very significant when considering the terrestrial effects of cosmic rays.

The most profound questions with the most surprising answers are often the simplest to ask. One is: why is the climate always changing? Historical and archaeological evidence of global warming and cooling that occurred long before the Industrial Revolution, and the geological traces of far greater variations before human beings even existed, require natural explanations. Regrettably, the global sense of urgency about understanding current climate change does not extend to profound investigations of natural influences. As a result, a small group of physicists has enjoyed a near monopoly in one unfashionable but highly productive line of research for 15 years.

Here the surprising answer about those never ending natural changes of climate is that galactic cosmic rays, atomic particles coming from the supernova remnants left by exploded stars, appear to play a major part. Although unseen, unfelt, generally harmless and therefore often ignored, cosmic rays are as much a part of our environment as sunlight or rainfall. About two particles go through our heads every second, but luckily the air is a good shield. When highly energetic protons and other atomic nuclei hit the Earth's outer atmosphere, traveling close to the speed of light, they react with nuclei in the gas molecules to make all sorts of atomic particles, most of which are stopped high overhead. Survivors reaching the ground include neutrons, which are counted at many ground stations to gauge the varying influx due to solar activity or lack of it. But neutrons have virtually no ionizing effect. Charged muons are the survivors mainly responsible for ionizing the air at low altitudes, and it is only thanks to the time dilation of Special Relativity that they live long enough to play this all-important role.

This ionization of the air by penetrating cosmic rays helps to form the aerosols, cloud condensation nuclei, required for water droplets to condense and create clouds in the lowest 3000 meters of the atmosphere. Here is where the relative weakness of cosmic rays near the Earth's surface becomes significant. At higher altitudes there is always plenty of ionization, whether the cosmic ray influx is high, when the sun is least active, or low when the solar influence is strongest. But at low altitudes, the solar effect is much stronger. Because of the scarcity of ionizing radiation, adding to it or subtracting from it becomes more influential.

As low-altitude clouds exert a strong cooling effect, increases or decreases in the cosmic ray influx and in cloudiness can significantly lower or raise the world's mean temperature. This is our central hypothesis, and it provides a ready explanation for the frequent coolings and warmings experienced by our ancestors over many thousands of years—most recently in the Medieval Warm Period, the Little Ice Age, and the Modern Warm Period of the 20th Century. Over 11,500 years, as revealed in the seafloor record,[1] repeated cold spells coincided with times of high cosmic-ray influx, recorded by high counts of radioisotopes, carbon-14 and beryllium-10, made when cosmic rays react with terrestrial atoms. Variations in solar irradiance measured by satellites suggest that they would be too small to explain the cooling and warming. An amplifier is needed, and the hypothesis is that the cosmic rays are not just a proxy for solar activity but a causal agent. The cold spells occurred whenever a magnetically weak sun failed to repel as many of the cosmic rays coming from the Galaxy as it does when it is most active.

EXPERIMENTAL RESULTS

Such explorations through time would be less meaningful if there were not clear evidence from experiments and observations that cosmic rays really do affect the clouds and the climate, here and now. Our experimental work has revealed the existence of a microphysical and chemical mechanism by which cosmic rays help to produce aerosols. It is a matter of ion-induced nucleation of micro-aerosols, small clusters of sulfuric acid and water molecules, which can then grow into the cloud condensation nuclei on which water droplets form.

In 2005, our team completed the world's first laboratory experiment on the role of cosmic rays in aerosol formation in the basement of the Rockefeller Complex on Juliane Maries Vej in Copenhagen.[2] We called the experiment SKY (*cloud* in Danish) and used natural cosmic-ray muons coming through the ceiling, supplemented by gamma-ray sources when we wanted to check the effect of increased ionization of the air. A 7-m^3 reaction chamber contained purified air and the trace gases that occur naturally in unpolluted air over the ocean. Ultraviolet lamps mimicked the Sun's rays, and instruments traced the chemical events. We were surprised by how quickly the ionizing radiation worked, in a split-second, suggesting that electrons released in the air act as efficient catalysts for building the micro-aerosols.

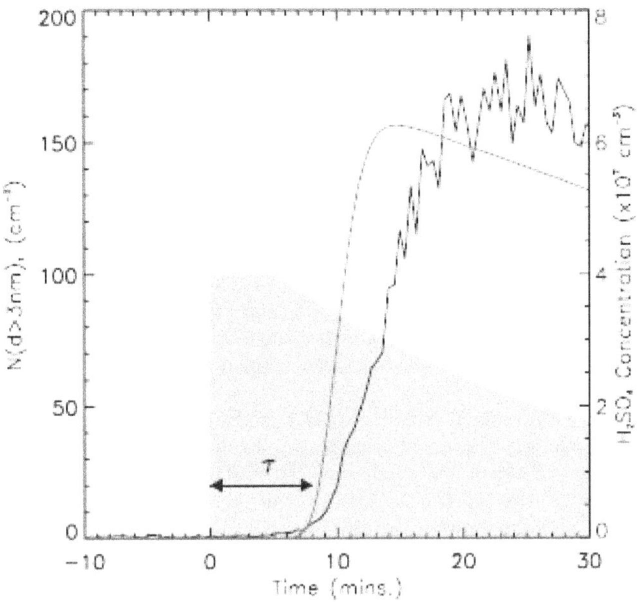

Fig. 1: A typical result from the SKY experiment showing the first half-hour of a four-hour run. It begins at t=0, when the ultraviolet lamps were activated to form H_2SO_4 molecules from SO_2 present in the chamber. This mimics the natural formation of H_2SO_4 from sulfur compounds in the real atmosphere. The ionization of the air by gamma-rays, simulating cosmic rays, promotes the formation of micro-aerosols—clusters of H_2SO_4 and H_2O molecules that become detectable when they reach 3 nanometers in diameter. The gray area and right-hand scale show the H_2SO_4 concentration, which reaches a maximum of 4 107 cm^{-3} before decaying. The gray curve and left-hand scale give the count of micro-aerosols concentration, peaking after 20-30 minutes. Subsequently the count declines to ~0 after four hours, because of exhaustion of H_2SO_4 and the loss of micro-aerosols to the chamber walls. The red curve shows a simple model with parameters that approximate to the micro-aerosol observations. The delay τ is the time taken for the aerosol concentration > 3 nm to reach a threshold of 10 cm^{-3}.

A small-scale (50-liter) version of the experimental chamber confirmed the SKY results at sea-level, before going 1.1 km underground in the deep Boulby Mine in England. Called SKY@Boulby its purpose was to investigate aerosol production in the absence of cosmic ray muons, blocked by the overlying rocks. With the equipment wrapped in lead, radioactivity was effectively excluded from the experiment as well, allowing the team to investigate aerosol formation both in the absence of ionization and with controlled ionization using radioactive sources, while experimenting with various concentrations of vapors.

We then adapted the SKY@Boulby chamber for exposure to a beam of energetic electrons from the ASTRID accelerator at Aarhus University, simulating cosmic rays. Again we detected the production of micro-aerosols, and we were able to show that the results were no different from what we obtained using gamma-rays to ionize the air.[3] Shortly after publication of the Aarhus results, independent support for our experimental findings came in the summer of 2011. A large international collaboration at the European Organization for Nuclear Research, CERN, announced findings of the CLOUD experiment which also uses a beam of accelerated particles.[4] These results were very satisfactory for our hypothesis, and confirmed and extended our own experimental results by showing an amplification of the formation of small aerosols by ions by a factor of ten or more.

Although the results of both the SKY and CLOUD experiments point to the existence of a micro-physical mechanism that links ions and aerosol formation, they do not reveal how important the mechanism is in the real atmosphere. An argument advanced against an important influence of cosmic rays on clouds is that any effect may be drowned in the real atmosphere by other sources of aerosols. So an important question is whether a significant fraction of the small size portion of ion-induced aerosols grows to cloud-seeding size. Following our own experiments in 2005, we looked to the real atmosphere for enlightenment.

COSMIC RAYS, AEROSOLS AND CLOUDS IN THE REAL ATMOSPHERE

It turns out that this question of aerosol growth and effects on low-level clouds can be answered with what one can call "natural experiments" where solar explosions (Coronal Mass Ejections) send out plasma clouds that shield the Earth and cause the cosmic ray flux to decrease within half a day. The largest of these Forbush decreases, as they are called, cut the cosmic ray ionization by about 10-15%, and obviously one should look for an impact on the Earth's aerosols and cloud cover following such events.

Our results published in 2009[5] show clear effects in cloud properties following the sudden decrease in cosmic ray ionization. In fact the whole chain from solar activity to cosmic rays ionization to aerosol impact to cloud properties could be observed for the stronger events. The effect is quite large—nearly 6% loss in cloud liquid water averaged over the 5 strongest events. That translates into a disappearance of about 3 billion tonnes of liquid water vanishing from the sky. The water remains there in vapor form, but unlike cloud droplets it does not get in the way of sunlight trying to warm the ocean.

An important preliminary step was to calculate changes in ionization in the lower atmosphere due to each Forbush decrease considered, using responses in about 130 neutron monitors world-wide and in the Nagoya muon detector to compute the changes in the primary cosmic ray spectrum in the Earth's vicinity. To see the effect of cosmic ray loss on aerosols in the atmosphere, we turned to the data from the solar photometers of the AERONET program, widely scattered around the world, which record changes in the intensity of sunlight at different frequencies. With fewer fine aerosols in the lower atmosphere to scatter it, more violet light reaches the photometers at the Earth's surface. As expected, in the aftermath of major Forbush decreases the AERONET data revealed falls in the relative abundance of fine aerosol particles which, in normal circumstances, could have evolved into cloud condensation nuclei.

The effects on low altitude liquid-water clouds were traced using three independent sources of satellite data. The Special Sounder Microwave Imager (SSM/I) observes changes in the cloud liquid water content (CWC) over the world's oceans. The Moderate Resolution Imaging Spectroradiometer (MODIS) on NASA's Terra and Aqua satellites (land and oceans) gives the liquid water cloud fraction (LWCF). The International Satellite Cloud Climate Project (ISCCP) provides data on infrared detection of low clouds (<3.2 km) over the oceans. The main results are summarized in Figure 2.

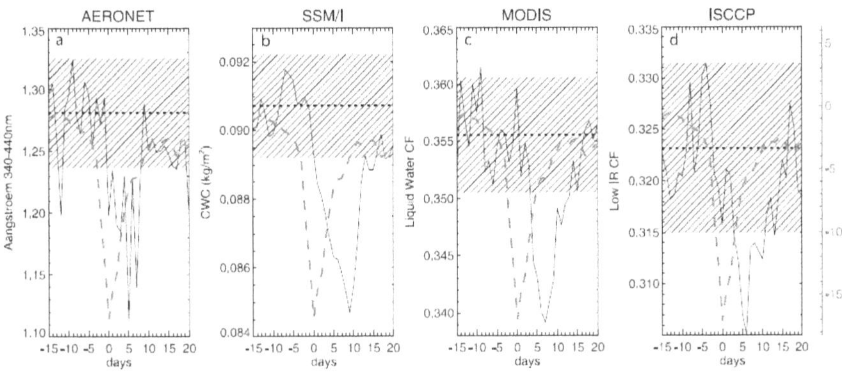

Fig. 2: The evolution of (a) fine aerosol particles in the lower atmosphere (AERONET), (b) cloud water content (SSM/I), (c) liquid water cloud fraction (MODIS), and (d) low IR-detected clouds (ISCCP) is here averaged for the five strongest Forbush decreases that their data sets have in common (i.e. between 2000 and 2005). The broken red curve is the average of cosmic ray neutron counts at the Climax (Colorado) station. In (a) each data point is the daily mean from about 40 AERONET stations world-wide, using stations with more than 20 measurements a day. The broken horizontal lines denote the mean for the first 15 days before the Forbush minimum, and the hatched zones show ±1 sigma for the data, estimated from the average variance of a large number of randomly chosen periods of 36 days of each of the four data sets. Observe that first the count of fine aerosols went to a minimum after about 5 days and the cloud indicators did so a few days later. That small delay was due to the fact that the aerosols in short supply were not yet large enough to act as cloud condensation nuclei.

There have been a few papers on this subject which disagree with our 2009 results, but on close inspection there is no contradiction. We have always emphasized that meteorological "noise" due to the natural daily variability of the weather tends to mask the effects of relatively weak Forbush decreases and the critics have relied on weak events to deny any effect of cosmic ray variations. But the fact that the detectable impact is proportional to the magnitude of the decreases is support for our hypothesis rather than the reverse.

Recent support for our results comes from a study that uses routinely observed changes the diurnal temperature range (DTR) as a proxy for cloud variations.[6] Reductions in cloud cover tend to have a warming effect by day and cooling at night, thereby increasing the DTR. Advantages in this approach are (1) it is not confined to the satellite era for data on cloudiness, and (2) it extends the analysis to continents (where there are most weather stations) in contrast with the primarily oceanic data used in our 2009 report. When Aleksandar Dragić and his colleagues at the Institute of Physics in Belgrade pick out the 13 strongest Forbush Decreases between 1954 and 1995 (with more than 10% reduction in cosmic rays) and look at meteorological data from 184 stations, they find that in the days following the fall in cosmic rays, the DTR in Europe increased on average by 0.6 deg. C—an impressive result.

There is, therefore, strong support for the existence and importance of the experimentally found link between cosmic ray ionization, aerosols and clouds, in response to variations in solar activity. It continues to offer the simplest explanation of why the frequent coolings and warmings during the last 11,500 years have followed solar-induced variations in the cosmic ray flux, through to the Medieval Warm Period and the Little Ice Age. Even in the Modern Warm Period of the 20th Century cosmic rays continued to have a significant impact.

Discussion of the solar influence on climate via cosmic rays is complicated by lesser solar effects due to changes in total irradiance or ultraviolet intensities. An unambiguous perspective comes from a totally different cause of much larger changes in the cosmic ray flux, namely the variation over millions of years in the rate of supernova explosions in the vicinity of the Solar System. Strong evidence links the changing flux to the alternations between "hothouse" and "icehouse" conditions over geological time. But that is another story.

REFERENCES

1. G. Bond et al. *Science*, 294:2130–6, 2001.
2. H. Svensmark et al. *Proceedings of the Royal Society A*, 463:385–96, 2007 (released online 2006).
3. M.B. Enghoff et al. *Geophysical Research Letters*, 38:L09805, 2011.
4. J. Kirkby et al. *Nature*, 476, 429-433, 2011.
5. H. Svensmark, T. Bondo and J. Svensmark. *Geophysical Research Letters,* 36: L15101, 2009.
6. A. Dragić et al. *Astrophysics and Space Sciences Transactions*, 7:315–318, 2011.

THE ROLE OF THE SOLAR FORCING IN THE 20TH CENTURY CLIMATE CHANGE

NIR J. SHAVIV
Hebrew University of Jerusalem, Mount Scopus Campus
Jerusalem, Israel

ABSTRACT

A long list of empirical results strongly suggests that solar variations play an important role in climate change. We begin by trying to understand why such variations are crucial if we are to understand 20th century climate change, and how it is related to the value of the climate sensitivity. The latter is, of course, necessary if we are to predict future climate change. We then review some of the aforementioned evidence, and in particular, the data which can be used to quantify the size of the solar climate link. We will end by building a more consistent picture for 20th century global warming, while mentioning a few words on the expected 21st century climate change.

INTRODUCTION: WHY DO WE NEED TO UNDERSTAND THE CLIMATE ROLE OF THE SUN?

In the standard picture advocated by the IPCC, most of the global warming observed over the 20th century is attributed to the increase in anthropogenic greenhouse gasses (AGHG). Indeed, when one considers the observed increase in temperature and the increase in AGHG, it is very tempting to do so. However, we have to remember that there are many uncertainties—primarily the unknown radiative forcing changes and unknown climate sensitivity—which imply that most of the warming is not necessarily human.

When the Earth's energy budget changes, that is, when the net radiative forcing changes, so does the climate equilibrium. Climate sensitivity is simply the ratio between the temperature change needed to reach the new equilibrium and the changed radiative forcing. Thus, loosely speaking, the temperature change over the 20th century is the product of the changed energy balance, presumably mostly anthropogenic GHG, and the climate sensitivity. It is somewhat more complicated because it takes many decades for the climate system to adjust.

Here comes the problem, we very poorly know the net radiation forcing imposed by humans over the 20th century and we very poorly know the climate sensitivity. It turns out that the Achilles heel for both is clouds. We do not know the net radiative forcing because human activity increased the amount of atmospheric aerosols which "seed" clouds and cool (e.g., IPCC AR4 WG1, 2007). Unfortunately there is a very large uncertainty on size of the effect. Clouds are also very important to the determination of climate sensitivity. The reason is that the climate feedback through clouds, namely, by how much does the cloud cover change when the global temperature changes, is not really known. It was shown by Cess et al. (1989) that it is by far the dominant source of uncertainty. Twenty-odd years later, the situation is virtually the same.

For these reasons, there is no unique prediction for how large the anthropogenic warming should have been over the 20th century—multiplying two very uncertain numbers gives an even more uncertain temperature change.

Because theory cannot uniquely predict the 20th century warming, it can be attributed to human activity only because of *indirect* lines of argument. First, 20th century warming is unprecedented. This was based on the famous "hockey stick". However, it was shown by McIntyre and McKitrick (2003) to be false, as was later corroborated by the climategate e-mails. The second argument is that climate modelers cannot explain the warming without including the anthropogenic contributions to the net radiative forcing, in particular that of the GHGs.

Now we see why the climate role of the sun is so important. Because solar activity increased over the 20th century, if it has an effect on the climate it should have contributed a net positive forcing and it may have been responsible for some of the 20th century warming. This would then diminish the role of anthropogenic activity.

More quantitatively, if the sun has contributed a positive radiative forcing, then the total radiative forcing change over the 20th century is necessarily larger as well. As we shall see below, the sun does have a large effect on the climate, and it is roughly twice as large as the anthropogenic forcing alone. This implies that in order to explain the same observed 20th century warming, we require a climate sensitivity which is only half as large. In fact, the range of sensitivities required to explain 20th century warming is just below the often quoted IPCC range of 1.5 to 4.5°C increase per CO_2 doubling.

Needless to say, a lower climate sensitivity is very important if we are to predict 21st century temperature increase. For a given emissions scenario, such as a "business as usual" one, the warming should be correspondingly smaller.

EVIDENCE FOR A SOLAR/CLIMATE LINK

One of the most interesting aspects of our sun is that it is not entirely constant. The variations that it exhibits appear in the total irradiance of the sun, i.e., primarily in the visible and infrared bands (by as much as 0.1%). But they also appear in components other than the total emitted flux. These include very large *relative* changes in the magnetic field, the sunspot numbers, the strength of the solar wind, and the amount of UV, to name a few.

The basic variation is an activity cycle of about 11 years, which arises from quasi-periodic reversals of the solar magnetic dipole field. On longer time scales (of decades to millennia) there are irregular variations which modulate the 11-year cycle. For example, during the middle ages and during the latter half of the 20th century, the peaks in the 11-year cycles were notably strong, while these peaks were almost absent during the Maunder minimum. On the other hand, eruptions may appear on the time scale of days. Today there is evidence linking solar activity to the terrestrial climate on all of these time scales.

Since the work of Jack Eddy in the 1970's, many empirical results show a clear correlation between different climatic reconstructions and different solar activity proxies on the time scale of decades or longer. Eddy realized that there is a correlation between solar activity and the European climate over the past millennium (Eddy, 1976). For example, the little ice age in Europe took place while the sun was particularly inactive, during the Mounder minimum. The medieval warm period, on the other hand, occurred while the sun was as active as it was in the late 20th century. Since then, many empirical results show a correlation between different climatic reconstructions and different solar activity proxies.

One of the most beautiful results is that of a multi-millennial correlation between the temperature of the Indian Ocean as mirrored in the ratio between different oxygen isotopes in stalagmites in a cave in Oman, and solar activity, as reflected in the cosmogenic carbon 14 isotope (Neff et al. 2001).

Another impressive result over the same time scale is that of Bond et al. (2001), where the solar activity was compared with the Northern Atlantic climate, as recorded on the ocean floor through ice rafted debris. Many other correlations exist at other locations in the world.

One way to see that this solar-climate link is global and it affects the global temperature is to look at *borehole* data (Huang et al. 1997). This reveals that the solar variations give rise to changes which are as large a 1°C between low and active states of the sun.

Over the 11-year solar cycle, it is much harder to see climate variations. There are two reasons for that. First, if we study the climate on short time scales, we find that there are large annual variations (for example, due to the el-Niño oscillation) which introduce cluttering "noise", hindering the observation of solar related signals. Second, because of the large ocean heat capacity, it takes decades until it is possible to see the full effects of given changes in the radiative budget, including those associated with solar variability. It is for this reason that climates of continental regions are typically much more extreme than their marine counterparts.

If, for example, a given change in solar forcing is expected to give rise to a temperature change of ½°C after several centuries, then the same radiative forcing varying over the 11-year solar cycle is expected to give rise to temperature variations of only 0.05-0.1°C or so (e.g., Shaviv, 2005 and references therein). This is because on short time scales, most of the energy goes into heating the oceans, but because of their very large heat capacity, large changes in the ocean heat content do not translate into large temperature variations.

Nevertheless, if the global temperature is carefully analyzed (for example, by folding the global temperature of the past 120 years over the 11-year solar cycle), it is possible to see variations of about 0.1°C in the land temperature, and slightly less in the ocean surface temperature (e.g., Shaviv, 2005 and references therein).
Moreover, as we shall see in the next section, it is in fact possible to see the large amount of heat going into the oceans every solar cycle.

We therefore conclude that the sun has a large effect on the climate. Although the link itself is not the topic of this article, it should be mentioned that the leading contender is through solar modulation of the cosmic ray flux reaching the Earth (Svensmark, 1998). This is now supported by a range of both empirical and experimental results.

For example, on the time scale of days, the sun can undergo flaring activity which is caused from the reconnection of magnetic loops. These flares are accompanied by a strong solar wind "gust" which later causes a decrease in the cosmic ray flux for several days. If the cosmic ray flux has an effect on clouds, then cloud properties should change following these events, known also as Forbush decreases. Several results indicate that clouds are affected during Forbush decreases. In particular, Svensmark et al. (2009) have shown the cosmic ray mechanism at work during these decreases. Not only was a cloud decrease signal clearly observed, the intermediate step of affecting the aerosol size distributed was detected as well.

QUANTIFYING THE SOLAR CLIMATE LINK

Having established that the sun has a large effect on the climate, we can proceed to quantify the size of the link. In particular, we are interested in the radiative forcing associated with solar variability. This is important if we are to assess its role in 20th century climate change.

As mentioned above, looking for the temperature response over the 11-year solar cycle is tricky because of the large heat capacity of the oceans and the climate variability on short time scales. Nevertheless, we can use the large ocean heat capacity to our advantage, since it implies that short term variations in the energy balance will translate into heat content variations in the Oceans without affecting other components, that is, without any internal feedbacks operating. This implies that the 11-year cycle variations in the heat content in the oceans can be straightforwardly used to calculate the radiative forcing imposed by the sun.

Presently, the ocean heat content can be derived from three independent data sets. First, there is the direct measurement of the heat content, as measured by small temperature changes down to depths of 700m since 1955, over the whole ocean. The second data set is the surface sea temperature, while the third is that of tide gauge records of the sea level. Each of the three data sets has its advantages and disadvantages. For example, the data over the ocean volume are the most direct measurements of the heat content, however, because the oceanic volume was only poorly covered (at different locations and different times), there are problems of incompleteness. The sea surface data is of higher quality, but one requires some modeling in order to translate the small sea surface temperature variations to changes in the heat content. The third record, of tide gauges, is of relatively high quality. It suffers however from contamination. Although the dominant contribution to sea level change rate over the several year time scale is that of thermal expansion, some contribution is due to changes in the total land ice volume. The tide gauge record can be seen in Figure 1.

Nevertheless, all the three records consistently reveal that the amount of heat going into the oceans every solar cycle is about 6 to 7 times larger than the changes expected from just the variations in the total irradiance (Shaviv, 2008). In absolute terms, it is a variation of about 1 W/m^2.

The number we obtain this way is very interesting. First, the leading contender to explain the solar/climate link is through cosmic ray flux modulation of the atmospheric ionization, which in turn affects the cloud cover. This implies that the radiative forcing change associated with the cloud cover variations over the 11-year solar cycle should too be of order 1W/m^2—the amount of heat going into the oceans. Within the radiative forcing uncertainties of clouds, this is indeed the observed variations (Shaviv, 2005). The different estimates for the forcings are summarized in Figure 2.

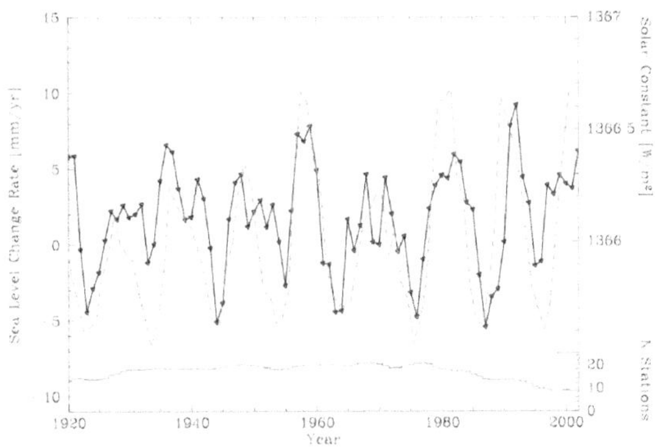

Fig. 1: The sea level change rate. On short time scales, it originates predominantly from changes in the oceanic heat content. Using this data, the derived changes in the energy budget over the solar cycle correspond to 1 W/m², almost an order of magnitude more than can be expected from changes in the solar irradiance (see Shaviv, 2008).

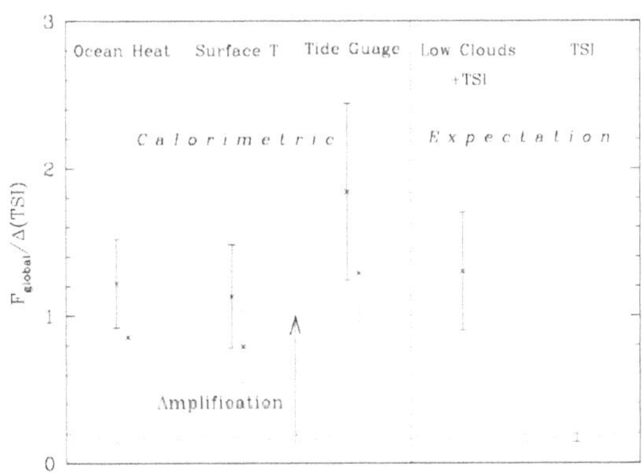

Fig. 2: Summary of the "calorimetric" measurements and expectations for the average global radiative forcing (Shaviv, 2008). The solid error bars are the global radiative forcing obtained while assuming that similar forcing variations occur over oceans and land. The dotted error bars assume that the radiative forcing variations are only over the oceans. These measurements should be compared with two different expectations. The TSI is the expected flux if solar variability manifests itself only as a variable solar constant. The "Low Clouds+TSI" point is the expected oceanic flux based on the observed low altitude cloud cover variations, which appear to vary in sync with the solar cycle. Since the TSI cannot explain the observed flux going into the ocean, an amplification mechanism is required.

Second, because the forcing variations are large, it is comparable to the net anthropogenic changes in the radiative forcing over the 20th century. This implies that one has to consider solar variability when trying to understand 20th century global warming.

20TH CENTURY CLIMATE CHANGE—THE FULL PICTURE

Now that we have quantified the size of the link, we can proceed to estimate the effect that the sun had over the 20th century. Since the increased solar activity between the first half and second half of the 20th century is comparable to the variations between solar minim and solar maximum, over the 11-year cycle, we can expect the radiative forcing to be similar, i.e., around 1 W/m^2. For comparison, the IPCC estimates the net anthropogenic forcing to have been 0.6 to 2.4 W/m^2 (IPCC AR4 WG1, 2007), but the solar forcing that modelers typically include is only the changes in the solar irradiance, which are of order 0.1 to 0.2 W/m^2.

To better understand the climate change, we can employ a simple "box" climate model, one which includes temperatures for the land, ocean mixed layer, and diffusion into the deep ocean. We can then ask the question, what are the allowed climate variables, including the different couplings, sensitivity, radiative forcings, and so forth which can consistently explain the 20th century global warming. The answer is that if we allow the sun to have contributed more than changes in the solar irradiance, we find that 20th century warming can be much better explained than present global circulation models which exclude a large solar effect. In fact, the residual in the fit between model and observations is twice smaller! This can be seen in Figure 3.

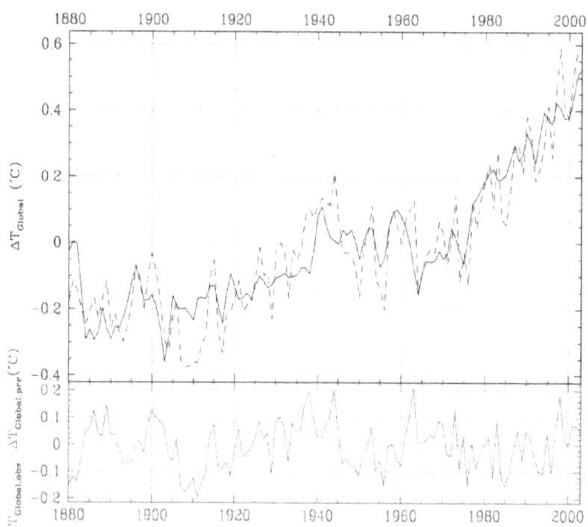

Fig. 3: A comparison between the observed global temperatures and the temperatures modeled using an energy balance model with a diffusive ocean. The small residual, which is twice smaller than obtained in typical global circulation models, can be obtained if we allow the sun to have large effect on the climate, and that the climate to have a low climate sensitivity. (From Ziskin & Shaviv, 2011).

The numbers that this fit gives is a net solar contribution of 0.8±0.4 W/m^2, and a climate sensitivity of 0.95±0.35°C increase per CO_2 doubling. These values are consistent with previous empirical determinations of the solar effect and of the climate sensitivity.

SUMMARY

We have seen that there is ample evidence to prove that the sun has a large effect on the climate. This is important because it allows us to present a much more consistent picture to explain the observed 20th century global warming—one in which model fit the predictions much better. In this picture, the sun has contributed a net radiative forcing which is comparable to the anthropogenic contribution. As a consequence, the same 20th century warming can be explained with a smaller climate sensitivity. It also implies that for a given emissions scenario the predicted 21st century warming should be correspondingly smaller, typically around 1 to 1.5°C, for a "business as usual scenario".

REFERENCES

1. Bond et al. (2001) "Persistent Solar Influence on North Atlantic Climate During the Holocene", *Science*, 294:2130-2136.
2. Cess et al. (1989) "Interpretation of Cloud-Climate Feedback as Produced by 14 Atmospheric General Circulation Models", *Science,* 245:513-516.
3. Eddy, J.A. (1976) "The Maunder Minimum", *Science*, 192:1189–1202.
4. Huang et al. (1997) "Late Quaternary temperature changes seen in worldwide continental heat flow measurements", 24:1947-1950.
5. IPCC AR4 WG1, Solomon et al. (2007) "Climate Change 2007: The Physical Science Basis, Contribution of Working Group I to the Fourth Assessment Report of the Intergovernmental Panel on Climate Change", Cambridge University Press.
6. McIntyre, S. and R. McKitrick (2003) "Corrections to the Mann et. al. (1998) Proxy Data Base and Northern Hemisphere Average Temperature Series". *Energy and Environment* 14:751–771.
7. Neff et al. (2001) "Strong coherence between solar variability and the monsoon in Oman between 9 and 6 kyr ago", *Nature* 411:290.
8. Shaviv, N.J. (2005) "On climate response to changes in the cosmic ray flux and radiative budget", *Journal of Geophysical Research*, 110(A8):8105–8105.
9. Shaviv, N.J. (2008) "Using the oceans as a calorimeter to quantify the solar radiative forcing", *J. Geophys. Res.*, 113:A11101, doi:10.1029/2007JA012989.
10. Svensmark, H. (1998) "Influence of cosmic rays on earth's climate", *Phys. Rev. Lett.*, 81:5027–5030.
11. Svensmark, H., Bondo, T. and Svensmark J. (2009) "Cosmic ray decreases affect atmospheric aerosols and clouds", *Geophys. Res. Lett.*, 36:L15101.
12. Ziskin, S. and Shaviv, N.J. "Quantifying the Role of Solar Radiative Forcing over the 20th Century", accepted to *Astrophysics and Space Science*.

PLANETARY TEMPERATURES/CLIMATE ACROSS GEOLOGICAL TIME SCALES

JAN VEIZER
Department of Earth Sciences, University of Ottawa
Ottawa, Canada

Based on present understanding of stellar evolution, our planet should have been a frozen ice ball up to about 1 billion years (Ga) ago.[1] Yet, geological evidence argues for presence of running water as far back as we have a record, up to 3.8 or even 4.2 Ga ago. This limits planetary surface temperatures to 0–100 degrees Celsius range. The usual solution is to assume a massive greenhouse, mostly by carbon dioxide and/or methane, that would compensate for the "faint young sun".[2] The carbon isotope record, however, suggests that the fundamentals of the planetary carbon cycle—and life—may have been established as early as 3.5 Ga ago, an observation difficult to reconcile with the postulated levels of greenhouse gases in the atmosphere. These constraints also mitigate against huge swings in their concentrations that were postulated for initiation and termination of the "snowball" episodes.[3] Unfortunately, for the Precambrian (older than 0.5 Ga), no other proxies or complementary constraints are presently available.

During the Phanerozoic, additional constraints (stomata in leaves, soils) enable broad estimates of atmospheric carbon dioxide levels and they were most of the time higher (up to 20x) than at present.[4] The broad patterns show that climate correlates poorly with such estimates, but co-varies with variations in cosmic ray flux.[1,5] The temporal resolution of the records, millions of years (Ma), is not sufficient for a claim of causative sequence. Nevertheless, for time intervals where the record with superior resolution is available, such as 55.8 Ma ago, the Paleocene-Eocene Thermal Maximum,[6] the temperature signal appears to precede carbon signal, arguing for climate being the cause of variability in the carbon cycle.

The latest 1 Ma brings forward additional proxy records, such as ice cores for climate and concentrations of atmospheric greenhouse gases and cosmogenic nuclides for impact of cosmic rays (solar activity).[7,8] While ice cores demonstrate conclusively that on multi-millennial time scales climate and greenhouse gases correlate, they again show that the shifts in climate precede the trends in gas concentrations.[9] Finally, after termination of the last major glacial episode, some 10,000 years ago, atmospheric greenhouse gases varied only marginally while the climate variations were considerable. These climate swings correlate well with coeval records of cosmogenic nuclides, arguing for the role of sun as the primary driver of the planetary climate.[7,8,10]

1. Veizer, J. (2005) "Celestial climate driver: a perspective from four billion years of the carbon cycle." *Geoscience Canada*, 32:13-28.and Veizer, J. (2006) "Celestial climate driver: a perspective from four billion years of carbon cycle." *34th Int. Seminar on Nuclear War and Planetary Emergencies*. World Scientific. Erice, Italy, 117-144.
2. Kasting, J.F. (2005) "Methane and climate during the Precambrian era." *Precambrian Res.*, 137:119-129.

3. Crowley, T.J., Hyde, W.T. and Peltier, W.R. (2001) "CO_2 levels required for deglaciation of a 'near-snowball' Earth." *Geophys. Res. Lett.*, 28:283-286.
4. Berner, R.A. and Kothavala, Z. (2001) "Geocarb III: a revised model of atmospheric CO_2 over Phanerozoic time." *American J. Science*, 301:182-204.
5. Shaviv, N.J. and Veizer, J. (2003) "Celestial driver of Phanerozoic climate?" *GSA Today*, 13/7:4-10.
6. McInerney, F.A. and Wing, S.L. (2011) "The Paleocene-Eocene thermal maximum: a perturbation of carbon cycle, climate, and biosphere with implications for the future." *Ann. Rev. Earth Planet. Sci.*, 39:489-576.
7. Scherer, K., Fichtner, H., Borrmann, T., Beer, J., Desorgher, L., Flückiger, E., Fahr, H.-J., Ferreira, S.E.S., Langner, U., Potgieter, M.S., Heber, B., Masarik, J., Shaviv, N. and Veizer, J. (2006) "Interstellar-terrestrial relations: the variable cosmic environments, the dynamic heliosphere, and their imprints on terrestrial archives." *Space Science Reviews*, 127/1-4:327-465.
8. Kirkby, J. (2008) "Cosmic rays and climate." *Survey in Geophysics*, 28:333-375.
9. Monnin, E., Indermühle, A., Dallenbach, A., Flückiger, J., Stauffer, B., Stocker, T.F., Raynaud, D., and Barnola, J.M. (2001) "Atmospheric CO_2 concentrations over the last glacial termination." *Science*, 291p. 112-114.
10. Veizer, J. (2011) "The role of water in the fate of carbon: implications for the climate system." *43rd Int. Seminar on Nuclear War and Planetary Emergencies*. World Scientific. Erice, Italy, 313-327.

SYNCHRONIZATION OF CLIMATE MODES FOLLOWED BY AN INCREASE IN THEIR COUPLING STRENGTH LEADS TO MAJOR CLIMATE SHIFTS

ANASTASIOS A. TSONIS AND KYLE SWANSON
Department of Mathematical Sciences, University of Wisconsin-Milwaukee
Milwaukee, USA

ABSTRACT

Theoretical and experimental results[1-3] have established that a network of coupled nonlinear oscillators may synchronize, at some appropriate coupling strength. Once a synchronous state is in place, an increase in the coupling strength may lead to a bifurcation, which destroys the network. Here we show that this mechanism is present in the climate system and is responsible for major climate shifts. First, we construct a network of observed climate modes in the period 1950–2005 and investigate their collective behavior. The results indicate that this network synchronized at around 1975. The synchronous state was followed by a steady increase in the coupling strength until the system bifurcated and the network was destroyed, upon which a new climate state emerged. This significant event is known as the great climate shift of the 1970s. We have also discovered the evidence for such type of behavior in a 100-yr climate simulation using a state-of-the-art model. During this period, the network has synchronized three times. In one of the cases, the synchronous state was followed by a decrease in coupling strength and no climate shift was observed, consistent with the theory. In the other two cases, the synchronous state was followed by an increase in coupling strength and in both these cases a major shift occurred. Thus, the combination of theory, observations and model simulations consistently suggest this is a new mechanism for climate shifts.

One of the most important and mysterious events in recent climate history is the climate shift in the mid-1970s.[4] In the northern hemisphere 500-hPa atmospheric flow the shift manifested itself as a collapse of a persistent wave-3 anomaly pattern and the emergence of a strong wave-2 pattern. The shift was accompanied by sea-surface temperature (SST) cooling in the central Pacific and warming off the coast of western North America.[5] The shift brought sweeping long-range changes in the climate of northern hemisphere. Incidentally, after "the dust settled," a new long era of frequent El Niños superimposed on a sharp global temperature increase has begun. Understanding the dynamics of the above phenomena is essential for our ability to make useful prediction of climate change. A major obstacle to this understanding is the extreme complexity of the climate system, which makes it difficult to disentangle causal connections leading to the observed climate behavior. Here we present a novel approach—the so called network approach,—which reveals an important new mechanism in climate dynamics and explains several aspects of the observed climate variability in the late 20th century. The new methodology also provides a promise for an improved prediction of major climatic shifts.

The network approach to complex systems is a rapidly developing methodology, which has proven to be useful in analyzing such systems' behavior.[6,7] In this approach, a complex system is presented as a set of connected nodes; the degree of connectivity between the pairs of nodes may vary. The collective behavior of all the nodes and links (the topology of the network) describes the dynamics of the system and offers new ways to investigate its properties. We construct a network from six major climate indices for the purpose of studying the underlying network topology and its relation to climate dynamics. The indices represent the Pacific North American (PNA) teleconnection pattern, the Pacific Decadal Oscillation (PDO), the North Atlantic Oscillation (NAO), the El Niño/Southern Oscillation (ENSO), and the Western Pacific (WP) and Tropical North Atlantic (TNA) indices.[8-12] Arguably, these indices represent the dominant modes of climate variability, with time scales ranging from months to decades and regional-to-global spatial scales, and we treat them as nonlinear sub-systems of the grand climate system exhibiting complex dynamics.[13-16] Monthly-mean values in the interval 1950–2004 are available for all indices.

In our approach, the six climate indices are assumed to be connected to each other with the strength related to the cross-correlation coefficient between them; the correlation coefficient is a standard measure of connectivity. A commonly used measure of the network's strength as a function of time is the mean distance $d(t)$[17-19]

$$d(t) = \frac{2}{N(N-1)} \sum_{d_{ij}^t \in D^t} d_{ij}^t \qquad (1)$$

Here t denotes the time in the middle of a sliding window of width Δt, $N=6;\ i,j=1,\ldots, N$, and

$$d_{ij}^t = \sqrt{2(1-|\rho_{ij}^t|)},$$

where ρ_{ij}^t is the cross-correlation coefficient between nodes i and j in the interval $[t-\Delta t/2,\ t+\Delta t/2]$. The above formula uses the absolute value of the correlation coefficient because the choice of sign of indices is arbitrary. The distance can in fact be interpreted as a measure of the synchronization of the network's components. Synchronization between nonlinear (chaotic) oscillators occurs when their corresponding signals converge to a common, albeit irregular, signal. In this case, the signals are identical and their cross-correlation is maximized. Thus, a distance of zero corresponds to a perfect synchronization and a distance of $\sqrt{2}$ signifies a set of non-connected uncorrelated nodes.

Figure 1 shows the distance as a function of time for a window length of $\Delta t = 11$ years, with tick marks corresponding to the year in the middle of the window. The correlations (and thus distance values for each year) were computed based on the annual-mean indices constructed by averaging the monthly indices over the period of November–March. The dashed line parallel to the time axis in Figure 1 represents the 5% significance level associated with the null hypothesis that the observed indices are sampled from a population of a 6-dimensional AR-1 process driven by a spatially (cross-index) correlated Gaussian noise; the parameters of the AR-1 model and the covariance matrix of the noise are derived from the full time series of the observed indices. This test assumes that the variations of the distance with time seen in Figure 1 are due to sampling associated with a finite-length (11-yr) sliding window used to compute the local distance values. Retaining overall cross-correlations in constructing the surrogates makes this test very stringent. Nevertheless, we still find that the event in the mid 1970s is associated with distance variations that fall below the 5% significance level. We therefore conclude that this feature is not likely to be due to sampling limitations but it represents a statistically significant synchronization event. Note that the window length used in Figure 1 is a compromise between being long enough to estimate correlations but not too long to "dilute" transitions. Nevertheless, the observed synchronization in 1975 is insensitive to the window size in a wide range of $7 \leq \Delta t \leq 15$.

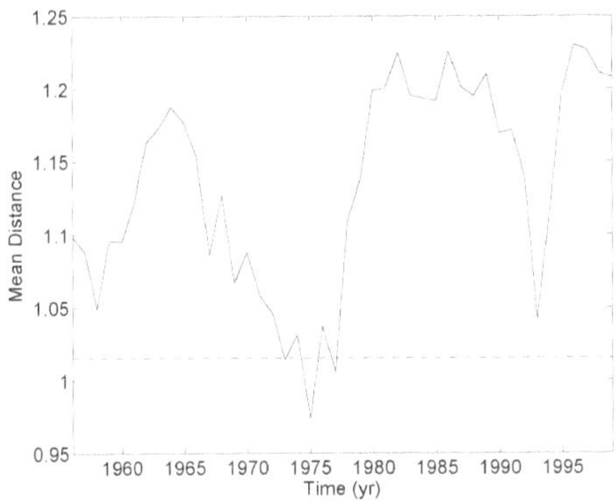

Fig. 1: The distance (see definition in text) of a network consisting of six observed major climate modes as a function of time. This distance is an indication of synchronization between the modes with smaller distance implying larger synchronization. The parallel dashed line represents the 5% significance level associated with a null hypothesis of spatially correlated red noise (see text). From this figure it follows that the event in 1975 represents a statistically significant synchronization.

The climate shift of 1970s occurred right after this synchronous state was in place. The question thus arises as to whether or not this shift represents a bifurcation induced by increasing of the coupling strength between the nodes, as the theory of synchronized chaos would predict.[1-3] Although linear measures such as the correlation are commonly used to describe the extent to which two time series are coupled, the correlation is not necessarily sufficient. For example, two time series could be coupled but be lagged in phase, and the correlation could vanish. More generally, in nonlinear time series analysis it is desirable to use the phase between two time series, and more specifically, how well changes in that phase can be predicted as a measure of how strongly two time series are coupled.[20] If future changes in the phase can be predicted using only information about the current phase, the series may safely be considered strongly coupled. Unfortunately, in climate data analysis we are limited by the fact that data are sparse and the phase might not be well known. To circumvent this difficulty in this paper, we use symbolic dynamics.

For any given time series point, we can define a symbolic phase by examining the relationship between that point and its nearest two neighbors in time. As shown in Figure 2, if the 3 points are sequentially increasing, we can assign to the middle point a phase of 0, while if they are sequentially decreasing, a phase of π. Intermediate values then follow. Notice that this procedure is totally non-parametric, as it does not compare the actual values of the points aside from whether a point is larger or smaller than its neighbors. The advantage of this approach is that it is blind to ultra-low frequency variability, i.e., decadal scale and longer. The symbolic phase was then constructed for each of the six climate time series considered here. The symbolic phase ϕ_n is represented as the trigonometric pair $Z_n = (\cos\phi_n, \sin\phi_n)$, and changes in this pair from year to year are represented as $\Delta Z_{n,n+1} = (\cos\phi_{n+1} - \cos\phi_n, \sin\phi_{n+1} - \sin\phi_n)$. It is straightforward to construct a least squares estimator of $\Delta Z_{n,n+1}$ of the type

$$\Delta \mathbf{Z}^{est}_{n,n+1} = \mathbf{M}\mathbf{Z}_n \tag{2}$$

where $\mathbf{Z}_n$ denotes a vector of all 15 possible phase pair combinations between our climate indices and **M** is the least squares predictor matrix calculated using all years. A measure of the coupling then is simply

$$\left\| \Delta \mathbf{Z}^{est}_{n,n+1} - \Delta \mathbf{Z}_{n,n+1} \right\|^2,$$

where strong coupling is associated with small values of this quantity, i.e., good prediction. This quantity is plotted in Figure 3. It is apparent that the coupling measured in this manner increases over the interval 1975–1980, implying that the coupling increased while synchronization (Figure 1) was high, thus leading to a bifurcation of the climate system.[1–3]

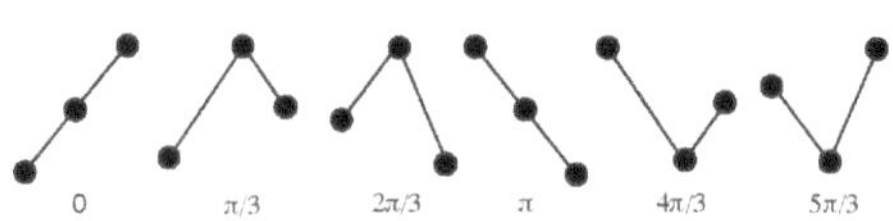

Fig. 2: The six states for the symbolic phase construction. The points in each triplet correspond to three consecutive points in a time series, and their relative vertical positions to each other indicate their respective values.

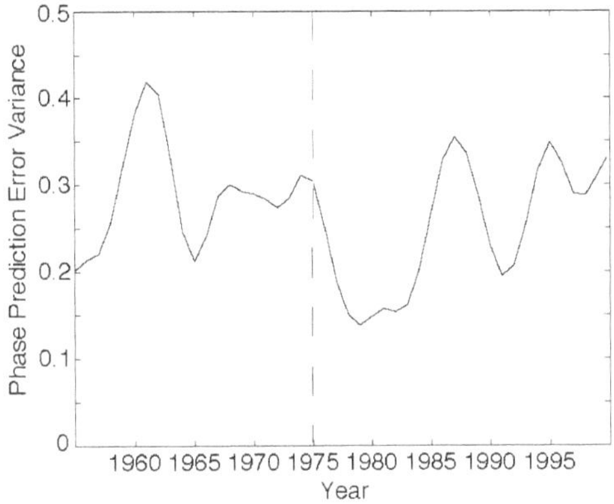

Fig. 3: Coupling strength for the observed network of six major climate modes as a function of time. Smaller phase prediction error variance implies stronger coupling. The vertical dashed line corresponds to the time of synchronization.

It is of interest to see if this type of behavior is found in a state-of-the-art coupled ocean/atmosphere model. The particular model we examine here is the GFDL CM2.1 coupled ocean/atmosphere model, run in support of IPCC 2007.[21] The particular scenario we examine is the SRESA1B, which is a "business as usual" scenario with CO_2 levels stabilizing at 720 ppmv at the close of the 21st century.[22] Figure 4a shows the global-mean temperature during the 21st century for this simulation; the increase due to radiative forcing is apparent. However, there is a curious jump in the temperature over the period 2035-2045, with the temperature gapping up by roughly 0.5° over that period, before settling onto a warming trend quite similar to the trend prior to 2030. This gap up coincided with a complete collapse of the El Niño oscillation at around this time. Figure 4b shows that the model went from a very aggressive series of alternating El Niño/La Niña events prior to 2035 to very subdued fluctuations for the period 2040-2070. It is apparent from these two figures that a very significant dynamical climate event occurred around 2035, probably pushing the climate onto a totally new attractor.

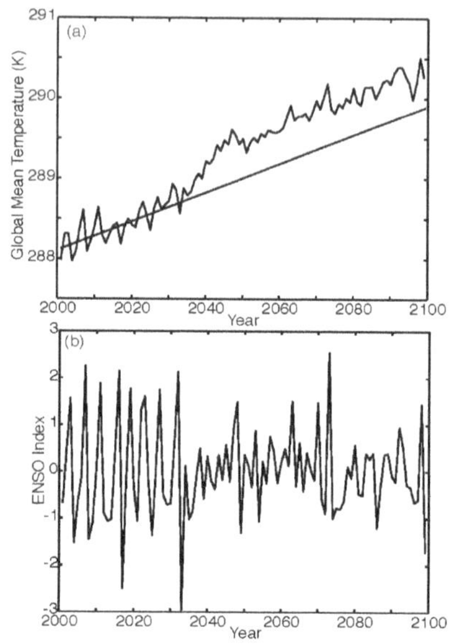

Fig. 4: (a) Global mean temperature for the GFDL CM2.1 SRESA1B coupled climate model simulation. The line indicates the 2001–2030 extrapolated trend. (b) Normalized ENSO index constructed from the leading EOF of SST in the tropical Pacific.

In order to examine this from the standpoint of synchronicity and coupling, we constructed from the simulated fields the same six modes used above and estimated the distance and coupling strength as a function of time. Figure 5a shows that the climate becomes synchronized at around 2020, 2030, and 2070; these times are highlighted by the vertical dashed lines. The synchronous state in 2020 is followed (see Figure 5b) by a decrease in coupling and no major changes are observed. On the other hand, the synchronous state around 2030 is followed by an increase in coupling strength and then, just as in the observed case of the 1970's, the climate system appears to undergo a bifurcation marked by a major climate transition. The synchronization around 2070 is also followed by an increase in the coupling strength, although the increase is not as substantial as it was around 2035, and the climate undergoes another transition, marked by a super El Niño event (Figure 4b) followed by a different El Niño/La Niña variability, and with the global temperature trend again breaking, gapping downward this time by about 0.25°. It is interesting to speculate on the similarity of the GCM event in 2035 and the observed event in the 1970's. The standard explanation for the warming post-mid 1970's is that the radiative effect of greenhouse gases overcame shortwave reflection effects due to aerosols. However, comparison of the GCM event and this event suggests an alternative hypothesis, namely that the climate shifted to a different state after the 1970's event that simply has a warmer climate, and the warming since that time has been the climate adjusting to the new state.

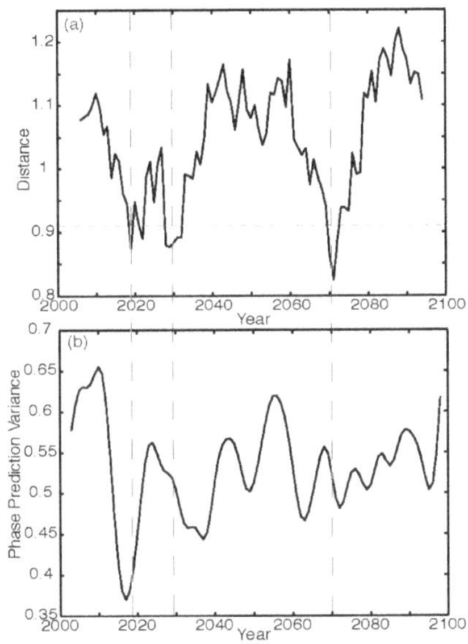

Fig. 5: (a) Distance and (b) coupling for the climate simulation network of the same six climate mode. The vertical dashed lines indicate the times of synchronization. The horizontal dashed line is similar to that in figure 1 but for the model network.

The above theoretical, observational, and modeling results suggest that the following intrinsic mechanism of the climate system leading to major climate shifts. First, the major climate modes tend to synchronize at some coupling strength. When this synchronous state is followed by an increase in the coupling strength between the modes, the network becomes unstable and fluctuations grow to lead the system to a bifurcation, after which climate emerges in a new state. The whole event marks a significant shift in climate. This scenario is consistent with studies that have suggested that in the case of the shift in the 1970s such a fluctuation may have been a sea surface temperature anomaly in the tropical Pacific.[4,23] Combination of climate-network-based diagnosis with general circulation models able to simulate major climate modes may enhance our potential to predict such climatic events.

ACKNOWLEDGEMENTS

We thank J.B. Elsner for his critical comments and suggestions. AAT and KLS are supported by NSF grant ATM-0438612, SK is supported by DOE grant DE-FG-03-01ER63260 and by NASA grant NNG-06-AG66G-1. GFDL climate model data was made available through the GFDL Data Portal at http://nomads.gfdl.noaa.gov/CM2.X.

REFERENCES

1. Heagy, J.F., Pecora, L.M. and Carroll, T.L. (1994) "Short wavelength bifurcations and size instabilities in coupled oscillator systems." *Phys. Rev. Lett.* **74**:4185-4188.
2. Pecora, L.M. and Carroll, T.L. (1990) "Synchronization in chaotic systems." *Phys. Rev. Lett.* **64**:821-824.
3. Pecora, L.M., Carroll, T.L., Johnson, G.A. and Mar, D.J. (1997) "Fundamentals of synchronization in chaotic systems, concepts, and applications." *Chaos* **7**:520-543.
4. Graham, N.E. (1994) "Decadal scale variability in the 1970s and 1980s: Observations and model results." *Clim. Dyn.* **10**:60-70.
5. Miller, A.J., Cayan, D.R., Barnett, T.P., Craham, N.E. and Oberhuber, J.M. (1994) "The 1976-77 climate shift of the Pacific Ocean." *Oceanography* **7**:21-26.
6. Albert, R.and Barabasi, A-L. (2002) "Statistical mechanics of complex networks." *Rev. Mod. Phys.* **74**:47-101.
7. Strogatz, S.H. (2001) "Exploring complex networks." *Nature* **410**:268-276.
8. Barnston, A.G. and Livezey, R.E. (1987) "Classification, seasonality, and persistence of low-frequency atmospheric circulation patterns." *Mon. Wea. Rev.* **115**:1083-1126.
9. Hurrell, J.W. (1995) "Decadal trends in the North Atlantic oscillation regional temperature and precipitation." *Science* **269**:676-679.
10. Mantua, N.J., Hare, S.R., Zhang, Y., Wallace, J.M. and Francis R.C. (1997) "A Pacific interdecadal climate oscillation with impacts on salmon production." *Bull. Amer. Meteor. Soc.* **78**:1069-1079.
11. Thompson, D.W.J. and Wallace, J.M. (1998) "The Artic Oscillation signature in the wintertime geopotential height and temperature fields." *Geophys. Res. Lett* **25(9)**:1297-1300.
12. Wallace, J.M. and Gutzler, D.S. (1981) "Teleconnections in the geopotential height field during the northern hemisphere winter." *Mon. Wea. Rev.* **109**:784-812.
13. Tsonis, A.A. and Elsner, J.B. (1989) "Chaos, strange attractors and weather." *Bull. Amer. Meteor. Soc.* **70**:16-23.
14. Lorenz, E.N. (1991) "Dimension of weather and climate attractors." *Nature* 353:241-244.
15. Elsner, J.B. and Tsonis, A.A. (1993) "Nonlinear dynamics established in the ENSO." *Geophys. Res. Lett.* **20**:213-216.
16. Neelin, D.J. and Latif, M. (1998) "El Niño dynamics." *Physics Today* **51**:32-36.

17. Mantegna, R.N. (1999) "Hierarchical structure in financial markets." *Eur. Phys. J.* B **11**:193-197.
18. Onnela, J.-P., Chakraborti, A., Kaski, K., Kertesz, J. and Kanto, A. (2003) "Dynamics of market correlations: Taxonomy and portfolio analysis." *Phys. Rev. E* **68**:056110.
19. Onnela, J.-P., Saramaki, J., Kertesz, J. and Kaski, K. (2005) "Intensity and coherence of motifs in weighted complex networks." *Phys. Rev. E* **71**:065103.
20. Smirnov, D.A. and Bezruchko, B.P. (2003) "Estimation of interaction strength and direction from short and noisy time series." *Phys. Rev. E* **68**:046209.
21. GFDL CM2.1 development team. (2006) "GFDL's CM2 global coupled climate models, Parts 1-4." *J.Clim.* **19**:643-740.
22. IPCC: Climate change 2001- The scientific basis. Contribution of working group I to the third assessment report of the Intergovernmental Panel on Climate Change. Houghton, J. H. et al., (eds) (Cambridge University Press 2001).
23. Graham, N.E., Barnett, T.P., Wilde, R., Ponater, M. and Schubert, S. (1994) "On the roles of tropical and mid-latitude SSTs in forcing interannual to interdecadal variability in the winter Northern Hemisphere circulation." *J. Clim.* **7**:1500-1515.

OVERCOMING CHAOTIC BEHAVIOR OF CLIMATE MODELS

S. FRED SINGER
Science and Environmental Policy Project
Arlington, USA

CHRISTOPHER W. MONCKTON
The Viscount Monckton of Brenchley
Carie, Rannoch, UK

ABSTRACT

A fundamental issue in climate science is Attribution—determining the relative importance of human and natural causes. The task generally involves comparing temperature trends from observations and from greenhouse (GH) models. However, a problem arises from the chaotic uncertainties inherent in the (non-linear) model calculations. Modelers try to overcome this problem by forming an "ensemble-mean" of a number of "simulations" (runs) of the same model. Here we conduct a synthetic experiment, and use two distinct procedures to demonstrate that no fewer than about 20 runs (of 20-yr length of an IPCC General-Circulation Model) are needed to place useful constraints upon chaos-induced model uncertainties.

INTRODUCTION

As Lorenz (1963, 1975: Zichichi, 2007 and also Giorgi, 2005) have demonstrated, climate models, using non-linear partial differential equations, generate results highly sensitive to the initial conditions. To reproduce an identical result in successive "simulations" (runs), the parameters describing the model's initial state must be given to a precision that is unattainable in practice. The Intergovernmental Panel on Climate Change (IPCC-TAR; Meehl et al. 2001) acknowledges that, mathematically speaking, the climate is a "complex, non-linear, chaotic object" and that, therefore, "the long-term prediction of future climate states is not possible." Accordingly, any comparison of modeled with observed temperature trends cannot be done satisfactorily without understanding the chaotic behavior of a climate model.

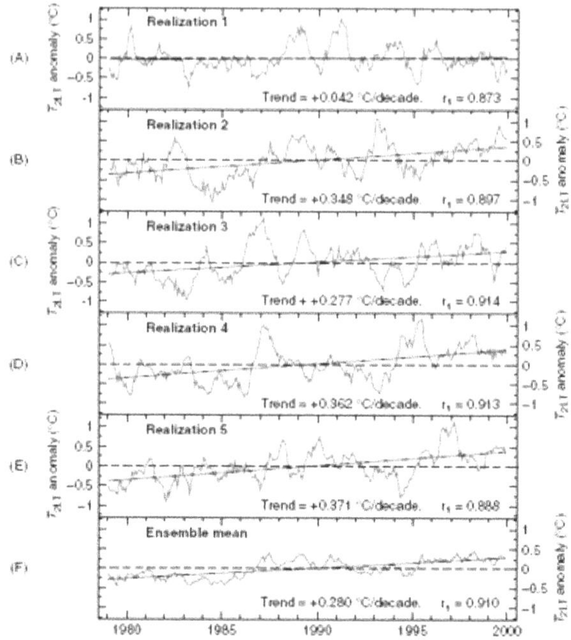

Fig. 1: Illustrating the chaotic nature of model trends, using the results of 5 runs (sometimes referred to as "realizations" or "simulations") from 1979 to 1999 of a particular GCM (from Japan's MRI), as presented in figure 1 of Santer et al. 2008. The OLS trends of the five runs range from +0.042 to +0.371 (K per decade). The range of trends would likely to be even larger if more runs had been carried out. None of the five trends, nor the 'ensemble-mean' shown, represents the 'true' model trend. As discussed in the text, one needs to show that the cumulative ensemble-mean approaches an asymptotic value as the number of runs increases.

One consequence is that successive runs of the same climate model can yield very different values for warming trends. These trends may vary by an order of magnitude or more, and even their sign may vary. For example, the Japanese MRI model carried out five runs for IPCC (Santer et al. 2008): the individual trends range from 0.042 to 0.371 K/decade (Figure 1)—and this error interval ('spread') would have been even greater if more runs had been performed.

Modelers, therefore, carry out several runs ("simulations") and then publish the 'model-ensemble-mean,' *E*, the arithmetic average of the individual trend values generated by the several runs. Only rarely do we learn the results of the individual simulations that are components in *E*. Yet how do we know that, say, five runs are sufficient to produce a reliable value of *E* to compare with an observed trend?

The present paper addresses a single but crucial aspect of the impact of chaoticity on the performance of general-circulation models: the strong dependence of the error-bars in temperature-trend projections on the number of simulations that are run on a particular model. We suggest that it may be possible to overcome the "chaoticity barrier" by performing a sufficient number of runs.

METHOD

The objective of this enquiry was to establish how many simulation runs of a GCM, at minimum, are necessary to provide reasonable constraints on the value of *E*. For this investigation, it would have been desirable to use climate models which had each been run at least 20 times. However, financial and time constraints on modelers mean there are no ready examples of such multiple runs. Therefore, we developed a synthetic approach to the problem.

A single, unforced ('constant forcing') control run of 1000 years' duration was obtained from the Program for Climate Model Diagnosis and Intercomparison (PCMDI) at Lawrence Livermore National Laboratory.

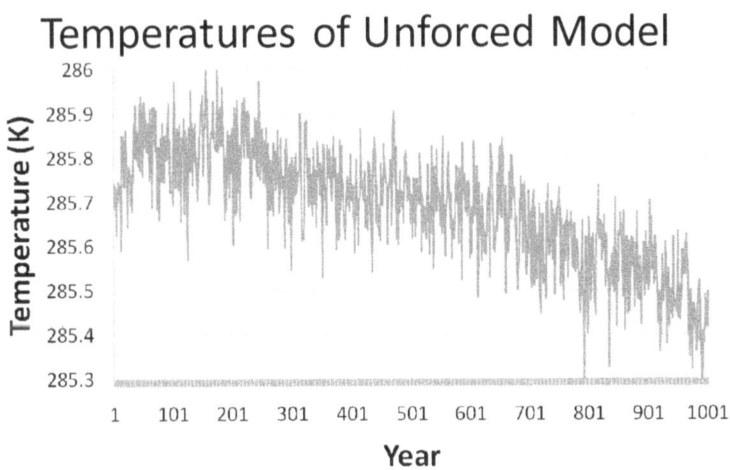

Fig. 2: Temperature values of an unforced 1000-year climate model control run. Source: PCMDI.

First, the temperatures were plotted against time in years (Figure 2) to check for inevitable drift from what should be a straight, zero-slope, horizontal line. Next, the temperature series was divided into 25 segments, each of 40 years' duration (and also into 50 segments of 20-year runs). For each segment, trend values T_1 ... T_{25} were determined. This procedure is analogous to, and (considering the chaoticity of the climate object) equivalent to, 25 separate runs of a GCM over a single 40-year time-interval. Another advantage of using an unforced model is that the *true* trend is known in advance —namely, zero (except for drift effects).

It was then possible to observe where this cumulative ensemble mean, E_{cum}, approaches an asymptotic value that may be termed the 'true' trend. Results (Figure 3) indicate that about 10 runs of the model seem to be sufficient for 40-yr runs (and 20 runs of 20-yr length).

First procedure: the cumulative ensemble mean E_{cum}.

In the first of two procedures, a cumulative ensemble mean E_{cum} was derived by adding the trend value of an additional run to the previous values, Eq. (1), so as to determine a new value of E. Finally, the cumulative trend was plotted as a function of n, the number of trend values used (Figure 3):

$$E_{\text{cum}} = 1/\text{n} \sum_{i=1}^{n} T_i \,, \tag{1}$$

Control experiments for the first procedure

- To investigate the influence of drift, which is seen to exist in the 1000-yr model run (Figure 2), we have also carried out the same procedure for two additional time periods (Years 200-1000 and 400-1000), for which the drift appears to be more uniform—or at least does not change its sign. For 40-yr runs, the asymptote of the cumulative ensemble mean E_{cum} is reached again after at least 10 runs.
- In addition to the 25 time-segments, each of 40 years, starting in years 1, 41, 81, ... , a further 24 trend values may be determined by starting the 40-year time-segments in years 11, 51, 91, ... , with still further trends determinable by starting in years 21, 61, 101, ... , and at 31, 71, 111, ..., for a total of 97 trend values. All of these were found on examination to behave in the same way. Lagged auto-correlation does not seem to be significant.
- We also checked against a possible influence of contiguity by selecting non-contiguous segments to form cumulative ensemble-means.
- Finally, by starting in each year on the interval (1, 961), it is possible to obtain 961 (partly overlapping) segments of 40-year length. When all 961 trend values are plotted, they are found to form a Gaussian distribution.

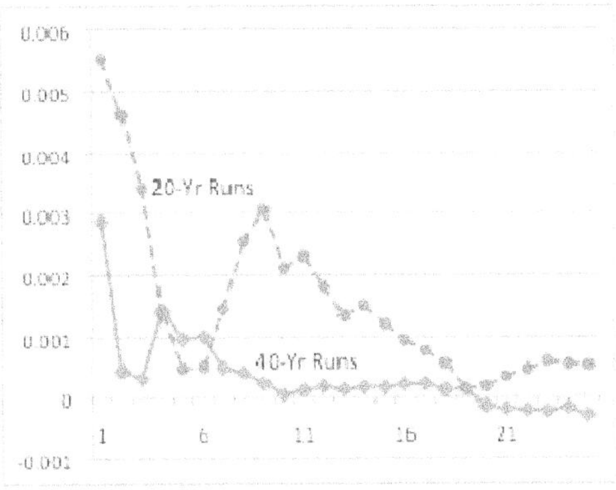

Fig. 3: Procedure #1: Cumulative ensemble-means of trend values as a function of n, the number of runs (of length 20 or 40 years). The cumulative ensemble mean, E_{cum}, is seen to reach an asymptotic value close to zero as the number of runs exceeds about 20 (for a run-length of 20 years)—and about 10 (for a run-length of 40 years). In the absence of model drift, the asymptotic value would presumably be zero.

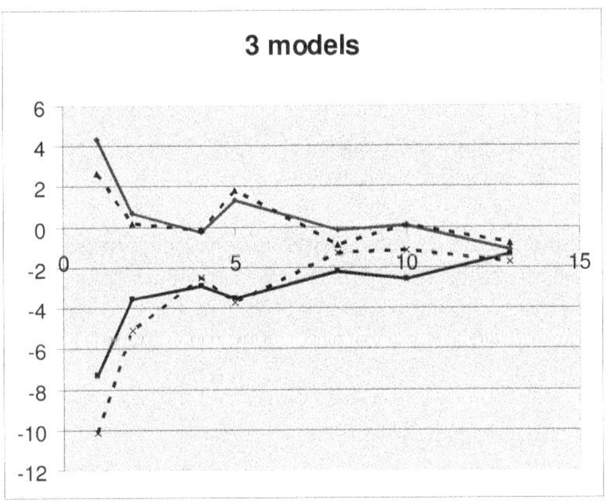

Fig. 4: Procedure #2: Error interval ("spread" = T_{max} - T_{min}) of ensemble-means of trend values of 3 (synthetic) models as a function of n, the number of runs. (All trend values shown on the y-axis were multiplied by 1000.) For a run length of 40 years, the spread is seen to collapse towards zero as the number of runs exceeds 10. Similar results are obtained for the cases of 4 models and 5 models. The dashed lines result from a different method of selecting trend values (see text).

Second procedure: constraining the error-interval

In the second of two procedures, the interval on which the values of E fall ('spread'), was investigated for (assumed) three synthetic models as a function of the number of runs. The result is shown in Figure 4, with the spread plotted against n, the number of simulation runs. For a run length of 40 years, the trend interval is seen to approach zero for $n > 10$ (Figure 4).

Details for the second procedure

- The time-series is truncated by removing years 1-200, to minimize possible effects of drift.
- The remaining 800 years were divided into 40 segments of length 20 years, whereupon 40 trend values $T_1 \ldots T_{40}$ were determined.
- We assume we have 3 models, each with the same number of runs n. We therefore assign n trend values, arbitrarily selected from the 40 values of T, to each of the three models.
- For n=1, the error interval among the 3 trend values is simply $(T_{max} - T_{min})$.
- For n=2, a series of ensemble-mean trend values was constructed thus: $T'_1 = ½(T_1+T_2)$; $T'_2 = ½(T_3+T_4)$; $T'_3 = ½(T_5+T_6)$. The trend interval among the 3 values was then determined as $(T'_{max}-T'_{min})$.
- The procedure was repeated for n = 4, 5, 8, 10, and 13 respectively. Results are plotted in Figure 4.
- To test sensitivity, the procedure was next repeated for an assumed 4 models (permitting n-values up to 10), and then for 5 models. Similar results were obtained.
- As a further check, we repeated all procedures by starting the selection of T-values with T_{40} (instead of T_1) and proceeding backwards. Those max and min trends are indicated by dashed lines in Figure 4.

Discussion

We have demonstrated that the ensemble-mean (E) trend obtained from a multi-run model is more reliable than the trend obtained from a model that is run only once. In general, ten or more 40-yr runs may be necessary to form a reliable E. Thus it may be possible to overcome the "chaoticity barrier" of climate models identified by IPCC in the 20CEN intercomparison.

Sensitivity Analysis of Segment Length

Our initial choice of a segment length of 40 years was arbitrary. We now investigate the effects of segment length on the convergence of the ensemble mean value. We find convergence after 5 runs for model runs of length 80 years and after 20 runs (see Figure 3) for a segment length of 20 years (which is typical of the models in the IPCC compilation; see Figure 1). Empirically, it appears that convergence is achieved in 400 run-years—i.e., (20 x 20), (10 x 40), and (5 x 80). We have not discovered a theoretical explanation for this useful result.

Few modelers have the resources to carry out ten or more runs of a particular general-circulation model. They frequently report a temperature trend based on only a single run. For example, the IPCC's compilation of 22 '20CEN' models has five models with just one run, five with 2 runs, and only seven with 4 or more runs. The run lengths varied between 20 and 24 years. Modeler should be encouraged to report not only the details of the forcings and parameterizations used in their particular models but also the results of each run and its length (in years).

Multi-Model Averaging

Many investigators when considering a group of models compound the problem by simply using the average of the ensemble-mean trends of all the models to compare with an observed trend. This procedure, however, is defective in that it gives equal weight to single-run models and multi-run models, and leads to greatly enhanced uncertainty of the multi-model trend.

The "spread" in trend values trend values discussed in Procedure #2 is akin to the "range' of extreme values of a Gaussian distribution. But "range" is an improper statistical metric; it increases as the number of independent data points increases—while the Standard Deviation of their distribution decreases. For example, it can be shown that the wide model uncertainty displayed as a grey region in Figure 6 of Santer et al. (2008) and labeled as a "2-sigma Standard Deviation" is actually an artifact, caused by the presence of five single-run models in the IPCC compilation of models. A compilation comprising only multi-run models would, therefore, help to constrain chaotic uncertainty, and would provide a more reliable means of comparing the consistency of modeled with observed trends.

ACKNOWLEDGEMENTS

We are grateful to Roger Cohen, Curtis Covey, Robert Levine, Craig Loehle, Christopher Monckton, and Ronald Stouffer for useful discussion and to Garrett Harmon and Will McBride for technical assistance in preparing the figures.

REFERENCES

1. Giorgi, F. (2005) "Climate Change Prediction." *Climatic Change* 73:239-265: DOI: 10.1007/s10584-005-6857-4.
2. Lorenz, Edward N. (1963) "Deterministic nonperiodic flow." *Journal of the Atmospheric Sciences*, 20:130-141.
3. Lorenz, E.N. (1975) "The physical bases of climate and climate modeling," in *Climate Predictability*, #16 in GARP Publication Series, pp. 132-136, World Meteorological Organization.
4. Meehl, G., et al. (2001) "Climate Change: The Scientific Basis" in *Fourth Assessment Report of the Intergovernmental Panel on Climate Change,* Cambridge University Press, London.
5. Santer, B.D., Thorne, P.W., Haimberger, L., Taylor, K.E., Wigley, T.M.L., Lanzante, J.R., Solomon, S., Free, M., Gleckler, P.J., Jones, P.D., Karl, T.R., Klein, S.A., Mears, C., Nychka, D., Schmidt, G.A., Sherwood, S.C., Wentz, F.J. (2008) "Consistency of modeled and observed temperature trends in the tropical troposphere." *Int. J. Climatol.*: doi:1002/joc.1756.

6. Zichichi, A. (2007) "Meteorology and climate: Problems and expectations." *Pontifical Council for Justice and Peace,* The Vatican, 26-27 April.

SESSION 6

WATER & POLLUTION

FOCUS: CONTAMINANTS OF EMERGING CONCERN (CEC)

CECS IN WATER: CHALLENGES AND FUTURE SOLUTIONS

DEBORAH L. SWACKHAMER, PH.D.
Science, Technology, and Public Policy, Hubert H. Humphrey Institute of Public Affairs, Environmental Health Sciences, University of Minnesota, St Paul, USA

After World War II, the ability to synthesize organic chemicals for industrial and household product use was established and embraced by society in the boom of peacetime. Plastic polymers were commercialized, and an explosion of new products entered the market to improve the standard of living and modernize society. Unfortunately, along with this progress also came environmental contamination from the manufacture, use, and disposal of these products. This paper will discuss what chemicals are of concern in the environment today, what their sources are, what happens to them in the environment, and what we know (and don't know) about the exposure and toxic effects to humans, fish and wildlife. This will be followed by a summary of what policy measures might be considered to address these concerns.

INTRODUCTION TO CECS

Chemicals of emerging concern (CECs) are (1) those chemicals that have been in commerce for some time, but have only recently been identified as environmental contaminants because of advances in analytical detection; (2) chemicals that have been in commerce for some time, have been regulated in the past for a toxic endpoint, but have now been associated with a more sensitive toxic endpoint; and (3) relatively newly produced chemicals that are now showing up in the environment. These chemicals of emerging concern generally are not regulated. The overarching concern is their potential to affect cell signaling processes in humans and animals that may disrupt a wide variety of the body's regulatory systems (reproduction, growth, metabolism, development, behavior, etc; see below).

THE "LIST" OF CECS—WHAT ARE THEY

There are more than 85,000 chemicals registered for commercial use in the United States (U.S.) under the Toxics Substances Control Act (TSCA, under the authority of the U.S. Environmental Protection Agency), which contains many of the same compounds on the Canadian list of 58,000 chemicals required under the Canadian Environmental Protection Act (CEPA). The U.S. Food and Drug Administration (FDA) regulates another 12,000 pharmaceuticals, but they are regulated for therapeutic medicinal use and not for exposure or harm as environmental contaminants. Chemicals on the TSCA and CEPA lists are simply registered for use; acceptance on the list is based on minimal data. Chemicals that are actually regulated in the U.S. (i.e., have had health standards developed for them) make up a small subset of these 100,000 chemicals. The Clean Water Act regulates 126 priority chemicals; the Safe Drinking Water Act regulates approximately 90 contaminants; the Clean Air Act regulates 187 toxic air pollutants; and the "Superfund" Act addresses 276 hazardous substances. The global Stockholm POPs (Persistent Organic Pollutants) Treaty bans 17 chemicals. Many of these lists overlap. The point being made here is that the number of chemicals that have been addressed in a regulatory framework is very small compared to the possible universe of chemicals of concern.

Of this universe of approximately 100,000 chemicals, approximately half of them are high molecular weight polymers that generally are too large to cross cell walls and cause a toxic response. Also, only small-to-medium molecular weight chemicals having specific physical chemical properties such that they would get into the environment (air, water, food) and subsequently allow for exposure are of concern. However, these lists do not include degradation byproducts or chemicals imported products. A well-cited study used computer screening of physical chemical properties to evaluate what chemicals in this "universe" might be predicted to be persistent in the environment and bioaccumulative (meaning they are taken up into foodwebs).[1,2] They concluded that approximately 600 chemicals fit these criteria, and should be of high priority for further examination of their toxicity and potential risk. Many organizations, including federal and state agencies, have also attempted to make "lists", and it appears that several hundreds of chemicals fall into this general category of CECs. There are very limited data on their physical chemical properties or toxicological effects.

WHERE DO CECS COME FROM?

The general category of CECs contains a wide variety of chemical classes, functionalities, and structures. They include natural and synthetic hormones, a variety of industrial chemicals, plasticizers, surfactants, musks and fragrances, pharmaceuticals, current use pesticides, and naturally-occurring phytoestrogens. The hormones include endogenous reproductive hormones such as 17β-estradiol and synthetic ones such as ethynyl estradiol (birth control pills), steroid hormones, thyroid hormones, and growth hormones. Industrial chemicals include flame retardants, perfluoroalkyl contaminants, cyclic siloxanes, and others. Plasticizers are additives to polymer formulations to make them clear, opaque, malleable, hard, etc., and include a number of phthalate compounds as well as the now-infamous bis-phenol A (BPA). Surfactants are added to soaps and detergents to improve their performance; the class of chemicals of most concern is the alkylethoxylates and their stable degradation products, the alkyl phenols. Musk and fragrances are added to consumer products to make them more attractive. Pharmaceuticals include those for human and veterinary use, prescription and over-the-counter drugs. Some current use pesticides have been shown to be endocrine disruptors, including atrazine and metalochlor.[3] Phytoestrogens are compounds in certain legumes that have estrogenic properties, such as the compound genestien found in soy products. Examples of these chemicals are shown in Figure 1.

Historically, hazardous chemicals were managed by regulating point sources—the banning of a particular industrial chemical from or used by a given manufacturer, for example. However, many CECs today are coming from consumer product use and disposal (as well as manufacture) and thus they enter the environment as non-point source pollutants as well as from point sources. They largely collect in our waste streams (usually water as opposed to solid) and get funneled to and concentrated in wastewater treatment plants. By far the largest body of literature on the occurrence of CECs is from studies of wastewater treatment plant effluent.

HOW DO THEY CAUSE HARM?

The various CECs enter the blood stream from environmental exposure, and can be transported throughout the body. They enter certain cells, and bind to chemical receptors on cell membranes or within the cell nucleus, thus initiating a variety of cell responses that were not intended. The potential impacts on humans and animals include reproductive disruption (e.g., estrogen and androgen receptors), metabolic and growth disorders (thyroid receptors, cortisol receptors), adrenal system (stress) disruption, immune system disorders, hepatic and renal disorders, and central nervous system impacts (neurotransmitter receptors).[4,5]

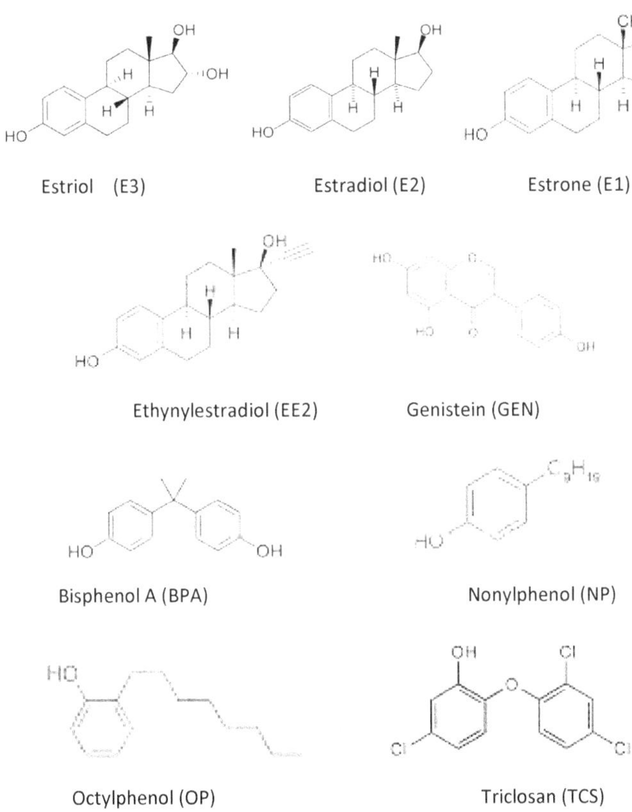

Fig. 1: Chemical structures of representative CECs.

EVIDENCE FOR OCCURRENCE

The iconic book "Our Stolen Future" by Theo Colborn and colleagues[5] raised public awareness of the specific problem of endocrine disrupting chemicals (EDCs). This book was the layman's story of how Dr. Colborn had synthesized a great deal of previously unconnected research to develop the hypothesis that there were chemicals acting as EDCs in the environment, and then brought together some of the best animal and wildlife toxicologists in the country to share their data to test her hypothesis. The data strongly supported her hypothesis, and provided ample evidence that wild fish and other wildlife were being exposed and, in many cases, adversely impacted by EDCs. But what these chemicals were was still not clear.

In 2002, Kolpin et al.[6] published their landmark paper in *Environmental Science & Technology* (still the most highly cited article published in *ES&T*) documenting CEC occurrences throughout the U.S. This reconnaissance study, done by USGS scientists, examined 139 streams in 30 states downstream of likely CEC sources such as wastewater treatment plants, intensive livestock areas, and urban areas, for 95 different chemicals. The researchers found at least one contaminant in 80% of the samples, and 82 of the 95 target analytes were found in one or more samples.

Benotti et al.[7] published a survey done in 2006-2007 of 19 water utilities that provide drinking water to 138 million people (over a third of the U.S. population). They looked for 55 chemical contaminants, and found 11 with substantial frequency. Nine of the frequently detected contaminants were pharmaceuticals, including betablockers, anticonvulsants, anti-inflammatories, antianxiety drugs, antibiotics, and antilipidemics. Also found with great frequency were a flame retardant and the herbicide atrazine. The concentrations were at low relative to therapeutic doses, but there are no data on the potential or real impacts of constant, long term exposure to these compounds at the concentrations found. Snyder and colleagues have recently reviewed pharmaceuticals and EDCs in drinking water and highlight current challenges to drinking water management.[8-10]

In addition to these studies that looked at specific compounds, there have been hundreds of studies that have assessed exposure of wild fishes, birds, and turtles to environmental estrogens by measuring vitellogenin in blood of males of the species. Vitellogenin is the protein precursor to the development of egg yolk, and although males have the genes to produce this protein, it should not be expressed unless a signal is received from an activated estrogen receptor. Thus this easily measured protein is an ideal indicator of environmental estrogen exposure in egg-laying wild fish, birds, and animals.[10] Studies from the UK, U.S., EU, China, Japan, and elsewhere have documented exposures that span the globe.[11-15]

It is also documented that exposures to some CECs may be greatest from unexpected exposure pathways. Typically, exposures from air, food, and water are the pathways that are of most importance. However, for some consumer product additives such as flame retardants, the dominant exposure may be housedust.[16] Thus, infant exposure may be greatest when they are playing on the floor of the home.

The National Health and Nutrition Examination Survey (NHANES) is conducted by the National Center for Health Statistics, Centers for Disease Control and Prevention. This program was designed to collect data and to assess the health and nutritional status of adults and children in the United States. NHANES has documented many CECs in the U.S. population (http://www.cdc.gov/nchs/nhanes.htm). Several researchers have begun to mine these data and examine correlations of CECs in blood and urine with health outcome epidemiology data (e.g.,[17]).

EVIDENCE FOR ADVERSE EFFECTS

A tragic demonstration of the adverse effects of CECs came from the widespread use of DES (diethylstilbesterol) to prevent miscarriages. DES was first synthesized in the UK in 1938, and was found to have estrogenic properties. It was approved for human use in the mid-1940s, and reached peak use in 1953. In the U.S., an estimated 5-10 million persons were exposed to DES during the period 1941-1971, including women who were prescribed DES while pregnant and the female and male children born of these pregnancies. Tragically, a double blind study in 1950s showed it had no effect on preventing miscarriages. In mid 1970s, studies linked mothers' exposure to daughters in the age group of 14 -22 with clear cell adenocarcinoma of the vagina; a rare cancer of the vagina and cervix.[18] These daughters have suffered from reproductive and other health problems, in addition to an elevated cancer incidence.[19,20]

A landmark set of studies on the reproductive and endocrine disruption of alligators from Lake Apopka, FL, was conducted by Louis Gillette.[21-23] Following federal and state protections as endangered species, the alligator populations rebounded in all lakes in Florida with the notable exception of Lake Apopka. Gillette and colleagues found that there was high mortality of males, and that the few males present had decreased penis size, low levels of testosterone, elevated levels of estrogen, and intersex (the presence of both gonadal and ovarian tissue in the same animal). After a series of laboratory and field studies, the cause of the endocrine disruption was found to be exposure to the anti-androgen Dicofol, a now-banned pesticide similar to DDT that had spilled into the lake from a manufacturing facility.

Since then, many other studies have documented adverse effects to colonial nesting and fish-eating birds worldwide. The effects include reproduction impacts, malformations, as well as behavior changes (e.g., female-female pairing).[24-31]

A recent survey by the U.S. Geological Survey[32] examined 15 species of fish from 111 sites at the downstream portions of nine major river basins in the U.S. The study found that at 34% of the sites, intersex was found in the fish, and concentrated in 4 of the 15 species of fish sampled. Typical background levels of intersex in fish are estimated to be 2%. This study highlights the widespread nature of the exposure and impact of EDCs, and illustrates one of the complicating aspects of the impact indicated by species sensitivity.

Demonstrating cause and effect in human populations is far more difficult. Typically, relationships are inferred from controlled laboratory studies with animal models, combined with correlative observations in epidemiological studies in a weight-of-evidence approach. Considerable controversy exists in using these studies in any regulatory context.

CHALLENGES TO UNDERSTANDING CECS

Our regulatory and policy framework for managing chemicals has been predicated on a risk assessment approach. Risk assessment evaluates the exposure and dose-response of a specific hazard (chemical) to determine what a "safe" level of that chemical is based on a probabilistic risk determination. The process addresses hazards one-chemical-at-a-time, and requires considerable data before a determination of risk can be made. In addition to the paucity of data on both exposures and dose-response of individual CECs, there are many other subtle but complicating challenges to understanding and thus managing CECs.

Assessing Exposure.

For exposure assessment, there are two general approaches used. The first is to use analytical chemistry to measure concentrations of chemicals in a given media (water, food, etc.). This is complicated by the fact that CECs often are present at low concentrations relative to analytical detection limits. Also, the instrumentation needed for many CECs is liquid chromatography coupled with tandem mass spectrometry, and these instruments are very expensive and require specialized operators and thus are not found routinely in environmental analytical laboratories. Finally, these analytical techniques "look under the lamppost", and can only measure a pre-determined list of compounds that may not include all possible CECs.

The second approach for assessing exposure is to use *in vitro* bioassays that measure total estrogenicity, without identifying the specific compounds that contribute to this estrogenicity. Generally, these are screening assays involving estrogen receptor competition that give semi-quantitative results. They can be fouled by particulates or dissolved organic matter in water samples, and so cannot be used on complex samples such as wastewater influent.[33]

Assessing Dose-Response.

The well-known phrase, "dose makes the poison", applies here but not in the conventional manner that is the underpinning of traditional toxicology. Most compounds that are currently regulated have a linear dose-response curve (monotonic)—the greater the dose of exposure, the greater the adverse impact. A major complication in assessing CECs is that many of their dose-response curves have been shown to be either U shaped or inverse-U shaped. This is thought to be related to the fact that the mode of action is receptor mediated, or that there are different modes of action occurring at different exposure doses. For example, for the inverse-U shaped curve, it is thought there is an optimal mid-level dose that interacts with the available receptors to cause an effect—lower doses have less of an effect, and greater doses saturate the receptors and decrease the effect. For U-shaped dose-response curves, very low doses and very high doses may have an adverse impact, but lesser effects are seen at intermediate doses. This may be due to two different receptor responses occurring, or two different modes of action (e.g.,[34]). This complication is greater than it may seem, as not only do different chemicals have different dose-response curves, the curve will vary by animal, by genius and species of that animal, and by when in the life cycle of the animal they were exposed (e.g.,[35]). A particular confounding effect is that different doses of a given chemical to a given animal can lead to different effects. This also implies that low dose exposures may be far more significant than previously thought.[36] Imagine the number of experiments that would be necessary to acquire all the necessary toxicological data for just one CEC—for relevant species, different doses, and for different life cycle exposures—the number becomes astronomically large and impracticable when considering the potential universe of hundreds of CECs.

These issues are further complicated by the fact that there are many types of endocrine receptors, including the estrogen, androgen, thyroid, cortisol, neurotransmitter, and others. They are found in many tissue types, and there are multiple versions of them—in the human body, there are 3 types of estrogen receptors, and they are found in different ratios and abundance in all different tissue types in the body. Thus the effect may vary depending on which target cell type receives the contaminant.

These complications are all related to assuming receptor mediated responses. However, the endocrine system consists of a cascade of responses—a chemical messenger (or contaminant) passes through a cell wall and enters the nucleus, binds to a receptor, and then interacts with the DNA to up-regulate or down-regulate specific genes. This results in mRNA transcription and a production of a specific protein, causing a biological response. See Figure 2 for a picture of this process. At the point of receptor binding, the CEC can mimic the andogenous hormone (agonist) or block its normal binding (antagonist). In addition to receptor binding, the contaminant can also interfere at any point in this cascade[37]—and we do not yet have the needed bioassays to evaluate these other types of interactions.

The resulting effects are wide-ranging, and can include multiple interferences in the reproductive hormone system, thyroid system, steroidogenic system, adrenal system, hepatic system, renal system, and central nervous system.[4,38-40]

Yet another level of complexity is added by the fact that these effects do not occur in isolation. "Cross-talk" is ubiquitous across signaling pathways, and there is now evidence that disruption of one signaling pathway (e.g., thyroid) can disrupt another (e.g., fatty acid metabolism).[41-43]

Finally, there are two additional, insidious confounding aspects to understanding CEC effects. First, many of these effects have been shown to be transgenerational, meaning that the effects are not manifested in the individual who is exposed, but are expressed in their offspring and subsequent generations. This is the landmark aspect of the DES story, and has been shown to be true for other chemical exposures. For example, a longitudinal epidemiology study of children born to women who consumed highly contaminated Great Lakes fish showed no effects in the mothers, but showed a series of effects in their children from birth through adolescence. The effects were all related to their dose in the womb, and not to exposures after birth.[44-49] Numerous animal studies have established transgenerational effects from a variety of CECs.[50-55]

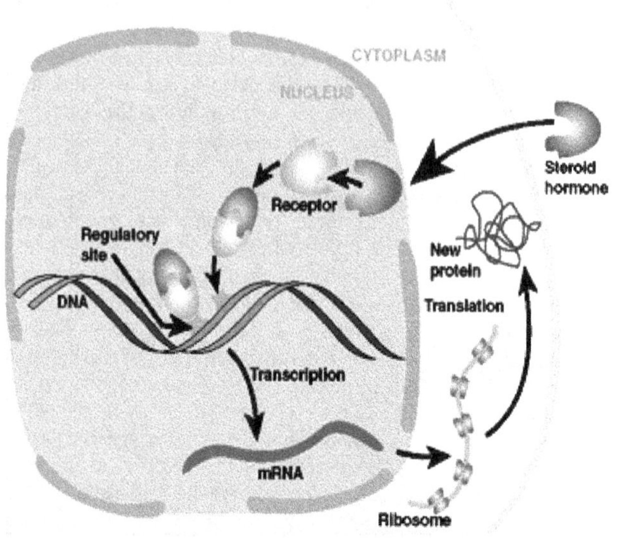

Fig. 2: How chemical messengers react with receptors to create a biological response. From www.ehormone.tulane.edu/

The other confounding aspect is the relationship of these chemical exposures to epigenetic disease. This phenomenon results when there are changes to DNA from chemical exposure (or that occur with age) that are not genetic, but affect the susceptibility of an individual to getting a disease from chemical exposure. A growing body of literature is demonstrating that diabetes, obesity, and other diseases can be caused by epigenetic factors, and may be linked to environmental chemical exposures.[37,54,56-64]

A NEW PATH FORWARD

Clearly, given these confounding factors, assessing each chemical one-by-one will not be successful at protecting human health and ecosystems. There are a number of strategies that have been brought forward in recent years, and many provide a ray of hope that society can get ahead of this problem.

Precautionary Hazard Approach:

This approach would regulate chemicals based on hazard identification and exposure, but not on risk. Thus chemicals would be banned and restricted on the potential for harm without calculating the specific dose response and assessing risk. This is called 'precautionary' as it is a highly conservative approach.

Burden of Proof:

In the U.S., chemicals in commerce are assumed to be reasonably safe, until a problem is discovered and EPA can show that they are harmful and need to be restricted or banned (PCBs are an example). The burden is on the government, and only after a chemical is in the environment and has shown to cause harm. An alternative approach is for industry to demonstrate that a chemical is not harmful in its intended use or as an environmental contaminant prior to being used in commerce. This is essentially the approach being used by the EU and its **R**egistration, **E**valuation, **A**uthorization and Restriction of **Ch**emical substances program[65] which went into effect in 2007.

21st Century Toxicological Approaches:

If a risk-based approach is used, it will need to be transformed considerably to account for cumulative exposures, exposures to mixtures, and the use of modeling instead of extensive animal studies.[66] Of promise in this regard is the active use of computer simulation models (*in silico* models) that predict toxicological activity from chemical structure and elemental composition, using verification data from existing laboratory and animal studies. The U.S. EPA has a world-class Computational Toxicology program that is working to develop and verify such models.[67] EPA has also developed computer models that simulate organ response, such as the Virtual Liver. Improved data bases and data management are also key to the success of such approaches.

21st Century Risk Evaluation:

There are also improvements being developed in risk assessment, including evaluating cumulative risk (risk over a life time), multiple stressor risk (cumulative impact of being exposed to mixtures of chemicals, combined with other non-chemical stressors such as temperature, age, etc.), and evaluation of groups of chemicals by 'lumping' them based on mode of action. For example, one could do risk assessment at the rigor needed for regulation of all compounds that cause an effect by binding to one of the thyroid receptors. Recent research supports that the effects are additive, provided they are caused by the same mode of action.[37]

Green Chemistry and Manufacturing:

This approach[68] aims at getting at the root cause of the problem, the manufacture and use of toxic chemicals in industry. The transition from using toxic chemicals to using safer chemicals solves several problems at once: it reduces the cost of waste storage, handling, and disposal to industry; it improves safety for industrial workers; it reduces the human and environmental costs caused by adverse effects downstream. It should also stimulate innovation and economic growth, provided the right incentive structure is put in place to encourage industry to move in that direction and assist with the upfront investments needed to stay competitive. There is increased momentum within the business world to embrace green chemistry.[69]

Addressing CECs is a daunting task, but one that must be done to achieve a sustainable and healthy society. Many of the approaches summarized are being actively explored and adopted, and have the potential to overcome the challenges we face today.

REFERENCES

1. Howard, P.H. and C.G. Muir Derek. *Environmental Science & Technology*, 2010. **44**: p. 2277-2285.
2. Muir Derek, C.G. and P.H. Howard. *Environmental Science & Technology*, 2006. **40**: p. 7157-7166.
3. Tillitt, D.E., D.M. Papoulias, J.J. Whyte, and C.A. Richter. *Aquat. Toxicol,* 2011. **99**(2): p. 149-159.
4. Diamanti-Kandarakis, E., E. Palioura, S.A. Kandarakis, and M. Koutsilieris. *Hormone Metabolism Research*, 2010. **42**(8): p. 543-552.
5. Colborn, T., D. Dumanoski, and J.P. Meyers, *Our Stolen Future: Are We Threatening our Fertility, Intelligence, and Survival? A Scientific Detective Story*. 1996, New York, NY: Dutton Publishers.
6. Kolpin, D.W., et al. *Environ. Sci. Technol.*, 2002. **36**(6): p. 1202-1211.
7. Benotti, M.J., et al. *Environmental Science & Technology,* 2009. **43**(3): p. 597-603.
8. Snyder, S.A. and M.J. Benotti. *Water Science and Technology*, 2010. **61**(1): p. 145-154.
9. Snyder, S.A., et al. *State of Knowledge of Endocrine Disruptors and Pharmaceuticals in Drinking Water*. 2009, Water Intelligence Online, IWA Publishing.
10. Sumpter, J.P. and S. Jobling. *Environ. Health Perspect.*, 1995. **103(Suppl 7)**: p. 173-178.
11. Furuichi, T., K. Kannan, J.P. Giesy, and S. Masunaga. *Water Res.*, 2004. **38**: p. 4491-4501.
12. Gracia, T., et al. *Environmental Science and Pollution Research International*, 2008. **15**: p. 332-343.
13. Jobling, S., et al. *Environ. Sci. Technol.*, 1998. **32**: p. 2498-2506.
14. Pojana, G., et al. *Environ. Toxicol. Chem.*, 2004. **23**(8): p. 1874-1880.
15. Ternes, T.A. *Water Res.*, 1998. **32**: p. 3245-3260.
16. Stapleton, H.M., et al. *Environ. Sci. Technol.*, 2005. **39**(4): p. 925-931.
17. Elobeid, M.A., et al. *Int. J. Environ. Res. Public Health.* **7**: p. 2988-3005.
18. Herbst, A.L., H. Ulfelder, and D.C. Poskanzer. The New England Journal of Medicine, 1971. **284**(15): p. 878-81.
19. McLachlan, J.A., R.R. Newbold, M.E. Burow, and S.F. Li. *Apmis*, 2001. **109**(4): p. 263-272.
20. Newbold, R.R. *Toxicology and Applied Pharmacology*, 2004. **199**(2): p. 142-150.
21. Guillette Jr., L.J., et al. *Environ. Health Perspect.*, 1994. **102**: p. 680-688.
22. Guillette Jr., L.J., et al. *Environ. Health Perspect.*, 1995. **103 (Suppl 4)**: p. 31-36.
23. Guillette Jr., L.J., et al. *General Comparative Endocrinology*, 1995. **101**: p. 32-34.
24. Feyk, L.A. and J.P. Giesy. *Xenobiotic modulation of endocrine function in birds*. 1998.
25. Ottinger, M.A., et al. *Brain Res. Rev.*, 2008. **57**(2): p. 376-385.
26. Ottinger, M.A., et al. *J. Environ. Sci. Health, Part C: Environ. Carcinog. Ecotoxicol. Rev.*, 2009. **27**(4): p. 286-300.

27. Ottinger, M.A. and K.M. Dean. *J Toxicol Environ Health B Crit Rev*, 2011. **14**(5-7): p. 413-22.
28. Touart, L.W. and M.A. Ottinger. *Target Organ Toxicol. Ser.*, 2010. **27**(Endocrine Toxicology): p. 313-323.
29. Bosveld, A.T.C. and M. van den Berg. *Toxicology*, 2002. **181-182**: p. 155-159.
30. Crews, D., et al. *Pure and Applied Chemistry*, 2003. **75**(11-12): p. 2305-2320.
31. Carere, C., et al. *Ann Ist Super Sanita*. **46**(1): p. 81-8.
32. Hinck, J.E., et al. *Aquat. Toxicol*, 2009. **95**(1): p. 60-70.
33. Swackhamer, D.L., et al. *Submitted for publication*, 2011.
34. Weltje, L., F.S. vom Saal, and J. Oehlmann. *Hum. Exp. Toxicol.*, 2005. **24**(9): p. 431-437.
35. Braw-Tal, R. *Pediatr Endocrinol Rev*, 2010. **8**(1): p. 41-6.
36. Welshons, W.V., et al. *Environ. Health Perspect.*, 2003. **111**(8): p. 994-1006.
37. Tabb, M.M. and B. Blumberg. *Mol. Endocrinol.*, 2006. **20**(3): p. 475-482.
38. Latini, G., et al. *Mini-Rev. Med. Chem.*, 2010. **10**(9): p. 846-855.
39. Miodovnik, A., et al. *NeuroToxicology*. **32**(2): p. 261-267.
40. Weiss, B. J. *Neurol. Sci.*, 2011. **305**(1-2): p. 11-21.
41. Swedenborg, E., J. Ruegg, S. Maekelae, and I. Pongratz. *J. Mol. Endocrinol.*, 2009. **43**(1,2): p. 1-10.
42. Mu, X., et al. *Environ. Toxicol. Chem.*, 2005. **24**(1): p. 146-152.
43. Iguchi, T. *Kagaku to Seibutsu*, 2005. **43**(1): p. 61-66.
44. Fein, G.G., et al. *Journal of Pediatrics*, 1984. **105**: p. 315-320.
45. Jacobson, J.L., S. W. Jacobson, *New methodologies for assessing the effects of prenatal toxic exposure on cognitive functioning in humans*, in *Toxic Contaminants and Ecosystem Health: A Great Lakes focus*, M.S. Evans, Editor. 1988, John Wiley and Sons: New York. p. 373-388.
46. Jacobson, J.L., et al. *Am. J. Public Health*, 1984. **74**: p. 378.
47. Jacobson, J.L. and S.W. Jacobson. *J. Great Lakes Res.*, 1993. **19**(4): p. 776-783.
48. Jacobson, J.L. and S.W. Jacobson. *N. Engl. J. Med.*, 1996. **335**: p. 783-789.
49. Jacobson, J.L. and S.W. Jacobson. *Developmental effects of PCBs in the fish eater cohort studies*. in *PCBs, Recent Advances in Environmental Toxicology and Health Effects*. 2001: Unversity Press of Kentucky, Lexington, KY.
50. Anway, M.D. and M.K. Skinner. *Prostate* (Hoboken, NJ, U. S.), 2008. **68**(5): p. 517-529.
51. Nilsson, E.E., M.D. Anway, J. Stanfield, and M.K. Skinner. *Reproduction* (Bristol, U. K.), 2008. **135**(5): p. 713-721.
52. Porta, M., D.-H. Lee, and *E. Puigdomenech. Occup Environ Med*, 2009. **66**(3): p. 141-2.
53. Walker, D.M. and A.C. *Gore. Nat. Rev. Endocrinol.*, 2011. **7**(4): p. 197-207.
54. Skinner, M.K., M. Manikkam, and C. Guerrero-Bosagna. *Reprod. Toxicol.*, 2011. **31**(3): p. 337-343.
55. Skinner, M.K., M. Manikkam, and C. Guerrero-Bosagna. *Trends Endocrinol. Metab.*, 2010. **21**(4): p. 214-222.
56. Anway Matthew, D. and K. Skinner Michael. *Reprod Biomed Online*, 2008. **16**(1): p. 23-5.
57. Baccarelli, A. and V. Bollati. *Curr Opin Pediatr*, 2009. **21**(2): p. 243-51.

58. Birnbaum Linda, S. and P. Jung. *Health Aff (Millwood),* 2011. **30**(5): p. 814-22.
59. Crews, D. *Hormones* (Athens), 2010. **9**(1): p. 41-50.
60. Newbold, R. and H.K. Kinyamu. *Reprod. Toxicol.* (3rd Ed.), 2010: p. 317-331.
61. Patisaul, H.B. and H.B. Adewale. *Front. Behav. Neurosci.*, 2009. **3**(June): p. No pp given.
62. Hatch, E.E., J.W. Nelson, R.W. Stahlhut, and T.F. Webster. *Int. J. Androl.*, 2010. **33**(2): p. 324-332.
63. Chen, J.-Q., T.R. Brown, and J. Russo. B*iochim. Biophys. Acta, Mol. Cell Res.*, 2009. **1793**(7): p. 1128-1143.
64. Grun, F. and B. Blumberg. *Mol. Endocrinol.*, 2009. **23**(8): p. 1127-1134.
65. REACH. Available from:http://ec.europa.eu/environment/chemicals/reach/reach_intro.htm.
66. (NRC), N.R.C., ed. *Toxicology for the 21st Century.* 2007, National Academy of Sciences Press: Washington, DC. 196.
67. EPA, U. *Computational Toxicology Program.* Available from: http://www.epa.gov/ncct/.
68. Anastas, P.T. and J.C. Walker, *Green Chemistry: Theory and Practice.* 2000, USA: Oxford University Press. 152.
69. Berns, M., et al. *The Business of Sustainability*, in *MITSloan Management Review*. 2009, MIT: Hollywood, CA.

FATE OF CONTAMINANTS OF CONCERN IN CONVENTIONAL AND ADVANCED WATER TREATMENT AND MANAGED AQUIFER RECHARGE SYSTEMS

PROFESSOR JÖRG E. DREWES
Environmental Science and Engineering, Colorado School of Mines
Golden, USA

ABSTRACT

Drinking water sources that receive wastewater discharge or that are augmented in part through wastewater recycling can contain wastewater-derived trace organic compounds at concentrations that might cause potential adverse effects to aquatic life or human health. Conventional drinking water treatment processes with the exception of chemical oxidation are rather inefficient in removing CECs. Managed aquifer recharge (MAR) systems, such as riverbank filtration (RBF) and soil-aquifer treatment (SAT), are characterized as multi-objective and low carbon-footprint treatment processes that could either substitute above-ground treatment systems or be employed in conjunction with engineered water treatment processes to mitigate CECs. With respect to CECs, previous research has identified that adsorption, biotransformation, and dilution are the most important attenuation mechanisms during subsurface treatment, however, the environmental conditions (i.e., microbial adaptation, redox conditions, availability of organic substrate) under which this attenuation is occurring or could potentially be enhanced are not well understood. This paper will highlight recent findings regarding the efficiency of conventional and advanced engineered water treatment processes as well as managed natural treatment system to attenuate CECs.

INTRODUCTION

Over the past decades, an increasing number of scientific studies began reporting the presence of organic compounds at trace concentrations in surface and groundwater under the impact of wastewater discharge. These trace organic chemicals of emerging concern (CECs) include endocrine disrupting chemicals (EDCs), pharmaceutically active compounds and personal care products (PPCPs), household chemicals, food additives, and emerging disinfection by-products. These trace organic compounds can occur at concentrations that might cause potential adverse effects to aquatic life or human health. A study published by the U.S. EPA in 1980 estimated that about 15 million people in the United States were served by surface supplies containing at least 10 percent wastewater at low flow conditions and 4 million people use municipal supplies that contain 100 percent wastewater during low flow conditions (Swayne and Boone et al. 1980). With increasing water demand, dwindling water resources in many communities, and increasing interest in planned potable reuse applications over the last 30 years, the proportion of wastewater-impacted drinking water supplies in the U.S. has likely risen significantly. With the recent recognition that chemicals of emerging concern might be present at elevated concentrations in these water supplies, mitigation strategies are needed that are effective in reducing the exposure to these chemicals both to aquatic life and humans.

While water treatment technologies are available to attenuate the majority of contaminants detectable in water, these processes are often energy intensive and may result in residual disposal challenges. Although, past studies regarding treatment efficiency have been mostly empirical with little focus on modeling process attenuation mechanisms. This presentation will discuss the current state-of-the-art to use engineered and natural treatment system to mitigate these trace organic chemicals and current attempts to develop a better process understanding for design and operation of these treatment processes.

FATE OF CECS DURING CONVENTIONAL WASTEWATER TREATMENT

Trace organic chemicals are discharged directly into sewage contributing to the mix of organic compounds present in municipal wastewater. Some of the most prevalent CECs found in wastewater treatment plants are pharmaceutical residues and personal care products (PPCPs), household chemicals (HHC), industrial chemicals, synthetic and natural hormones, pesticides and disinfection by-products (Ternes, 1998; Daughton et al. 1999; Drewes and Shore, 2001; Heberer, 2002, Glassmeyer et al. 2005). Until very recently, relatively little was known about the efficiency with which CECs are removed during wastewater and biosolids treatment processes. The two prevalent mechanisms for the CEC indicator removal during secondary treatment are biotransformation of the parent compound and sorption. The removal efficiencies for strongly adsorbing, biologically labile, and volatile organic compounds range from 25 percent for primary treatment and 80-98 percent for secondary treatment (Tchobanoglous, 2003).

No systematic and comprehensive work has described the dimensions of CEC issues in wastewater treatment, including origins, distributions, fate and transport. Various approaches have been proposed to estimating CEC concentrations in sewage. Some of these studies have focused on closed systems such as hospitals (Kuemmerer et al. 1997; Heberer and Feldmann, in press), utilization of prescription rate data in combination with per-capita sewage volume (Stuer-Lauridsen et al. 2000, Huang et al. 2001, Sedlak and Pinkston, 2001), production data of high-production volume chemicals (Drewes et al. 2008), and physicochemical information and mass balances approaches (Ternes et al. 2004; Khan and Ongerth, 2004) to predict the presence of CEC in conventional wastewater. There are a number of operational factors which are likely to influence the biological removal of CEC in activated sludge systems such as biochemical oxygen demand (BOD_5), suspended solids (SS) loading, hydraulic residence time (HRT), solids retention time (SRT), food-microorganism ratio (F/M ratio), mixed liquor suspended solids (MLSS), pH, and temperature. Frequently, however, these operational details are lacking in the literature on the fate and transport of CECs during wastewater treatment. Figure 1 provides a summary of frequently detected CECs in secondary/tertiary treated effluents in the United States.

Mechanisms of CEC Removal: Sorption

Findings from multiple studies have demonstrated that adsorption onto suspended solids, aerobic and anaerobic biodegradation, chemical (abiotic) degradation (via processes such as hydrolysis), and volatilization are the primary removal mechanisms for CEC in wastewater. Physicochemical properties will influence whether a CEC will remain in the aqueous phase (like many of the acidic, neutral and basic pharmaceuticals) or interact with solid particles, such as estrogens or certain antibiotics, which have a higher potential to be adsorbed to sewage sludges. Sorption and volatilization are physical processes and their relevance for specific contaminants can be predicted using physicochemical property information. Hydrophobic contaminants may partition onto primary or secondary sludge solids and the tendency to accumulate in sludge solids can be assessed using the octanol-water partition coefficient (K_{OW}).

Mechanisms of CEC Removal: Volatilization

Volatilization losses of CEC during wastewater treatment can be estimated using the following empirically defined categories based on Henry's Law constant (H_c) and K_{OW} (Rogers, 1996):

$H_c > 1*10^{-4}$ and $H_c/K_{OW} > 1*10^{-9}$	high volatilization potential
$H_c < 1*10^{-4}$ and $H_c/K_{OW} < 1*10^{-9}$	low volatilization potential

Mechanisms of CEC Removal: Biotransformation

Biotransformation of CEC might occur during secondary treatment, which involves both aerobic (trickling filters, activated sludge treatment) or anaerobic (biological P-removal; sludge digestion) unit operations. The removal efficiencies of CECs during primary and secondary treatment have been highlighted in previous review articles (Drewes and Shore, 2001; Drewes, 2007; Snyder et al. 2008). Lessons learned from these reviews are that many studies published in the peer-reviewed literature regarding CEC removal during conventional wastewater treatment have utilized either simplistic batch reactors receiving synthetic feed water that poorly represents municipal wastewater qualities, employed spiked target compounds orders of magnitude above ambient levels, conducted experiments that did not allow full adaptation of biocommunities, or fail to report the operational boundary conditions of laboratory-, pilot- or full-scale reactors employed in these studies. Findings of these studies do allow some estimation of key removal pathways of CECs during conventional wastewater treatment and the relationship between operational parameters and physicochemical characteristics of the compounds, but the knowledge base is insufficient to develop accurate and predictive fate models.

A recent study funded by the Water Environment Research Foundation (WERF) explored the relationships between operational parameters and attenuation of CECs during conventional wastewater treatment and the use of quantitative structural property relationship (QSPR) to describe their fate. The overall goal of this study was to better assess the performance of secondary treatment processes for CEC removal based on selected CEC indicator compounds and to advance our ability to identify operational strategies that may lead to a higher efficiency for reduction. The functional relationships between operational parameters (such as SRT and HRT) and CEC removal established in full-, pilot-, and lab-scale sample events formed the basis for efforts to calibrate CEC fate models that help improve our ability to predict CEC removal during conventional wastewater treatment. In this study, both key operational parameters, SRT and HRT, generally showed a positive correlation with the overall removal of CEC during secondary treatment.

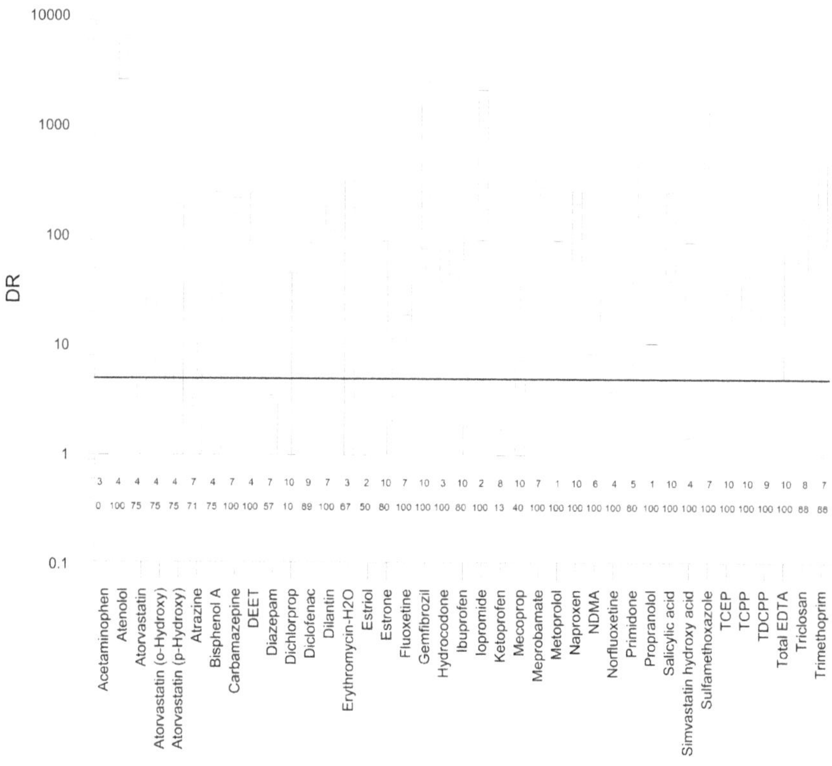

Fig. 1: Box and whisker plots of the detection ratio of trace organic chemicals in U.S. treated wastewater effluents. The total number of WWTP effluents evaluated and detection frequency (%) for a given compound are listed horizontally and numerically within the graph. (adopted from Dickenson et al. 2011, Water Research 45, 1199)

FATE OF CECS DURING CONVENTIONAL AND ADVANCED WATER TREATMENT

Fate of CECs during Conventional Drinking Water Treatment

Several studies reported that pharmaceuticals (i.e., sulfamethazine, trimethoprim), steroids (i.e., estrone), personal care products (i.e., DEET) and other compounds (i.e., chlorinated flame retardants such as TCEP) are poorly removed during conventional drinking water treatment processes using ferric or alum coagulation (Adams et al. 2002;

Ternes and Meisenheimer et al. 2002; Ballard and MacKay, 2005; Westerhoff and Yoon et al. 2005; Vieno and Tuhkanen et al. 2006).

Stackelberg et al. (2004) evaluated the occurrence of PPCPs within a drinking water treatment plant (DWT) within a highly urbanized area where more than 50 wastewater treatment plants discharge effluent to the two streams that provide source water for the 235 million L/d DWT plant. The process train includes the following processes in order: coagulation/flocculation/sedimentation ($FeCl_3$ = 20-45 mg/L), disinfection (NaClO; contact time of 200-300 minutes), GAC filtration (after 2 months operating time; contact time of 1.5 to 3 min), and 2nd disinfection (chlorine residual ~ 1.2 mg/L). The average concentrations before and after the respective treatment processes are presented in Figure 2. Based upon findings of this study, clarification had little impact on the removal of PPCPs, whereas chlorination and GAC filtration were more effective removal mechanisms depending on the individual compound. In the finished water all compounds were reduced to various degrees, however some compounds were detected in the finished water indicating incomplete degradation or removal. For example, carbamazepine and DEET were detected in every finished water sample and cotinine and tonalide were detected in 75 percent and 50 percent, respectively, of finished-water samples.

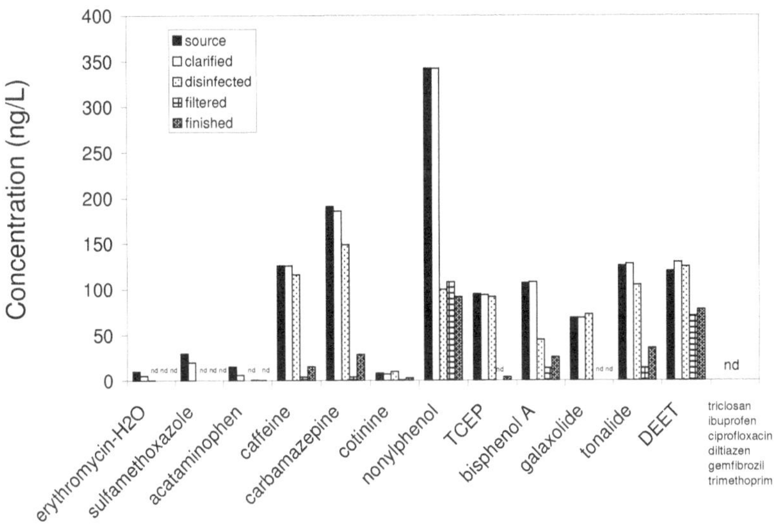

Fig. 2: Average concentrations of select CECs before, during, and after full-scale drinking water treatment. Data is from Stackelberg et al. (2004)

Fate of CECs during Disinfection Processes

Since disinfection processes are oxidation processes, these unit operations have the potential to transform CECs. In drinking water treatment systems, chlorine, chlorine dioxide, ozone, and UV are frequently used as disinfectants. Chlorine, chlorine dioxide and ozone oxidations of CECs are selective for certain chemical structures and functional groups. Among these three oxidants, ozone is the most reactive.

Chlorine can be used to remove CECs to concentrations below the detection limit (Adams and Wang et al. 2002; Gallard and Leclercq et al. 2004; Lee and Kamata et al. 2004; Boyd and Zhang et al. 2005; Dodd and Shah et al. 2005; Westerhoff and Yoon et al. 2005; Bedner and MacCrehan 2006). Westerhoff et al. (2005) observed removal exceeding 95 percent of the following compounds: acetaminophen, diclofenac, estrone, 17b-estradiol, 17a-ethynylstradiol, naproxen, sulfamethoxazole, triclosan, erythromycin-H_2O, carbamazepine, gemfibrozil, and trimethoprim. Pinkston and Sedlak (Pinkston and Sedlak, 2004) also observed significant reactivity of chorine with gemfibrozil, naproxen, acetaminophen, atenolol, metoprolol, and propranolol. Westerhoff et al. (2005) observed that compounds with primary or secondary amines (i.e., diclofenac, sulfamethoxazole, trimethoprim) and compounds with phenols (i.e., estrone, 17b-estradiol, 17a-ethynylstradiol, and acetaminophen) or substituted phenols (i.e., triclosan) were very reactive with chlorine. Alum et al. (2004) and Lee et al. (2004) also reported bisphenol A and nonylphenol, phenolic compounds, were very reactive with chlorine. However, chlorine substituted aromatic rings (i.e., diazepam) were less reactive with chlorination because the chlorine atom is electron-withdrawing (Westerhoff and Yoon et al. 2005). Westerhoff et al. (2005) also observed chlorine was inefficient at removing ibuprofen, DEET, iopromide, and TCEP. These compounds have either electron-withdrawing groups or lack conjugated carbon bonds.

Chlorine dioxide is generally a stronger oxidant than free chlorine. Huber et al. (2005) observed appreciable removals of sulfamethazine, sulfamethoxazole, estrone, 17b-estradiol, 17a-ethynylstradiol, roxithromycin, erythromycin-H_2O, and diclofenac by chorine dioxide. However, caffeine, clofibric acid, gemfibrozil, ketoprofen, naproxen, iopromide were recalcitrant to chlorine dioxide oxidation. Compared to chlorine, the rate constants for the reactions of chlorine dioxide with 17a-ethynylstradiol and naproxen are higher than chlorine rate constants (Huber and Korhonen et al. 2005). Ozonation is a strong oxidant and very effective in the transformation of EDCs and PPCPs (Adams and Wang et al. 2002; Andreozzi and Marotta et al. 2002; Ternes and Meisenheimer et al. 2002; Huber and Canonica et al. 2003; Alum and Yoon et al. 2004; Huber and Göbel et al. 2005; McDowell and Huber et al. 2005; Westerhoff and Yoon et al. 2005). However, Huber et al. (2005) observed X-ray contrast media (i.e., iopromide, deteriorate) were only partly oxidized. Westerhoff et al. (2005) reported that ozone oxidized 20 EDCs, PPCPs and other compounds by more than 80 percent except for iopromide, TCEP and ibuprofen. TCEP is an aliphatic compound with polar (chlorine) functional groups and ibuprofen has an electron-withdrawing functional group on the aromatic ring. Ozone will have higher compound removals compared to chlorine. For nucleophilic sites, ozone is a better oxidant. Also, during ozonation, hydroxyl radicals will form, which are powerful oxidants that react nonselectively with most organic compounds. Ultraviolet irradiation is a very efficient disinfection process. UV can also be used for treatment of CECs with chromophores that adsorb at UV wavelengths. However, typical UV doses required for disinfection (i.e., <5-30 mJ/cm^2) are orders of magnitude lower than those used for destructive treatment of CECs. Even at higher doses of 3,000 mJ/cm^2 only 50 to 80 percent of the pharmaceuticals (i.e., trimethoprim, sulfamethazine) could be removed (Adams and Wang et al. 2002). Therefore, trimethoprim, sulfamethazine will have low susceptibility to typical UV dosages. Rosenfeldt and Linden (2004) also reported poor removal (<22 percent) of 17b-estradiol, 17a-ethynylstradiol, and bisphenol A at UV doses of 3,000 mJ/cm^2. Thus, UV at disinfection dosages is inappropriate to remove CECs.

Fate of CECs during Advanced Water Treatment

Several advanced water treatment processes can efficiently remove most CECs. The most common processes currently employed are activated carbon treatment, advanced oxidation processes (AOP), and high-pressure membranes (reverse osmosis, nanofiltration). It is noteworthy that none of these processes represent a 100 percent barrier against contaminants of concern and for complete removal processes combinations employing different types of processes ("multiple barriers") are required. The major drawback of these advanced processes is their carbon footprint. These processes are energy intense (e.g., AOPs, high-pressure membranes) and generate residual streams (e.g., activated carbon, high-pressure membranes), which are difficult to dispose off, especially in inland settings.

Activated Carbon Treatment

Activated carbon can be used to remove CECs completely or partially from drinking water (Fuerhacker and Dürauer et al. 2001; Tanghe and Verstraete, 2001; Adams and Wang et al. 2002; Ternes and Meisenheimer et al. 2002; Yoon and Westerhoff et al. 2003; Vieno and Tuhkanen et al. 2005; Westerhoff and Yoon et al. 2005). Activated carbon can be used in the following two types: powdered activated carbon (PAC) and granular activated carbon (GAC). PAC is usually added in contact basins (1-5 hours) prior to coagulation at dosages of 5-50 mg/L. At typical plant PAC dosages, PAC can effectively remove pharmaceuticals (i.e., trimethoprim), steroids (i.e., 17b-estradiol), personal care products (i.e., triclosan), and other compounds (i.e., bisphenol A) (Adams and Wang et al. 2002; Yoon and Koyanagi et al. 2003; Westerhoff and Yoon et al. 2005).

The performance of activated carbon adsorption depends upon the properties of the activated carbon sorbent (i.e., surface area, pore size distribution, surface charge, oxygen content) and on the properties of the CEC (i.e., shape, size, charge, and hydrophobicity). Activated carbon with the point of zero charge near the ambient pH will have less ionization and potentially have a greater affinity for hydrophobic compounds (Yoon and Westerhoff et al. 2003). Activated carbon adsorption can readily remove CECs except for a limited number of polar compounds, such as the iodinated contrast media, sulfamethoxazole or the artificial sweetener sucralose, that are insufficiently removed by activated carbon.

Advanced Oxidation Processes (AOP)

AOPs can form hydroxyl radicals, which can non-selectively attack and transform CECs. AOPs include UV/hydrogen peroxide, ozone/hydrogen peroxide, and UV/ozone. Zwiener and Frimmel (2000) observed removals exceeding 90 percent for diclofenac, ibuprofen, and clofibric acid at ozone and hydrogen peroxide doses greater than 3.7 and 1.4 mg/L, respectively (O_3:H_2O_2 = 2.5 mg/mg; 10 minute contact time). For a UV/H_2O_2 process, Rosenfeldt and Linden (2004) observed removal exceeding 90 percent removals of bisphenol A, 17b-estradiol, and 17a-ethynylstradiol at a UV dose of 1,000 mJ/cm^2 and a hydrogen peroxide dose of 15 mg/L. Huber et al. (2003) observed that the ozone/hydrogen peroxide AOP process considerably increased the removal of ibuprofen from 40 to 80 percent. Westerhoff et al. (2005) observed that an addition of a small amounts of H_2O_2 (0.025 mg of H_2O_2/mg of O_3) prior to ozonation generally improved the extent of CEC oxidation by 5-15 percent as compared to ozone alone. In this study testosterone exhibited more than 20 percent higher oxidation in the presence of hydrogen peroxide. AOPs are very effective treatment processed for oxidizing CECs, however compared to ozone, AOPs provide only small increase in removal efficiency.

High-pressure Membranes (Reverse Osmosis and Nanofiltration)

High-pressure membranes consists of nanofiltration (NF) or reverse osmosis (RO). NF/RO membrane operations can physically remove CECs and greater than 90 percent removal can be achieved by RO and tight NF membranes. Some of the mechanisms affecting the permeation of CECs during RO and NF treatment are fairly well understood, such as (1) physical sieving of solutes larger than the molecular cutoff (MWCO) of a membrane. Other mechanisms of rejection, such as (2) electrostatic repulsion and (3) hydrophobic-hydrophobic interactions between membrane and solute, have been studied mainly at the bench-scale (Kiso and Nishimura et al. 2000; Kiso and Kon et al. 2001; Kiso and Sugiura et al. 2001; Ozaki and Li 2002; Ngheim and Schaefer et al. 2004). The following key solute parameters have been identified by Bellona et al. (2004) in a comprehensive literature review to primarily affect solute rejection during NF/RO: molecular weight, molecular size (i.e., length and width), acid dissociation constant (pKa), hydrophobicity /hydrophilicity (i.e., log K_{ow}), and diffusion coefficient. Bellona et al. (2004) also identified key membrane properties that affect rejection, which included MWCO, pore size, surface charge, hydrophobicity/hydrophilicity and membrane surface morphology.

Xu et al. (2005) and Nghiem et al. (2005) examined the rejection of pharmaceuticals due to electrostatic repulsion and they observed that the membrane surface charge of NF and RO membranes determines the rejection of ionic hydrophilic solutes, such acidic pharmaceuticals (i.e., naproxen, diclofenac, ibuprofen, ketoprofen, gemfibrozil) and dissociated antibiotics (i.e., sulfamethoxazole). Speciation of acidic pharmaceuticals as a function of pH may result in a dramatic change of rejection (Bellona and Drewes, 2005; Nghiem and Schaefer et al. 2005; Xu, Drewes et al. 2005). Also, fouling can affect the rejection of charged compounds due to the foulant altering the membrane surface characteristics (i.e., surface charge). Xu et al. (2006) observed that the transport of ionic organic compounds was hindered as a result of electrostatic exclusion likely due to a more negative membrane surface charge caused by the foulant.

For neutral pharmaceutical compounds, intrinsic physicochemical properties (i.e., size, polarity, hydrophobicity) of the compounds can affect their rejection. Xu et al. (2005) performed bench-scale experiments and observed the solute size and MWCO of the membrane affected the rejection of hydrophilic and hydrophobic nonionic compounds, such as neutral pharmaceuticals (i.e., carbamazepine, primidone) and steroids (i.e., EE2). Steric exclusion is the major mechanism for retention of uncharged and non-adsorptive organic solutes. When electrostatic repulsion is absent, compounds with a sufficiently high dipole moment (polarity) will approach the membrane pores in a particular orientation, which can affect the separation process (Nghiem, Schaefer et al. 2005). The influence of molecular polarity is significant of compounds that are cylindrical in shape, such as sulfamethoxazole (Nghiem, Schaefer et al. 2005). Nghiem et al. (2005) reported the lower retention of neutral sulfamethoxazole as compared to neutral carbamazepine (NF-270, pH 3.5) was due to the lower dipole moment and more bulky structure of carbamazepine even though they approximately had similar molecular weights. Note the molecular polarity interaction mechanism remains steric in nature.

Yoon et al. (2006) and Nghiem et al. (2004) reported that the last rejection mechanism, hydrophobic-hydrophobic interactions, was responsible for the tendency of hydrophobic estrogens to adsorb onto hydrophobic nanofiltration membrane surfaces. Nghiem et al. (2004) observed steroid (i.e., 17b-estradiol, estrone, testosterone) rejection decreased rapidly initially, which was caused by the hormones adsorbing onto the membrane surface and subsequent increased permeate concentration due to diffusion across the skin layer. Ng and Elimelech (2004) also reported that 17b-estradiol rejection with time was much more severe when colloidal fouling occurred. They speculate that the cake layer formed due to fouling hinders the back diffusion of 17b-estradiol from the membrane surface back to the bulk solution. Consequently, buildup of 17b-estradiol at the membrane surface provided a larger concentration gradient for its diffusion across the RO membrane. However, partial rejection or evidence for steroid partitioning with time was not observed during sampling of full-scale RO installations (Drewes and Bellona et al. 2005), which is likely due to the formation of a fouling layer while RO membranes are employed on surface or recycled water. This fouling layer can act as a second barrier for hydrophobic compounds. In general, RO and *tight* NF membranes are very effective treatment processes for removing CECs from drinking water and only a few compounds, such as N-nitrosodimethylamine, chloroform or acetaminophen exhibit poor rejection.

FATE OF CECS IN MANAGED AQUIFER RECHARGE SYSTEMS

Riverbank Filtration (RBF)

Riverbank filtration (RBF) is a natural process that has been used for public and industrial water supply in Europe for more than a century and for nearly half a century in the United States. The process has been shown to be effective in moderating the peak concentrations of various contaminants present in river or lake water (for lake bank filtration) and in many cases involving removal of contaminants to below detection levels. The performance of riverbank filtration systems depends upon well type and pumping rates, travel time of surface water between the river and the pumping wells, source water quality, site hydrogeologic conditions, biogeochemical reactions in sediments and aquifer, and the quality of background groundwater. Studies from Europe and the United States indicate significant removal of dissolved organic carbon, nitrate, pesticides, pathogens, and turbidity, and recent studies also indicate efficient removal of select CECs during RBF.

Soil-aquifer Treatment (SAT) and Artificial Recharge and Recovery (ARR)

Subsurface systems, such as RBF or soil-aquifer treatment (SAT), demonstrated high removal efficiencies for CECs (Drewes and Shore, 2001; Heberer, 2002; Montgomery-Brown and Drewes et al. 2003; Verstraeten and Heberer et al. 2003;

Kreuzinger and Clara et al. 2004; Mansell and Drewes et al. 2004). Mansell et al. (2004) and Mansell and Drewes (2004) reported that steroidal hormones (such as 17b-estradiol, estriol, and testosterone) were efficiently removed after short travel times. Montgomery-Brown et al. (2003) extensively studied fate and transport of alkylphenolpolyethoxylates (APECs) and nonylphenol during SAT and findings of this study indicate a significant removal after travel times of approximately two weeks. Riverbank filtration, however, showed some limitations in removing organic contaminants. Certain trace pollutants have been regularly found in the product water of RBF systems, among them urotropin (aliphatic amine) and 1,5-naphthalindisulfonate (aromatic sulfonate) (Brauch, Sacher et al. 2000). Recent studies investigating the fate of PPCPs reported that antiepileptic drugs (e.g., carbamazepine, primidone), some blood-lipid regulator (e.g., clofibric acid), antibiotics (e.g., sulfamethoxazole), X-ray contrast media (e.g., iopromide), fragrances (e.g., galaxolide and tonalide), perfluorinated chemicals (e.g., PFOS, PFOA), and artificial sweenteners (e.g., sucralose) were present in both river water and bank filtered water (Kuehn and Mueller, 2000; Drewes and Fox et al. 2001; Drewes and Shore, 2001; Drewes and Heberer et al. 2003; Kreuzinger and Clara et al. 2004; Schmidt and Lange et al. 2004). A partial reduction in concentration was only achieved under certain redox conditions and through dilution with local groundwater.

Enhancing the Efficiency of Natural Treatment Processes

In subsurface systems, biodegradable organic carbon is often promoting the removal of river-borne pollutants. Heterotrophic denitrification is often limited due to a lack of organic carbon which originates in part from infiltrated river water and organic matter fixed in the sediments of the aquifer material (Grischek and Hiscock et al. 1998). It has also been reported that biodegradable organic matter plays a dominant role in the degradation of certain trace organic pollutants through a co-metabolic degradation pathway. During RBF, co-metabolic removal of trace pollutants has been observed for certain aromatic and aliphatic amines (Boernick and Eppinger et al. 2001; Paul and Boernick et al. 2001) and halogenated organics (Drewes and Jekel, 1996). Recent studies by Rauch et al. (2004) suggest that quantity and composition of NOM affect the removal efficiency of organic micropollutants during soil infiltration in groundwater recharge systems. Hydrophilic and colloidal carbon fractions have been found to stimulate soil microbial activity (Rauch and Drewes 2004) and thereby promoting the metabolic or cometabolic removal of certain organic micropollutants (Rauch and Munoz et al. 2004). Recent work by the author also suggest that certain environmental stress conditions will favor the establishment of a biocommunity that is capable to remove CECs more efficiently. The communities are more diverse and capable of expressing enzymes that are not present under "unstressed" conditions.

CONCLUSIONS

In order to treat impaired water sources to remove CECs from water requires a sequence of diverse treatment processes that can tackle the wide range of physicochemical properties of CECs. It is noteworthy that there is not a single treatment process that provides a 100 percent barrier to CECs in drinking water or the receiving environment.

Conventional drinking water treatment consisting of coagulation/flocculation followed by sedimentation and filtration commonly employed for surface water is not capable of removing CECs occurring at the parts-per-trillion level. Removal of CECs in particular EDCs like steroid hormones, however, can be expected during drinking water disinfection using chlorine, chlorine dioxide, and ozone. No removal is achieved by UV radiation applied at disinfection dosages. Especially, steroid hormones are well destructed during chlorine disinfection. Stronger oxidants, such as ozone or advanced oxidation processes (AOP), can remove a wide range of CECs, but exhibit limited removal regarding chlorinated flame retardants (e.g, TCEP, TCPP) or the mosquito repellent DEET. Activated carbon can remove a wide range of hydrophobic CECs including steroid hormones but has limitations regarding polar compounds, such as x-ray contrast agents. Low-pressure membranes, such as MF and UF, are ineffective in removing CECs from water. High-pressure membranes, such as NF or RO, however, can remove a wide range of CECs. Problematic for high-pressure membranes are low-molecular weight organics, such as N-nitrosamines or certain pharmaceuticals (i.e., acetaminophen, phenacetine). Natural processes, such as riverbank filtration (RBF) or soil-aquifer treatment (SAT), can be employed either as an additional treatment step for wastewater prior to discharge to the environment or as a pre-treatment to subsequent drinking water treatment. RBF and SAT both provide a significant barrier for a majority of CECs, which the exception of recalcitrant compounds that are not amendable to a biological attack. By modifying the environmental conditions, natural treatment systems can be tuned to increase the attenuation of CECs.

REFERENCES

1. Adams, C., Y. Wang, K. Loftin, and M. Meyer. (2002) "Removal of antibiotics from surface and distilled water in conventional water treatment processes." *Journal of Environmental Engineering*, 128(3):253-260.
2. Andreozzi, R., R. Marotta, G. Pinto, and A. Pollio. (2002) "Carbamazepine in water: persistence in the environment, ozonation treatment and preliminary assessment on algal toxicity." *Water Research*, 36(11):2869-2877.
3. Ballard, B.D., and A.A. Mackay. (2005) "Estimating the Removal of Anthropogenic Organic Chemicals from Raw Drinking Water by Coagulation Flocculation." *Journal of Environmental Engineering (Reston, VA, United States)*, 131(1):108-118.
4. Bedner, M., and W.A. Maccrehan. (2006) "Reactions of the amine-containing drugs fluoxetine and metoprolol during chlorination and dechlorination processes used in wastewater treatment." *Chemosphere*, 65:2130-2137.
5. Bellona, C., J.E. Drewes, P. Xu, and G. Amy. (2004) "Factors affecting the rejection of organic solutes during NF/RO treatment—A literature review." *Water Research*, 38:2795-2809.
6. Bellona, C., and J.E. Drewes. (2005) "The role of membrane surface charge and solute physico-chemical properties in the rejection of organic acids by NF membranes." *Journal of Membrane Science*, 249:227-234.
7. Boyd, G.R., S. Zhang, and D.A. Grimm. (2005) "Naproxen removal from water by chlorination and biofilm processes." *Water Research*, 39:668-676.

8. Daughton, C. G., and T. A. Ternes. (1999) *Environ. Health Perspect.* 107:907.
9. Dodd, M.C., A.D. Shah, U. Von Gunten, and C.-H. Huang. (2005) "Interactions of fluoroquinolone antibacterial agents with aqueous chlorine: reaction kinetics, mechanisms, and transformation pathways." *Environmental Science & Technology*, 39(18):7065-7076.
10. Drewes, J.E., and M. Jekel. (1996) "Simulation of groundwater recharge with advanced treated wastewater." *Water Science and Technology*, 33(10-11):409-418.
11. Drewes, J.E., P. Fox, and M. Jekel. (2001) "Occurrence of iodinated X-ray contrast media in domestic effluents and their fate during indirect potable reuse." *Journal of Environmental Science and Health Part a- Toxic/Hazardous Substances & Environmental Engineering*, 36(9):1633-1645.
12. Drewes, J.E., and L.S. Shore. (2001) "Concerns about pharmaceuticals in water reuse, groundwater recharge, and animal waste." *ACS Symposium Series*, 791:206-228.
13. Drewes, J.E., T. Heberer, T. Rauch, and K. Reddersen. (2003) "Fate of pharmaceuticals during groundwater recharge." *Ground Water Monitoring & Remediation*, 23(3):64-72.
14. Drewes, J.E., C. Bellona, M. Oedekoven, P. Xu, T.-U. Kim, and G. Amy. (2005) "Rejection of wastewater-derived micropollutants in high-pressure membrane applications leading to indirect potable reuse." *Environmental Progress*, 24(4):400-409.
15. Drewes, J.E., D. Sedlak, S. Snyder, E. Dickenson. (2008a). Final Report. WRF. Alexandria, VA.
16. Fuerhacker, M., A. Dürauer, and A. Jungbauer. (2001) "Adsorption isotherms on 17ß-estradiol on granular activated carbon (GAC)." *Chemosphere*, 44:1573-1579.
17. Gallard, H., A. Leclercq, and J. Croué. (2004) "Chlorination of bisphenol A: kinetics and by-products formation." *Chemosphere.*, 56(5):465-473.
18. Glassmeyer, S. T., et al. (2005) *Environ. Sci. Technol.* 39:5157.
19. Heberer, T. (2002) "Occurrence, fate, and removal of pharmaceutical residues in the aquatic environment: a review of recent research data." *Toxicology Letters*, 131(1-2):5-17.
20. Heberer, T. and D. Feldmann. (2005) *J. Hazardous Material.*
21. Huang, C., J.E. Renew, K.L. Smeby, K.E. Pinkston, and D.L. Sedlak. (2001) *Wat. Res. Update* 120:30.
22. Huang, C., and D. Sedlak. (2001) *Environ. Toxicol. Chem.* 20:133.
23. Huber, M.M., S. Canonica, G.-Y. Park, and U. Von Gunten. (2003) "Oxidation of pharmaceuticals during ozonation and advanced oxidation processes." *Environmental Science & Technology*, 37(5):1016-1024.
24. Huber, M.M., S. Korhonen, T.A. Ternes, and U. Von Gunten. (2005a) "Oxidation of pharmaceuticals during water treatment with chlorine dioxide." *Water Research Water Research*, 39:3607-3617.
25. Huber, M.M., A. Göbel, A. Joss, N. Hermann, D. Löffler, C.S. Mcardell, A. Reid, H. Siegrist, T.A. Ternes, and U. Von Gunten. (2005b.) "Oxidation of pharmaceuticals during ozonation of municipal wastewater effluents: a pilot study." *Environmental Science & Technology*, 39(11):4290-4299.

26. Khan, S.J., and J.E. Ongerth. (2004) *Chemosphere* 54:355.
27. Kiso, Y., Y. Nishimura, T. Kitao, and K. Nishimura. (2000) "Rejection properties of non-phenylic pesticides with nanofiltration membranes." *Journal of Membrane Science*, 171(2):229-237.
28. Kiso, Y., Y. Sugiura, T. Kitao, and K. Nishimura. (2001a) "Effects of hydrophobicity and molecular size on rejection of aromatic pesticides with nanofiltration membranes." *Journal of Membrane Science*, 192(1-2):1-10.
29. Kiso, Y., T. Kon, T. Kitao, and K. Nishimura. (2001b) "Rejection properties of alkyl phthalates with nanofiltration membranes." *Journal of Membrane Science*, 182(1-2):205-214.
30. Kreuzinger, N., M. Clara, B. Strenn, and B. Vogel. (2004) "Investigation on the behaviour of selected pharmaceuticals in the groundwater after infiltration of treated wastewater." *Water Science and Technology*, 50(2):221-228.
31. Kuemmerer, K., T. Steger-Hartmann, and M. Meyer. (1997) *Wat. Res.* 31:2705.
32. Lee, B.-C., M. Kamata, Y. Akatsuka, M. Takeda, K. Ohno, T. Kamei, and Y. Magara. (2004) "Effects of chlorine on the decrease of estrogenic chemicals." *Water Research*, 38(3):733-739.
33. Mansell, J., J.E. Drewes, and T. Rauch. (2004) "Removal mechanisms of endocrine disrupting compounds (steroids) during soil aquifer treatment." *Water Science and Technology*, 50(2):229-237.
34. Mansell, J., and J. Drewes. (2004) "Fate of steroidal hormones during soil-aquifer treatment." *Ground Water Monitoring & Remediation*, 24(2):94-101.
35. Mcdowell, D.C., M.M. Huber, M. Wagner, U. Von Gunten, and T.A. Ternes. (2005) "Ozonation of carbamazepine in drinking water: identification and kinetic study of major oxidation products." *Environmental Science & Technology*, 39(20):8014-8022.
36. Montgomery-Brown, J., J. Drewes, P. Fox, and M. Reinhard. (2003) "Behavior of alkylphenol polyethoxylate metabolites during soil aquifer treatment." *Water Research*, 37:3672-3681.
37. Ngheim, L.D., A.I. Schaefer, and M. Elimelech. (2004) "Removal of natural hormones by nanofiltration membranes: measurement, modeling, and mechanisms." *Environmental Science & Technology*, 38:1888-1896.
38. Nghiem, L., A. Schaefer, and M. Elimelech. (2005a) "Nanofiltration of hormone mimicking trace organic contaminants." *Separation Science and Technology*, 40(13):2633-2649.
39. Nghiem, L.D., A.I. Schaefer, and M. Elimelech. (2005b) "Pharmaceutical retention mechanisms by nanofiltration membranes." *Environmental Science & Technology*, 39:7698-7705.
40. Ozaki, H., and H. Li. (2002) "Rejection of organic compounds by ultra-low pressure reverse osmosis membrane." *Water Research*, 36:123-130.
41. Paul, S., H. Boernick, T. Grischek, and E. Worch. (2001) "The relevance of aliphatic amines for drinking water production from river Elbe bank filtrate." *Vom Wasser*, 97:125-133.
42. Pinkston, K.E., and D.L. Sedlak. (2004) "Transformation of aromatic ether- and amine-containing pharmaceuticals during chlorine disinfection." *Environmental Science & Technology*, 38(4019-4025).

43. Rauch, T., and J.E. Drewes. (2004) "Assessing the removal potential of soil-aquifer treatment systems for bulk organic matter." *Water Science and Technology*, 50(2):245-253.
44. Rauch, T., J. Munoz, and J.E. Drewes. (2004) "Metabolic removal of organic micropollutants in artificial groundwater recharge systems using water of impaired quality." In *WEFTEC*. Edited by New Orleans, Louisiana.
45. Rosenfeldt, E., and K.G. Linden. (2004) "Degradation of endocrine disrupting chemicals bisphenol A, ethinyl estradiol, and estradiol during UV photolysis and advanced oxidation processes." *Environmental Science & Technology*, 38(20):5476-5483.
46. Schmidt, C., F. Lange, and H.-J. Brauch. (2004) Assessing the impact of different redox conditions and residence times on the fate of organic micropollutants during riverbank filtration: 4th International Symposium on Pharmaceutically Active Compounds and Endocrine Disrupting Chemicals.
47. Sedlak, D.L., and K.E. Pinkston. (2001)*Wat. Res. Update* 120:56.
48. Snyder, S.A., P. Westerhoff, Y. Yoon, and D.L. Sedlak. (2003) "Pharmaceuticals, personal care products, and endocrine disruptors in water: implications for the water industry." *Environmental Engineering Science*, 20(5):449-469.
49. Sontheimer, H. (1991) *Trinkwasser aus dem Rhein?* Sankt Augustin, Germany: Academia Verlag.
50. Stuer-Lauridsen, F., et al. (2000) *Chemosphere* 40:783-793.
51. Swayne, M., Boone, G., Bauer, D. and Lee, J., (1980) Wastewater in Receiving Waters at Water Supply Abstraction Points EPA-60012-80-044. Washington, D.C.: U.S. Environmental Protection Agency.
52. Tanghe, T., and W. Verstraete. (2001) "Adsorption of nonylphenol onto granular activated carbon." *Water Air and Soil Pollution*, 131:61-72.
53. Tchobanoglous, G., et al. (2003) *Wastewater Engineering Treatment and Reuse*. 4th Edition. McGraw Hill.
54. Ternes, T.A., M. Meisenheimer, D. Mcdowell, F. Sacher, H.-J. Brauch, B. Haist-Gulde, G. Preuss, U. Wilme, and N. Zulei-Seibert. (2002) "Removal of pharmaceuticals during drinking water treatment." *Environmental Science & Technology*, 36(17):3855-3863.
55. Ternes, T.A., N. Herrmann, M. Bonerz, T. Knacker, H. Siegriest, and A. Joss. (2004) *Wat. Res.* 38:4075.
56. Ternes, T.A. (1998) *Wat. Res.* 32:3245.
57. Verstraeten, I.I., J.D. Carr, G.V. Steele, E.M. Thurman, K.C. Bastian, and D.F. Dormedy. (1999) "Surface-water/groundwater interaction: Herbicide transport into municipal collector wells." *Journal of Environmental Quality*, 28(5):1396-1405.
58. Verstraeten, I.I., T. Heberer, J. Vogel, T.F. Speth, S. Zuehlke, and U. Duennbier. (2003) "Occurrence of endocrine-disrupting and other wastewater compounds during water treatment with case studies from Lincoln, Nebraska and Berlin, Germany." *Practice Periodical of Hazardous,Toxic, and Radioactive Waste Management*, 7(4):253-263.

59. Vieno, N., T. Tuhkanen, and L. Kronberg. (2006) "Removal of pharmaceuticals in drinking water treatment: effect of chemical coagulation." *Enivronmental Technology*, 27:183-192.
60. Vieno, N.M., T. Tuhkanen, and L. Kronberg. (2005) "Seasonal variation in the occurrence of pharmaceuticals in effluents from a sewage treatment plant and in the recipient water." *Environmental Science & Technology*, 39(21):8220-8226.
61. Westerhoff, P., Y. Yoon, S. Snyder, and E. Wert. (2005" "Fate of endocrine-disruptor, pharmaceutical, and personal care product chemicals during simulated drinking water treatment processes." *Environmental Science & Technology*, 39(17):6649-6663.
62. Xu, P., J.E. Drewes, C. Bellona, G. Amy, T.-U. Kim, M. Adam, and T. Heberer. (2005) "Rejection of emerging organic micropollutants in nanofiltration-reverse osmosis membrane applications." *Water Environment Research*, 77(1):40-48.
63. Xu, P., J.E. Drewes, T.-U. Kim, C. Bellona, and G. Amy. (2006) "Effect of membrane fouling on transport of organic contaminants in NF/RO membrane applications." *Journal of Membrane Science*, 279:165-175.
64. Yoon, B.O., S. Koyanagi, T. Asano, M. Hara, and A. Higuchi. (2003a) "Removal of endocrine disruptors by selective sorption method using polydimethylsiloxane membranes." *Journal of Membrane Science*, 213(1-2):137-144.
65. Yoon, Y., P. Westerhoff, S.A. Snyder, and M. Esparza. (2003b) "HPLC-fluorescence detection and adsorption of bisphenol A, 17b-estradiol, and 17a-ethynyl estradiol on powdered activated carbon." *Water Research*, 37(14):3530-3537.
66. Yoon, Y., P. Westerhoff, S.A. Snyder, and E.C. Wert. (2006) "Nanofiltration and ultrafiltration of endocrine disrupting compounds,pharmaceuticals and personal care products." *Journal of Membrane Science*, 270:88-100.
67. Yoon, Y., P. Westerhoff, S.A. Snyder, E.C. Wert, and J. Yoon. (2007) "Removal of endocrine disrupting compounds and pharmaceuticals by nanofiltration and ultrafiltration membranes." *Desalination*, 202(1-3):16-23.
68. Zwiener, C., and F.H. Frimmel. (2000) "Oxidative treatment of pharmaceuticals in water." *Water Research*, 34(6):1881-1885.

AGRICULTURAL SOURCES OF CONTAMINANTS OF EMERGING CONCERN AND ADVERSE HEALTH EFFECTS ON FRESHWATER FISH

DONALD E. TILLITT AND HERBERT T. BUXTON
Biochemistry and Physiology, Columbia Environmental Research Center
Columbia, USA

ABSTRACT

Agricultural contaminants of emerging concern (CECs) are generally thought of as certain classes of chemicals associated with animal feeding and production facilities. Veterinary pharmaceuticals used in animal food production systems represent one of the largest groups of CECs. In our review, we discuss the extensive increase in use of antibiotics in animal feeding operations (AFOs) around the world. AFOs are a major consumer of antibiotics and other veterinary pharmaceuticals and over the past decade there has been growing information on the occurrence, release, and fate of CECs from animal food production operations, including the application of pharmaceutical-containing manure to agricultural fields and releases from waste lagoons. Concentrations of CECs in surface and ground water in proximity to AFOs correspond to their presence in the AFO wastes. In many cases, the environmental concentrations of agriculturally-derived CECs are below toxicity thresholds. Hormones and hormone replacement compounds are a notable exception, where chemical concentrations near AFOs can exceed concentrations known to cause adverse effects on endocrine-related functions in fish. In addition, some agricultural pesticides, once thought to be safe to non-target organisms, have demonstrated endocrine-related effects that may pose threats to fish populations in agricultural regions. That is, we have pesticides with emerging concerns, thus, the concern is emerging and not necessarily the chemical. In this light, one must consider certain agricultural pesticides to be included in the list of CECs. Even though agricultural pesticides are routinely evaluated in regulatory testing schemes which have been used for decades, the potential hazards of some pesticides have only recently been emerging. Emerging concerns of pesticides in fish include interference with hormone signaling pathways; additive (or more than additive) effects from pesticide mixtures; and adverse population-level effects at concentrations below predicted toxicity thresholds. Consequently, there is a need to evaluate the environmental concerns related to pesticide exposures to fish populations based on current biological and toxicological techniques. This presentation reviews some of the agricultural chemicals that have emerged as contaminants of concern and potentially threaten fish populations in agricultural watersheds.

INTRODUCTION

Agricultural productivity of both crops and livestock operations have greatly increased over the past half-century, due in part to advances in chemistry, including the use of pesticides and pharmaceuticals. Crop protection through the use of pesticides has increased yields, controlled weeds, and reduced insect damage; while increased use of pharmaceuticals in animal food production has reduced disease, enhanced animal health, and increased growth rates. The animal food production industry in the United States (U.S.) has transformed over this time; going to more confined animal feeding operations (AFOs) with greater animal numbers and higher densities (USEPA, 2005). The higher animal densities at AFOs are achieved through the use of veterinary pharmaceuticals, such as antibiotics, that protect against disease and at sub-therapeutic concentrations are used as growth promoters. Antibiotic use in livestock and poultry farms in the U.S. has steadily increased over the past half century to an estimated 8-13,000,000 Kg annually (AHI, 2000; Mellon et al. 2001, USFDA, 2010). In addition to antibiotics, veterinary pharmaceuticals routinely used in animal production facilities of developed countries include nutritional supplements, anti-fungal agents, growth promoters, hormones, anthelminthic agents, coccidiostats, anti-protozoals, sedatives, anti-bloating agents, and endectocides, among others (Boxall et al. 2003). The intestinal absorption of these chemicals in livestock and poultry can be low and often as much as 70-80% of the active agent passes through the animal in an original form (Elmund et al. 1971; Aga et al. 2003). Thus, the release of large amounts of these veterinary pharmaceuticals from agricultural systems into the environment has been a growing concern (Mellon et al. 2001; Boxtall et al. 2003; Sarmah et al. 2006).

Collectively, veterinary pharmaceuticals are among the growing number of chemicals of emerging concern (CECs) along with personal care products, human pharmaceuticals, over-the-counter drugs, new pesticides, as well as industrial chemicals such as the brominated flame retardants. Veterinary pharmaceuticals are a major aggregation of chemicals derived from agricultural practices and considered among the CECs. However, we must also consider chemicals in which ecological concerns are "emerging", as opposed to the emergence of the use of a particular chemical. In this sense, there are risks from some of the long-used agricultural pesticides that have only emerged in recent years as our understanding of more subtle effects on fish and wildlife populations has grown.

Our presentation will provide an overview of CECs coming from agricultural practices. Where possible, we try to provide quantitative information of usage patterns of the chemicals and amounts. However, the countries with the greatest amounts of agricultural practices using these chemicals, such as AFOs, do not always have records or policies of sharing this information. Thus, the extent of the risk or potential hazard is not known. Second, we review some of the potential adverse ecological risks of CECs on fish populations in agricultural watersheds.

AGRICULTURAL SOURCES OF EMERGING CHEMICALS OF CONCERN

Farming practices have drastically changed in developed countries around the world. Intensive animal production systems have been developed for the rapid and economical generation of animal protein within the livestock and poultry industries. In the U.S., animal production over the past three decades has shifted from widely distributed small animal operations to a greater reliance on large, concentrated AFOs which produce larger amounts of livestock and poultry under more confined conditions at higher densities (Mellon et al. 2001; USEPA, 2005). A reliance on veterinary pharmaceuticals has been a key factor that has allowed farmers to house livestock and poultry at higher densities, while maintaining animal health and increasing growth rates.

A variety of medicines, nutrients, and dietary supplements fall into the category of veterinary pharmaceuticals, and potentially CECs. Improved growth and maintenance of healthy individuals in AFOs has been achieved only through the intensive use of antibiotics, anti-parasitic agents (internal and external), growth promoters, hormones, analgesics, and other dietary supplements (Boxall et al. 2003). Of these chemicals, antibiotics are by far used most frequently and in the greatest amounts as dietary supplements in modern AFO feed and watering regimes (Sarmah et al. 2006). Antibiotic use in agricultural production of animals has increased over the past decades in the U.S. and it also common practice in most countries, at least where information is available (Table 1). Antibiotics have been used as a prophylaxis for disease at therapeutic doses of the given antibiotic, but also at non-therapeutic doses (often low mg/Kg range) to enhance the growth rates and improve the feed conversion efficiencies in cattle, swine, and poultry operations (Sarmah et al. 2006). The enhanced use of antibiotics in the food production, particularly for growth promotion, has been controversial due to concerns regarding development of antibiotic resistant microbes (USEPA, 2005; GAO, 2011). However, there has not been worldwide consensus on the use of antibiotics for non-disease-related purposes in animal food production. For example, European countries have generally had treatment regimes that included lower proportions of their total agricultural antibiotics for growth promotion (Table 1). Indeed, the use of antibiotics for growth promotion in livestock and poultry was banned in Sweden (1986), restricted in Germany (1996), voluntarily restricted in Denmark (1998), and ultimately banned as a growth promoter in swine in all of the European Union (EU) in 2006 (Sarmah et al. 2006). The methods of data estimation and collection for antibiotic use in agriculture, as well as the dissemination of that information, are not consistent across countries. Even within the U.S., there are gravely different reports of antibiotic treatment patterns among therapeutic and non-therapeutic usages (Mellon et al. 2001; AHI, 2000). Consequently, it is challenging to prioritize hazards and estimate risk derived from the use information of agricultural veterinary pharmaceuticals across the world. There is generally even less information available on the sales and usage patterns of the other veterinary pharmaceuticals used in modern food industrial practices (Boxall et al. 2003; Sarmah et al. 2006). Thus, the public and environmental health hazards associated with the non-antibiotic CECs are even more difficult to predict and evaluate.

Table 1. Estimated therapeutic (disease treatment and prevention), non-therapeutic (growth promotion), and total antibiotic usage rates (Kg/year) in agricultural practices by country.

Country	Disease Treatment/Prevention	Growth Promotion	Total	Reference
China			5,500,000	Zhao et al. 2010
			162,000,000	Luo et al. 2011
United States	1,800,000	7,300,000	9,100,000	Swartz 1989
	6,700,000	1,400,000	8,100,000	AHI 2000
	940,000	12,500,000	13,400,000	Mellon et al. 2001
			13,100,000	U.S. FDA 2010
European Union	3,500,000	1,500,000	5,000,000	Sarmah et al. 2006
United Kingdom	420,000	28,000	448,000	Sarmah et al. 2006
Korea			230,000	Oh et al. 2006
New Zealand	8,800	26,200	35,000	Sarmah et al. 2006
Kenya			14,600	Sarmah et al. 2006

The release of these veterinary pharmaceuticals from AFOs and other animal production facilities occurs through distinct pathways. Veterinary care pharmaceuticals given to livestock and poultry, often in their feed or water, can have very low rates of adsorption into the animal (Elmund et al. 1971; Aga et al. 2003) resulting in large amounts of the parent compound in the waste products (urine and feces). The pharmacokinetics of the various medicines and supplements in cattle, swine or poultry can differ greatly, but as much as 80-90% of the active compound may pass through the animals (Boxall et al. 2003; Sarmah et al. 2006; Luo et al. 2011). The waste products from AFOs, namely, manure, urine, litter materials, and waste water slurries, contain large amounts of veterinary pharmaceuticals and represent the main route by which these chemicals are released into the environment. The significance of this fact becomes apparent when the sheer mass of waste products produced by livestock is considered. In the U.S. an estimated 1.2-1.37 billion tons of manure is produced annually in agriculture (USEPA, 2005), while estimates in China are approximately 2.17 trillion tons of manure produced per year (Zhao et al. 2010). Routine waste management strategies for manure and certain slurries of waste from AFOs include storage and land application to agricultural fields in order to recycle nutrients and amend the soil. Consequently, two major pathways of chemical transport of CECs from agricultural farms are from: 1) surface water runoff into lakes, streams, and rivers; along with 2) leaching of more hydrophilic (less adsorbed) chemicals into groundwater. The leaching of CEC from manure and other animal waste products generated by AFOs into groundwater occurs from both the land application to agricultural fields (Weiss et al. 2008; Song et al. 2010) and the on-site storage of the manure in large piles (Bartelt-Hunt et al. 2011; Huang et al. 2011). Therefore, recent studies to determine the release and chemical fate of veterinary pharmaceuticals around AFOs have compared concentrations of CECs in manure, surface waters, and groundwater (Weiss et al. 2008; Matsui et al. 2008; Kim et al. 2008; Huang et al. 2011). For example, the concentrations of amprolium, carbadox, monesin, and tylosin were measured in surface waters nearby farms that applied manure agriculturally (Song et al. 2010). Amprolium and monesin were frequently detected in surface waters, while carbadox and tylosin were rarely detected, which corresponded to the soil sorption coefficients determined in the laboratory (Song et al. 2010). Another route of release of CECs from AFOs and other agricultural food production facilities is from the waste water and slurry materials generated from these operations. Liquid wastes from AFOs are often applied to agricultural fields, but generally are stored in waste lagoons prior to the application (Mellon et al. 2001; USEPA, 2005). The retention of the wastes in the lagoons is subject to the integrity of the lagoon itself. If the lagoon is not large enough for the operational capacity of the AFO wastes, direct discharges can occur either inadvertently (spill ways, dam rupture) or intentionally (spray irrigation or emergency dumping) and can result in dramatic effects on receiving waters, such as fish kills. Also, AFO waste lagoons can be subject to leaching of chemicals into the underlying groundwater (see review by Pan et al. 2009). The transport of CECs from AFOs by volatilization or release as aerosols may be important fate processes, but these types of mechanism have not received much attention in the literature (Boxall et al. 2003).

The ultimate environmental disposition of veterinary pharmaceuticals from animal production facilities is dependent on the physical and chemical properties of the molecules and the properties of the environment to which they are released. Because there are hundreds to thousands of separate chemicals in this category called "veterinary pharmaceuticals," there are a variety of chemical structures and resulting physical-chemical properties (Blackwell et al. 2007; Aust et al. 2010). So, the importance of different environmental transport and transformation processes are not always the same for each chemical in this class. That said, many of the veterinary pharmaceuticals can be grouped based on common structural attributes. In general, many of these chemicals have moderate to high water solubility and low to moderate octanol-water partition coefficients (Sarmah et al. 2006; Pal et al. 2010). Sorption coefficients (K_{OC}) vary greatly and tend to predict mobility in soils and the propensity for a chemical to volatilize (Song et al. 2010), along with the Henry's Law constant (vapor pressure relative to water solubility). Degradation rates of veterinary pharmaceuticals have a wide range of values, again depending on the structure of the molecule and the nature of the environment, and can range from days to months (Sarmah et al. 2006), even within a single chemical class such as the sulfonamides (Aust et al. 2010).

Documentation of the occurrence, environmental fate, and concentrations of veterinary pharmaceuticals has increased globally over the past decade. Sporadic reports of agricultural pharmaceutical CECs occurred in the 1980's and as methods for chemical analysis improved over the 1990's, more complete and broader scale reconnaissance was possible (Kolpin et al. 2002). Most monitoring efforts for veterinary pharmaceuticals were initially conducted in the U.S., Europe, and New Zealand (Kolpin et al. 2002; Boxall et al. 2003; Sarmah et al. 2006, Zuccato et al. 2006). Concentrations of veterinary pharmaceuticals in surface waters were often less than 1 ug/L with a few exceptions, and generally in the low to mid-nanograms per liter (see review in Boxall et al. 2003), while concentrations in groundwater are routinely even lower (Weiss et al. 2008; Hu et al. 2010). With increasing frequency over the past decade, more environmental assessments of the occurrence and fate of veterinary pharmaceuticals associated with AFOs or in general monitoring, have been conducted in other parts of the world. In particular, agricultural CECs in Asia have received an increased amount of attention, including investigations in Korea (Oh et al. 2006; Kim et al. 2008), Vietnam (Managaki et al. 2007) and China (Zhao et al. 2010; Pan et al. 2010; Zhao et al. 2010; Luo et al. 2011, Wei et al. 2011).

ADVERSE HEALTH EFFECTS ASSOCIATED WITH VETERINARY PHARMACEUTICALS

Initial concerns regarding the release of veterinary pharmaceuticals from agricultural animal production facilities into the environment were largely associated with the potential for development of antibiotic resistant strains of bacteria. Farmers in the U.S. (and others of the part of the world) increased usage rate of antibiotics both for the control of disease and for non-therapeutic, promotion of growth in livestock and poultry (Mellon et al. 2001; Boxall et al. 2003; USEPA, 2005; Sarmah et al. 2006). The prevalence of antibiotic resistant bacteria increased with the widespread use of sub-therapeutic doses of antibiotics used for growth promotion and increasing feed efficiency in animal food production (USEPA, 2005). European countries, including Sweden, Denmark, the Netherlands, the United Kingdom and Germany, were among the first countries to initiate legislation to limit particular antibiotics, and in 2006 the European Union banned the use of antibiotics as growth promoters in swine (Sarmah et al. 2006). Sweden had banned non-therapeutic use of antibiotics in animal food production in 1986 and subsequently reported fewer detections of antibiotic resistant bacteria compared to surrounding countries over the period from 1986-1995 (Sarmah et al. 2006). Potential hazards associated with linkages between antibiotic use in animal feeding operations and increases in drug resistant bacteria are beyond the scope of this presentation, but are reported with increasing frequency (see review USEPA, 2005).

Ecotoxicological assessments of the hazards posed by veterinary pharmaceuticals have largely been focused on adverse outcomes as predicted by invertebrate sensitivities to these CECs (Boxall et al. 2003). That is due to the fact that most toxicological information on agricultural CECs comes from laboratory testing of *Daphnia magna* (e.g. cladoceran zooplankton), while much less information exists for fish species. Initial hazard evaluations of potential ecological effects indicated that the handful of veterinary pharmaceuticals with toxicity thresholds determined using controlled laboratory exposures (mostly antibiotics) indicated little potential risk to aquatic organisms, with the exception of ciprofloxacin that has toxicity thresholds of less than 10 *ug*/L and environmental concentrations up to approximately 50 *ug*/L (Boxall et al. 2003). Toxicity thresholds compiled in this report were generally from the low *ug*/L to the high mg/L, while environmental concentrations were typically < 1ug/L (Boxall et al. 2003). A recent review of evidence for effects of pharmaceuticals (mostly human-use) on fish health concluded that most of these chemicals had environmental concentrations 1-3 orders of magnitude less than toxicity thresholds (Corcoran et al. 2010). Toxic effects of these human-use pharmaceuticals in fish often occurred at *ug*/L to mg/L concentrations, while environmental concentrations based on literature reports were routinely in the ng/L range (Corcoran et al. 2010). Diclofenac and ibuprophen were notable exceptions in this review of environmental concentrations and toxicity thresholds in fish, with toxicity thresholds of these non-steroidal anti-inflammatory drugs (NSAIDs) at or just below 1 *ug*/L for interruption of prostaglandin synthesis, histological alterations in gills, liver and kidneys, and reproductive impairment (Corcoran et al. 2010).

Steroid hormones and hormone replacements are the veterinary pharmaceuticals that provide the greatest environment risk to fish and potentially other aquatic organisms. Many hormone receptors, including steroid hormone receptors and metabolic pathways are highly conserved in vertebrate taxa, including fish, birds, and mammals (Gunnarsson et al. 2008). Conservation of drug target genes in zebrafish is approximately 86% as compared to humans, making it understandable that drugs are likely to have highly specific interactions with the intended receptor, even across taxa (Gunnarsson et al. 2008). Thus, steroid hormones intended to control reproductive cycles (e.g. ethinyl estradiol, EE2) or anabolic steroids used for growth promotion in cattle (e.g. trenbolone) can interfere with homologous receptors in fish. EE2 has been linked to disruption of reproductive processes and abnormal gonad development in feral fish populations in the UK (Jobling et al. 1998). The source was not agricultural, but the linkage of population-level adverse outcomes to this steroid hormone replacement drug is a leading example of adverse ecological outcomes that can occur from CECs. Controlled exposures of an entire lake to EE2 at approximately 5 ng/L resulted in reproductive disruption and subsequent collapse of fish populations (Kidd et al. 2007). Trenbolone acetate is a synthetic androgen used in approximately 90% of cattle feeds in the U.S. (Balter, 1999). Trenbolone is fairly stable in manure (Shiffer et al. 2001) and has been found in feedlot runoff and adjacent streams at concentrations great enough to induce androgenic responses (Durhan et al. 2006) and concentrations known to reduce fecundity and alter secondary sexual characteristics in fish (Ankley et al. 2003). Thus, steroid hormones and synthetic steroid hormone replacements can have dramatic effects on fish and fish populations; therefore, this class of veterinary pharmaceuticals represents a potential risk posed by AFOs to aquatic resources.

AQUACULTURE AS A SOURCE OF CHEMICALS OF EMERGING CONCERN

Veterinary pharmaceuticals are used in aquaculture for disease prevention. In the U.S., certain antibiotics are registered for disease prevention in aquaculture by the Federal Drug Administration, U.S. Department of Agriculture. Chemicals registered for use in aquaculture must be tested for efficacy and pharmacokinetics in fish. However, very few studies have been conducted on the occurrence and fate of fish health chemicals in raceways and waste effluents from fish hatcheries. Dietze et al. (2005) surveyed the effluents of 13 fish hatcheries in seven states from the U.S. for three commonly used antibiotics (sulfadimethoxine, oxytetracycline, and ormetoprim). The antibiotics were detected in the effluents, but at low rates of detection.

AGRICULTURAL CHEMICALS WITH EMERGING CONCERNS

Agricultural practices are one of the few activities in which chemicals are intentionally released into the environment, particularly chemicals that are designed to be toxic. Over the years, environmental issues with pesticides have come to our attention. The toxicity of some pesticides to non-target species, when it has occurred, has often been a surprise and has brought about paradigm shifts in environmental toxicology. The best example of that was DDT, an insecticide designed to persist in the agricultural fields, a trait thought to be particularly useful at the time it was first produced. The process of bioaccumulation was elucidated by the environmental fate of DDT, after it became clear that secondary poisoning of birds was occurring from dietary exposures to DDT. Prior to these events, the concept of a chemical contaminant increasing concentration at higher levels of the foodweb was not understood. Now, nearly every hazard assessment scheme for chemicals, and most certainly pesticides, includes some sort of evaluation of the bioaccumulation potential. Thus, even though agricultural pesticides are routinely evaluated in regulatory testing schemes and may have been used for decades, their potential hazards may come to light based on a more comprehensive understanding of the fundamental biological processes, molecular mechanisms, or previously unrecognized critical biological processes. Pesticides and other agricultural chemicals once thought to be harmless, have, in certain cases, more recently been observed to cause adverse effects in fish and wildlife populations. Emerging concerns of pesticides in fish include interference with hormone signaling pathways (i.e., endocrine disruption), additive (or more than additive) effects from pesticide mixtures, and adverse outcomes at concentrations below predicted toxicity thresholds. Consequently, there is a need to evaluate the environmental concerns related to pesticide exposures to fish populations based on current biological and toxicological knowledge.

Atrazine is an agricultural pesticide that has come under greater scrutiny as more information on the persistence and ecotoxicological effects has been revealed. Atrazine use as an herbicide has increased steadily since its introduction forty years ago. Currently, atrazine is the pesticide with the second highest use rate in the U.S., at approximately 70 million pounds annually (Keily et al. 2004). Atrazine is a broad spectrum pre-emergent herbicide used to control weeds mainly on corn and sorghum crops in the U.S. The use of atrazine has been banned in the EU in 2004 due to persistence of the pesticide in surface and groundwater. Toxicological concerns related to atrazine have emerged as more information has come available on its ability to disrupt endocrine-related functions in fish and amphibians. In fish, studies indicate atrazine adversely affected endocrine-mediated olfactory function (Moore and Lower, 2001) and reproductive function (egg production) at environmental concentrations (Tillitt et al. 2010). The endocrine-related effects of atrazine, or other pesticides for that matter, were not part of the registration scheme and hazard evaluation process until recent years. In the U.S. in 1996, the Food Quality Protection Act mandated an endocrine disruption screening program be developed and implemented for testing and chemical registration purposes. It was the fish reproduction test developed under this program (Ankley et al. 2001) that was used to discover the endocrine-related effects of atrazine on fish reproduction (Tillitt et al. 2010). Based on a 0.5 *ug*/L toxicity threshold for reproductive impairment in fish, there are large areas of the U.S. in which the surface waters have concentrations of atrazine that could threaten fish reproduction (Gilliom et al. 2006). Thus, our knowledge of endocrine disruption in fish and the appropriate screening methods for that functional endpoint were important for elucidating potential hazards of atrazine; hazards which had previously gone undetected.

CONCLUSIONS

Agricultural practices have changed over the past half-century and have resulted in enhanced use of chemicals for crop protection, animal health, and animal production. Veterinary pharmaceuticals used in animal food production systems represent one of the largest sources of chemicals of emerging ecological concern. Among the veterinary pharmaceuticals, antibiotic use has become extensive throughout animal production facilities around the world. Release of antibiotics and other veterinary pharmaceuticals from animal food production operations has been documented over the last decade and occurs from application of pharmaceutical-containing manure and liquid wastes to agricultural fields and unintentional releases from waste lagoons. Concentrations of these chemicals in surface and ground water in proximity of AFOs correspond to their presence in the AFO wastes. In many cases, the environmental concentrations of agriculturally-derived CECs are below toxicity thresholds. Hormones and hormone replacement compounds are a notable exception, where chemical concentrations near AFOs can be above those required to cause adverse effects on endocrine-related functions in fish. Finally, some agricultural pesticides, once thought to be safe, have demonstrated endocrine-related effects that may pose threats to fish populations in agricultural regions. These emerging concerns were only revealed through application of the evolving endocrine screening protocols which offer greater opportunities to predict and document the effects of agricultural CECs on non-target ecological receptors such as fish.

REFERENCES

1. Aga, D.S., R. Goldfish, P. Kulshrestha (2003) "Application of ELISA in determining the fate of tetracyclines in land-applied livestock wastes." *Analyst* 128:658–662.
2. Ankley G.T., K.M. Jensen, M.D. Kahl, J.J. Korte, E.A. Makynen (2001) "Description and evaluation of a short-term reproduction test with the fathead minnow (Pimephales promelas)." *Environ Toxicol Chem* 20:1276–1290.
3. Ankley, G.T., Jensen, K.M., Makynen, E.A., Kahl, M.D., Korte, J.J., Hornung, M.W., Henry, T.R., Denny, J.S., Leino, R.L., Wilson, V.S., Cardon, M.C., Hartig, P.C., Gray, L.E. (2003) "Effects of the androgenic growth promoter 17-beta-trenbolone on fecundity and reproductive endocrinology of the fathead minnow." *Environ. Toxicol. Chem.* 22:1350-1360.
4. Animal Health Institute (2000) *Survey indicates most antibiotics used in animals are used for treating and preventing disease*, Washington, DC: Animal Health Institute, February 2000.
5. Aust, M.O., T.B. Sören, J. Seeger, F. Godlinski, R. Meissner, P. Leinweber. (2010) "Sulfonamides leach from sandy loam soils under common agricultural practice." *Water Air Soil Pollut.* 211:143–156
6. Balter, M. (1999) "Scientific cross-claims fly in continuing beef war." *Science* 284:1453-1455.
7. Bartelt-Hunt, S., D.D. Snow, T. Damon-Powell, D. Miesbach (2011) "Occurrence of steroid hormones and antibiotics in shallow groundwater impacted by livestock waste control facilities." *J. Contam. Hydrol.* 123(3-4):94-103.
8. Blackwell, P.A., P. Kay, A.B.A. Boxall (2007) "The dissipation and transport of veterinary antibiotics in a sandy loam soil." *Chemosphere* 67(2):292-299.
9. Boxall, A.B.A., D.W. Kolpin, B. Halling-Sorensen, J. Tolls (2003) "Are veterinary medicines causing environmental risks?" *Environ. Sci. Technol.* 37(15):286A-294A.

10. Corcoran, J., M.J. Winter, C.R. Tyler (2010) "Pharmaceuticals in the aquatic environment: A critical review of the evidence for health effects in fish." *Crit. Rev. Toxicol.* 40(4): 287-304.
11. Dietze, J.E., E.A. Schribner, M.T. Meyers, D.W. Kolpin (2005) "Occurrence of antibiotics in water from 13 fish hatcheries, 2001–2003." *Intern. J. Environ. Anal. Chem.* 85(15):1141–1152.
12. Durhan, E.J., Lambright, C.S., Makynen, E.A., Lazorchak, J., Hartig, P.C., Wilson, V.S., Gray, L.E., Ankley, G.T. (2006) "Identification of metabolites of trenbolone acetate in androgenic runoff from a beef feedlot." *Environ. Health Perspect.* 114S, 65–68.
13. Elmund, G.K., S.M. Morrison, D.W. Grant, M.P. Nevins (1971) "Role of excreted chlortetracycline in modifying the decomposition process in feedlot waste." *Bull. Environ. Toxicol.* 6:129–132.
14. Gilliom, R.J., Barbash, J.E., Crawford, C.G., Hamilton, P.A., Martin, J.D., Nakagaki, N., Nowell, L.H., Scott, J.C., Stackelberg, P.E., Thelin, G.P., and Wolock, D.M., (2006) The Quality of Our Nation's Waters—Pesticides in the Nation's Streams and Ground Water, 1992–2001: U.S. Geological Survey Circular 1291, 172 p.
15. Government Accountability Office (2011) Antibiotic resistance: Data gaps still remain despite HHS taking steps to improve monitoring: Government Accountability Office report GAO-11-406, 60 p. http://www.gao.gov/n
16. Gunnarsson, L., A. Jauhiainen, E. Kristiansson, O. Nerman, D.G.J. Larsson (2008) "Evolutionary conservation of human drug targets in organisms used for environmental risk assessments." *Environ. Sci. Technol.* 42:5807–5813.
17. Hu, X., Q. Zhou, Y. Lu (2010) "Occurrence and source analysis of typical veterinary antibiotics in manure, soil, vegetables and groundwater from organic vegetable bases, northern China." *Environ. Pollut.* 158(9):2992-2998.
18. Huang, C.H., J.E. Renew, K.L. Smeby, K. Pinkston, D.L. Sedlak (2001) "Assessment of potential antibiotic contaminants in water and preliminary occurrence analysis." *J. Contemp. Water Res. Educ.* 120(1):30-40.
19. Jobling, S., Nolan, M., Tyler, C.R., Brighty, G., Sumpter, J.P. (1998) "Widespread sexual disruption in wild fish." *Environ. Sci. Technol.* 32:2498–2506.
20. Kidd, K.A., P.J. Blanchfield, K.H. Mills, V.P. Palace, R.E. Evans, J.M. Lazorchak, R.W. Flick (2007) "Collapse of a fish population after exposure to a synthetic estrogen." *Proc. Nat. Acad. Sci.* 104(21): 8897–8901.
21. Kiely, T., D. Donaldson, and A. Grube (2004) Pesticide industry sales and usage – 200 and 2001 market estimates. U.S. Environmental Protection Agency, Office of Prevention, Pesticides, and Toxic Substances. EPA-733-R-04-001, 33 p.
22. Kim, Y., K. Choi, J. Jung, S. Park, P.G. Kim, J. Park (2007) "Aquatic toxicity of acetaminophen, carbamazepine, cimetidine, diltiazem and six major sulfonamides, and their potential ecological risks in Korea." *Environ. International* 33:370-375.
23. Kolpin, D.W., E.T. Furlong, M.T. Meyer, E.M. Thurman, S.D. Zaugg, L.B. Barber, H.T. Buxton (2002) "Pharmaceuticals, hormones and other waste water

contaminants in U.S. streams 1999–2000. A national reconnaissance." *Environ. Sci. Technol.* 36:1202–1211.

24. Luo, Y., L. Xu, M. Rysz, Y. Wang, H. Zhang, P. J. J. Alvarez (2011) "Occurrence and transport of tetracycline, sulfonamide, quinolone, and macrolide antibiotics in the Haihe River basin, China." *Environ. Sci. Technol*: 45(5):1827-1833.
25. Managaki, S., A. Murata, H. Takada, B.C. Tuyen, N. Chiem (2007) "Distribution of macrolides, sulfonamides, and trimethoprim in tropical waters: Ubiquitous occurrence of veterinary antibiotics in the Mekong Delta." *Environ. Sci. Technol.* 41(23):8004-8010.
26. Matsui, Y., T. Ozu, T. Inoue, T. Matsushita (2008) "Occurrence of a veterinary antibiotic in streams in a small catchment area with livestock farms." *Desalination* 226(1-3): 215-221.
27. Mellon, M., C. Benbrook, and K.L. Benbrook (2001) Hogging It: Estimates of Antimicrobial Abuse in Livestock. Cambridge, MA:Union of Concerned Scientists Publications.
28. Moore, A., Lower, N. (2001) "The impact of two pesticides on olfactory-mediated endocrine function in mature male Atlantic salmon (*Salmo salar* L.) parr." *Compar. Biochem. Physiol.* Part B. 129:269-276.
29. Oh, S. J., J. Park, M.J. Lee, S.Y. Park, J.H. Lee, K. Choi (2006) "Ecological hazard assessment of major veterinary benzimidazoles: Acute and chronic toxicities to aquatic microbes and invertebrates." *Environ. Toxicol. Chem.* 25(8):2221-2226.
30. Pal, A., K.Y. Gin, A.Y. Lin, M. Reinhard (2010) "Impacts of emerging organic contaminants on freshwater resources: Review of recent occurrences, sources, fate and effects." *Sci. Total Environ.* 408:6062–6069.
31. Pan, B., P. Ning,P. Xing (2009) "Part V—Sorption of pharmaceuticals and personal care products." *Environ. Sci. Poll. Res.* 16(1):106-116.
32. Sarmah, A. K.; M.T. Meyer, A.B.A. Boxall (2006) "A global perspective on the use, sales, exposure pathways, occurrence, fate and effects of veterinary antibiotics (VAs) in the environment." *Chemosphere*, 65(5):725-759.
33. Song, W., Y. Ding, C.T. Chiou, H. Li (2010) "Selected veterinary pharmaceuticals in agricultural water and soil from land application of animal manure." *J. Environ. Qual.* 39(4):1211-1217.
34. Schiffer, B., Daxenberger, A., Meyer, K., Meyer, H.H.D. (2001) "The fate of trenbolone acetate and melengestrol acetate after application as growth promoters in cattle: environmental studies." *Environ. Health Perspect.* 109:1145-1151.
35. Swartz, M.N. (1989) Committee on Human Risk Assessment of Using Subtherapeutic Antibiotics in Animal Feeds, Institute of Medicine, Division of Health Promotion and Disease Prevention. Human Health Risks with the Subtherapeutic Use of Penicillin or Tetracyclines in Animal Feed. Washington, DC: National Academy Press.
36. Tillitt, D.E., D.M. Papoulias, J.J.Whyte, and C.A. Richter. (2010) "Atrazine reduces reproduction in fathead minnow (*Pimephales promelas*)." *Aquatic Toxicol.* 99(2):149-159.
37. United States Environmental Protection Agency (USEPA). (2005) Detecting and Mitigating the Environmental Impact of Fecal Pathogens Originating from

Confined Animal Feeding Operations: Review. Office of Research and Development National Risk Management Research Laboratory Cincinnati, OH 45268 EPA/600/R-06/021 September 2005. 175 p.

38. United States Food and Drug Administration (USFDA). (2010) 2009 Summary report on *Antimicrobials sold or Distributed for Use in Food-Producing Animals.* Center for Veterinary Medicine. December 9, 2010 release at: http://www.fda.gov/downloads/ForIndustry/UserFees/AnimalDrugUserFeeActADUFA/UCM231851.pdf

39. Wei, R., F. Ge, S. Huang, M. Chen, R. Wang. (2011) "Occurrence of veterinary antibiotics in animal wastewater and surface water around farms in Jiangsu Province, China." *Chemosphere* 82:1408–1414.

40. Weiss, K., W. Schüssler, M. Porzelt (2008) "Sulfamethazine and flubendazole in seepage water after the sprinkling of manured areas." *Chemosphere* 72:1292–1297.

41. Zhao, L., Y.H. Dong, H. Wang (2010) "Residues of veterinary antibiotics in manures from feedlot livestock in eight provinces of China." *Sci. Total Environ.* 408:1069–1075.

42. Zuccato, E., S. Castiglioni, R. Fanelli, G. Reitano, R. Bagnati, C. Chiabrando, F. Pomati, C. Rossetti, D. Calamari (2006) "Pharmaceuticals in the environment in Italy: Causes, occurrence, effects and control." *Environ. Sci. Poll. Res.* 13(1):15-21.

EPIDEMIOLOGY OF CHEMICALS OF EMERGING CONCERN

DR. JOE M. BRAUN
Department of Environmental Health, Harvard University
Boston, USA

Over the last several decades, epidemiological studies have determined that gestational and childhood exposure to a wide range of environmental chemicals can adversely affect human health. The developing fetus and child are especially vulnerable to the effect of chemicals.[1] Metals (Pb and methylmercury), organophosphate pesticides, tobacco smoke, air pollutants, and polychlorinated biphenyls (PCBs) have the potential to adversely affect a wide range of child health outcomes, including growth, pubertal development, neurodevelopment, and even mortality.[2-11] A growing body of animal and human studies suggests that fetal and early life exposure to these "old" CECs, as well as new CECs, are associated with adverse health effects. Biomonitoring and exposure assessment studies indicate that humans are exposed to a numerous old and new CECs.

There is now considerable evidence that environmental chemical exposure during early life can adversely affect later life health. These findings are consistent with the Developmental Origins of Health and Disease Hypothesis (DOHaD), which posits that adverse conditions during early life can "program" the fetus to be predisposed for poor health.[12-13] While the toxicity of chemicals like Pb, PBCs, and methylmercury have been extensively studied, >80,000 chemicals are used in commerce, >3,000 are produced in excess of 1 million pounds annually, and few have been subjected to toxicity testing.[14-15] Of particular concern is the potential for many new CECs to be introduced into the environment every year and little known about their potential to affect human health.

SUSCEPTIBILITY OF THE FETUS AND CHILD

The fetus and child are especially vulnerable to the impact of environmental chemicals due to the rapid development of multiple organ systems that rely on the synchronization of time-dependent events. The disruption of critical early-life developmental processes may manifest later in the form of clinical disease. Even subtle shifts in the distribution of continuous traits like blood pressure or IQ can have profound impacts on the burden of disease at the population level.[16-17] Children receive higher doses of environmental chemicals than adults because of their higher consumption of food, water, and air per unit body mass; larger surface area to volume ratio; and higher gastrointestinal absorption.[1] In addition, the developing fetus and child have immature detoxification pathways for some chemicals.[18-19] Combined with their higher doses, the developing fetal or child organ systems are exposed to a greater concentration of biologically active compounds than an adult.

MEASURING CHILD HEALTH OUTCOMES

Children may unwillingly serve as proverbial "canaries in the coal mine" because of their increased susceptibility to environmental chemicals. Child health outcomes like body mass index, IQ, and behavior predict later life health outcomes and may be sensitive to the impact of CEC expsoure.[20-23] While the observed effects of environmental chemicals on child health outcomes may be small, trajectories of optimal or adverse development manifest early in life and small changes in an early marker can have profound effects on later life morbidity and mortality.[24] Clinical diseases related to disrupted growth and development can have profound public health impacts.[25-26]

Several prospective cohorts in the United States (U.S.) are underway to study the impact of environmental chemical exposure on a variety of infant, child, and adolescent health outcomes.[27-28] These cohorts offer the most robust way to quantify the impact of environmental chemicals on human health, since exposure measures are prospectively collected before the occurrence of the health outcome of interest. These studies are measuring a wide range of CECs in multiple biological matrices and longitudinally assessing growth/adiposity, neurodevelopment, puberty, and respiratory outcomes.

EMERGING CHEMICALS OF CONCERN FOR HUMAN HEALTH OUTCOMES: BISPHENOL A

Recent concern has been raised over the ability of bisphenol A (BPA) to increase the risk of adverse health outcomes in humans.[29-30] BPA is an illustrative example of a CEC that has been in commerce over 50 years, but a vast animal literature on its adverse health effects has accumulated in the last decade.[31-32] However, there are a relatively small number of epidemiological studies quantifying potential health risks in humans.[33] There is intense debate regarding the toxicity of BPA, and the lack of human studies has hampered efforts to make decisions about the impact of BPA on human health. While hundreds of animal and in vitro studies have examined the toxicity of BPA, at the time of this writing, there are only 32 human studies examining the impact of BPA exposure on human health.[33] Of these 32 studies, only 7 (22%) are prospective cohorts and only 9 (28%) have examined infant/child health outcomes.

BPA was originally investigated as a synthetic estrogen in the 1930s. In the 1950s, polymer chemists discovered BPA could be used to manufacture polycarbonate plastics and resins.[34] As a result, BPA is used in a variety of commercial products including thermal receipts, resin linings in canned foods, polycarbonate plastics, and cigarette filters.[35] Exposure is thought to occur primarily through dietary ingestion, but individuals involved in the manufacture or handling of BPA containing products may have higher exposure.[36-40] Biomonitoring results from the United States (U.S.) and Europe demonstrate that pregnant women are almost universally exposed to BPA.[41-44] Urinary concentrations of BPA have been found to be higher in children than adults.[45]

STUDYING THE IMPACT OF BPA EXPOSURE ON HUMAN HEALTH

As the epidemiological literature related to BPA expands, systematic reviews and meta-analyses of similar outcomes can be conducted, allowing policy makers to assess the impact of BPA exposure on human health. The current literature suggests that BPA has the potential to impact some human health outcomes, but not others. There is some evidence that BPA may adversely impact semen quality and reproductive function in adult men.[38,46-49] These findings seem to be consistent with results from some animal studies.[50] Two studies in 6-9 year old girls show inconsistent and inconclusive effects of childhood BPA exposure on pubertal development.[27,51] Future studies will need to address the role of fetal and neonatal BPA exposures on these health endpoints; however, the study of outcomes with long latencies and potentially small risks will require large (n>1000) prospective cohorts to be followed over long periods of time.

Results from two prospective birth cohorts suggest that maternal urinary BPA concentrations are associated with child behavior.[41,52] Animal studies have found that behavioral endpoints like anxiety, aggression, and hyperactivity may be affected by gestational BPA exposure.[50,53] Consistent with these prior studies, new results from a study of 237 mother-child dyads presented at this symposium suggests that gestational, but not childhood, BPA exposures are associated with impaired behavioral and emotional regulation among 3-year old female children.[41,54] These findings illustrate the importance of studying whether there are periods of vulnerability in human development that would make exposure to BPA and other CECs more detrimental.

Two important factors warrant additional consideration for epidemiological investigations of BPA and other non-persistent CEC exposure assessment and study design. BPA is ubiquitous in the environment and exogenous contamination of samples is of concern.[55] Furthermore, because of its short biological half-life, a single spot urine measurement will be insufficient to classify BPA exposure beyond a time period of days or weeks.[35,56] Variability of urinary BPA concentrations due to diet, time of day, and time since last urine void are additional considerations that will require standardized sample collection procedures or 2-stage exposure modeling techniques.[57]

Cross-sectional studies may have limited utility for examining the toxicity of BPA, especially for adult diseases like cardiovascular disease and metabolic disorders, which have long latency periods and etiologies originating years or decades earlier. Contemporaneous urinary BPA concentrations may not reflect the relevant etiologic window for the development of many child and adult diseases. However, cross-sectional studies should not be entirely dismissed since they can inform the design of future studies and may provide relevant information for some health outcomes (e.g., steroid production or gene expression). Future studies will need to collect relevant biomarkers of exposure before the onset of disease and ensure that factors associated with exposure variability are minimized. New biomarkers of exposure that can better characterize exposure during critical periods of development are needed, as well as biomarkers that serve as sensitive markers of exposure to individual or mixtures of CECs.

FUTURE DIRECTIONS FOR CEC RESEARCH

The current approach to evaluating the toxicity of CECs in humans and animals has understandably examined single chemicals in isolation without considering the sheer number of chemicals we are exposed to at any one moment.[44] Approaches to study mixtures of chemicals have been considered for some time,[58] but are logistically and methodologically difficult to implement. Future epidemiological studies will need to consider the impact of chemical mixtures. This will require identifying comparable endpoints used by toxicologists and incorporating animal toxicity data into exposure assessment and modeling. This work has the potential to radically change our thinking about the way CECs impact human health.

CONCLUSIONS

Epidemiological studies over the last half century have shown that environmental agents have the ability to impact human health. From Pb-induced deficits in IQ to increased mortality among smokers, we have learned that environmental factors impact human health across the lifespan.[3,59] Recent work showing that gestational and early childhood factors can impact adult health has renewed interest in identifying modifiable factors, including CECs, that may be responsible for disease/disability and are amenable to public health interventions. Epidemiology will continue to serve an important role by quantifying these effects in human populations and will need to incorporate new methods to address the complex relationships between CECs and human health.

REFERENCES

1. Selevan, S.G., Kimmel, C.A., Mendola, P. (2000) "Identifying critical windows of exposure for children's health." *Environmental health perspectives.* Jun 2000;108 Suppl 3:451-455.
2. Axelrad, D.A., Bellinger, D.C., Ryan, L.M., Woodruff, T.J. (2007) "Dose-response relationship of prenatal mercury exposure and IQ: an integrative analysis of epidemiologic data." *Environmental health perspectives.* Apr 2007;115(4):609-615.
3. Lanphear, B.P., Hornung, R., Khoury, J., et al. (2005) "Low-level environmental lead exposure and children's intellectual function: an international pooled analysis." *Environmental health perspectives.* Jul 2005;113(7):894-899.
4. Oken, E., Huh, S.Y., Taveras, E.M., Rich-Edwards, J.W., Gillman, M.W. (2005) "Associations of maternal prenatal smoking with child adiposity and blood pressure." *Obes Res.* 13:2021-2028.
5. Eubig, P.A., Aguiar, A., Schantz, S.L. (2010) "Lead and PCBs as risk factors for attention deficit/hyperactivity disorder." *Environmental health perspectives.* Dec 2010;118(12):1654-1667.
6. Kalkbrenner, A.E., Daniels, J.L., Chen, J.C., Poole, C., Emch, M., Morrissey, J. (2010) "Perinatal exposure to hazardous air pollutants and autism spectrum disorders at age 8." *Epidemiology.* Sep 2010;21(5):631-641.
7. Akinbami, L.J., Lynch, C.D., Parker, J.D., Woodruff, T.J. (2010) "The association between childhood asthma prevalence and monitored air pollutants in metropolitan areas, United States, 2001-2004." *Environ Res.* Apr 2010;110(3):294-301.
8. Woodruff, T.J., Darrow, L.A., Parker, J.D. (2008) "Air pollution and postneonatal infant mortality in the United States, 1999-2002." *Environ Health Perspect.* Jan 2008;116(1):110-115.
9. Bouchard, M.F., Chevrier, J., Harley, K.G., et al. (2011) "Prenatal Exposure to Organophosphate Pesticides and IQ in 7-Year Old Children." *Environmental health perspectives.* Apr 21 2011.
10. Rosas, L.G., Eskenazi, B. (2008) "Pesticides and child neurodevelopment." *Curr Opin Pediatr.* Apr 2008;20(2):191-197.

11. Selevan, S.G., Rice, D.C., Hogan, K.A., Euling, S.Y., Pfahles-Hutchens, A., Bethel J. (2003) "Blood lead concentration and delayed puberty in girls." *The New England journal of medicine.* Apr 17 2003;348(16):1527-1536.
12. Barker, D.J. (2004) "The developmental origins of adult disease." *Journal of the American College of Nutrition.* Dec 2004;23(6 Suppl):588S-595S.
13. Gillman, M.W. (2005) "Developmental origins of health and disease." *The New England journal of medicine.* Oct 27 2005;353(17):1848-1850.
14. Grandjean, P., Landrigan, P.J. (2006) "Developmental neurotoxicity of industrial chemicals." *Lancet.* Dec 16 2006;368(9553):2167-2178.
15. Landrigan, P.J. (2010) "What causes autism? Exploring the environmental contribution." *Curr Opin Pediatr.* Apr 2010;22(2):219-225.
16. Needleman, H.L. (1990) "What can the study of lead teach us about other toxicants?" *Environmental health perspectives.* Jun 1990;86:183-189.
17. Bellinger, D.C. (2004) "What is an adverse effect? A possible resolution of clinical and epidemiological perspectives on neurobehavioral toxicity." *Environmental research.* Jul 2004;95(3):394-405.
18. Holland, N., Furlong, C., Bastaki, M., et al. (2006) "Paraoxonase polymorphisms, haplotypes, and enzyme activity in Latino mothers and newborns." *Environ Health Perspect.* Jul 2006;114(7):985-991.
19. Ginsberg, G., Rice, D.C. (2009) "Does rapid metabolism ensure negligible risk from bisphenol A?" *Environmental health perspectives.* Nov 2009;117(11):1639-1643.
20. Lammi, N., Moltchanova, E., Blomstedt, P.A., Tuomilehto, J., Eriksson, J.G., Karvonen, M. (2009) "Childhood BMI trajectories and the risk of developing young adult-onset diabetes." *Diabetologia.* Mar 2009;52(3):408-414.
21. Williams, S.M., Goulding, A. (2009) "Early adiposity rebound is an important predictor of later obesity." *Obesity (Silver Spring, Md.* Jul 2009;17(7):1310.
22. Nagin, D., Farrington, D. (1992) "The stability of criminal potential from childhood to adulthood." *Criminology.* 30:235-260.
23. Whalley, L.J., Deary, I.J. (2001) "Longitudinal cohort study of childhood IQ and survival up to age 76." *BMJ.* Apr 7 2001;322(7290):819.
24. Baker, J.L., Olsen, L.W., Sorensen, T.I. (2008) "Weight at birth and all-cause mortality in adulthood." *Epidemiology (Cambridge, Mass.* Mar 2008;19(2):197-203.
25. Leibson, C.L., Katusic, S.K., Barbaresi, W.J., Ransom, J., O'Brien, P.C. (2001) "Use and costs of medical care for children and adolescents with and without attention-deficit/hyperactivity disorder." *JAMA.* Jan 3 2001;285(1):60-66.
26. Russell, R.B., Green, N.S., Steiner, C.A., et al. (2007) "Cost of hospitalization for preterm and low birth weight infants in the United States." *Pediatrics.* Jul 2007;120(1):e1-9.
27. Wolff, M.S., Teitelbaum, S.L., Pinney, S.M., et al. (2010) "Investigation of Relationships between Urinary Biomarkers of Phytoestrogens, Phthalates, and Phenols and Pubertal Stages in Girls." *Environmental health perspectives.* Jul 2010;118(7):1039-1046.
28. Dietrich, K.N., Eskenazi, B., Schantz, S.L. (2005) "Principles and practices of neurodevelopmental assessment in children: lessons learned from the Centers for

Children's Environmental Health and Disease Prevention Research." *Environmental health perspectives.* 113(10):1437-1446.

29. Meeker, J.D., Sathyanarayana, S., Swan, S.H. (2009) "Phthalates and other additives in plastics: human exposure and associated health outcomes." *Philos Trans R Soc Lond B Biol Sci.* Jul 27 2009;364(1526):2097-2113.
30. Thompson, R.C., Moore, C.J., vom Saal, F.S., Swan, S.H. (2009) "Plastics, the environment and human health: current consensus and future trends." *Philos Trans R Soc Lond B Biol Sci.* Jul 27 2009;364(1526):2153-2166.
31. Chapin, R.E., Adams, J., Boekelheide, K., et al. (2008) "NTP-CERHR expert panel report on the reproductive and developmental toxicity of bisphenol A." *Birth Defects Res B Dev Reprod Toxicol.* Jun 2008;83(3):157-395.
32. vom Saal, F.S., Akingbemi, B.T., Belcher, S.M., et al. (2007) "Chapel Hill bisphenol A expert panel consensus statement: integration of mechanisms, effects in animals and potential to impact human health at current levels of exposure." *Reproductive toxicology (Elmsford, N.Y.* Aug-Sep 2007;24(2):131-138.
33. Braun, J.M., Hauser, R. (2011) "Bisphenol A and children's health." *Curr Opin Pediatr.* Feb 2 2011.
34. Vogel, S.A. (2009) "The politics of plastics: the making and unmaking of bisphenol a 'safety'." *American journal of public health.* Nov 2009;99 Suppl 3:S559-566.
35. Braun, J.M., Kalkbrenner, A.E., Calafat, A.M., et al. (2011) "Variability and Predictors of Urinary Bisphenol A Concentrations during Pregnancy." *Environmental health perspectives* ;119(1):131-137.
36. Wilson, N.K., Chuang, J.C., Morgan, M.K., Lordo, R.A., Sheldon, L.S. (2007) "An observational study of the potential exposures of preschool children to pentachlorophenol, bisphenol-A, and nonylphenol at home and daycare." *Environmental research.* Jan 2007;103(1):9-20.
37. von Goetz, N., Wormuth, M., Scheringer, M., Hungerbuhler, K. (2010) "Bisphenol A: How the Most Relevant Exposure Sources Contribute to Total Consumer Exposure." *Risk Anal.* Jan 29 2010.
38. Li, D., Zhou, Z., Qing, D., et al. (2010) "Occupational exposure to bisphenol-A (BPA) and the risk of self-reported male sexual dysfunction." *Human reproduction (Oxford, England).* Feb 2010;25(2):519-527.
39. Fu, P., Kawamura, K. (2010) "Ubiquity of bisphenol A in the atmosphere." *Environ Pollut.* Oct 2010;158(10):3138-3143.
40. Kaddar, N., Harthe, C., Dechaud, H., Mappus, E., Pugeat, M. (2008) "Cutaneous penetration of bisphenol A in pig skin." *Journal of toxicology and environmental health.* 71(8):471-473.
41. Braun, J.M., Yolton, K., Dietrich, K.N., et al. (2009) "Prenatal bisphenol A exposure and early childhood behavior." *Environmental health perspectives.* Dec 2009;117(12):1945-1952.
42. Wolff, M.S., Engel, S.M., Berkowitz, G.S., et al. (2008) "Prenatal phenol and phthalate exposures and birth outcomes." *Environmental health perspectives.* Aug 2008;116(8):1092-1097.
43. Ye, X., Pierik, F.H., Hauser, R., et al. (2008) "Urinary metabolite concentrations of organophosphorous pesticides, bisphenol A, and phthalates among pregnant

women in Rotterdam, the Netherlands: the Generation R study." *Environmental research.* Oct 2008;108(2):260-267.

44. Woodruff, T.J., Zota, A.R., Schwartz, J.M. (2011) "Environmental Chemicals in Pregnant Women in the U.S.: NHANES 2003-2004." *Environmental health perspectives.* Jan 14 2011.
45. Calafat, A.M., Ye, X., Wong, L.Y., Reidy, J.A., Needham, L.L. (2008) "Exposure of the U.S. population to bisphenol A and 4-tertiary-octylphenol: 2003-2004." *Environmental health perspectives.* Jan 2008;116(1):39-44.
46. Meeker, J.D., Ehrlich, S., Toth, T.L., et al. (2010) "Semen quality and sperm DNA damage in relation to urinary bisphenol A among men from an infertility clinic." *Reproductive toxicology (Elmsford, N.Y.* Jul 23 2010;30(4):532-539.
47. Mendiola, J., Jorgensen, N., Andersson, A.M., et al. (2010) "Are environmental levels of bisphenol a associated with reproductive function in fertile men?" *Environmental health perspectives.* Sep;118(9):1286-1291.
48. Li, D.K., Zhou, Z., Miao, M., et al. (2010) "Relationship between Urine Bisphenol-A (BPA) Level and Declining Male Sexual Function." *J Androl.* May 13 2010;31(5):500-506.
49. Li, D.K., Zhou, Z., Miao, M., et al. (2011) "Urine bisphenol-A (BPA) level in relation to semen quality." *Fertil Steril.* Feb 2011;95(2):625-630 e621-624.
50. WHO/FAO. *Joint FAO/WHO Expert Meeting to Review Toxicological and Health Aspects of Bisphenol A Summary Report*: World Health Organization and Food and Agriculture Organization of the United Nations;;2010.
51. Wolff, M.S., Britton, J.A., Boguski, L., et al. (2008) "Environmental exposures and puberty in inner-city girls." *Environmental research.* Jul 2008;107(3):393-400.
52. Miodovnik, A., Engel, S.M., Zhu, C., et al. (2011) "Endocrine disruptors and childhood social impairment." *Neurotoxicology.* Mar 2011;32(2):261-267.
53. Tian, Y.H., Baek, J.H., Lee, S.Y., Jang, C.G. (2010) "Prenatal and postnatal exposure to bisphenol a induces anxiolytic behaviors and cognitive deficits in mice." *Synapse (New York, N.Y.* Feb 18 2010;64(6):432-439.
54. Braun, J., Kalkbrenner, A., Yolton, K., et al. (2011) "Impact of early life bisphenol A exposure on behavior and executive function in children." *Pediatrics.* Under Review.
55. Calafat, A.M., Needham, L.L. (2009) "What additional factors beyond state-of-the-art analytical methods are needed for optimal generation and interpretation of biomonitoring data?" *Environmental health perspectives.* Oct 2009;117(10):1481-1485.
56. Mahalingaiah, S., Meeker, J.D., Pearson, K.R., et al. (2008) "Temporal variability and predictors of urinary bisphenol A concentrations in men and women." *Environmental health perspectives.* Feb 2008;116(2):173-178.
57. Ye, X., Wong, L.Y., Bishop, A.M., Calafat, A.M. (2011) "Variability of Urinary Concentrations of Bisphenol A in Spot Samples, First-morning Voids, and 24-Hour Collections." *Environmental health perspectives.* Mar 15 2011.
58. NRC. *Phthalates and Cumulative Risk Assessment The Task Ahead.* Washington, DC: National Academies Press. 2008

59. Doll, R., Peto, R., Boreham, J., Sutherland, I. (2004) "Mortality in relation to smoking: 50 years' observations on male British doctors." *BMJ.* Jun 26 2004;328(7455):1519.

SESSION 7

ENERGY

FOCUS: ENERGY EFFICIENCY

ENERGY EFFICIENCY IN POWER GENERATION, INDUSTRY AND BUILDINGS: A SURVEY PAPER

DR. HISHAM KHATIB
World Energy Council
Amman, Jordan

ABSTRACT

Reducing demand for energy through energy efficiency measures and conservation is the cheapest and most effective way for enhancing energy security, providing a stable basis for economic planning as well as cutting carbon emissions.

This is a survey paper that deals with many aspects pertaining to energy efficiency, particularly in the electricity generation sector. Efficiency in the transport sector is dealt with by another paper in this session, so also aspects pertaining to sustainability. The electrical power generation sector is chosen for special emphasis due to its present low efficiency and the growing importance of electricity as the major energy carrier.

Energy efficiency can be defined as percentage of total energy input to a machine, equipment or a facility that is consumed in useful work and not wasted as useless heat. The value of the global energy efficiency is debatable. Presently, however, it does not exceed 40%; that of the average electricity generation is even lower at 35%. These are extremely low figures and demonstrate the strides that can be achieved in this quarter. Most of the energy losses in delivering final energy to consumers are in the power system, and that is why this sector receives a lot of emphasis in this paper. Significant losses in converting delivered primary energy into useful energy in final use also occur in all sectors of the economy, mainly in industry, transport and buildings.

Energy efficiency has to be differentiated from energy conservation or demand side management. Whereas conservation can be achieved through behavioral changes, education and pricing, energy efficiency can be mainly improved by technology, investment, codes and practices also energy (and carbon) pricing, introducing edits and enforcing regulations. Energy intensity varies from one country and region to another depending on economic output, technology achievements, price of energy products, and weather. Many barriers exist that are delaying the deployment of energy efficient technologies.

In the global quest for curbing emissions, energy efficiency figures out to be the most effective and cheapest means to reduce carbon emissions. In many cases it is a win-win situation where benefits can be attained at minimum (or no) cost, with short pay-back periods. However this means overcoming many barriers, one of which is the proliferation of subsidies in the developing world. Any environmental strategy that does not have energy efficiency and conservation as a center of interest will be missing the target.

INTRODUCTION

Reducing demand for energy, through energy efficiency measures and conservation, is the cheapest and most effective way for enhancing energy security, providing a stable basis for economic planning as well as cutting carbon emissions.

Energy efficiency is the most influential means of achieving sustainable energy future. It will significantly economize on energy use, reduce investments in energy infrastructure, reduce costs and maximize consumer welfare; as important it cuts off on emissions. In its World Energy Outlook 2010, the International Energy Agency (IEA), adopted the BLUE Map scenario which sets a goal of halving global energy-related CO_2 emissions by 2050, compared to 2005 levels, and sets out the least-cost pathway to achieve that goal; through the deployment of existing low-carbon technologies (IEA-WEO, 2010).

In 2050, with business as usual, global CO_2 emissions are expected to reach 57 Gt of CO_2. The Blue Map emissions aim at reducing them to only 14 Gt. Although this has become a rather unrealistic goal, however, what is of interest to us here is that the most effective means of achieving these emissions reduction through energy efficiency is end-use fuel and electricity generation. In the IEA 450 Scenario, energy efficiency is capable, by 2035, of achieving up to 48% of the abatement goal, more than twice that of renewables or CCS. The total reductions in CO_2 emissions as a direct result of energy efficiency measures are almost equal to the combined containment by all other means (CCS, renewables, fuel switching, nuclear, etc., see Figure 1).

This is a survey paper that deals with many aspects pertaining to energy efficiency, particularly in the electricity generation sector and also industry and buildings. Efficiency in the transport sector is dealt with by another paper in this session, as are aspects pertaining to sustainability. The electrical power generation sector is chosen for special emphasis due to its present low efficiency and the growing importance of electrification as we are going to see later.

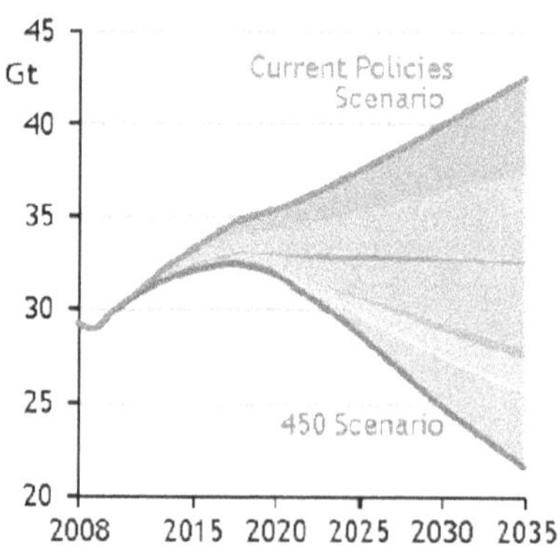

	Abatement		
	2020	2030	2035
Efficiency	71%	49%	48%
- End-use (direct)	*34%*	*24%*	*24%*
- End-use (indirect)	*33%*	*23%*	*23%*
- Power plants	*3%*	*2%*	*1%*
Renewables	18%	21%	21%
Biofuels	1%	3%	3%
Nuclear	7%	9%	8%
CCS	2%	17%	19%
Total (Gt CO_2)	3.5	15.1	20.9

Fig. 1: IEA 450 Scenario for Reducing Global CO2 Emissions (IEA–WEO 2010).

ENERGY EFFICIENCY

Let us first of all start by defining what we mean by "energy efficiency". According to the IEA—something is more efficient if it delivers more services for the same energy input, or the same services for less energy input. For example, when a compact florescent light (CFL) bulb uses less energy than an incandescent bulb to produce the same amount of light, the CFL is considered to be more energy efficient (IEA, Governance). Energy efficiency is also defined as "percentage of total energy input to a machine or equipment that is consumed in useful work and not wasted as useless heat" (ABB, 2010). The EU Directives define efficiency as "a ratio between an output of performance, service, goods or energy and an input of energy" (ESD Article 3 b).

The exact value of global energy efficiency is a debatable figure. Of the 12,500 million tons of oil equivalent (Mtoe) of primary energy consumed in 2010, 8,600 Mtoe are delivered as total final energy to users and consumers. There are estimated 7,500 Mtoe lost, mostly as low- and medium-temperature heat (WEA). This means that globally the energy efficiency ratio does not exceed 40%, which is extremely low and demonstrates the future strides that can be achieved in this quarter. A large proportion of the energy losses in delivering final energy to consumers are in the electricity generation sector, and that is why this sector receives a lot of emphasis in this paper. Losses in converting delivered energy into useful energy in final use also occur in all sectors of the economy, mainly in industry, transport and building.

Energy efficiency has to be differentiated from energy conservation. Energy conservation means more caution, better economy and care in the use of energy mainly through cutting waste and unnecessary use. Whereas conservation can be achieved through behavioral changes, education and pricing, energy efficiency can be mainly improved by technology, investments, codes and practices as well as energy pricing. Energy efficiency also needs to be differentiated from energy management which means timing energy use in a manner that minimizes cost and maximizes useful utilization of efficient facilities.

Energy is consumed for production, services and comfort. Its extent of use varies from one country and region to another depending on technology achievement, prices of energy products, economic output and weather. Primary energy intensity of a country is its total primary energy consumption per unit of the country's gross national product (GDP). Primary energy intensity measures the total amount of energy required to generate one unit of GDP. GDP is expressed at constant exchange rate or purchasing power parity (ppp) to remove the impact of inflation and changes in currency rates. Using purchasing power parity rates instead of exchange rates to convert GDP in the same currency (e.g., $) makes it possible to account for differences in general price levels: it increases the value of GDP in regions with low cost of living (case of developing countries) and, therefore, decreases their energy intensities.

Prior to the age of increasing energy prices, i.e., pre-1973, the growth of energy consumption used almost to mirror that of global economic growth, i.e., elasticity of energy demand versus that of economic growth was almost 1:1, with little improvement in global energy intensity. However, in recent years, and due to rising energy prices, environmental awareness and resource depletion, as well as technology oriented efficiency programs, significant trends took place in reducing energy intensity. Global energy intensity decreased by 1.4% per year over the last two decades and is continuing to do so (see Figure 2). Over the last twenty years the world economy grew at an average of 3.2% annually, while primary energy consumption grew from 8,800 Mtoe to 12,500 Mtoe i.e., an annual average of 1.9% only. This lead to these significant improvements is energy intensity.

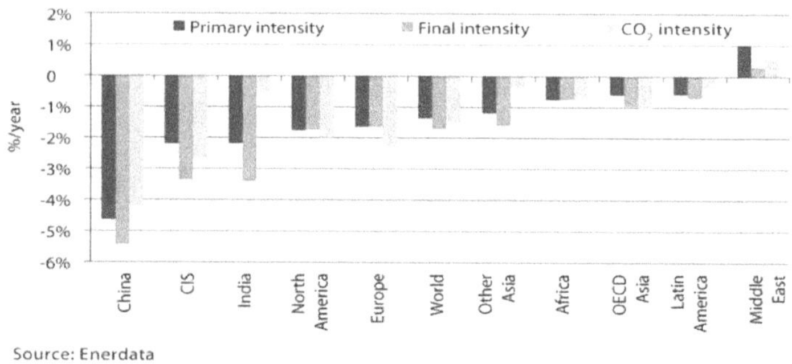

Fig. 2: Energy and CO_2 intensity trends (1990-2009) (% / year) Global Energy Efficiency Trends (ABB 2010).

Most reductions in energy intensity occurred in OECD countries. One of the reasons of the decrease of the energy intensity in OECD countries is that the heavy and high energy intensity industries (steel and cement industries, aluminum smelters, etc.) moved out of these countries into places where energy is still cheap (like China and India) and where there is less national emphasis on environmental and emission issues. OECD economies are emphasizing the services sector which is less energy intensive, while many developing countries are building and expanding their industrial base with emphasis on heavy industries and utilization of local natural resources while exporting industrial products to OECD countries. Correspondingly the true country figure of energy intensity should be modified to reflect imports to OECD countries of industrial material that carry a high energy bill (more on that later in the Industry sector).

EFFICIENCY IN THE ELECTRICITY SECTOR

Energy losses occur in every activity in the electricity sector, not only in the utilization of electricity, but mainly in the electricity supply industry—electricity production (generation), its transmission to consumer centers as well as its distribution to users and consumers. It is mainly in electricity generation where the majority of losses occur. Till today the average efficiency of electricity generation worldwide is around 35%, with almost two thirds of calorific content of primary fuel input into electricity generators is lost as waste heat (Khatib, 2003). Efficiency of electricity generation which was as low as 25% in the mid-twenties significantly improved to 32% in 1990, 34% in 2,000 and 35%-36% today and is continuously but slowly improving mainly through the increasing use of combined cycle facilities firing natural gas.

Figure 3 (IEA-ETP, 2010) shows how the efficiency of coal firing power stations was held steady at almost 34% over the last twenty years, while that of gas firing facilities improved from 34% in 1990 to 43% in 2010 through the wide introduction of combined cycle gas turbine plant (CCGT) facilities firing natural gas.

Electrical efficiency is most important because the world is electrifying. Energy services are now being increasingly offered in the form of electricity, rather than in any other form—mechanical or human, Whereas energy demand is expected to grow at an annual rate of 1.4% over the next 25 years, that of electricity is expected to grow at a rate of 1.7%, reflecting this growing electrification; that is why improving efficiency of electricity production figures so high in any effort to improve global energy efficiency.

Efficiency of the electricity generation depends on the generating facility. Hydroelectric production has an efficiency of over 90%; whereas vintage thermal plant firing coal has efficiencies of no more than 25%. Modern steam plant, firing coal, is becoming the preferred means of generation, particularly in China and India, where thermal plants represent 80% of production due to availability of cheap coal. Steam generation, mainly firing coal, presently represents 40% of world electricity generation and future improvements in efficiency of electricity production depend on technology advancement in this quarter. The major losses mainly occur in the thermal plant itself, in condensing steam; also in auxiliaries and generators and these can be as high as 5-6% of the generated electricity.

There are two approaches to improve steam power generation efficiency: one is through increasing live steam parameters (pressure and temperature) to develop supercritical (SC) and ultra-supercritical (USC) technologies; another is by system integration, a typical example is Integrated Gasification Combined Cycle (IGCC). In the next 10 years, SC and USC will be built in significant numbers, and these new plants are likely to remain in use until 2050 for electricity production from coal because of their flexibility and general advantages of lower cost, reliability, high availability, maintainability, and operability. IGCC currently looks promising in its ability to produce deep CO_2 reductions at least cost while maintaining high generation efficiency.

Pulverized coal combustion (PCC) is currently the predominant technology for generating electricity from coal. It accounts for more than 97% of the world's coal-fired capacity. Most existing plants operated at less than SC steam conditions, with best examples reaching 39% efficiency.

New pulverized coal power plants—utilizing SC and USC—operate at increasingly higher temperatures and pressures and therefore achieve higher efficiencies than conventional units. Supercritical power generation has become the dominant technology for new plants in industrialized countries. Now considerable efforts are underway in Untied States, Europe, Japan, also China, to develop 700 C-class Advanced Ultra-Supercritical (A-USC) steam turbines. If successful, this will raise the efficiency of the A-USC units to about 50% by 2020 (Lui, 2010).

Natural gas (NG) presently accounts for more than 20% of electricity production. Increasingly availability of NG and its prices are favoring the utilization of NG in high efficiency combined cycle gas turbines (CCGT) where efficiencies as high as 60% are becoming increasingly possible. This CCGT trend and improvements in the efficiency of thermal plant through the increasing use of critical and super critical steam turbines means that the efficiency of electricity generation is destined to continue to improve, although slowly, in the future.

There are also major losses in the grid—the transmission network as well as distribution lines. The transmission network that delivers bulk electricity from major power stations into bulk supply substations, can incur losses as high as 1–3 % of transmitted energy (depending in the length and voltage of the network). Other 6–10% losses occur in the distribution networks (also depending on the network configuration—length and conductors). Therefore, world average transmission and distribution (T&D) losses are around 8–9% of energy sent out from the power station (IEA-ETP, 2010). When this is taken into account, of the primary energy input into the power plant only 30–32% of this energy reaches consumers, i.e., less than one third. This demonstrates the efficiency dilemma of the electrical power system.

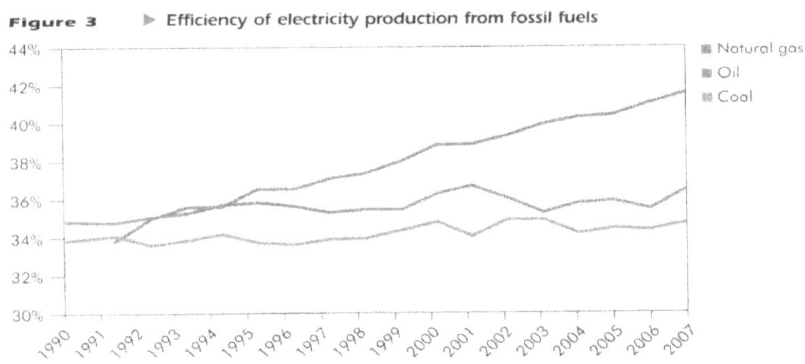

Fig. 3: Development of Efficiency of Thermal Generating Plant (ETP 2010)

Carbon Emissions from Electricity Generation

Thermal power stations firing coal are major emitters of CO_2 with around 750–1,000 grams CO_2/kWh in 2010. With electricity generation of 20,500 TWh in that year, total CO_2 emission from the electricity sector amount to around 12,000 Mtons, this is 40% of global carbon emissions. With each one percentage point improvement in efficiency of electricity generation, CO_2 emissions can be reduced by as high as 350Mt annually.

It has to be realized that considerable progress has been achieved in reducing emissions from thermal plant during recent years. Past vintage plant of the late twentieth century burning coal operated at an efficiency of 25% while emitting 1.30 tons

CO_2/MWh. New thermal generation operating at an efficiency of 45% and burning coal is emitting only 0.720 tons CO_2 /MWH, while a new CCGT burning NG has an efficiency of 60 % and emits only 0.320 tons CO_2 /MWH. This is only one quarter of the emissions per unit of electricity one quarter of a century ago.

Nuclear plants, due to security reasons have to operate at relatively moderate temperatures and pressures. Correspondingly its efficiency tends to be at 35% or lower. This is a low efficiency value compared to modern thermal plant. However a nuclear plant has the advantage of no emissions.

ENERGY EFFICIENCY IN INDUSTRY

Industry is the major energy user by sector, slightly ahead of power generation, buildings and transport. It is also a major user of electricity. In 2011 it is expected to use one quarter of global electricity production compared to one fifth in 1990. Due to advancement in technology major improvements took place in the efficiency of energy use in the industrial sector during recent years. The energy required per unit of value added (industrial energy intensity) has decreased in all regions since 1990. As a result of the globalization of industrial activities, energy intensity levels are converging. At world level, it fell by 1.6 percent / year between 1990 and 2009 (1 percent/year between 2000 and 2009).

Industrial energy intensity is lowest in Europe and OECD Asia, where it stands 40 percent below the world average since 1990. Due to the migration of heavy industries to the developing world, the European industrial energy intensity has decreased by 2.5 percent/year, while North America and OECD Asia show reductions of around 1.5 percent/year over the period. Energy intensity is slightly higher in North America than in Europe and OECD Asia: it stands just 13 percent below the world average and 50 percent above the EU average. Industrial cogeneration (CHP) is aiding the improvement in energy use in industry.

As mentioned above, one of the main reasons for improvements in the industrial energy intensity of Europe and North America is the shift of much of the heavy industrial production from these countries into China, India and few other developing countries which are rich in cheap energy resources.

Energy intensive industries are: the steel industry (20% of global energy use in industry), chemical industry (14%), cement industry (11%), paper industry (6%) and the aluminum industry. These heavy industries account for over than 55% of global energy use in industry. (ABB, 2010)

Advancement in technology in heavy industries (like: increasing use of electric process in steel, change from heavy into light chemicals, the dry process in cement production and recycling in the aluminum industry, etc.), was the main reason for improvements in efficiency in energy use in industry.

The energy required per unit of industrial value added has been decreasing in all regions; as a result of the globalization of industrial activities, energy intensity levels are converging. The global economic crisis in 2009 had a strong impact on industrial trends, especially in developed countries, where energy-intensive industries were severely hit.

In the steel industry, which is the main industrial energy consumer, significant progress has been made over the last 20 years thanks to the spread of the electric process.

However, the main producers, in cheap energy regions, still use outdated processes like open-hearth furnaces (Russia and Ukraine) or small-sized plants and low quality ore (China).

In the chemical industry, the energy consumption per unit of value added decreased in all main producing countries except in the United States, which is the world's main producer of chemical products and also the country with the highest energy intensity.

The spread of dry processes and kilns using pre-heaters and pre-calciners let to reductions in the average energy consumption per ton of cement (specific energy consumption) in several large producing countries. The sharpest drop in specific energy consumption was achieved in China, thanks to the replacement of small cement plants by larger facilities.

Nevertheless, a great deal still remains to be done, as shown by the significant energy savings potentials identified in this study. The energy efficiency improvement potential can be as high as 40 percent in the case of steel and 20 percent in other industries.

ENERGY EFFICIENCY IN BUILDINGS

The building sector consumed the equivalent of 2760 Mtoe of primary energy in 2007; this is almost to 23% of global primary energy use. However buildings accounted for almost 40% of the world's final energy consumption. Almost 70% of this was consumed in the residential sector and the other 30% in the services buildings sector (commercial and public). The expected growth in energy consumption in buildings, at an average of 1.2% annually, is expected to mirror that of global average energy growth. However most of this increase is expected to take place in commercial buildings due to the global rapid growth of services.

Energy usage in buildings is quite different from that in industry or the power sector in that globalization and technology do not significantly assist in converging efficiency of use between different countries and communities. There are at least one billion households worldwide, however, their use of energy varies widely accordingly to income, affluence and weather. Whereas major consumption of energy in building in OECD countries goes for heating, that in developing countries is accounted for by cooling. With increasing affluence, cooling is becoming a major user of energy in buildings in developing economies. Biomass still plays a prominent role in energy consumption in buildings in developing countries, while that of electricity is rapidly growing. Lighting accounts to almost one fifth of energy consumption in buildings.

Most of today's buildings, particularly in OECD countries, were built during the cheap energy era with modest emphasis on energy efficiency. Also the need for cheap and popular housing in developing countries is delaying the introduction of energy efficient technologies in buildings. However many buildings can benefit out of passive buildings design, particularly in DCs, in which the buildings are oriented in such a way so that heating and lighting in winter is assisted by solar exposure while in summer solar heat is avoided as much as possible. Of course quality of building materials and utilization of insulation plays a major role in energy efficiency of buildings; these are still modest in most of the developing world due to emphasis on providing cheap housing, most of the time, at the expense of long term economy in energy use.

Public policy, particularly in OECD countries and in growing economies, is promoting energy efficiency in buildings through (ETP, 2010):

- Building codes
- Certificate of energy efficiency
- Government promoting efficiency in public buildings
- Training and certification of building experts
- Fiscal and financial incentives (not only in new buildings but also in improvements of existing buildings)

Combined lighting, space heating and space cooling account for almost half of the energy consumption in buildings in the industrialized world. Significant slow improvements in efficiency of lighting in almost all countries are taking place through the increasing adoption of compact florescent lighting (CFL) fixtures.

Improving energy efficiency in buildings and correspondingly reducing emissions require effort on three fronts: better control and management of energy use, better design for new construction and retrofits of existing buildings. In its 2007 study McKinsey & Company found that the largest potential reduction in CO_2 emission at costs below \$50 per ton of CO_2 emitted are in the power sector (27%), followed closely by the buildings sector (24%). (Creyts et al. 2007)

Aggressive measures are taking place to improve energy efficiency in buildings with design changes that harness passive heating and daylight (Levins et al. 2007). However the barrier to better energy efficiency in the building sector are larger than other sectors due to the very large stock of existing buildings worldwide with low energy efficiency, shortage of financial and other incentives to improve these, also the high demand for cheap housing in DCs where affordability rather than efficiency is the paramount consideration. Correspondingly improving energy efficiency in buildings, in the future, may be slower than that in industry and power generation.

ENERGY GOVERNANCE

To achieve the benefits of energy efficiency calls not only on technological advancement but also on institutional, legal, public awareness and similar coordination arrangements. This is called "energy governance" and is defined as "combination of legislative frame works, funding mechanisms, institutional arrangements and coordination mechanism, which work together to support implementation of energy efficiency strategies, policies and programs". Such energy governance is needed to overcome the many barriers to energy efficiency. These barriers are market, financial, institutional and technological, as well as lack of awareness and information. The Table of Annex One details these barriers.

To these must be added another major barrier which is "energy subsidies". The existence of subsidies is a major barrier to the attainment of the goals of energy efficiency and this is dealt with in detail below.

Dealing with Barriers

It is not intended here to go into full details of how to deal with barriers to energy efficiency. These are detailed elsewhere (IEA-Energy Governance).

It is enough to mention here the importance of regulatory, pricing, fiscal and financial measures in overcoming such barriers. The most important is setting minimum energy performance standards (MEPS) for equipment, apparatus, buildings and vehicles with obligatory energy audits to ensure attainment of MEPS as well as appliances labeling and enforce building certification. Pricing has to aim at avoiding unnecessary subsidies with an increasing tariff so that higher consumption is penalized by higher unit tariff that curbs overuse. Fiscal and financial measures need to provide incentive to energy efficiency investments, by providing revolving funds and contingent financing facilities.

Awareness and dissemination of information is most important in facilitating adoption of energy measures, particularly among the public. This is lacking in many cases and requires public information campaigns and promotions, enforcing the already mentioned appliances labeling, also demonstration and training. The energy service companies (ESCOs) can play a major role in this regard. They can provide the services demanded by industry as well as the public in certification, audits, training and promotion as well as disseminating public information. They also can be intermediaries in facilitating funding. Such ESCOs are now common in industrialized countries but are missing in most developing economies where their services are most badly needed.

Energy Subsidies as a Major Impediment to Energy Efficiency

A subsidy implies selling an energy product (refined fuel or electricity) at prices lower than the product's opportunity cost. The opportunity cost is the price which the product will have obtained had it been exported instead of direct burning or utilization in local electricity production. We have to note that energy cost includes, beside the fuel's opportunity cost, the amortization of investment as well as operational and maintenance costs.

If we apply this definition, subsides proliferate energy use in most of the developing world. Electricity tariff subsidies are more widespread than subsides in selling refined products like petroleum for cars. This is due to the fact that electricity is a basic commodity that is widely used in all aspects of the economy as well as the society, particularly among limited income groups. This forces governments to provide it (like water) at prices lower than actual cost.

The dangers of subsidies manifest themselves in over use of the cheap resources more than is required by actual needs. Instead of spending the country's limited resources on human development, subsidies lead to waste, inefficiency and over use. This is besides the harmful environmental implications of this over use.

There is always the call for the need to provide energy in subsidized manner to certain limited income consumer groups as well as industries. If there is real need for this, then the subsidy should be made in cash and not in energy bills. Electricity subsidies lead to inefficient use and waste. Correspondingly it is essential to charge fair electricity prices that cover actual cost to avoid waste. Subsidizing the industry and economy can be in cash payment or other fiscal means, if need be.

Energy subsidies proliferates mostly in oil/energy exporting countries where refined products are provided at only a fraction of what they command in the export market and electricity tariffs are so low thus encouraging waste and over use. In few instances, in oil and gas rich countries, electricity is also provided free!!

The IEA (IEA-WEO, 2010) estimate that the value of fossil-fuel consumption subsidies, including electricity generation, in 2009, amounted to $312 billion. Most of these subsidies encouraged wasteful over use with social, economical and environmental damage. Most important they were an impediment to achieving the goals of energy efficiency. The value of the energy products market now approaches $5 trillion annually. Subsidies account to almost 6% of this market value. If market pricing can reduce this by a half almost 3% of world energy consumption can be saved by phasing out subsidies.

CONCLUSIONS

Due to increasing cost of primary energy and its products, also environmental awareness, the efficiency of utilizing energy is improving worldwide year after another. Major efficiency improvements happened in OECD countries in recent years through lowering energy intensity in their economies. However this was assisted by the migration of heavy industries to developing economies. In spite of all technological advances and regulatory arrangements, energy efficiency worldwide does not exceed 40%. This demonstrates the still wide scope for improvements in this regard.

The efficiency of the electricity sector demands particular attention. Of the primary energy input to power stations less than one third of the calorific input reaches consumers, in many countries only one quarter. The wider introduction of combined cycle plant and higher parameters in PC plants are gradually improving the electricity sector efficiency. This is particularly important because the world economy is increasingly electrifying.

In the global quest for improving energy security and reducing environmental impacts, improvements in energy efficiency are the cheapest and easiest means to attain results. Any global sustainable energy strategy which does not have improvements in energy efficiency as its center of interest will be missing the correct emphasis.

ANNEX ONE

BARRIERS TO ENERGY EFFICIENCY

Barrier Examples

Market

- Market organization and price distortions prevent customers from appraising the true value of energy efficiency.
- Split incentive problems created when investors cannot capture the benefits of
- improved efficiency
- Transaction costs (project development costs are high relative to energy savings).

Financial

- Up-front costs and dispersed benefits discourage investors
- Perception of EE investments as complicated and risky, with high transaction costs
- Lack of awareness of financial benefits on the part of financial institutions.

Information and awareness

- Lack of sufficient information and understanding, on the part of consumers, to make rational consumption and investment decisions.

Regulatory and institutional

- Energy tariffs that discourage EE investment (such as declining block prices).
- Incentive structures encourage energy providers to sell energy rather than invest in cost-effective energy efficiency.
- Institutional bias towards supply-side investments.

Technical

- Lack of affordable energy efficiency technologies suitable to local conditions.
- Insufficient capacity to identify, develop, implement and maintain EE investments.

(Source: IEA-Energy Efficiency Governance Hand Book second Edition–2010)

REFERENCES

1. ABB—Global Energy Efficiency Trends—ABB product 2010.
2. Creyts, J., A. Derkach, S. Nyquist, K. Ostrowski, and J. Stephenson. 2007. "Reducing U.S. Greenhouse Gas Emissions: How Much at What Cost?" U.S. Greenhouse Gas Abatement Mapping Initiative Executive Report. December 2007. McKinsey & Company. Available online at *http://www.mckinsey.com /clientservice/ccsi/pdf/US_ghg_final_report.pdf.*
3. Enerdata - Global Energy Intelligence. More information about Enerdata available www.enerdata.net
4. ESD—Energy Services Directive, Article 3b, 2006.
5. ETP 2010—IEA Energy Technology Perspectives, 2010.
6. Lui, L. & G. S. Gallagher, Energy Policy Journal: "Catalyzing Strategic transformation to a low-carbon economy—A CCS roadmap for China", Col. 38, Jan 2010, pp 59-74.
7. IEA-Energy Governance: IEA Energy Efficiency Governance Handbook—Second Edition 2010.
8. IEA-WEO 2010: The International Energy Agency—World energy Outlook 2010.
9. Khatib, H., "Economic Evaluation of Projects in the Electricity Supply Industry", I.E.E., London 2003.

10. Lee, J.J., et al. "Historical and Future Trends in Aircraft Performance, Cost and Emissions", Annual review of *Energy and Environment*, 2001, Vol. 26, pp. 167-200.
11. Levine, M. et al. "Residential and Commercial Buildings," in *Climate Change 2007: Mitigation,* Contribution of Working Group III to the Fourth Assessment Report of the Intergovernmental Panel on Climate Change, Cambridge, UK: Cambridge University Press, 2007, pp. 404–405.
12. WEA—World Energy Assessment "Energy End-Use Efficiency" Eberhard Joehen, UNDP 2000.

ENERGY EFFICIENCY AND SUSTAINABILITY

MICHAEL JEFFERSON
International Business and Sustainability Center, London Metropolitan Business School, London, UK

ABSTRACT

At the 1992 Rio Earth Summit, great emphasis was placed on energy efficiency in the Opening Session. That message, and indeed the subject of energy more generally, largely disappeared in the forty chapters and 600 pages of Agenda 21 that emerged from Rio. This situation has largely remained in subsequent UN deliberations. Closer focus on climate change, deforestation, and poverty has failed to produce significant benefits.

By 2010, world carbon dioxide emissions from the use of fossil fuels had risen over 46% from 1990 levels. Renewable energy projects have been pursued with scant regard for efficiency or costs to users. Still 2.7 billion people—some 40% of the World's population—rely overwhelmingly on traditional biomass; 1.5 billion others have no electricity supply; and a further 1 billion only have sporadic supply. Fossil fuels continue to provide nearly 85% of the World's primary energy, while renewable energy sources provide about 13%—of which traditional biomass accounts for 10.2% and all other renewable resources for only 2.8%. Of this 2.8%, hydropower accounts for 2.3%, wind power for 0.2%, and direct solar energy and geothermal for 0.1% each.

Not one of the eight Millennium Development Goals (MDG) produced in 2000 refers to energy. Then in 2005 UNDP, together with the World Bank and ESMAP, produced "Energy Services for the Millennium Development Goals", which pointed out that "failure to include energy considerations in national MDG strategies and development planning frameworks will severely limit the ability to achieve the MDGs." It also stated: "Increased energy efficiency—whether during generation/production, transport/ transmission, or end-use—can have wide-ranging benefits." In 2010, the UN Advisory Group on Energy and Climate Change (AGECC) report "Energy for a Sustainable Future" referred to improving energy access and strengthening energy efficiency as the "two priorities". AGECC's Chairman (and Head of UNIDO) stated that "a vast potential for energy efficiency improvements across the supply and delivery chain remains largely untapped." Also in 2010, the UN's General Assembly proclaimed 2012 the International Year for Sustainable Energy. Will there be a new, and more effective, focus on energy and efficiency in its provision and use?

INTRODUCTION

There are many fields of human endeavour where policies, measures, and actions are marked by internal contradictions. The field of energy as it relates to sustainable development is no exception.

It is often claimed that universal access to energy is a vital step for sustainable development. This should not be accepted at face value. For example, many poor families in India can access volumes of primary energy relatively easily—from fuel wood and animal dung—and the volumes of these in relation to needs at face value might seem greater than their requirements. The problem is, of course, that transforming these to useful energy involves great losses, possible adverse health effects, and both poor and limited final energy services. In other words, this is a very inefficient use of energy resources. Also, by using large quantities of fuel wood it can undermine sustainable use of land and, by using human energy and time in its collection, undermines sustainable human development.

Then there is the question of whether a particular form of energy or transforming technology is regarded as socially desirable. For example, in South Africa it was noticed as far back as the 1980s by ESKOM that local people had a strong preference for grid-connected electricity, rather than solar panels for electricity or for water heating, not because the latter failed to provide a superior service compared with what had hitherto been available, but because grid-connection was regarded as a superior and more socially desirable form of transmission. Here the link to energy efficiency may be more tenuous, yet the link to sustainable development perceived by customers nevertheless real.

These two examples illustrate the point that people do not want energy as such but the services it offers—lighting and heating, cooking, transportation, motive power—with as much comfort, affordability, and reliability as feasible. In other words, people expect, and certainly aspire to, the comfort, affordability, and reliability that they associate with modern forms of energy. These features they commonly associate with the efficiency of their provision.

In recent years the search for low carbon technologies and introduction of lower emissions targets among industrialised countries, especially many of those designated as Annex I Countries under the UN Framework Convention on Climate Change, has led to many initiatives which seem counter-intuitive. The push for wind energy to the point where ever-larger wind turbines are placed where mean wind speeds are relatively low, capacity factors achieved are well below those widely claimed, and the problems of variability loom large. Supporting solar PV in areas where solar irradiation levels are poor unavoidably means that the results will be modest, and almost certainly an unwise use of resources. The drive to expand biomass and biofuel availability with scant regard for their wider implications—such as impacts on food availability and prices, water requirements, associated emissions, loss of biodiversity and species' habitats. Talk of tidal range technologies (estuarine barrages) without adequate recognition of their wider ecological consequences. In these fields severe adverse impacts on sustainable development may be inherent, and are insufficiently taken into account. Some may consider other renewable energy technologies—large hydropower schemes, wave and tidal stream, Concentrating Solar Power (CSP), for example—can also have serious negative consequences. Here we come to matters of personal judgement and greater uncertainty. But for this author the latter have fewer negative implications for sustainable development, having taken into account the likely eventual scale of operations and net benefits. One reason for this relative equanimity is that these other sources of renewable energy have higher power densities than wind power, biomass, and even solar PV unless situated in a sunny location.

All renewable energy sources have lower power densities than the fossil fuels or nuclear energy. The latter, in turn, are trivial in relation to the power densities of such phenomena as tornadoes, thunderstorms, earthquakes, or even hurricanes and monsoons. Fossil fuels as primary energy have power densities in the range 1-10 kW/per square metre, although the entire cycle of fuel, transportation, generation and transmission can reduce this to about 600 W/square metre in some coal-fired power stations. By contrast, wind power, geothermal, and the more efficient hydro-electricity schemes can achieve 5–15 W/per square metre. Flat plate solar heat collectors in areas of high solar irradiation can achieve up to 80 W/square metre; solar PV up to 50 W/square metre in areas of high solar irradiation. Biomass is closer to 1 W/square metre.[1] Thus although renewable energy sources are sometimes claimed to be 'free', which before any investment may be true, their efficiency is generally low and highly variable. Land usage, intermittency, state of technology, and cost of investments and subsidies all come into the equation.

THE ROLE OF THE UN

Maurice Strong, Secretary-General to the UN Conference on Environment and Development held in Rio in 1992, stressed in his Opening Speech:

> "Nowhere is efficiency more important than in the use of energy. The transition to a more energy-efficient economy that weans us off our overdependence on fossil fuels is imperative to the achievement of sustainable development."[2]

Yet the subject of energy efficiency, and of energy more generally with its links to sustainable development, failed to warrant its own chapter in what became known as Agenda 21. Instead, energy issues were confined to Chapter 9: "Protection of the Atmosphere". This eight-page chapter proved to be the most difficult and protracted to negotiate in Rio after the discussions on finance. It has been claimed: "The Arab Group maintained that the chapter was not only duplicative of the work of the Climate Change negotiations, but that it placed an over-emphasis on energy efficiency and conservation".[3] While some may cavil at this claim, it would surely have been wiser to have two separate chapters, one covering protection of the atmosphere and the other on the full range of energy issues relating to sustainable development. The result of bringing the two topics together into one chapter was a loss of focus on the many sound statements on energy in the chapter, not least the call under Objectives for "more efficient energy production, transmission, distribution and use." (para. 9.11). This is not to claim that, if there had been this tighter focus (looking at the doubtful outcome to date of the two international treaties opened for signature in Rio—on climate change and biodiversity), developments since would have been very different.

Under Activities, inter-governmental, governmental, non-governmental, and private sector bodies were urged to co-operate in identifying and developing commercially viable and environmentally sound energy sources to promote sustainable development; to promote integrated energy, environment and economic policy decisions for sustainable development; to promote R&D and the transfer and use of improved energy-efficient technologies; and to support environmentally sound energy from new and renewable energy sources. Cost-effective policies, measures, standards, labelling, education and awareness-raising were all to be encouraged in order to promote energy efficiency. The chapter then paid particular attention to the transport and industrial sectors. All worthy sentiments, but the effectiveness of the message was lost among all the other messages being conveyed.

INPUT FROM ENERGY SPECIALISTS

Much of the world's energy sector was well aware of the dangers at the time. The World Energy Council had established a Commission in 1990 to try and get a balanced and proportionate view on the importance of energy supply and use in relation to sustainable development. In a report entitled "Energy for Tomorrow's World" (1993) a balanced overview was given of prospects and needs, including a 24-page chapter on energy efficiency and conservation. All the regional reports submitted to the Commission and the report's author stressed "the need to give very high priority to increasing the overall efficiency of energy use—indeed this was one of the most widely supported of all aims in the energy sector".[4] It was pointed out that there was huge technical potential for meeting the needs and aspirations of society with much less energy use. Moreover, it was pointed out that the aim should be to seek a theoretical minimum energy use for performing a required or desired task, and anything above that should be regarded as wasted energy. This concept of useful energy was, and is, termed "exergy". End-use energy efficiency in the USA was estimated to be only 2.5%, in other words in principle the same final services could have been obtained by the expenditure of only 1/40th as much energy as was actually used. Western Europe and Japan were considered to be more efficient, in the 4%-5% range, while Eastern Europe and the CIS were estimated to be in the 1.5% to 2% range. Overall, the word's energy efficiency rating was estimated to be then no greater than 3% to 3.5%.[5]

Examples of the potential for improving energy efficiency were drawn from the manufacturing, agricultural, residential and commercial, and transport sectors as well as from energy supply and conversion. This report was referred to by a number of energy specialists not linked with the World Energy Council as "The Bible" on energy for several years thereafter. In its assessment of the potential for improving energy efficiency "Energy for Tomorrow's World" did not overlook the challenge of what is called "the rebound effect"—that in improving energy efficiency resources will be freed up (to an unknown extent) to consume more energy.[6]

Then in 1998 there followed from the World Energy Council and senior staff members of the International Institute for Applied Systems Analysis (IIASA) "Global Energy Perspectives", where the scope for energy intensity improvements and technological change was thoroughly aired. This study also announced a departure from the traditional analytic separation of energy supply and demand, treating the energy system holistically from the primary sources to the services provided. It demonstrated and concluded that in 1990 global energy flows had been 9 Gtoe (gigatonnes oil equivalent), of which 3.3 Gtoe had provided useful services and 5.7 Gtoe (63%) had been "wasted or rejected".[7] This analytical framework has now become standard, although it does not plumb the future possibilities for efficiency improvements suggested by the principle of exergy.

A similar flow chart has been produced for energy use in the USA in 2008, by the Lawrence Livermore National Laboratory. Here 57.5% of primary energy inputs are diagnosed as 'rejected' (or wasted). Although there have been many suggestions that a disproportionate amount of energy is wasted in the USA compared to most other countries, the more important point would seem to be that little progress has been achieved in reducing the volumes and proportions of wasted energy in the world over the past twenty years. Indeed, Robert Ayres has suggested that official data often overstate the level of energy efficiency achieved.[8] Thus for the USA, he contrasts U.S. Energy Information Administration claims with his own findings:

Current U.S. Energy Efficiencies

	U.S.EIA	Ayres' est.
Electric power generation and distribution	33%	33%
Industrial sector	80%	30%*
Transportation services	27%	< 1%*
Residential and commercial buildings	80%	10%
TOTAL	42%	8%

* Regarded by Bob Ayres as consistent with other engineering studies.

UN AGENCIES RESPOND

These studies and findings did not go entirely unnoticed among some of those engaged in the UN system. One result was an initiative of a World Energy Assessment conducted by personnel from UNDP (UN Development Programme), UNDESA (UN Department of Economic and Social Affairs), and the World Energy Council. This 500-page study was published in September, 2000, and contained a major chapter on energy end-use efficiency (pps. 173-217), other material on supply-side efficiency, and in the chapter "Energy Policies for Sustainable Development" a section on "Raising Energy Efficiency" pointed out the potential and the need to exploit it as a matter of urgency (pps. 427-429).[9] Chapter 6, on "Energy end-use efficiency", provided a wide range of energy efficiency potentials from many countries and regions even by 2020.

Regrettably, the process of producing and promoting this major study was undermined somewhat from two quarters. UNDESA's role was not entirely helpful, and divisions and political manoeuvring within the World Energy Council even more so. Yet it remains a major contribution, even though in the same month that it was published the eight Millennium Development Goals (MDGs) were issued without a single reference to energy or energy efficiency. This was despite the MDGs intended purpose of promoting environmental sustainability and developing a global partnership for sustainable development. This could not have been for lack of space. The Millennium Declaration itself runs to 32 paragraphs over nine pages.

There will be some severe questions for the UN to answer when the Rio+20 meetings in 2012 seek to assess what has happened since the Rio Summit in 1992, and when in 2015 performance towards the MDGs is examined. But there have been some more hopeful responses from within the UN system in recent years—along with many simultaneous signs of unrealistic targets and failure to meet them, and disparate aims and conflicts with reality.

One of the most encouraging signs was the appearance of "Energy Services for the Millennium Development Goals", produced by UNDP, the World Bank, and ESMAP (Energy Sector Management Assistance Programme) in 2005. This report recognised that "much wider and greater access to energy services is critical in achieving all of the MDGs." It pointed out that "without scaling up the availability of affordable and sustainable energy services, not only will the MDGs not be achieved, but by 2030 another 1.4 billion people are at risk of being left without modern energy." It noted that "no MDG refers to energy explicitly", yet improved energy services are necessary for meeting all the Goals. The report recommended that the issue of energy services be placed on a par with other MDGs, that measures be taken to ensure reliable electricity supply, and that a sharper focus on the availability of modern energy services was essential for human welfare and increased productivity. In summary, the report concluded:

> "It is clear that energy services have an impact on all of the MDGs and associated targets. Access to energy services facilitates the achievement of these targets. Failure to consider the role of energy in supporting efforts to reach MDGs will undermine the success of the development options pursued, the poverty reduction targets, as well as the cost effectiveness of the resources invested."[10]

Perhaps the main weakness of this Report was its avoidance of any specific reference to energy efficiency. However, there were various references to the need for reliable supplies and specifically under "Challenges Facing Energy Institutions and Systems" (ibid, p. 66) to: "technical problems such as electric utilities with limited generation capacity and losses in transmission and distribution that nonetheless face high and growing demand". This topic will be returned to below in discussion of the concerns many people have over the reliability of some forms of renewable energy in terms of variability, and need for back-up of electricity supply from traditional sources and well-founded concerns about the implications of having this "spinning reserve".

Thus it was that in 2009. a UN Advisory Group on Energy and Climate Change was established under the chairmanship of the Director-General of UNIDO (UN Industrial Development Organization), Kandeh Yumkella. This Advisory Group identified two priorities: improving energy access and strengthening energy efficiency. The Chairman pointed out that the "vast potential for energy efficiency improvements across the energy supply and delivery chain remains largely untapped."[11] In the main body of the report it is stressed:

> "Energy efficiency is one of the few 'no-regret' policies that can offer a solution across challenges as diverse as climate change, energy security, industrial competitiveness, human welfare and economic development." (ibid, p. 18).

The Report estimated that in capturing all cost-effective energy efficiency measures the growth of global energy consumption could be reduced from currently forecast levels of 2,700-3,700 Mtoe (million tonnes oil equivalent) in 2030 to 700-1,700 Mtoe. The reduction would be split fairly evenly between high income countries and the rest of the World, much of the former coming from retro-fitting existing infrastructure in the developed economies. The large potential for improving efficiency in power generation, and reducing transmission and distribution losses, was covered as well as the potential for exploiting opportunities in industry, buildings and transport. By applying the best currently available technologies it was estimated that potential savings in the buildings sector is around 1,500-2,000 Mtoe; in power generation around 1,000 Mtoe; in the industrial sector some 600-900 Mtoe; and in the transport sector around 500 Mtoe.

These matters are still, for the most part, aspirational. As was pointed out at the Vienna Energy Forum in June, 2011, there are still 2.7 billion people (40% of the World's population) who rely overwhelmingly on traditional biomass. A further 1.5 billion have no electricity supply. A further 1 billion have only a sporadic supply of electricity. There are fears that, unless there is careful re-examination of the reliability of large-scale electricity generation from renewable energy sources not only will the aim of meeting the needs of all those currently without access to modern energy services by 2030 fail by a large margin. There are also concerns that a growing number of people in some industrialised countries will also be reduced to sporadic electricity supplies due to over-dependence on wind and solar energy.

An additional cause for concern is the little progress being made towards shifting the World away from reliance on fossil fuels. As the recent IPCC Special Report on Renewable Energy Resources has pointed out, 85% of the World's primary energy still comes from fossil fuels, and use of fossil fuels accounts for 56.6% of all anthropogenic greenhouse gas emissions. Renewable energy on the other hand accounts for 12.9% of global primary energy use. Unfortunately, this breaks down into: 10.2% traditional biomass; 2.3% hydro; others 0.4%. Of this latter 0.4%, wind power accounts for 0.2%, and solar energy and geothermal for 0.1% each.[12] Nuclear accounts for the remaining 2.1%, and in many countries is again under general threat because of the particular circumstances of the Fukushima incident. As already indicated, there are reasonable fears that in shifting from reliance on fossil fuels (albeit there now being enhanced potential to shift away from coal to gas) and nuclear power too hastily a loss of reliability will occur for many electricity users.

Not surprisingly, therefore, as pointed out at the Pre-Launch of the "Global Energy Assessment" in Vienna on June 21, 2011: "Efficiency improvement is the most cost-effective option."[13] There are also other encouraging signs. The year 2012 has been designated the 'International Year of Sustainable Energy for All', and the International Energy Agency (IEA) issued a report in 2009: "Towards a More Energy Efficient Future" which pointed out yet again the "increasing consensus that improving energy efficiency is often the most economic, proven, and readily accessible means of ensuring a better use of the world's energy resources." The report focussed on rising household energy use due to increasing ownership of small appliances; increased use of electricity in the services sector; passenger and freight transport increases; and electricity generation growth (which increased by 37% within IEA Member countries between 1990 and 2006 alone).[14] The IEA will focus part of its World Energy Outlook 2011 on how this could be achieved (although the last estimates of costs involved were U.S.$756 billion between now and 2030). Only time will tell whether these are just more fanciful targets or realistic ones.

RISKS FOR ENERGY EFFICIENCY AND SUSTAINABILITY

Time and space only allow a few issues to be touched on here. First, there is the question of accounting for greenhouse gas emissions. To date international convention makes no allowance for the large rise in manufactured goods imported into industrialised countries from such countries as China and India, where dependence on coal-fired electricity generation is high. Thus the UK can claim it has reduced its carbon dioxide emissions by some 20% between 1990 and 2010. In fact, if the emissions exported to such countries as

China, India and others where manufacturing capacity has in effect been transferred, then the UK's emissions of carbon dioxide actually rose some 16% between 1990 and 2010. This should not cause surprise. The UK Government's Chief Scientist to its Department of Energy and Climate Change (DECC) "called for more openness in admitting Britain's apparent cuts in greenhouse gases are an illusion". Robert Watson, former Chairman of the Intergovernmental Panel on Climate Change, said: "if emissions 'embedded' in imported goods are counted, UK emissions are up, not down." He is quoted as saying: "At face value UK emissions look like they have decreased 15% or 16% since 1990. But if you take in carbon embedded in our imports, our emissions have gone up about 12%. We've got to be more open about this." Not surprisingly perhaps, though disingenuously, the official DECC response was: "Our position is that greenhouse gas emissions in the UK have been cut by 22% since 1990." The year 2010 saw, according to UK Government statistics, carbon dioxide emissions rise 3%, hence the +16% figure given above.[15] This report led to calls for the UK to start measuring its full carbon footprint, but to no effect as yet. [One intriguing feature of the recent Vienna Energy Forum was that there was no mention of Austria's 10% increase in carbon dioxide emissions from fossil fuel use between 1990 and 2010.]

The UK is not, of course, alone in this situation. Germany, which has decided to abandon nuclear power from 2022, has led the reduction in greenhouse gas emissions in Western Europe since 1990 (assisted by re-unification). BP's Statistical Review gives Germany a 19.6% reduction in carbon dioxide emissions from fossil fuel use between 1990 and 2010 (and for the UK a 27.1% reduction). But Germany imports two-thirds of its energy. China accounts for over 8% of Germany's imports, with a value in excess of € 55 billion per year, well in excess of Germany's exports to China.

In 2009 Germany's primary energy use was: oil 34.6%; natural gas 21.7%; coal 22.5% (of which bituminous coal 11.1% and lignite 11.4%); nuclear 11.0%; hydro and wind 1.5%; and others 9.0%. The figures get even more interesting when Germany's electricity generation is considered. In 2008 the sources of Germany's electricity generation were: lignite 23.5%; bituminous coal 20.1%; nuclear 23.3%; natural gas 13.0%; wind 6.5%; biomass 3.7%; hydro 3.4%; wastes 0.8%; solar PV 0.7%; and others 5.0%. Germany announced in September, 2010, that renewable energy sources would provide 35% of its electricity requirements by 2020, and 80% by 2050. Furthermore, renewable energy sources would provide 18% of primary energy by 2020; 30% by 2030; and 60% by 2050. Although in 2010 it was claimed that nearly 17% of Germany's electricity was provided by renewable energy sources, a reality check is required on the targets being bandied about. Huge subsidies have gone into wind energy (mean wind speeds in the inner Länder are low and capacity factors achieved often below 12% in these areas), and into solar PV (Germany, even in the South, is not among the World's leading countries in terms of solar insolation). And although Germany has taken a leading position in support of the Desertec concept (placing large sets of parabolic mirrors in the Mediterranean region—notably in North Africa—and transmitting electricity from these Concentrating Solar Power plants by Ultra High Voltage Direct Current transmission to Northern Europe), socio-political uncertainties in North Africa have recently been compounded by the so-called 'Arab Spring'.

The UK has apparently eagerly followed the German example in terms of emissions reduction and renewable energy targets, though not in the abandonment of nuclear power. Thus the UK has called for the EU to adopt a 30% reduction in greenhouse gas emissions from 1990 levels by 2020, and an 80% reduction by 2050. As we have seen, the adoption of rational emissions accounting standards would quickly scupper that.

Another area where UK data and targets require close examination is the efficacy of renewable energy. Capacity (or load) factors achieved in electricity generation are provided by the UK's Department of Energy and Climate Change:

UK Capacity Factors Achieved by Electricity Generating Type—2010

Combined Cycle Gas Turbines	61%
Nuclear Power Stations	60%
Conventional coal-fired stations	41%
Onshore wind	22%
Offshore wind	30%
Hydro	30%
Biomass (including co-firing)	53%
Landfill gas	57%

Source: UK Digest of Energy Statistics, July, 2011, pps. 146 and 214.

These figures raise a number of questions. At a relatively trivial level, it may be noted that the UK's nuclear sector once averaged a capacity factor of around 80%. Given its low power density, the figure for biomass may be considered rather misleading. More intriguing are the claims that UK wind energy developments "typically" achieve a capacity factor of 30%, and at least one author has claimed 35% is closer to modern reality. Let us consider the claims for wind energy more closely.

There is another source for capacity factor assessment in the UK's wind energy sector, the data supplied by the operators themselves to Ofgem (Office of Gas and Electricity Markets). In 2009 the simple average capacity factor achieved by onshore wind energy developments in England was 21.16%; in Scotland 28.46%; in Wales 23.86%; and in Northern Ireland 24.06%. This suggests a capacity factor achieved in the UK of no more than 25%. In 2010 the figures were even lower: in England 18.7%; in Scotland a sharp reduction to 23.75%; in Wales 18.75%; and in Northern Ireland 17.57%. The 2010 data cover 313 onshore developments—142 in England, 89 in Scotland, 43 in Northern Ireland, and 39 in Wales. The average UK capacity factor was clearly below 20% in 2010. These figures contrast with the 30% capacity factor claimed as "typical" both by the industry and by the UK Government in its Companion Guide to PPS 22: Planning for Renewable Energy, and which is used to hoodwink the public (domestic and business electricity customers in the UK subsidise wind energy—the current programme estimated to cost them £100 billion over its lifetime) into believing the exaggerated claims made by both industry and government.[16] One of the problems with the official statistics is that they appear to have omitted large numbers of developments in the past, although this has been addressed more recently.[17] Nevertheless, the capacity factors claimed for wind energy in the UK appear to be greatly exaggerated. There is little official discussion of the estimated costs of building 17 new gas-fired plants (some £10 billion) to provide a 'spinning reserve' in the event that there is insufficient wind. There is a high likelihood that by the early 2020s there will be days when the UK will be able to supply over 20% of its electricity requirements from wind energy—only to be confronted a few days later with wind energy scarcely able to provide 2%. This seems neither efficient nor sustainable. For as long as one of the leading UK wind energy personalities continues to believe reference to capacity factors is "bizarre pseudo-science" and "absolute nonsense" it may prove difficult to pursue a rational path forward. Similarly, if the industry and government are in denial about the existence of aerodynamic modulation from wind turbines, visual intrusion, and adverse effects on residential property values (England's population density now exceeds 400 people per square kilometre) social acceptability will remain a major issue.

But in the context of efficiency perhaps the most remarkable statement that has emerged is the following from the British Wind Energy Association (now RenewableUK): "the meaning of efficiency is a redundant concept to apply to wind energy, where the fuel is free."[18] This appears to completely overlook the wider implications of variability in the relevant resource (wind or solar especially), and the importance of selecting optimal sites. It appears to overlook the industry's claims that they have spent billions of £s in investments, and the cost of subsidies. It is estimated that the current UK programme will cost domestic and business electricity customers £100 billion by 2030 in subsidies alone. The cost of that subsidy per worker employed in the UK wind energy sector in the year to March, 2010, was £57,000 (compared to the median public sector income of 29,000 and median private sector income of £25,000).[19] The nearest comparable figures are for Spain's subsidies to wind power and solar PV, which have now largely collapsed due to the global financial and Eurozone crises, of around € 8 billion between 2000 and 2008, with an average subsidy per renewable energy sector worker of € 571,138. [19, pps. 92—93]

It is widely claimed that the UK has the best wind resource in Europe, and that claim can be substantiated for Scotland but not for the rest of the UK. It is also not the case that "calm conditions never extend across the entire UK" or that "low wind speed conditions extending across 90% or more of the UK during winter occur around one hour every five years".[20] For example, in October, 2007 there were at least eleven days during which calm conditions existed across over 90% of the UK, and many hours during the winter of 2010/2011 when wind energy output was almost zero—despite the need for heating.

The variability of wind places large question marks around the UK's official support for onshore wind power in particular, and the reliance on building at least 10,000 more turbines as the main route to achieving emissions reductions from the electricity generating sector. It is highly likely that, with coal-fired power stations being decommissioned under the EU's Large Combustion Directive and withdrawal of existing nuclear power stations (most of them before new plants will come on stream), there will be severe disruptions to electricity supplies in the UK before 2016.

There are question marks over the use of other sources of renewable energy where the impacts on the environment may be severely adverse, or their contribution marginal or inefficient because of the relative lack of resource (e.g., lack of solar insolation). Among the most disturbing cases are planning decisions which permit the simple burning of palm oil in proposed electricity generating plants in rural England, even when the Planning Inspector involved is informed about associated emissions (on average 33 tonnes of carbon dioxide are estimated to be emitted for every tonne of palm oil produced in Indonesia, before its transportation to the generating plant). In such cases the Planning Inspector claims that palm oil is defined as a renewable source of energy under UK Planning Guidance (Planning Policy Statement 22), apparently regardless of deforestation and species' habitat loss, and approves the burning of palm oil.

There is also a growing literature on the impacts of biomass and biofuel production on food availability and prices, on water requirements, and on biodiversity. The technical potential of second- and third-generation biofuel technologies is interesting, but the time horizon and scale of output provide little ground for optimism in the short and medium term. There are ecological concerns about estuarine barrages (the La Rance scheme "destroyed the local ecology" according to its developer). Unless and until political stability is assured in North Africa Concentrating Solar Power is unlikely to take off. In short, the potential for large-scale efficient and sustainable energy schemes on the supply side seems far more modest than usually claimed.

The distortions which have arisen in the renewable energy sector are partly the result of excessive or poorly targeted subsidies (the € 110 billion subsidy programme for the UK wind energy industry, funded by electricity customers, has already been noted). This is an important issue. Yet the subsidies still going to fossil fuels—globally over U.S.$550 billion per year (IEA estimate for 2008)—and with organisations like the World Bank involved (over U.S.$ 3 billion in 2008, with lending to the coal sector 256% up on 2007) and countries such as Germany still involved (Germany's coal sector enjoys € 16 billion of subsidy in the period 2005-2012, and no cut in subsidies to the coal sector is proposed before 2018), are surely even less acceptable.

CONCLUSION

The past twenty years have seen an unfortunate path of evolution in the energy sector. Largely ignored, or mixed up with other aims and targets, energy as a general topic and energy efficiency as a particular one, have failed to get sufficient recognition or rational support in policies, measures and investments. Part of the problem has been the uncertain path trodden by UN agencies and the UN system more generally. Part has arisen at national levels. There remain disconcerting levels of subsidies to fossil fuels and perverse incentives for renewable sources. There are signs of greater recognition of energy efficiency as being the most cost-effective route to tackling a range of pressing sustainability issues. Whether these signs will be translated into effective action and desired results remains an open question. One ongoing challenge will be the possible "rebound" effects of resources freed up by energy efficiency improvements being ploughed back into higher energy use.
August 22, 2011.

* Corresponding author's e-mail addresses: m.jefferson@londonmet.ac.uk and michael.jefferson@buckingham.ac.uk

REFERENCES

1. Vaclav Smil (2008) "Energy in Nature and Society: General Energetics of Complex Systems", MIT Press, pps. 309-310.

2. Stanley P. Johnson (1993) "The Earth Summit", Graham & Trotman/Martinus Nijhoff. p. 53.
3. Stanley P. Johnson (1993) "The Earth Summit", Graham & Trotman/Martinus Nijhoff. p. 213.
4. World Energy Council (1993) "Energy for Tomorrow's World", Kogan Page/St. Martin's Press. p. 109.
5. *Supra* 2, pps. 113-114, after Robert U. Ayres.
6. *Supra* 2, p. 110.
7. N. Nakicenovic, A. Grübler, A. McDonald (eds.) (1998) "Global Energy Perspectives", IIASA/WEC. Cambridge University Press. Box 5.1, p. 64.
8. Robert U. Ayres: "Energy Efficiency", a presentation to the Vienna Energy Forum, June 22, 2011. See also: Robert U. Ayres and Benjamin Warr: "The Economic Growth Engine: How Energy and Work Drive Material Prosperity", International Institute of Applied Systems Analysis, Laxenburg, Austria, 2010.
9. UNDP/UNDESA/WEC: "World Energy Assessment: Energy and the challenge of sustainability", UNDP, 2000.
10. UNDP/World Bank/ESMAP: "Energy Services for the Millennium Development Goals", 2005. p. 32.
11. UN: "Energy for a Sustainable Future—the Secretary-General's Advisory Group on Energy and Climate Change (AGECC): Summary Report and Recommendations", 28 April, 2010.
12. Intergovernmental Panel on Climate Change: "Special Report on Renewable Energy Sources and Climate Change Mitigation", June, 2011 ('Final Release' before official publication).
13. Ged Davis, Co-President, the Global Energy Assessment, Vienna, June 21, 2011. The "Global Energy Assessment" is expected to be published in October, 2011, and to be 1,300 pages in length.
14. See: www.iea.org/publications/free new Desc.asp?PUBS ID=2142
15. See: http://www.bbc.co.uk/news/science-environment-11172230 (accessed May 25, 2011) and various UK Press reports on September 3, 2010. 'Embedded' emissions are alternatively described as 'embodied' emissions.
16. "Planning for Renewable Energy: A Companion Guide to PPS 22", 2004, at para. 34, page 165.
17. See the rebuttal of such claims in: Michael Jefferson: "Accelerating the transition to sustainable energy systems" in 'Energy Policy', Vol. 36, Issue 11, November 2008, at page 4122. David J.C. MacKay has since confirmed this rebuttal in "Sustainable Energy—without the hot air", UIT Cambridge, 2009.
18. Renewable UK (formerly British Wind Energy Association), http://www.bwea.com/ref/faq.html. The same website (Frequently Asked Questions) dissembles in its claim that low frequency noise [aerodynamic modulation] is "so low that they lie below the threshold of perception." Under "Top Myths about Wind Energy" the same website dissembles in claiming that (under 'Myth 12') that: "There is currently no evidence in the UK showing that wind farms impact house prices." http://bwea.com/energy/myths.html.
19. John Constable. "The Green Mirage", *Civitas*, July, 2011.

20. See DECC: "Energy Trends" Special Feature on Renewable Electricity, September, 2010, pps. 25-36. The author complained about a similar Special Feature on Load Factors in the September, 2008, issue (at pps. 43-44) which may have a bearing on the wider database.

THE ENERGY EFFICIENCY POTENTIAL OF GLOBAL TRANSPORT TO 2050

DAVID L. GREENE
Center for Transportation Analysis Energy Division, Oak Ridge National Laboratory, Oak Ridge, USA

ABSTRACT

Transport accounts for 27 percent of total global energy use, 26 percent of global energy-related CO_2 emissions, and obtains 95 percent of its energy from petroleum (IEA, 2010, p. 255). Carbon dioxide from the combustion of petroleum comprises approximately 95 percent of transport's GHG emissions. The sector's share of global primary oil consumption is expected to rise from slightly more than 50% today to 60 percent by 2035, with an approximate doubling of energy use. Transportation activity is inextricably linked to energy use because the nature of transportation is physical work—the application of force over a distance—and energy is the ability to do work. On the other hand, energy is used inefficiently to move people and commodities. For example, the overall energy efficiency of passenger transportation by motor vehicles in the United States today is only about 1 percent. Despite this, transportation is widely regarded as the most difficult and most costly sector for GHG mitigation and for substitution of alternative sources of energy. It is shown that given appropriate public policies to overcome the deficiencies of transportation markets and plausible technological progress, it should be possible to hold global transport energy use to approximately today's level through 2050.

INTRODUCTION

Transporting people and materials is essential to human society. Modern economies move approximately 15 trillion ton kilometers of freight each year, and passengers travel 40 trillion kilometers by motorized transport (IEA, 2010, p. 273). The essence of transport is moving a physical object over a distance, which requires the application of force and therefore energy. Today, over 95 percent of the energy used globally in motorized transport is obtained from one fossil form of energy: petroleum. The economic and social importance of transportation and its near total dependence on petroleum, combined with the geographical concentration of petroleum resources and their control by nation states has led to serious concerns about energy security among many oil importing economies. Combustion of petroleum fuels to power the world's transportation systems also causes significant damage to the global environment. Carbon dioxide accounts for nearly all of the global warming potential of greenhouse gases (GHGs) emitted by transportation vehicles, which comprise approximately 15 percent of global anthropogenic GHG emissions.

Global resources of conventional petroleum are finite and the rate of global consumption is now large (1.5 to 3 percent) relative to best estimates of remaining ultimately recoverable conventional petroleum resources (1 to 2 trillion barrels). Unconventional fossil resources that can be converted into liquid hydrocarbon fuels at costs the world's transportation system has already demonstrated it is willing to pay are vast. Indeed, a transition to unconventional petroleum resources in the forms of oil sands and extra-heavy oil has already begun. The transition will almost certainly result in greater emissions of GHGs per unit of energy than conventional petroleum. Perhaps more importantly, it will begin the exploitation of vast additional fossil carbon that could be added to the atmosphere. Energy security, environmental and sustainability concerns about petroleum have led many of the world's economies to promote biofuels, electricity and hydrogen as alternatives to petroleum. However, the energy density, relatively low cost and convenience of petroleum fuels, as well as the technology "lock-in" of the internal combustion engine poses major economic barriers to technologies that rely on alternative sources of energy. To date, increasing energy efficiency has been the most effective strategy for restraining the growth of transport petroleum consumption.

Road vehicles produce the most transportation and use the greatest quantities of energy. Of the 103 exajoules of energy used by transport each year, cars, trucks and buses account for approximately about 75 percent (EIA, 2010, Table 14; IEEJ, 2010). In 2007, the OECD countries accounted for approximately 60 percent of transport energy use. The U.S. Energy Information Administration (2001, Table 14) projects that by 2035, non-OECD countries will account for 55 percent of total transport energy use. Air transport is second in passenger-kilometers and energy use, using 11 exajoules, nearly all of which is kerosene for subsonic turbine engine aircraft. Ninety percent of the 9 exajoules of energy used for waterborne transport is used in international shipping. The IEA (2010, p. 294) projects that energy use for water transport will more than double by 2050. Rail freight energy use is expected to increase by 25 to 30 percent by 2050 (IEA, 2010, 289).

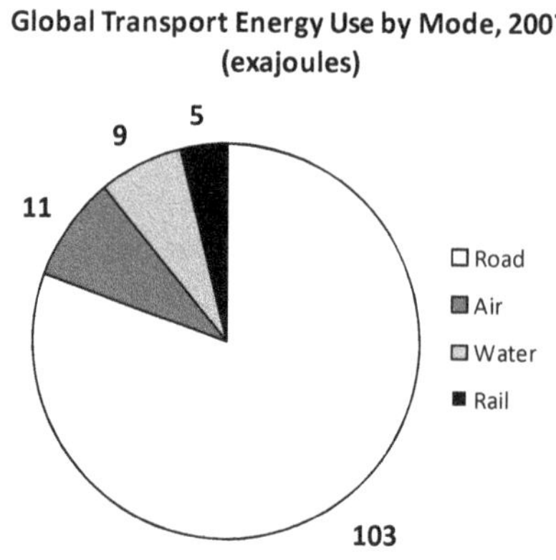

Fig. 1: Global Transport Energy Use by Mode.

Energy use for transport is expected to grow substantially in coming decades as a result of rapid motorization in the developing world. Between 1970 and 2007, global energy use in transportation more than doubled, increasing at an average rate of more than 2 percent per year, with the majority of the growth occurring in road transport (IEA, 2010, p. 257). The world stock of light-duty vehicles increased 60 percent over the same period, reaching 780 million vehicles by 2007. The IEA projects the world's motor vehicle park will increase to 2.2 billion vehicles by 2050, with most of the increase taking place in the developing economies (IEA, 2010, p. 281).

Assuming business as usual, the International Energy Agency projects a doubling of transport energy use by 2050 (IEA, 2010, p. 255).

In OECD countries, energy efficiency improvements have succeeded in restraining the growth of transport energy use and GHG emissions but not in reducing it. Given an expected slowing of the growth of transport demand in these economies, future efficiency improvements are expected to lead to actual reductions in energy use. This will not be the case for the world's developing economies, where motorization of transport lags far behind the developed economies. In these economies, improving energy efficiency will be a key strategy for restraining growth.

In OECD and non-OECD countries alike, meaningful government policies will be necessary to achieve the full potential of proven and advanced technologies to improve energy efficiency. To a large degree this is because markets tend to significantly undervalue potential energy savings. It is also because markets do not take into account the external costs of fossil fuel consumption. Nor do today's markets adequately consider the energy needs of future generations.

This paper considers the potential to increase the energy efficiency of global transport through technological improvements to transport vehicles and their propulsion systems. In so doing, it draws on recent studies from the EU and U.S. The technological potential for energy efficiency improvement is very large, on the order of a 60 to 70 percent reduction in energy required to produce a given quantity of transport services However, realizing that potential will depend on both continued advances in technology and effective public policies.

MEASUREMENT

The energy efficiency of transport is most frequently measured in terms of energy use per, 1) passenger-kilometer (pkm), 2) ton-kilometer (tkm) or 3) vehicle-kilometer (vkm). None of these is a perfect measure. Transport, like other economic services, has many dimensions. Among the more important are speed, reliability, safety, comfort, accessibility and cost. Failure to account for these other important attributes can lead to erroneous conclusions about the substitutability of one mode of transport for another. For example, although waterborne transport would be far less energy intensive and less costly, fresh flowers are most often shipped by air.

Evaluation of the potential to increase energy efficiency is most often done within modes producing very similar services and at costs that can approximately be recouped by the future value of fuel saved. This is the approach followed in this paper. There is likely some potential to increase overall transport energy efficiency by shifting passenger and freight transport from more to less energy intensive modes. However, when like services are considered, these opportunities turn out to be far more limited. There are also opportunities to improve the energy efficiency of transport systems through operational changes: increasing load factors, reducing circuity and improving the way vehicles are operated. The available evidence suggests that such improvements are usually on the order of 10 percent or less, are difficult to achieve on a system wide basis, and are as easily lost as gained (Greene and Plotkin, 2010, p. 32). The focus in this paper is on improvements to vehicle efficiency by technological means, although system efficiency improvements are considered in the air, rail and waterborne transport modes.

Most often, the energy efficiency of transport is measured by the energy directly used to move vehicles and their payloads. However, a substantial amount of energy is embodied in transportation vehicles and their infrastructure. For road vehicles, the energy embodied in vehicle manufacture and disposal comprises roughly 10 to 15 percent of total lifecycle energy use (Bandivedekar et al. 2008, Table 3). Because embodied energy is a minority of energy use and does not vary greatly across modes and vehicle types, it is usually neglected in the measurement of transport energy efficiency and that less than optimal practice is followed in this paper, as well.

EFFICIENCY POTENTIAL TO 2050

In this section the potential of technology to increase the energy efficiency of transportation vehicles is summarized, drawing on a number of recent assessments.

Passenger Cars and Light Trucks

The potential to improve the energy efficiency of passenger transport by light-duty vehicles is great in large measure because current energy efficiency is so low. The flow of energy from the tank of a modern 2.5 liter engine, 5 passenger automobile is shown in Figure 2. Engine losses, including thermodynamic losses, friction and pumping losses account for 74 percent of the chemical energy put into a vehicle's fuel tank. Various accessories, including oil and water pumps, fuel injection and cooling fan, require 6 percent of the total energy available. Of the remaining 20 percent available at the engine's crankshaft, 4 percent is lost to transmission inefficiency and other driveline losses. This leaves 16 percent to overcome the forces opposing the motion of the vehicle. Over the urban drive cycle, 4 percent goes to overcoming aerodynamic drag, 5 percent to overcoming rolling resistance and 7 percent to overcoming inertia (converted to low grade heat during braking). In highway driving a much greater percentage of total tractive energy is used to overcome aerodynamic drag which increases per unit of distance traveled with the square of velocity.

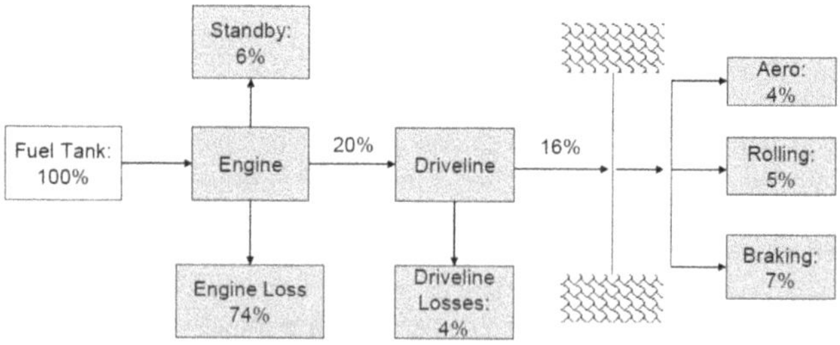

Fig. 2: Representative Energy Flows for a 2.5-liter, 5-passenger Automobile over the U.S. Urban Drive Cycle (Bandivadekar et al. 2008).

In the United States, the typical occupancy of a passenger car is 1.7 persons. The average weight of a passenger vehicle is approximately 1,820 kg. If the average occupant weighs 70 kg, and if the total energy requirement is proportional to loaded vehicle mass, then the overall energy efficiency of transporting the passengers is approximately 1 percent.[1]

$$(70 \bullet 1.7)/1820 = 0.06\ ;\ 0.06 \bullet 0.16 \approx 1\%$$

The U.S. example is extreme and the calculation is an oversimplification but the main point remains valid: there is enormous opportunity to increase the energy efficiency of light-duty road vehicles. Increasing vehicle occupancy, although theoretically possible, has proven to be difficult to sustain at significantly higher levels and at a widespread scale (Greene and Plotkin, 2010). Because of this, it appears that most of the energy efficiency improvement for light-duty vehicles will come from improvements to the vehicle itself.

Advances in the technology of internal combustion engine (ICE) light-duty vehicles could reduce fuel consumption rates over 2005 levels by about 45 percent (about an 80 percent increase in fuel economy) by 2030 (Kasseris and Heywood, 2007). Key technologies include advanced lightweight materials (20 percent mass reduction), gasoline direct-injected or diesel lean-burn engines with electronically boosted turbo-charging, camless valves, variable compression ratios, and continuously variable or manual-automatic transmissions. Advanced hybrid vehicles with advanced batteries and electrical components and downsized engines could double or even triple fuel economy over 2005 levels (Kasseris and Heywood, 2007). Policies will be necessary, however, to insure that technological advances are applied to increasing energy efficiency rather than to increasing vehicle weight or power (Cheah et al. 2007).

The U.S. EPA (2010, p. 3-10) estimates that the mass of U.S. automobiles could be reduced by 20 percent at a cost of U.S.$0.25 per kg or less, and up to 35 percent at costs below U.S.$0.50/kg. A one percent reduction in mass increases fuel economy by approximately 0.67 percent, provided that the vehicle's engine is downsized to maintain constant acceleration performance. A 35 percent mass reduction, by itself, could improve fuel economy by 25 percent. Researchers at the Massachusetts Institute of Technology estimated that aerodynamic drag could be reduced at the rate of 1 percent per year for the next 25 years and that tire rolling resistance could be decreased at the rate of 1.65 percent per year, resulting in cumulative reductions of 25 percent for drag (C_d=0.21) and 33 percent for rolling resistance (C_r = 0.006) (Bandivadekar et al. 2008).

In the driveline, the MIT researchers predicted that the efficiency of transmissions could increase from 84 percent in 2007 to 94 percent by 2035 (Bandivadekar, 2008). They estimated that a combination of friction reduction, smart cooling systems, variable valve lift and timing, cylinder deactivation, variable compression ratio engines and direct gasoline injection could reduce the fuel consumption of spark ignition engines at the rate of 0.5 percent per year (with a 12.5 percent increase in specific power). Adding turbo-charging with electric boosting to gasoline direct injection, as some manufacturers are

[1] The author first heard this example presented by Professor John Heywood of the Massachusetts Institute of Technology.

already doing, brings a spark-ignited gasoline engine's efficiency close to that of a diesel engine.

Hybridization (HEV), adding an electric motor for tractive power, regenerative braking and a 1-2 kWh battery for energy storage, enables a 40 percent reduction in fuel consumption over a conventional vehicle via engine downsizing, operation at more efficient engine-speed and load combinations, re-use of braking energy, engine shut-off during deceleration and idle, and all-electric operation under limited conditions. Adding 5-10 kWh of battery storage and a larger electric motor enables much greater use of electrical power and access to grid electricity in a plug-in hybrid vehicle (PHEV). PHEVs allow further engine downsizing and greatly expanded "all-electric" operation with a potential 70 percent reduction in fuel consumption. The substantially greater energy efficiency of all-electric drive can be achieved with battery electric vehicles (BEV) and hydrogen fuel cell vehicles (FCVs) (Bandivadekar et al. 2008).

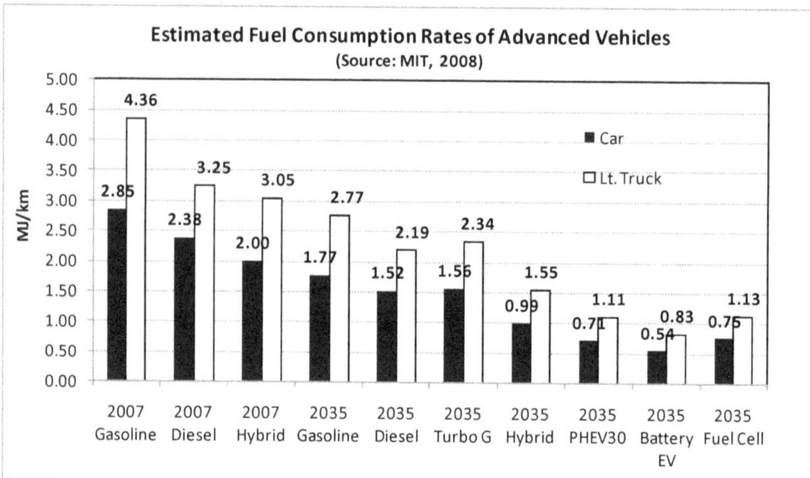

Fig. 3: Estimated Fuel Consumption Rates of Advanced Vehicles (Bandivadekar et al. 2008).

Estimates of potential energy efficiencies for road vehicles in the EU indicate that similar efficiency gains are possible (McKinsey & Co., 2010). Typical energy efficiency for a gasoline-powered new light-duty vehicle in the EU is approximately 20 percent. The study estimated that by 2030, the energy efficiency of gasoline-fueled internal combustion engine vehicles could reach 30 percent, while diesels could achieve 35 percent (Figure 3). Battery electric vehicles were judged to have the greatest energy efficiency potential: 68 percent from plug to wheels, followed by hydrogen fuel cell electric vehicles (FCV) at 56 percent. The well-to-wheel efficiencies were lower and varied considerably according to the primary energy source (Figure 4). These estimates pertain to powertrain efficiencies and do not include the potential for trimming loads via reductions in aerodynamic drag, rolling resistance and vehicle mass. A mix of one-third battery electric vehicles (BEV) and two-thirds fuel cell vehicles (FCV) would reduce the rate of fuel consumption to one third the current rate. Even if only a fraction of the reductions in road load available to U.S. vehicles could be achieved for vehicles sold in the EU, it should be possible to reduce fuel consumption rates by 70 percent by 2050.

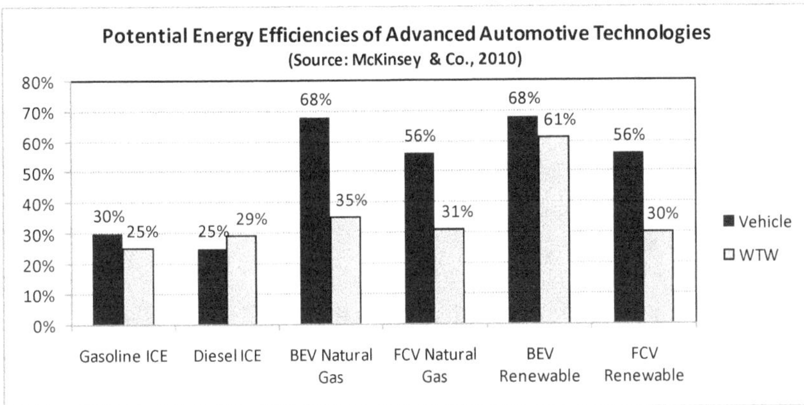

Fig. 4: Potential Energy Efficiencies of Advanced Automotive Technologies.

Energy efficiency improvements of this magnitude are likely to come at increased cost. Estimates of the costs of fuel economy improvements of up to 400 percent for U.S. vehicles are shown in Figure 5 (Bandivadekar, 2008). In comparison to the costs and potential estimated for 2005 technology, the improvement potential is enormously greater and the marginal costs are uniformly lower. Still, unless energy prices are higher than they are today, fuel economy increases beyond the 2035 hybrid will be difficult to justify to consumers based on the private benefits alone. Policies to reflect the societal value of further reducing energy use will be necessary.

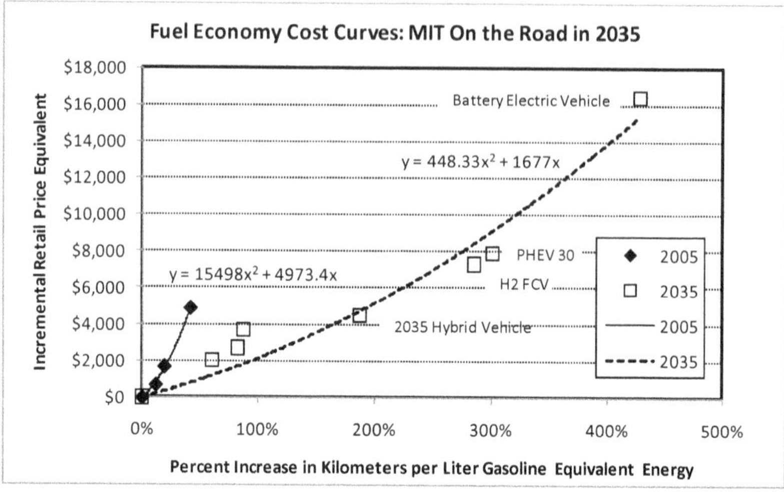

Fig. 5: Fuel Economy vs. Cost: Current and Advanced Automotive Technologies.

<u>Heavy-duty vehicles</u>

Key energy efficiency technologies for heavy-duty vehicles include advanced diesel engines with a 30 percent improvement in efficiency, hybridization to achieve a 60 percent fuel economy improvement for vehicles operating predominantly in stop-and-go service, advanced aerodynamics to achieve a 20 percent drag reduction for trucks and buses in long-haul service, reduction in parasitic losses of 50 percent, and advanced materials for a tare weight reduction of 15 percent to 20 percent (U.S. DOE, 2006). Applying all these advanced technologies appropriately could produce a 40 percent reduction in energy intensity (67 percent increase in fuel economy) for heavy trucks and buses. The IEA (2010, p. 288) estimates that global freight truck energy efficiency could be increased by 30 to 40 percent by 2030 and 40 to 50 percent by 2050. The U.S. National Research Council estimated that the fuel consumption (liters/100 km) of tractor trailers, which account for 80 percent of heavy truck energy use in the U.S., could be reduced by over 40 percent by 2020 through a variety of measures at a cost of just over $1 per gallon of fuel saved (NRC, 2010, p.132). Other types of heavy-duty vehicles could achieve similar reductions at higher costs, with hybridization playing the largest role for trucks having low-speed, stop-and-go duty cycles, and engine improvements and reductions in rolling and aerodynamic resistance more important for over-the-road vehicles.

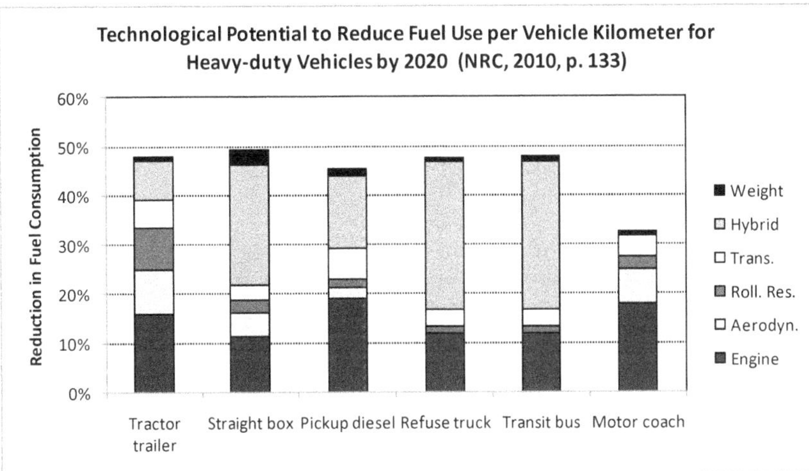

Fig. 6: Technological Potential to Reduce Fuel Use per Vehicle Kilometer for Heavy-duty Vehicles in the U.S. by 2020 (NRC, 2010, p. 133).

Air transport

Commercial aircraft are the second largest energy users in the transport sector, accounting for 11 percent of transport energy use and 12 percent of transport's GHG emissions. All of the energy used by commercial aircraft is derived from petroleum and 99 percent is jet fuel (McCollum, Gould and Greene, 2009, Table 1). The energy efficiency of air transport can be increased by aerodynamic improvements that increase lift-to-drag ratios, material substitution to reduce empty weight, increased efficiency in engine thermodynamics and propulsion, and increased aircraft size (seats/aircraft). Despite impressive progress over the past half century, substantial potential to increase energy efficiency remains.

The European Commission's Advisory Council on Aeronautics Research in Europe set a goal of reducing the fuel consumption per passenger kilometer of new aircraft by 50 percent by 2020 (IATA, 2009, p. 14). Of this, 15 to 20 percentage points were estimated to be available via improvements to engines, 20 to 25 percent via improvements to airframes and 6 to 10 percent via operational improvements to the air traffic management system, more direct routing and more energy efficient flight profiles. Similar estimates have been made by others (Schafer et al. 2009). The U.S. National Science and Technology Council's Aeronautics Science and Technology Subcommittee set a goal of a 70 percent reduction in fuel consumption per passenger kilometer within 25 years (NTSC, 2007, p. 47). Considering a list of technologies that could be retrofitted to existing aircraft and technologies that could be implemented on updated versions of current generation aircraft, IATA estimated that energy consumption could be reduced by about 20 percent (2009, Tables 3-3 and 3-4). Advanced technologies applicable to the next generation of aircraft designs were considered sufficient to reduce fuel consumption rates by 40 percent by 2020. By 2050, radically different concepts such as the blended wing-body and hydrogen powered long-distance aircraft might be able to reduce fuel burn per passenger mile by as much as 70 percent (Greene and Plotkin, 2010, Table 6). MIT researchers concluded that a "Double-bubble" fuselage design could achieve a 70 percent reduction in fuel burn in domestic operations while the blended wing-body, better suited to international flights, could produce more than a 50 percent reduction in fuel burn by 2035 (de la Rosa Blanco and Greitzer, 2010).

Waterborne transport

Globally, water-based transport accounts for 9 percent of transport energy use (IEA, 2009, p. 339). The energy efficiency of water transport could be increased via improved diesel engine efficiency, improved propeller designs, reducing hydrodynamic drag and wave resistance (U.S. DOT, 2010, p. 3-97). The IEA estimates that ship fuel consumption could be reduced up to 30 percent by advanced engine and propulsion systems, up to 20 percent by retrofitting and better maintaining existing vessels and up to 40 percent by means of operational changes, including speed reduction (IEA, 2009, pp. 359-360). Speed reduction is controversial, however, since its economic costs are not well understood. Combining these measures, the IEA estimates that ship energy use per ton-km could be reduced by as much as 40 percent by 2030 and 60 percent by 2050. The International Maritime Organization estimated a CO_2 abatement potential for marine vessels of 15 to 30 percent by 2020 at less than \$100 per ton CO_2 (IMO, 2009, p. 263).

Rail transport

The importance of the rail mode in passenger and freight transport varies greatly among the world's nations. In the U.S., rail carries 40 percent of total ton-kms while in Europe it transports only 17 percent. Rail accounts for less than 1 percent of passenger kms in the U.S. but is far more important in other OECD countries. Because of its energy efficiency, rail transport uses only about 2 percent of transport energy (Davis et al. 2010; Noreland, 2008).

Rail energy intensity could be reduced by 18-24 percent through a combination of more efficient locomotive (diesel) engines, switching from DC to AC propulsion, regenerative braking, and reduction in the empty weight of rolling stock (IEA, 2009, p. 299; U.S. DOT, 2010, p. 3-89). Researchers at Argonne National Laboratory estimated that rail energy intensity could be reduced by up to 50 percent by 2050 through a combination of operational and technological measures (Stodolsky, 2002).

BEHAVIOR, MARKETS AND POLICIES

Every major automobile manufacturing country in the world has either fuel economy or CO_2 emission standards, or both. This is true despite the fact that many also have motor fuel taxes that double or triple the production costs of the fuels. The explanation lies in fundamental human behaviors that deviate from the economic theory of rational behavior. Because of this, strong public policy will be essential to realizing the energy efficiency potentials described above. At the same time, increasing energy efficiency reduces the costs of energy services, leading to increased consumption of energy services. This effect, termed "rebound", has been claimed by some to be so large that it causes technical improvements in energy efficiency to ultimately increase, rather that decrease energy use. In fact, the reverse is true, the larger and more widespread the improvements in energy efficiency, the smaller the rebound effect.

Behavioral economics, loss-aversion and fuel economy

Fundamental differences between the rational economic agents of economic theory and real-world consumers cause energy efficiency to be undervalued in the marketplace. The reason is the inherently uncertain value of future fuel savings from improved energy efficiency combined with the most well-established principle of behavioral economics: loss aversion (Della Vigna, 2009). Faced with a risky bet, such as to give up something today to get energy savings over future years, human beings typically exaggerate the probability of loss and weight potential losses about twice as much as gains. The result is systematic underinvestment in energy efficiency from firms' R&D decisions to consumers' purchases of energy using durable goods.

The energy efficiency of vehicles is determined by manufacturers' decisions about how to design vehicles and what technology to incorporate in them, and consumers' decisions about which vehicles to purchase. In a competitive market[2] manufacturers will attempt to design vehicles that maximize consumers' satisfaction, that is, to design vehicles the way consumers want them designed. In their decision making manufacturers face two key trade-offs:

1. Energy efficiency can be increased by reducing vehicle size and power.
2. Energy efficiency can be increased by adding more expensive technologies.

In either case, the consumer gives up something immediate and tangible (a higher price, a smaller vehicle) to obtain energy savings in the future. But future energy savings are uncertain. Prices vary, energy efficiencies vary from standardized ratings, vehicle lifetimes and vehicle use are uncertain, as well. The apparently high discount rates with which consumers around the world seem to evaluate energy savings from efficiency improvements have been termed the energy paradox (Jaffe and Stavins, 1994). Engineering-economic analyses of the potential for technological energy efficiency improvements typically reveal substantial opportunities for cost-effective improvements. Yet markets decline to adopt the energy efficient technologies.

Delucchi (2007) argued that consumers' apparently high discount rates could be explained by consumers making conservative assumptions about all these uncertainties. Greene et al. (2008) and Greene (2011) explained consumers' desire for short pay-back periods in terms of a fundamental tenet of behavioral economics: loss aversion. Perhaps the most well established principle of behavioral economics is that consumers evaluate the outcomes of a risky bet relative to their status quo, exaggerate the probability of losing and weight potential losses about twice as much as potential gains. Greene (2011) showed how a typically loss averse consumer would decline technology-based fuel economy improvements that appeared highly cost-effective based on their expected net

[2] This also applies to a monopolistically competitive market in which there are a sufficient number of producers to avoid oligopoly but firms are able to differentiate their products from those of their competitors. This is a more realistic characterization of the global motor vehicle market.

present value. Indeed, the loss-averse consumer would behave very much like a consumer with either very high discount rate or a pay-back period of 2-3 years. Figure 7 shows the probability distribution of net present value (value of fuel savings minus first cost) for an expected present value maximizing fuel economy increase from 28 to 35 miles per gallon (8.4 to 6.7 liters/100 km) for a U.S. passenger car, as derived by Greene (2011) from estimates of the cost of fuel economy improvement made by the U.S. National Academy of Sciences (NRC, 2002). The expected value is over $400 but there is a substantial probability the consumer would lose money if fuel prices were low, if the expected fuel economy benefit were not realized, and if other factors, such as vehicle lifetime, usage rates, etc. were unfavorable.

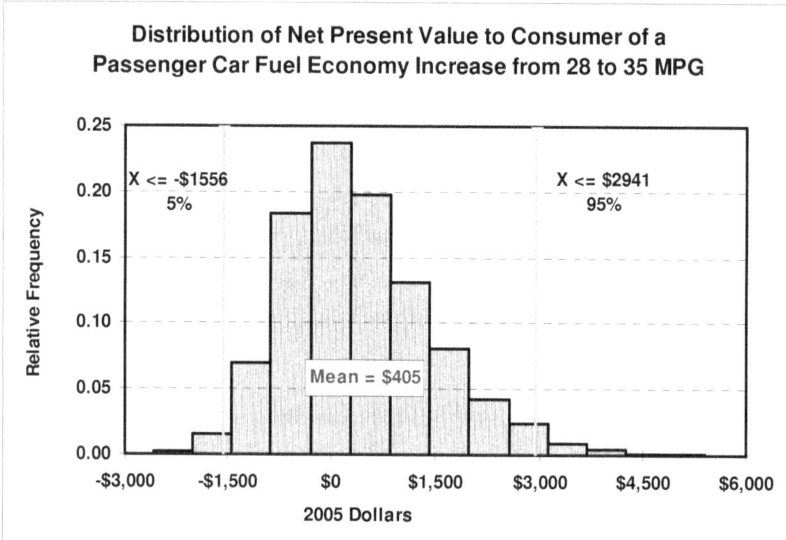

Fig. 7: Probability Distribution of Net Present Value of Optimal Fuel Economy Improvement for a Typical U.S. Passenger Car (Greene, 2011).

Weighting the distribution shown in Figure 6 by a typical loss aversion function obtained from Tversky and Kahneman (1992) produces a loss-averse value distribution with a negative expected value (Figure 8). This result is similar to the way a consumer with a required payback period of less than 3 years would evaluate future fuel savings in the absence of uncertainty. The implication is that consumers undervalue future fuel savings by more than a factor of two relative to expected value. The result is not only underinvestment in energy efficiency by consumers but underinvestment in fuel economy throughout the supply chain, including research and development.

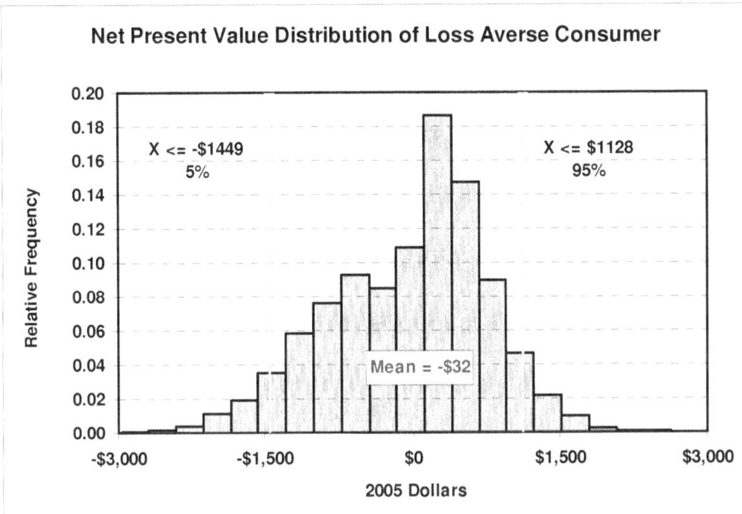

Fig. 8: Probability Distribution of Present Value of Fuel Economy Improvement for a Typical Loss-Averse Consumer (Greene, 2011).

There is nothing in the theory of loss-aversion that would limit this behavior to consumers. Loss aversion is apparently a fundamental characteristic of human behavior and has been observed in other primates as well (Shermer, 2009). Loss-aversion may explain why every major automobile manufacturing economy, from the U.S. to the EU to Japan and China, also has fuel economy or CO_2 emissions standards for automobiles (An et al. 2007). The fact that the behavior is not limited to consumers may explain why Japan, the U.S., the EU and China have also adopted or are in the process of adopting similar standards for heavy trucks. Indeed, energy efficiency standards may be needed for all types of transport vehicles if the potential efficiency improvements described above are to be fully realized.

The rebound effect: does energy efficiency reduce energy use?

Energy efficiency improvement reduces the quantity of energy required for a given energy service and thereby the cost of that service. The reduced cost of energy services leads to increased consumption of energy services, a phenomenon known as the "rebound effect" (Sorrell, 2007). Economists distinguish between direct rebound effects (the increased consumption of energy services due to the decrease in price as a consequence of improved energy efficiency) and indirect rebound effects (which result from the gain in income due to lower energy costs and the increased consumption as a consequence).

The idea that the rebound effect might be so strong that it would overcome increases in energy efficiency was first propounded by Jevons (1866). This argument has been very recently resurrected by Jenkins et al. (2011) who argue that the rebound effect is an "emergent" property of energy efficiency improvement. By this they mean that increased energy efficiency leads to a variety of more or less unpredictable innovations that ultimately cause increased rather than decreased energy use. Such an outcome is highly unlikely.

Let T[c(ep,x),w] be the demand for travel or transport as a function of the value of time, w, and the monetary cost of travel c(ep,x) which depends on fuel costs ep, where e is energy efficiency (in km/MJ) and p the price of fuel, and other costs, x. Differentiating travel demand with respect to energy efficiency, and rearranging terms to obtain an expression in terms of the price elasticity (percent change in quantity for a unit percent change in price) of travel and the energy cost share of travel yields the following.

$$\frac{dT}{de}\frac{e}{T} = \frac{dT}{dc}\frac{dc}{de}\frac{e}{T} = \frac{dT}{dc}\frac{c}{T}\left(\frac{ep}{c}\right) = -1\left(\frac{ep}{c}\right)$$

There is reasonable evidence that the elasticity of travel with respect to total monetary costs is approximately -1 (Goodwin et al. 2004). In that case, we obtain the simple rule of thumb that the direct rebound effect is the negative cost share of energy.

For road passenger vehicles in the U.S. the energy cost share of the monetary costs of travel has varied between 10 percent and 20 percent over the past 35 years as a function of the price of gasoline and the fuel economy of vehicles. Energy cost shares of air and truck transport are about 1.5 to 2 times as high. Cost shares in other OECD countries are also higher than in the U.S. because taxes on motor fuels are much higher than in the U.S. However, this is partly offset by higher vehicle ownership costs and lower annual kilometers of travel. As energy efficiency improves, the rebound effect will decrease because the energy cost share will decrease, all else equal. This relatively simple explanation agrees well with empirical estimates of the rebound effect for light-duty vehicles in the U.S. As Small and Van Dender (2007) demonstrated, studies based on pre-1990 data generally found the rebound effect to be on the order of 20 percent. Their study, which included data through 2004 and that of Greene (2011) which included data through 2007, found rebound effects in the last two decades on the order of 10 percent to 15 percent. These quantitative estimates mirror the trend of the energy cost share of automobile travel in the U.S. over the same period (Figure 9).

To the extent that increases in energy efficiency raise the initial costs of energy using durable goods, which increase in cost will, in the long run, partially offset reduce energy costs, further mitigating the rebound effect.

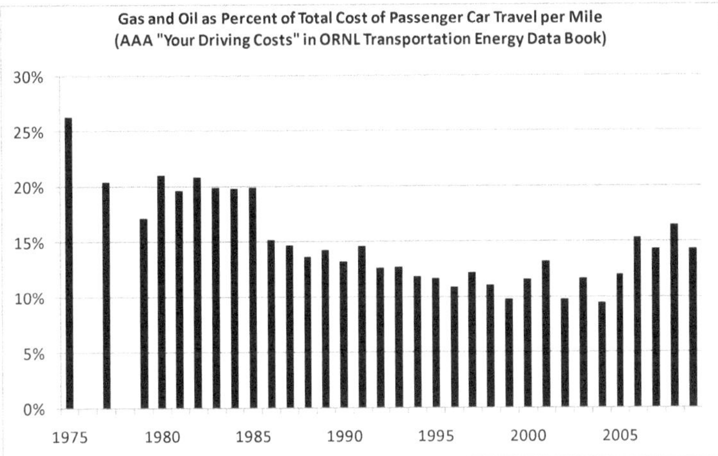

Fig. 9: Gas and Oil as a Percent of the Total Cost of Passenger Car Travel in the U.S. (Davis, Diegel and Boundy, 2010).

A similar line of reasoning applies to the indirect rebound effect. As economy-wide energy efficiency increases, the energy cost share of GDP declines. As the energy cost share of GDP declines the fraction of a dollar saved on energy that will be spent on energy declines. This simple relation contradicts the energy backfire argument. Indeed, the paradox is that the larger and more globally pervasive the energy efficiency improvements, the smaller the rebound effect. As Sweeney (2011) has pointed out, the backfire argument can only be possible if the economic value of consuming another unit of energy substantially exceeds its price, in other words, if markets are for some reason presently substantially under-utilizing energy. The "emergent" theory of an energy efficiency backfire rests on the idea that technical energy efficiency improvement, *per se*, will, through an unspecified mechanism, transform economic activity in a way that causes innovations in products and consumer preferences in related and unrelated areas such that energy consumption increases substantially relative to consumption of other goods and services. There is no evidence to support this proposition. It simply confuses the effects of innovation and economic growth with those of technical energy efficiency improvement. Improving energy efficiency reduces energy use, *ceteris paribus*. Rebound effects are real but take back only a fraction of the energy efficiency benefit.

CONCLUDING OBSERVATIONS

It will be difficult to hold global transportation energy use constant through 2050 through efficiency improvement alone. Assuming relatively moderate rates of growth in transportation demand, ranging from 1 percent per year for rail to 3 percent per year for air transport, energy use would increase by 240 percent in the absence of any improvement in energy efficiency. Using the potential efficiency gains described above and ignoring the rebound effect, energy use in 2050 would be about 20 percent below the 2007 levels. Assuming a 15 percent rebound effect brings 2050 transport energy use to approximately the same level as in 2007. This simple analysis implies that if global transport greenhouse gas emissions in 2050 are to be 50 percent below current levels, the carbon intensity of transport energy use would have to be cut in half, even if full advantage is taken of all opportunities to increase energy efficiency.

Table 1. Impact of Transport Energy Efficiency Improvement on Energy Use in 2050

	Energy Use	Growth Rate	Extrapolated Energy Use	Efficiency	Efficient Energy Use	Energy Use With Rebound
Mode	2007	%	2050	Improvement	2050	2050
Road	103	2.0%	241	70%	72	87
Air	11	3.0%	39	60%	16	18
Water	9	2.0%	21	50%	11	12
Rail	5	1.0%	8	50%	4	4
TOTAL	128		309		102	121

The efficiency improvements necessary to hold 2050 energy use at roughly today's levels will not occur in the absence of strong public policies. Absent such policies, markets will continue to under-invest in energy efficiency R&D, vehicle manufacturers will under-utilize available energy efficient technologies, and consumers will undervalue uncertain future fuel savings relative to their expected value. In the absence of strong public policies, markets will also ignore the external costs of energy use and fail to adequately account for the interests of future generations in sustainable energy.

REFERENCES

1. An, F., D. Gordon, H. He, D. Kodjak and D. Rutherford (2007) *Passenger Car Greenhouse Gas and Fuel Economy Standards: A Global Update*, International Council on Clean Transportation (ICCT), San Francisco.
2. Bandivadekar, A. et al. (2008) *On the Road in 2035*, Massachusetts Institute of Technology, Laboratory for Energy and the Environment Report No. LFEE 2008-05 RP, Cambridge, MA.
3. Cheah, L., C. Evans, A. Bandivadekar, J. Heywood. "Factor of Two: Halving the Fuel Consumption of New U.S. Automobiles by 2035" (LFEE 2007-04 RP, Laboratory for Energy and the Environment, Massachusetts Institute of Technology, Cambridge, Massachusetts, 2007).
4. Davis, S.C., S.W. Diegel and R. Boundy (2010) *Transportation Energy Data Book, Ed. 29*, ORNL-6984, Oak Ridge National Laboratory, Oak Ridge, Tennessee.
5. de la Rosa Blanco, E. and E.M. Greitzer (2010) "Subsonic Civil Transport Aircraft for a 2035 Time Frame", pp. 91-94 in, *ICAO Environmental Report 2010*, International Civil Aviation Organization, Montreal.
6. DellaVigna, S. (2009) "Psychology of Economics: Evidence from the Field", *Journal of Economic Literature.*

7. Greene, D.L. and S.E. Plotkin (2011) *Reducing Greenhouse Gas Emissions from U.S. Transportation*, Pew Center on Global Climate Change, Arlington, Virginia.
8. International Air Transport Association, 2009. *The IATA Technology Roadmap Report*, Switzlerland.
9. International Energy Agency (IEA) (2010) *Energy Technology Perspectives 2010: Scenarios and Strategies to 2050*, OECD, Paris.
10. International Energy Agency (IEA) (2009) *Transport Energy and CO_2 – Moving Towards Sustainability*, OECD, Paris.
11. International Energy Agency (IEA) (2008) *Energy Technology Perspectives 2008* (OECD, Paris, France, 2008).
12. International Maritime Organization (IMO) (2009) *Prevention of Air Pollution from Ships: Second IMO GHG Study*, MEPC 59/INF.10, London, UK.
13. Jaffe, A.B. and R.N. Stavins (1994) "The Energy Efficiency Gap: What Does It Mean", *Energy Policy*, vol. 22, no. 10, pp. 804-810.
14. Jenkins, J., T. Nordhaus and M. Shellenberger (2011) *Energy Emergence: Rebound & Backfire as Emergent Phenomena*, The Breakthrough Institute, Oakland, California.
15. Jevons, W.S. (1866) *The Coal Question*, Macmillan & Co., London.
16. E. Kasseris, J. Heywood (2007) "Comparative Analysis of Automotive Powertrain Choices for the Next 25 Years" (SAE Technical Paper Series 2007-01-1605, SAE International, Warrendale, Pennsylvania, 2007).
17. Lotus Engineering (2010) *An Assessment of Mass Reduction Opportunities for a 2017-2020 Model Year Vehicle Program*, International Council on Clean Transportation, Washington, D.C.
18. McCollum, D., G. Gould, D. Greene (2009) *Greenhouse Gas Emissions from Aviation and Marine Transportation: Mitigation Potential and Policies*, Pew Center on Climate Change, Arlington, Virginia.
19. McKinsey & Co. (2010) "A portfolio of power trains for Europe: a fact-based analysis", www.zeroemissionvehicles.eu .
20. National Research Council (NRC) (2010) *Technologies and Approaches to Reducing the Fuel Consumption of Medium- and Heavy-Duty Vehicles*, National Academies Press, Washington, D.C.
21. Noreland, J. (2008) *Modal Split in the Inland Transport of the EU: Freight and Passenger Transport up to 2006*, Eurostat 35/2008, EC, Luxembourg.
22. Shermer, M. (2009) *The Mind of the Market*, Henry Holt & Co., New York.
23. Small, K. and K. Van Dender (2007) "Fuel Efficiency and Motor Vehicle Travel: The Declining Rebound Effect", *The Energy Journal*, vol. 28, no. 1, pp. 25-51.
24. Sorrell, S. (2007) *The Rebound Effect: An Assessment of the Evidence for Economy-Wide Energy Savings from Improved Energy Efficiency*, Sussex Group for the Technology and Policy Assessment function of the UK Energy Research Centre, London, UK.
25. Stodolsky, F. (2002) "Railroad and Locomotive Technology Roadmap", ANL/ESD/02-6, Argonne National Laboratory, Argonne, Illinois.
26. U.S. Department of Energy, Energy Information Administration (EIA) (2010) *International Energy Outlook 2010*, DOE/EIA-0484(2010), Washington, D.C.

27. U.S. Department of Energy, Encrgy Efficiency and Renewable Energy (EERE) (2006) "21st Century Truck Partnership: Roadmap and Technical White Papers" (21CTP-0003, Washington, D.C. 2006, http://www1.eere.energy.gov/vehiclesandfuels/pdfs/program/21ctp_roadmap_2007.pdf).
28. U.S. Environmental Protection Agency, U.S. Department of Transportation, California Air Resources Board (2010) *Interim Joint Technical Assessment Report: Light-Duty Vehicle Greenhouse Gas Emission Standards and Corporate Fuel Economy Standards for Model Years 2017-2025*, http://www.epa.gov/otaq/climate/regulations/ldv-ghg-tar.pdf .
29. U.S. Department of Transportation (U.S. DOT) (2010) *Transportation's Role in Reducing Greenhouse Gas Emissions: Volumes 1 and 2*, Report to Congress, Washington, D.C.

SPECIAL SESSION:

LECTIO MAGISTRALIS

THE ROLE OF SCIENCE IN THE THIRD MILLENNIUM

ADDRESS OF THE RT. HON. IGNAZIO LA RUSSA, MINISTER OF DEFENCE OF ITALY

Presented by the Hon. Alessandro Pagano
Finance Commission—Deputy Chamber

Professor Zichichi, Professor Lee, distinguished scholars and researchers, eminent speakers, Ladies and Gentlemen,

One year after my last visit to Erice, I have the pleasure to present my contribution to your yearly international seminar on behalf of the Italian Defence and Armed Forces. In fact, more than others, they both rely on scientific understanding and technology improvements to ensure effectiveness, and responsiveness to our nation's needs.

I have been following all the activities developed by the 'Ettore Majorana' Centre for Scientific Culture with strong interest for some time now. The Centre has interpreted and promoted a new spirit of international scientific collaboration, which is well depicted by the motto "science without secrets and without borders."

It is a new, innovative, nonconformist approach—which I also appreciate. Over the last 50 years or so, the Centre in Erice hosted one hundred and nine thousand, two hundred and fifty (109,250) scholars from nine hundred and thirty two (932) laboratories and universities of one hundred and forty (140) nations.

It is a centre of international scientific excellence which relies strongly on two mandatory guiding principles: merit and 'scientific volunteering'.

As is well known, I am very interested in your studies about planetary emergencies, and not only in my position as member of the cabinet and minister, but also as a common citizen and father, that is to say someone who cares about the future of young and next generations, and of our planet as a whole.

My interest is also the result of a personal passion towards science and of the impression that the revolutionary experiment of Galileo Galilei on motion had on me in the primary school.

The key experiment involving balls of different weight falling from the Tower of Pisa and arriving at the same time—at least in vacuum—was a sort of revelation for me to understand how important experimentation, testing and scientific rigour are, especially when applied to everyday life.

Besides, although it is said that the experiment never took place and that Galileo Galilei may have reached his conclusions through reasoning rather than testing, I got the inherent message: "what appears to be evident per se may be both a useful and yet deceiving truth. Therefore, we must investigate apparent truths with rigour."

I made this approach mine and followed it on all occasions since then, both as a minister and a lawyer.

This is a guiding principle for a modern, consumer-oriented and technology-bound society. Otherwise, it may incur in two real risks.

If the logic of profit prevails, short-sightedness arises and therefore—either for convenience or advantage—we cannot see the real problems afflicting the core elements of our lives and planet: energy, water, environmental pollution, climate, terrorism, just to mention a few; in fact these are the very emergencies you study with such care and competence.

The second risk is that approximation prevails on scientific rigour. As Enrico Fermi said, there is a possibility that those who speak about scientific or technical issues have never dealt with any of those.

As a matter of fact, today, a greater number of people are interested in physics and science given their increased cultural level and the role played by modern media, such as the Internet.

And this is not necessarily a bad thing. On the contrary, it is a very positive fact that should be fostered—and I quote Professor Zichichi—"in order to bring science to every youngster's heart" and make it part of his or her cultural roots."

Again, Professor Zichichi reminds us that when one speaks, for example, of world energy crisis and planetary monitoring, rigour is much needed. The planet should be globally under control.

Hence, we should be capable to merge two necessities: an approach based on scientific rigour, on one side, and knowledge dissemination, on the other, with a view to raise awareness in and involve the society at large, so that support and consensus can be built around important projects, such as the establishment of the first Research Centre on Planetary Emergencies, which you care so much about.

Many times I have heard Professor Zichichi saying that if this was the era of science, there would be no planetary emergencies.

In order for all this to happen, we need to persuade the civil society that we need science in everyday life, especially in the third millennium.

Defence has always been attentive and sensitive to scientific progress, whose applications are part of the daily lives of men and women in uniform.

Today, we cannot imagine a modern and efficient military structure without a constant and active participation of Defence to scientific research.

I am referring to aerospace programmes and to the resulting modern vehicles and equipment, namely the modern space systems for Earth observation like COSMO-SkyMed, that are essential to deploy in the current international scenarios.

Investigating and finding out the details for every type of scenario, land, maritime and aero spatial, is a prerequisite for planning and conducting situation-based military operations, be they intended to defend peace and international security or to prevent or provide relief in case of natural disasters.

Real-time awareness of greater geographical trends is therefore essential to conduct analyses and operational planning, and is crucial for those who are required to fight international terrorism, natural disasters, emergencies and threats of whatever nature.

Defence, however—based on specific directions I have given—, is strongly committed to turn to better account the innovative means that science makes available to the common people. For example, Difesa Servizi, a company controlled by Defence, has been chosen to manage the photovoltaic systems installed on the roofs of defence buildings and barracks in order to produce energy from renewable sources. The project involves 64 military compounds for an extension of more than one thousand two hundred acres.

Among the planetary emergencies you have identified, those concerning earth monitoring, containment of terrorist attacks, cyber security, the world energy crisis, pollution and defence from cosmic objects are the more interesting to Defence.

These are very topical issues, which a larger public should be aware of. So, we need a collective effort where institutions and the scientific world play their part in order to overcome the cultural limitations that belong to yesterday's and today's societies.

Mankind and society have always manifested distrust and suspicion towards science.

Since the time of Galileo Galilei, the most innovative and revolutionary theories have been rejected with fear, and scientists have been prosecuted and convicted by the dominant culture of the time.

This clash between society and science has been investigated several times through studies and reflections, and all led to the same conclusion: science per se is not dangerous; the way it is used is.

This has not been enough—however—to get rid of the long standing reluctance to trust science.

Today is crucial—and I strongly mean it—that such reluctance can be overcome so that science can play its role in full, during daily public debates and investigations.

This is why I believe that what Professor Zichichi and his collaborators are doing at CERN, in Geneva, in full observance of Galilean research criteria, in the subnuclear and high energy physics field is a significant and important contribution to the progress in technological, civil and social issues.

This is why I will continue to follow your studies with renewed interest, while being increasingly convinced that we need—as I said last year—to bring science in everyday life.

I would like to thank all the participants for their contribution, all over the world, to these important and strategic issues.

IL RUOLO DELLA SCIENZA NEL TERZO MILLENNIO

INTERVENTO DEL MINISTRO DELLA DIFESA, ON. IGNAZIO LA RUSSA

Professore Zichichi, Professore Lee, scienziati, ricercatori, pregiatissimi relatori, Signore e Signori,

A distanza di un anno, è con vero piacere che, pur non potendo partecipare per sopraggiunti impegni istituzionali, invio il saluto della Difesa e delle Forze Armate, che costituiscono uno dei settori istituzionali maggiormente legati ad un continuo aggiornamento delle conoscenze scientifiche e delle applicazioni tecnologiche per assicurare uno strumento efficace e rispondente alle esigenze del Paese.

Da tempo seguo con molto interesse le attività del Centro di cultura Scientifica "Ettore Majorana" di Erice che si è fatto interprete e promotore di un nuovo spirito nella collaborazione scientifica internazionale che è ben sintetizzato dal motto: "una scienza senza segreti e senza frontiere".

Uno spirito nuovo, innovativo, anticonformista, come piace a me, che nel corso di mezzo secolo di attività ha richiamato ad Erice 109.250 scienziati provenienti da 932 Laboratori ed Università di 140 nazioni.

Un Centro di eccellenza scientifica internazionale saldamente fondato su due principi guida inderogabili: la meritocrazia e il "volontariato scientifico".

Come dicevo seguo con estremo interesse i vostri studi sulle Emergenze Planetarie e non solo in qualità di rappresentante di Governo e Ministro ma anche come comune cittadino e padre che ha a cuore il futuro delle giovani generazioni, di quelle che verranno e del nostro pianeta tutto.

Un interesse alimentato anche da una personale passione per il mondo scientifico che mi porto dietro fin dal periodo scolastico quando rimasi colpito dal famoso esperimento di Galileo Galilei che rivoluzionò l'idea di moto.

Fu infatti l'esperimento delle sfere di diverse dimensioni e peso lasciate cadere dalla torre di Pisa—che dimostrò come, indipendentemente dalla loro massa, tutti gli oggetti lasciati cadere dalla stessa altezza toccano terra nel medesimo istante, o almeno così avverrebbe nel vuoto—a farmi riflettere, già allora, sull'importanza della sperimentazione, della verifica, del rigore scientifico applicato alla vita di tutti i giorni.

Sebbene risulti che l'esperimento non abbia mai avuto luogo e che Galileo Galilei sia giunto alle note conclusioni attraverso un ragionamento piuttosto che per via sperimentale il messaggio che io recepii fu che "spesso ciò che appare evidente può risultare una verità tanto comoda quanto ingannevole e di conseguenza è bene sforzarsi sempre di approfondire la realtà apparente con il necessario rigore".

Un convincimento questo che ho fatto mio e che osservo scrupolosamente in ogni circostanza sia nella funzione di Ministro sia in qualità di avvocato.

E' questo un principio "bussola" che permette di far mantenere la rotta ad una società moderna, consumistica, tecnologica che altrimenti, nel campo del sapere, corre due concreti rischi.

Distratti dalla logica del profitto il primo rischio è quello della "miopia" che ci evita di vedere per comodità e convenienza dell'immediato i reali problemi che affliggono elementi essenziali della nostra vita e dell'intero pianeta: l'energia, l'acqua, l'inquinamento ambientale, il clima, il terrorismo; quelle Emergenze che Voi qui studiate con tanto scrupolo e competenza.

Il secondo rischio è che a prevalere sia l'approssimazione sul rigore scientifico. Come diceva Enrico Fermi il pericolo è che a parlare di scienza e di tecnica siano persone che non hanno mai fatto né l'una né l'altra cosa.

Oggi, infatti, in conseguenza del livello culturale mediamente più elevato e dei moderni strumenti di informazione, come internet, molte più persone si interessano di fisica e scienza.

Questo non è necessariamente un male, anzi, è un aspetto estremamente positivo che andrebbe maggiormente curato e sviluppato per "portare la scienza nel cuore dei giovani" e nelle loro radici culturali, riprendendo una efficace frase del Professore Zichichi.

Ma sempre il Professore Zichichi ci ricorda anche che quando si parla, ad esempio, di crisi energetica mondiale, di monitoraggio globale del pianeta c'è bisogno di rigore. Il pianeta deve essere globalmente messo sotto controllo.

Si tratta, pertanto, di riuscire a compendiare queste due necessità, quella dell'approccio basato sul rigore scientifico con quella della divulgazione della conoscenza, in modo da sensibilizzare e coinvolgere l'intera società perché si crei quella spinta e quel consenso generale che permettano di concretizzare al meglio progetti importanti, come quello a Voi caro, della realizzazione del primo Centro per le ricerche dedite alle Emergenze Planetarie.

Ho sentito tante volte dire dal Professore Zichichi che se vivessimo l'era della scienza, le Emergenze Planetarie non esisterebbero.

Perché ciò avvenga è necessario far capire alla società civile che, specie nel terzo millennio, c'è bisogno di scienza nella vita di tutti i giorni.

In questo senso la Difesa è da sempre particolarmente attenta e sensibile al progresso scientifico che trova costante applicazione nelle quotidiane attività degli uomini e delle donne con le stellette.

Oggi, non è pensabile uno strumento militare moderno ed efficiente senza una costante ed attiva partecipazione della Difesa alla ricerca scientifica.

Mi riferisco per esempio ai programmi del comparto aerospaziale, che permettono di progettare mezzi e strumenti indispensabili per operare negli odierni complessi scenari operativi internazionali come i moderni sistemi spaziali per l'osservazione della Terra (COSMO-SkyMed).

La lettura costante e dettagliata di cosa avviene in ogni tipo di scenario - terrestre, marittimo ed aerospaziale—è la condizione necessaria per pianificare organizzare e condurre un intervento militare che sia il più aderente e commisurato all'esigenza del momento: sia esso orientato alla difesa della pace e della sicurezza internazionale o alla prevenzione e al soccorso in caso di calamità naturali.

La conoscenza, quindi, degli andamenti "in tempo reale" delle grandi dinamiche geografiche diviene componente essenziale sia delle analisi sia della pianificazione operativa di chi è impegnato, giorno per giorno, a lottare contro il terrorismo internazionale, le calamità, le emergenze, le minacce di ogni natura.

Ma la Difesa, dietro mio preciso impulso, è attivamente impegnata anche nella valorizzazione degli strumenti innovativi che la Scienza mette a disposizione del cittadino. Penso alla recente convenzione per l'affidamento alla nostra società Difesa

Servizi SpA della gestione economica delle coperture di fabbricati e terreni in uso alla Difesa per la realizzazione di impianti per la produzione di energia da fonti rinnovabili, circa 500 ettari di 64 compendi militari presenti sul territorio nazionale.

A mio avviso tra le Emergenze Planetarie da Voi individuate e studiate che ritengo di estremo interesse per la Difesa rientrano il monitoraggio del pianeta, il contenimento degli atti terroristici così come la sicurezza informatica, la crisi energetica mondiale, i problemi dell'inquinamento e la difesa da oggetti cosmici.

Sono argomenti di grande attualità che dovrebbero essere portati a conoscenza di un più vasto pubblico per cui è necessario compiere uno sforzo collettivo, dove tutti, istituzioni e mondo scientifico, devono fare la propria parte per superare uno scoglio culturale che permea la società di oggi come quella di ieri.

Da sempre l'uomo e la società manifestano diffidenza e sospetto nei confronti della scienza.

Fin dai tempi di Galileo Galilei le teorie più innovative e rivoluzionarie venivano rifiutate, respinte con paura e gli scienziati condannati e perseguiti dalla cultura dominante di quei tempi.

Questa contrapposizione tra società e scienza è stata oggetto di tanti studi, riflessioni che hanno inevitabilmente portato sempre alla stessa conclusione: non è la scienza ad essere pericolosa ma il modo in cui la si utilizza.

Tutto questo non è però stato sufficiente ad eliminare completamente l'atavica ritrosia dell'uomo a fidarsi della scienza.

Oggi è però di vitale importanza, ne sono sinceramente convinto, che questa diffidenza venga superata affinché la scienza possa entrare a pieno titolo e con il ruolo centrale che le compete nei quotidiani dibattiti e approfondimenti pubblici.

Ecco perché ritengo che quanto sta facendo al CERN di Ginevra il Professore Zichichi con i suoi collaboratori, in ossequio ai criteri di ricerca di stampo galileiano, nel campo della Fisica Subnucleare delle alte energie, sia un lavoro prezioso e necessario sulla lunga ed indispensabile via del progresso tecnologico, civile e sociale.

E' con questa certezza che continuerò a seguire i vostri studi con rinnovato interesse nella convinzione che la strada da seguire sia, come dissi l'anno scorso, quella di far entrare la scienza nella vita di tutti i giorni.

Nel ringraziare tutti i partecipanti per l'impegno che pongono, in ogni parte del mondo, in questo importante e strategico settore rinnovo il mio personale saluto e l'auspicio di un proficuo risultato dei lavori del seminario.

SESSION 8

INFORMATION SECURITY

FOCUS: THE ROLE OF SCIENCE IN INFORMATION TECHNOLOGIES AND INTERNET TOOLS IN DEVELOPING COUNTRIES

CYBER REPRESSION: GOING WORSE. WHAT CAN BE DONE?

HENNING WEGENER
Ambassador of Germany (ret.),
Madrid, Spain

The World Federation established its Permanent Monitoring Panel on Information Security exactly ten years ago, impressed by the growing damage potential looming in cyberspace. These dangers were already very real at that time, but in view of the *current*, civilization-threatening dimension of the cyber threat and the over-riding role cyber issues are taking in concerned public debate, those PMP members of the first hour can almost be credited with a sense of premonition. More than before, we are today facing a truly planetary emergency. It affects developing countries no less than others.

The Panel, one of the first groups in civil society world-wide to take a multi-disciplinary approach to the cyber issue, has tried from the beginning, in the best tradition of Erice, to develop a comprehensive analysis and strategy for harnessing the threats in cyberspace, as is evident from the title of its first major document, "Toward a Universal Order of Cyberspace", and, throughout the last decade has concentrated on key issues of cyber security, especially cyberwar, cyberterrorism, and cyber conflict in general. Our analysis and recommendations have increasingly crystallized around the concepts of cyber stability and cyber peace, as evidenced by the Erice Declaration of 2009, and our latest publication, "The Quest for Cyber Peace", co-authored by the Secretary General of ITU.[1]

One important focus of our work has been the steady growth of massive Government interference in the freedom of digital information by censorship in the Internet, in violation of International Law,[2] and in contradiction with the sacred tenets of Erice where we have always defended the free flow of information as a prerequisite of civilized society.[3] The consequences of comprehensive censorship—cyber repression—are grave and cannot be overestimated. Citizens are cut off from important benefits of the information age, and receive a skewed view of world reality, condemning them to political immaturity. Massive cyber repression can alter the collective state of mind of a nation. The gravity of massive information suppression is at par with other variants of cyber crime and cyber conflict und thus rightly in the purview of our work. Moreover, cyber repression hits hardest in developing countries which tend to have less elaborate legal and judicial systems, and where authoritarian structures are more widely spread and

[1] www.itu.int/pub/S-GEN-WFS.01-1-2011

[2] In particular, in the Universal Declaration of Human Rights ("UDHR"), the International Covenant on Civil and Political Rights ("ICCPR") and the International Covenant on Economic, Social and Cultural Rights ("ICESCR") protect the right of anybody to receive and impart information of all types, regardless of frontiers and through any chosen medium.

[3] "All governments should make every effort to reduce or eliminate restrictions on the free flow of information, ideas and people" Erice Statement 1982. "All governments should recognize that international law guarantees the free flow of information and ideas; these guarantees also apply to cyberspace. Restrictions should be as necessary and accompanied by a process for legal review" Erice Statement on Principles for Cyber Stability and Cyber Peace, 2009

more persistent. Developing countries would be the greatest beneficiaries of the free acquisition of knowledge and free access to communication. Our work on cyber repression thus fits particularly well within the general theme of this session.

All governments need to have an eye on Internet contents for reasons of *ordre public.* There must be control to cut out child pornography, violation of intellectual property rights, and incitation to crime, racial hatred and anarchy. But in our earlier work we have been precise in providing a yardstick for alleviating the tension between freedom of expression and illegal repression. The criteria are the existence of clear legal prescription, respecting the International Covenants of Human Rights and Freedom of Expression, and access to independent legal review. Cyber Repression, as used in this presentation, lacks these essential prerequisites. No further definition is needed in this context. Cyber repression is by no means the preserve of developing countries that have inadequate democratic credentials. Many other countries are far from innocent.

The PMP has gone public with its indictment of cyber repression three times in the last years. At the 2005 session of the World Summit on the Information Society we submitted a document "Information Security in the Context of the Digital Divide".[4] Some of its recommendations are directed at the "Denial of information access through Internet filtering". The case against cyber repression has again been made in a book we published jointly with the EastWest Institute in 2010.[5] Finally, this mode of infringement of information access and information integrity is also featured in "The Quest for Cyber Peace", where it is clearly indicated that cyber repression is incompatible with cyber peace.[6]

If I undertake today to revisit the problem, there are alarming reasons to do so. Especially in the last two years the situation world-wide has deteriorated both quantitatively and qualitatively. That must spur our analysis, but also our thinking in action terms; our former operative recommendations must be strengthened.

One reason for the rapid, substantial worsening of the situation is the growth process in digital technologies; another is the recent emergence and phenomenal growth of the "new social networks". Since we first worked on the topic in 2004, the number of computers world-wide has more than doubled, and access to the Internet via broad band has grown even more steeply. There are now billions of mobile devices, especially in developing countries, increasingly on-line capable; at the same time, fixed and mobile technologies converge. The advent of the mobile Internet and interactive Web 2.0 options increasingly permeate the developing world as well.

At least since 2010, the New Media have firmly established their role, together with the Internet and mobile phones, as novel and mighty mobilization and news transmission tools. The new networks were initially designed for "social" purposes, to allow people networking and social interaction. But in the process they have developed unprecedented group dynamics, beyond mere communication and knowledge acquisition. The figures are astounding. By May 2011, Facebook had 600 million (!) members who communicate in 70 languages, helped by 50,000 servers. Twitter users now number close to 200 million (175 by September 2010) who emit corresponding millions of "tweets".

[4] Document WSIS-05/TUNIS/CONTR/01-E
[5] www.ewi.info
[6] All PMP documents are also available at its website www.unibw.de/infosecur

YouTube receives an average of 2 billion (!) visits daily (in exceptional moments up to 7 billion). The result is a true explosion of the numbers of stakeholders in the information process, and of the potential for collective action.[7]

I do not intend to join the current lively debate about to which extent these new processes, and especially the new social networks, have contributed to the current Middle Eastern revolutions, starting with the 2009 Iranian unrest. Some see the networks as the new "liberators", novel tools of democratization, profoundly reshaping our political processes and their instant communications; others are more sceptical and warn of "the Net illusion". But it is certain that the empowerment effect is considerable. And it is double, since the new networks can also be used by authoritarian governments to relay regime propaganda and enforce repression. The new media "between revolution and repression"![8]

The overall effect of the communication and information explosion of the last few years is in any event that governments wishing to establish or maintain control over their citizens and their access to information, and to stem instability and threats to their system, have to cope with a new dimension of the repression problem. This translates, in the practice of the last years, especially the last two years, into *more* repression and *new* repressive techniques.

The factual basis for observing this sad process exists. There are several organizations, including from academia, that operate observatories world-wide, identify and document cases, and report out in their comprehensive Internet pages. The most prominent ones—there are many others—admirable in their thoroughness, professional rigor, and spread of research, are Freedom House, Reporters without Borders, and the OpenNet Initiative.[9] They publish rankings of the governmental repressors on whom they collect information. Freedom House publishes annual reports on Freedom on the Net. On the basis of a detailed score table, the organization allots an International Freedom Status designation. In its 2011 report, it ranks 11 countries as "not free", and 19 as only "partly free". Reporters without Borders has a category of "Enemies of the Internet" where 10 countries are listed (they are essentially similar to the "not free" list of Freedom House), 16 other countries are placed "under surveillance". These lists comprise only the main offenders, on whom reliable information is available. Their geographical distribution is depicted in the following two maps. Censorship mostly concentrates on political and national security content, but many countries go beyond. Intensity varies, but on the basis of the existing material one can assume that governments of more than 60 countries practice some form of censorship and that more than 25% of the current world population—Reporters without Borders speaks of one third!—live under censorship, a staggering record.

Yet it is not so much the number of censoring countries—and thus of population—that has increased, but the intensity of control, and the number of new repressive techniques employed.

[7] And new interaction networks are springing up by the day, as for instance "google+".

[8] Reporters without Borders, 2011 Report

[9] www.freedomhouse.org; www.rsf.org; http://opennetinitiative.net
. To be mentioned also are the Global Network Initiative, http://globalnetworkinitiative.org and Amnesty International.

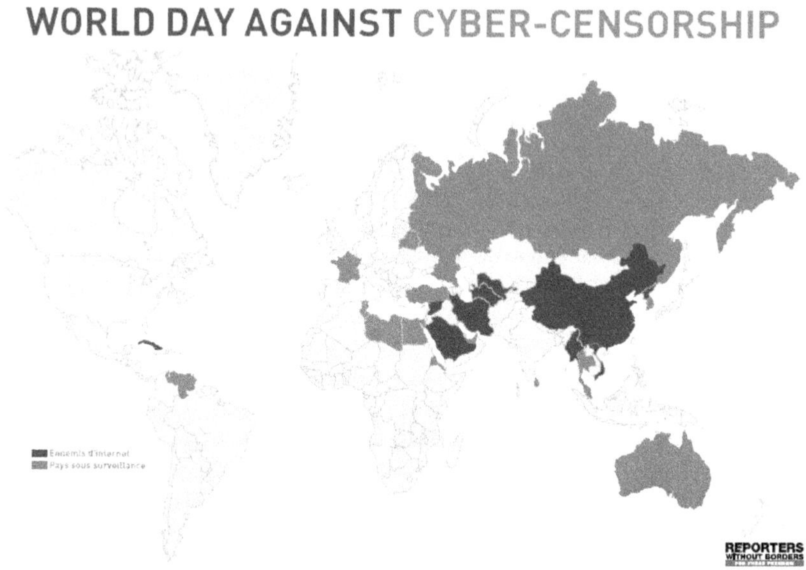
WORLD DAY AGAINST CYBER-CENSORSHIP
Pays sous surveillance
REPORTERS WITHOUT BORDERS

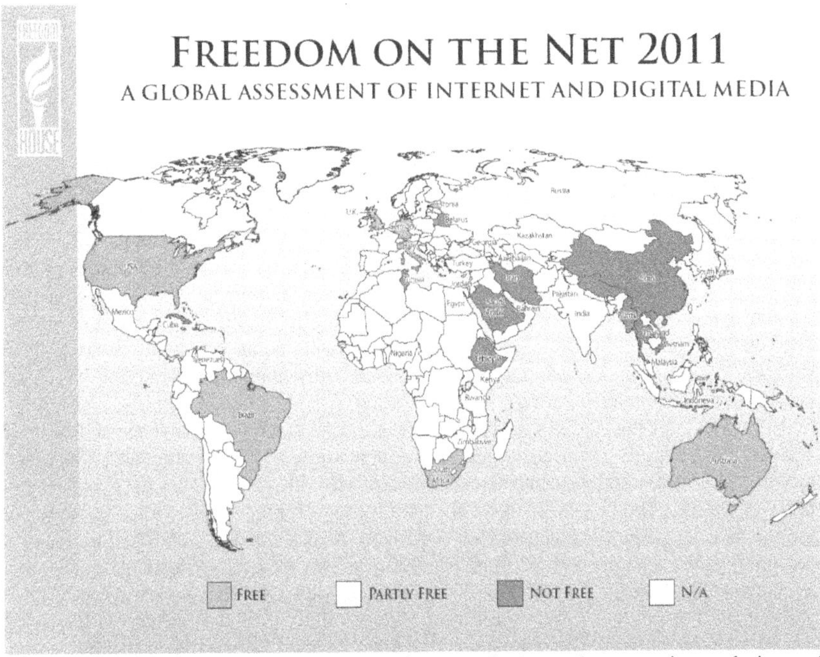

In our earlier work we concentrated our analysis on the censuring technique of content filtering by advanced routers, relying often upon foreign servers to install the corresponding software, and then block sites and practice surveillance; we were openly critical of what we saw as profit-oriented aiding and abetting in violation of ethical principles and international law. Filtering is still practiced at a massive scale by most repressive governments—and augmented in period of unrest and Government nervousness—but the shift away from the traditional *www* to new applications has also motivated a partial shift from content filtering and site blocking to new methods. At the same time, many developing countries are advancing in their own filter technologies and are taking the server structure into their own hands, thus depending less on foreign software technology imports and the readiness of foreign Internet service providers to obey their repressive orders. These new national infrastructures, often highly centralized, easily turn into Big Brother, Orwellian control mechanisms. However, the pressure on foreign service providers has also notably grown in some countries, thus heightening their ethical dilemmas.

Many repressor countries have responded to the growth in the number and nature of information stakeholders by a noticeable acceleration of reactions to the appearance of undesirable sites, and by institutionalizing their censoring agencies—in some cases these are huge, and technically competent. These countries tend to move from intermittent interferences to a much more intensive, agile and intrusive system of control. More sites are being controlled and interfered with, and sanctions become more severe.

Generally speaking, it can be observed that repressor governments move from whole-sale blocking and shutting down of certain sites to individual persecution. Massive, clearly directed DDoS attacks, espionage, defacing of individual pages, phishing, password theft, falsification of information and sites supplement general filtering. These attacks are often performed, at State behest, by hired non-State hackers. The harassment and terror effect of these individualized attacks is clearly designed to muzzle dissident opinions.

One increasingly frequent technique is the temporary slow-down in band-width speed (total, or geographically limited), with the effect that big data packets, photos and videos can no longer be received or sent. This strategy is increasingly applied in periods of turmoil—or expected turmoil—and political tension, like recently in the Middle East. As has been observed in countries like Iran, but also Egypt at the time of unrest, "connection speed has become the barometer of a country's political and social situation".[10] At times, this technique is accompanied by jamming or a shut-down of cell phone networks in relevant areas.

Blogs and the new social networks, as the perceived main agents of undesired political activity, are preferred targets of the new wave of cyber repression. Prominent bloggers are individually persecuted and castigated with shut-downs. Bloggers receive personal threats. The history of total, or just-in-time, blockings of Twitter, Facebook and the like is long; often these shut-downs are timed to coincide with periods of unrest. The technology and case history of these shut-downs will be dealt with in more detail in another presentation in this session. Repression of the New Media is particularly grave as, beyond filtering out information and limiting individual exchanges of opinion, it impedes the operation of a new universe of collective opinion-forming and mobilization.

There is often ruthless criminal persecution of cyber "offenders". Key repressor countries have a dedicated cyber police who intimidate and threaten, but also act. Arrests of bloggers and "netizens" in the key repressor countries continue unabated. Reporters without Borders keep track: as of March 2011 there were 119 perpetrators in jail, including 2011 Nobel peace prize winner Liu Xiaobo. In Iran, there have been the first condemnations to death. Many of the police actions are directed against users of the New Media, as they are perceived to be the principal virtual meeting place of adolescent political dissenters. In one of our earlier documents we unabashedly asked to what extent foreign software producers and service providers are—directly or indirectly—responsible for these "cyber criminals" to be denounced and condemned.

Another new feature of the cyber repression scene is that repressor governments do not content themselves with impeding undesirable information traffic and enact reprisals, but move to affirmative action. The new battle is about manipulation of information. Repressive governments increasingly and proactively use the new social networks for massive propaganda and distortion of facts to counter the Facebook and Twitter effect. Here again, the duplicity, the ambivalence of new technologies, in this case the social networks, shows clearly.

Our PMP has always thought that wailing and denouncing in the face of a common evil is not enough: we have always placed our analysis in an operational perspective, coupling it with precise recommendations, a time-proven Erice recipe. Many

---

[10] Reporters without Borders, March 2011 Report, p.5

of the afore-named organizations that are admirable in their analysis and voice their revolt, fail to proceed beyond that passive stance. However, the worsening situation of cyber repression makes action more necessary than ever

I will thus briefly review the action plan which has been part of our earlier documents, and will try to indicate where our operational options can be reinforced, and what else we can do.

Action can and should be taken by civil society—the articulate lobby for a Free Internet—by industry, by governments, and—most important—by the international community. Our earlier documents list the organizations that strive for freedom of expression in the Internet, denounce the wrongdoers, sensitize public opinion, and help the victims of cyber repression by providing evasion techniques to bypass censorship, anti-filter software and the like. This assistance is effective where the "traditional" filtering methods are used, but cannot cope with the new techniques repressive governments apply, especially when there are shut-downs, reduction of bandwidth speed, large-scale manipulation of information, or individual attacks and reprisals. Yet, the efforts of the Internet lobby are praiseworthy, and there should be more mobilization, especially in many countries not yet involved, to join or do likewise. The OpenNet Initiative and the Global Network Initiative are among those deserving particularly strong support.

Governments can do a lot. The U.S. Government is exemplary in its support for Internet freedom which includes the creation of the Global Internet Freedom Task Force and the provision of funds for assisting filter circumvention technologies. The European Union does not accept export of filter software technologies to repressor countries. Indeed there should be global export controls for this purpose.

The principal areas where more action should be deployed are the public arenas of the International Community: international fora and international organizations. Practically the entire world community has subscribed to the Universal Declaration of Human Rights and the two International Covenants, thus converting the principle of freedom of expression into international law of general validity. They all share a common responsibility. It is their duty collectively to respect, protect, promote and fulfil human rights, including the freedom of expression. We should therefore urge more strongly—and bring this to our governments—that the struggle against cyber repression be carried into international organizations that have a calling in this field. A call from the World Federation of Scientists to this effect could make a difference and produce impact.

In our earlier documents we have discussed the pros and cons of each one these relevant international bodies. We have also argued that an international complaint procedure of the "comply or explain"-type be introduced that—as direct sanctions are not feasible—would increasingly put international pressure and opprobrium on repressive Governments. In the interest of brevity, I will only refer to these texts and will merely list the organizations we have considered:

- The World Summit on the Information Society. We brought the case to the 2003/2005 First Summit. It did not take up our proposals for pertinent resolutions, but in its Geneva Declaration of Principles has strongly affirmed extant international law on freedom of expression. Governments should build on that to introduce the topic into the agenda for the forthcoming 2015 Second Summit.

- The Internet Governance Forum
- UNESCO as, by its founding act, the unique international guardian of freedom of information, and, in addition, mandated by the WSIS (and the UNGA) to deal with the topics of "Access to Information and Knowledge", and the "Ethical Dimension of the Internet"
- The UN Human Rights Council as the special body dealing with violations of the International Covenant on Civil and Political Rights which cyber repression violates; it could establish the suggested complaint procedure, and/or include the topic of Internet Freedom and censorship in the statutory Universal Periodic Review Process.
- Finally, the UN Human Rights Committee which is entrusted with periodic country reviews. These should in the future include Internet Freedom.

A POSTCRIPT: WIKILEAKS VS. GOVERNMENT CONTROLS

When this presentation was commissioned, the problem of Wikileaks vs. Government Control also figured in the assignment. As I have not treated it in the body of the text, a brief supplementary note on it might be in order.

Discussing Wikileaks, and especially the recent massive release of confidential government papers, predominantly from the U.S. Government, is not, properly speaking within the purview of this presentation: no repressive government has been at work, and developing countries and their legal systems are not involved. In fact, the leaks concern some of the freest countries in the world, where, in addition, legal procedures are available to enforce Freedom of Information (e.g., under the U.S. Freedom of Information Act).

But the Wikileaks case does, in fact, pose the question of the inherent tension between public order and security on the one side, and totally unbridled openness on the other. From the viewpoint of the Erice community, the Wikileaks perpetrators have promoted freedom of information and expression, and have thus done something positive. Most of the information, in addition, does not apparently affect State security or, in most cases, betray essential State secrets; most of the effect has been no more than embarrassing and damaging to the prestige, if not the vanity of those who would have preferred to preserve confidentiality. As some of the material uncovers illegal practices by a government—for example, corruption cases in Hungary or Tunisia—additional healthy effects may have occurred, and in some contexts opposition forces may have benefitted.

But this positive assessment has to be nuanced in two important ways. From an information security standpoint, there are probably massive elements of interference with the Internet and computer systems (illegal interception, data interference, system interference), punishable under the applicable national cybercrime legislation. If there was printed material—or any other material, like a CD Rom—involved, the perpetrators have committed theft. There may also be illegal acquisition if other protected interests have been affected (e.g., intellectual property rights). Furthermore, for some part of the material, State secrets may have been involved: this has been claimed by the Australian

Government, and the Espionage Act may apply for the U.S. if protected secrets have been disclosed within the country. A detailed analysis in the light of each national legal system would be necessary to determine whether penal laws apply. Government insiders with authorized access, like Pfc. Manning in the recent Wikileaks disclosure, may be subject to disciplinary or even penal sanction if classified information was revealed.

The other nuance is more of a philosophical nature. Society is made up of competing goods, this competition cannot be resolved in a once-for all way but requires the earnest search for acceptable equilibria. When State documents are involved the tension, referred to above, between security and individual liberty takes on a special twist. Statecraft, even beyond the formal qualification of State secrets, requires a margin of confidentiality to be effective; take international negotiations and their ever-changing tactical needs. Statecraft and total transparency are not compatible, and for statecraft not to be wrecked, some limits to transparency must be preserved. Call it maturity? Responsibility? Patriotism? In this perspective, Wikileaks cannot endure with its present absolute quest for total transparency. Already, it has done damage.

The consequences will possibly result in more damage. If Wikileaks does not mend its ways, Governments will become more restrictive in their information policy, clamp the stamp of "State Secret" on more, otherwise innocuous documents and restrict the freedom of information and opinion of which we, members of open democratic systems, are justly proud.

THE ROLE OF SCIENCE AND TECHNOLOGY AS EMPOWERMENT OF PERSON AND STATE

JODY R. WESTBY
Global Cyber Risk LLC, Information Security PMP, World Federation of Scientists, Washington, USA

INTRODUCTION

Information and communication technologies (ICTs) are empowering individuals and allowing them to exercise self-determination and impact their governments and foreign policy in an unprecedented manner. ICTs are also creating "global citizens" and enabling them to extend their influence to political elections and domestic agendas outside of their own countries. For example, the 2008 U.S. presidential election was clearly impacted by the gushing support that Europeans displayed for Barack Obama; his speech in Berlin was broadcast around the globe and widely distributed over the Internet. By declaring, "People of Berlin, people of the world, this is our moment. This is our time," he expanded the circle of influence upon American voters to the entire world.

A more recent phenomenon involves the role that social networking technologies are playing in the recent uprisings in the Middle East and North Africa (known as the "Arab Spring"), and the way they are changing the cultures of these countries. Individuals are also using ICTs to disclose information and impact global events. For example, Pfc. Bradley Manning's alleged theft of confidential U.S. diplomatic cables and Julian Assange's disclosure of them through his organization WikiLeaks provided a treasure trove of information that weakened and embarrassed governments and empowered protestors in numerous countries.

ICTs also strengthen the nation state. Satellites, the Internet, landline and mobile telephony, cable, radio, and broadcast have converged into a powerful weapon that can be manipulated by governments through their ownership of communication gateways, regulatory authority, and control over broadcast licenses and spectrum. They represent a serious capability in the correlation of forces that a nation can bring to bear against its adversaries. Government-owned media, such as Al-Jazeera (Qatar) and Al-Arabhiya (Saudi Arabia), enable countries to exercise a shadow power that broadens their sphere of influence and, in the case of Al-Jazeera, extends Qatar's diplomatic clout far beyond usual zones.

Individuals and governments alike have learned lessons from the Arab Spring uprisings. Similar to the Tunisian and Egyptian demonstrations, the recent riots in London were fueled and organized using Facebook, Twitter, and messaging systems. Also, equally similar to the actions taken by Egypt and other Middle Eastern governments, a frustrated U.K. government responded with proposals/threats to cut off social media communications, alarming activists and communications providers.

The interests of the nation state against the rights of the individual are colliding, with ICTs being the tool of choice to assert power by both sides. How governments are leveraging their control over communications, especially when used to restrict or cut off the individual right to freedom of expression, is raising serious legal and policy issues that merit a wide discussion. These are hard issues that cannot be dismissed with declarations that the Internet must never be shut down or that people must retain the right to freely communicate no matter what the circumstance. There may be situations when the Internet may need to be cut off for a period of time, just as the U.S. closed all air traffic on 9/11. For example, such action might be prudent if an exploit that came through Internet-connected control systems was causing water systems to be contaminated. Likewise, some instances of personal communication may be outside the right to freedom of speech, as noted by the U.S. Supreme Court in *Brandenburg v. Ohio,* when it ruled that freedom of speech does not extend "where such advocacy is directed to inciting or producing imminent lawless action and is likely to incite or produce such action."[1]

EMPOWERING THE INDIVIDUAL

Article 19 of the Universal Declaration of Human Rights (UDHR) guarantees the right to freedom of expression:

> Everyone has the right to freedom of opinion and expression; this right includes the right to hold opinions without interference and to seek, receive and impart information and ideas through any media and regardless of frontiers.[2]

Drafted in less than two years and passed with only eight United Nations (UN) member nations abstaining and none opposing, the UDHR stands as one of the lasting, positive outcomes of World War II. A member of the drafting sub-committee, Hernán Santa Cruz of Chile, later wrote about the UN's agreement on sweeping language at a time when the world was divided into Eastern and Western blocs:

> I perceived clearly that I was participating in a truly significant historic event in which a consensus had been reached as to the supreme value of the human person, a value that did not originate in the decision of a worldly power, but rather in the fact of existing—which gave rise to the inalienable right to live free from want and oppression and to fully develop one's personality. In the Great Hall…there was an atmosphere of genuine solidarity and brotherhood among men and women from all latitudes, the like of which I have not seen again in any international setting.[3]

Although the UDHR is not directly binding on UN member states, portions of it, including Article 19, have acquired legal force as customary international law.[4]

---

[1] *Brandenburg v. Ohio,* 395 U.S. 444 (1996), http://supreme.justia.com/us/395/444/case.html ("Freedoms of speech and press do not permit a State to forbid advocacy of the use of force or of law violation except where such advocacy is directed to inciting or producing imminent lawless action and is likely to incite or produce such action.").

[2] United Nations General Assembly Resolution 217A (III), Dec. 10, 1948, http://www.un.org/en/documents/udhr/.

[3] "The Universal Declaration of Human Rights: History of the Document," http://www.un.org/en/documents/udhr/history.shtml.

[4] *See, e.g., Barcelona Traction, Light and Power Company Limited Case* (Belgium v. Spain) (Second Phase), ICJ Rep. 1970 3 (Int'l Court of Justice); *Namibia Opinion*, ICJ Rep. 1971 16, Separate Opinion, Judge Ammoun (Int'l Court of Justice).

The right to freedom of expression is also guaranteed in the International Covenant on Civil and Political Rights (ICCPR), a treaty that has been ratified by 167 countries. Also in Article 19, the ICCPR language parallels the UNDR, but goes farther and specifies that the right applies regardless of whether the expression is made "orally, in writing or in print, in the form of art or through any other media of his choice."[5] In addition, the right to freedom of expression is set forth in the three major regional human rights treaties: the American Convention on Human Rights (Article 13),[6] the European Convention on Human Rights (Article 10),[7] and the African Charter on Human and Peoples' Rights (Article 9).[8]

The reach of the protection of the right to freedom of expression was succinctly set forth by the European Court of Human Rights:

> [F]reedom of expression...is applicable not only to "information" or "ideas" that are favourably received...but also to those which offend, shock or disturb the State or any other sector of the population. Such are the demands of pluralism, tolerance and broadmindedness without which there is no "democratic society....In its social dimension, freedom of expression is a means for the interchange of ideas and information among human beings and for mass communication.[9]

It is important to note that the guarantee of freedom of expression is particularly applicable to the media, including broadcast media, because it is the primary conduit of information to the public. Strong statements of protection for freedom of expression by the general media have been put forward by the Inter-American Court of Human Rights, the European Court of Human Rights, and the African Commission on Human and Peoples' Rights. Since the legal obligations for ensuring freedom of expression apply to member states, public media stations are particularly bound to uphold it.[10] The infringement of this right by government leaders for their own purposes is, therefore, particularly egregious.

The Arab Spring uprisings were ignited by a young fruit merchant who set himself on fire in Tunisia following police harassment,[11] and it spread to Egypt after a 28-year-old Egyptian man named Khaled Said died from a beating by police. When a young Google executive, Wael Ghonim, dedicated a Facebook page to Mr. Said that "went viral" and spread globally in a very brief period of time, it inspired the Egyptians' ouster of Mr. Mubarek.[12]

---

[5] UN General Assembly Resolution 2200A(XXI), Dec. 16, 1966, entered into force Mar. 23, 1976.

[6] Adopted Nov. 22, 1969, entered into force July 18, 1978.

[7] Adopted Nov. 4, 1950, entered into force Sept. 3, 1953.

[8] Adopted June 26, 1981, entered into force Oct. 21, 1986.

[9] *Handyside v. the United Kingdom,* Application No. 5493/72, para. 49, Dec. 7, 1976.

[10] Toby Mendel, *Public Service Broadcasting: A Comparative Legal Survey,"* United Nations Educational, Scientific and Cultural Organization, 2011, 2nd ed. at 9-12.

[11] Marc Fisher, "The spark that ignited a Revolution," *The Washington Post,* Mar. 27, 2011 at A1, A10-11.

[12] Andrew England and Heba Saleh, "Freed hero steps into a dead man's shoes," *Financial Times,* Feb. 10, 2011 at 4; Ernesto Londono, "Fatal beating became symbol of callous state," *The Washington Post,* Feb. 6, 2011 at A10.

There was no videotape of the scene when the Tunisian fruit vendor, Mohammed Bouazizi, argued with a retaliatory policewoman and then set himself on fire in protest of repeated abuses by police that kept him from earning a living. Following his death, however, his cousin filmed a small demonstration about the treatment of vendors that was held in front of the city hall and posted it on Facebook. Tunisia has the highest Internet usage of any Arab country, but it has been one of the most restrictive in censoring online content. Social networking sites, however, were a step ahead of government censors. One day after the Facebook posting, Al-Jazeera picked up the video clip and repeatedly broadcast it. A reporter for a Tunisian radio station summed up the electrifying effect of the video and story of Bouazizi: "Every family here has somebody who lost a job or was denied a raise or was called in by state security. Bouazizi just gave us the courage to let it out."[13]

A revolution was born. Former president Zein al-Abidine Ben Ali was forced from office after a 23-year stronghold, largely due to the Internet's ability to show Tunisia's youth how their lives differed from people in democratic societies. When U.S. diplomatic cables released by WikiLeaks confirmed long-standing rumors about corruption in the Tunisian government, particularly within the families of the president and his wife, a frustrated population turned angry.[14] Young, disenfranchised men and women shed the cloak of fear their government had wrapped around them, and used the Internet and social media to ignite a revolution that toppled their own regime and inspired a region.

The Egyptian uprising began with demonstrations of striking workers, which were largely leaderless and organized over the Internet.[15] The anger of the people was stoked by the Facebook page that Google executive Wael Ghonim posted of murdered Khaled Said[16] and news accounts detailing corruption and enrichment of those in power.[17] Mr. Said's problems were, ironically, born from the Internet. He was in his upstairs apartment working on his computer when his Bluetooth picked up a video clip that was being viewed by police officers in an Internet café beneath him. The video showed officers dividing up seized narcotics and cash. Mr. Said showed the video to friends and forwarded it to others. Two of the officers pictured in the video tracked Said and grabbed him outside of the Internet café. They proceeded to brutally beat him to death. Authorities then told his family that he was at the morgue and had died of a heart problems related to a drug overdose. Disbelieving the authorities, the family bribed a guard to take a photo of Said's body. When Egyptian prosecutors were confronted with the evidence of a horribly beaten young man, they were reluctant to pursue the case. Said's father held a news conference and set up a Facebook page, showing the police brutality.[18] Similar to the

---

[13] Marc Fisher, "The spark that ignited a Revolution," *The Washington Post,* Mar. 27, 2011 at A1, A10-11

[14] Roula Khalaf, "Time for Arab leaders to watch cable TV – and quake," *Financial Times,* Jan. 15-16, 2011 at 7.

[15] Heba Saleh and Roula Khalaf, "Opposition strives to exploit anger," *Financial Times*, Jan. 28, 2011 at 6.

[16] Andrew England and Heba Saleh, "Freed hero steps into dead man's shoes," *Financial Times,* Feb. 10, 2011 at 4.

[17] Andrew England, Michael Peel, and Heba Saleh, "Egyptian workers stage fresh strikes," *Financial Times,* Feb. 20, 2011 at 4.

[18] Ernesto Londono, "Fatal beating became symbol of callous state," *The Washington Post,* Feb. 9, 2011 at A10.

impact that Bouazizi's plight had on Tunisians, the killing of Said reminded Egyptians of the injustices that they themselves had suffered under Mubarek's regime, and it lit a revolt.

The protests that toppled Mubarek's 30-year regime were largely fueled by Facebook and Twitter and assisted by television. The messages and images sent around the world revealed the incredible courage of young men and women, who had been protesting peacefully, as they suddenly faced systematic and coordinated attacks from government-sponsored thugs and dodged Molotov cocktails, charging camels and horses with riders bearing whips and machetes, and vehicles wildly careening through crowds of protesters.[19] Journalists from CNN, BBC, The Washington Post, Associated Press, Danish TV2 News, and Swiss television also were assaulted.[20] These images enhanced the Egyptian public's distrust of the Mubarek regime, and they responded to the attacks by flooding Tahrir Square with an estimated 100,000 people chanting "Down with the regime" and calling for the Mubarek to leave.[21]

The Egyptian protesters also were emboldened by the WikiLeaks disclosure of confidential U.S. diplomatic cables that revealed the Egyptian army was hardly the cohesive force it presented to the public, and that it was not in favor of Mubarek's son succeeding him as president. Cables from high ranking U.S. diplomats revealed factionalism and mistrust within the military and referred to the Egyptian defense minister as "Mubarek's poodle." These diplomatic cables gave the Egyptian protesters hope that they were not alone in their discontent; the military also might stand with them.[22] Other cables disclosed by WikiLeaks revealed the true depths of the close relationship between the U.S. and Omar Suleiman, the former head of Egyptian intelligence and then-vice president, whom the Obama Administration had deemed central to an orderly transition of power from Mubarek to a new government. One cable referred to Suleiman as "the most successful element" of the U.S.-Egypt relationship.[23] Further revelations in the cables revealed the unfavorable views of Suleiman by Israel and his close ties with Iraqi Sunni insurgents.[24] All in all, the cables turned the protesters against Suleiman and surely contributed to the protestors' demands that he also go from office.

The use of the Internet and social networking sites, particularly Facebook and Twitter, has been instrumental in the dissemination of information among protestors across the Middle East and has facilitated their ability to organize. These technologies have become a fount of courage, hope, and inspiration and are fueling the determination

---

[19] Will Englund and Leila Fadel, "Mubarek supporters confront protestors," *The Washington Post,* Feb. 3, 2011 at A1, A11; *see also* Ahmed Alaidy, "I thought one day my grandson might see this. But here I am." *The Washington Post,* Feb. 6, 2011 at B4.

[20] Christopher Torchia, "Journalists are attacked in Cairo," Feb. 3, 2011 at A12.

[21] Michael Peel, Andrew England, and Steve Negus, "Anti-regime protests regain impetus," *Financial Times,* Feb. 5-6, 2011 at 2; *see also* Heba Saleh, "Besieged leaders struggle to find a strategy," *Financial Times,* Feb. 5-6, 2011 at 2.

[22] David Blair and Helen Warrell, "WikiLeaks cables portray army riven by factionalism," *Financial Times,* Feb. 5-6. 2011 at 2.

[23] Helen Warrell and David Blair, "Leaked cables reveal faith in Suleiman's ability," *Financial Times,* Feb. 7, 2011 at 3.

[24] *Id.*

of these people to risk their lives for freedom. Beyond these beneficial uses, however, ICTs have been used by ruthless and desperate leaders to attack peaceful protestors, spread propaganda, and stop the flow of communications for strategic advantage. In the days prior to the coordinated attacks on the Egyptian protestors, Mubarek tried to curb the demonstrations by shutting off the Internet. The individual response to the Egyptian shutdown of the Internet was broadened when the international hacking group known as Anonymous showed support for the protestors by orchestrating a distributed denial of service (DDOS) attack against important Egyptian government websites.[25]

To counter communication blockages and relay the events as they unfolded, the major media channels relied upon scientific-derived technologies, such as satellites and Google maps, to broadcast and report events, even when heavy censorship has been in place. The ability to bring instant information and visual depictions to the global population helped protect the human rights of the protestors, as images of brutality against unarmed citizens resulted in condemnation by leaders of nation states and international action by the United Nations and NATO.

As reporter Marc Fisher noted:

> In Tunisia and then in Egypt and across the region, people who had complained only to friends and family felt the fear that their rulers depended upon dissipate like air from a pierced balloon. The wizards who commanded seemingly omnipresent secret police forces were revealed to be just old men behind a curtain, running state security operations that didn't even know how to handle a virus of rebellious Facebook pages.[26]

Kathleen Parker brilliantly dubbed the Egyptian uprising as the "digitally inspired revolt" that was waged on a "digital battlefield" whose "front lines were manned with typists."[27] She further noted:

> The transformation taking place isn't only for Egypt but for mankind....Perhaps we are not doomed after all...Unarmed men and women inspired by tweets of freedom stared into the bullying armaments of dead ways. It was a stark image of the prolonged battle between good and evil....This time, enabled by what we casually call social media, evil may finally be outgunned.[28]

The common Western view that the only alternative to these autocratic regimes is Islamic fundamentalism is being tested by individuals who are charting a new path.[29]

The repercussions from the Egyptian and Tunisian uprisings spread like a tsunami across the Middle East and triggered responses from governments. Media around the

---

[25] rwaters, "Anonymous Strikes Egypt," Digital Forensic Investigator News, Feb. 4, 2011, http://www.dfinews.com/e-newsletter/february-4-2011-anonymous-strikes-egypt-photographer-faces-child-porn-charges-doj-seeks.

[26] Marc Fisher, "The spark that ignited a Revolution," *The Washington Post,* Mar. 27, 2011 at A1, A10-11.

[27] Kathleen Parker, "Tweets vs. tanks in Tahrir, *The Washington Post,* Feb. 13, 2011 at A25.

[28] *Id.*

[29] Roula Khalaf, "Time for Arab leaders to watch cable TV – and quake," *Financial Times,* Jan. 15-16, 2011 at 7.

world reported protests in Iran, Bahrain, and Yemen,[30] Libya, Morocco,[31] Syria, Jordan, Saudi Arabia, and Oman.[32] Rulers from country after country appeared before the media announcing concessions to their people, fearful that they might be next to topple. Jordanian King Abdullah fired his cabinet and appointed a new prime minister, Yemen's president announced he would not seek re-election,[33] Algeria's president announced that he would lift the 19-year emergency rule and allow opponents access to television and radio,[34] and Syrian officials said that bans on Facebook and YouTube would be lifted for the first time in three years.[35] Money also began to flow from kingdom coffers to the people. Kuwait's Sheikh Sabah Al Ahmed Al Sabah Kuwait's ruler said he would give grants of about U.S.$3,500 and food coupons to all one million Kuwaiti citizens,[36] and Saudi Arabia's King Abdullah announced he was allocating $35 billion to address affordable housing and youth unemployment.[37] In the midst of this, the Palestinian cabinet resigned and did not deny that the regional turmoil had been a factor.[38] The release of 1,600 documents on Palestinian concessions to Israel during peace talks that were made public by Al-Jazeera a couple of weeks earlier angered the Palestinian population.[39] The documents, posted on the Al-Jazeera website called Palestine Papers, included memoranda, emails, maps, minutes, and notes of high-level meetings from 1999 to 2010.[40]

Individuals also played a significant role in bringing news updates to the international press via their mobile phones, Skype, Facebook, Twitter, and email. For example, a Syrian who fled to Beirut, spends his days in his apartment collecting information from Syrians via Skype and then passing it on to media organizations that were expelled from Syria soon after the protests began. He retweets threats that he receives from the Syrian government and sends tips to Syrians on how to upload videos

---

[30] Sudarsan Raghavan, "Egypt's revolt stokes fires regionwide," *The Washington Post,* Feb. 15, 2011 at A1, A10.

[31] Sudarsan Raghavan, "Arab leaders use varying tactics to try to calm anger in the streets," *The Washington Post,* Feb. 21, 2011 at A9.

[32] "The Arab spring: the first three months," *Financial Times,* Apr. 2-3, 2011 at 3.

[33] Janine Zacharia, "An uncertain path to political change," *The Washington Post,* Feb. 6, 2011 at A1, A13; Tobias Buck, "King of Jordan promises reform, *Financial Times,* Feb. 2, 2011 at 3.

[34] Christian Lowe and Lamine Chikhi, "Snap analysis – Algeria to lift emergency rule soon," Reuters.com, http://www.reuters.com/article/2011/02/03/uk-algeria-emergency-repeal-snap-idUKTRE71268320110203.

[35] Bassem Mrouh, "Syrians see social-media sites; 3-year-old ban could be over," *The Washington Post,* Feb. 9. 2011 at A11.

[36] "Kuwait ruler orders to give money, food coupons to all citizens," albawaba, Jan. 17, 2011, http://www.albawaba.com/main-headlines/kuwait-ruler-orders-give-money-food-coupons-all-citizens.

[37] Deborah Amos and Steve Inskeep, "Saudi Arabia Uses Money to Address Protestor Issues," National Public Radio, Feb. 24, 2011, http://www.npr.org/2011/02/24/134017770/Saudi-Arabia-Update.

[38] Janine Zacharia, "Palestinian cabinet quits amid regional turmoil," *The Washington Post,* Feb. 15, 2011 at A12.

[39] Joel Greenberg, "Palestinian Erekat says he quit to set example of accountability," *The Washington Post,* Feb. 16, 2011 at A 10;

[40] Janine Zacharia, "Al-Jazeera will release 'Palestine Papers,'" *The Washington Post,* Jan. 24, 2011 at A7; *see also,* Tobias Buck, "Ramallah accuses al-Jazerra of plot to weaken Abbas," *Financial Times,* Jan. 25, 2011 at 6.

to YouTube.[41] Government actions captured on video also helped empower the Syrian protestors. One video showing Syrian security forces placing weapons on the bodies of dead protestors was posted on YouTube by activists to show how desperately the Syrian government wanted to mask the protests as an armed revolt.[42]

Another aspect of how ICTs can empower the individual was evidenced by a protest campaign launched by Qatari citizens against the country's primary telecommunications provider, Qtel. Using social media sites, such as Facebook and Twitter, the organizers asked Qtel subscribers to shut off their phones for an hour on July 7, 2011, as a campaign for better coverage and lower rates. Protesters tried to lessen the risk of getting jailed for their actions by clarifying that, "We are happy with the political set-up [of the country], we are not happy with the service."[43] The campaign eventually led to a meeting between the organizers and QTel officials. The organizers were emboldened by the role that their country has played in reporting on the Middle East uprisings through the Qatari government-owned television channel, Al-Jazeera,[44] which is an interesting twist on how government use of ICTs worked to empower its own citizens.

The Chinese government has historically engaged in heavy censorship of its citizens, including blocking Internet websites and filtering communications. The desire for individual expression is great, however, and Chinese citizens are increasingly innovative in using cutting edge technologies to circumvent government controls. A U.S.-based group of Chinese activists formed the Global Internet Freedom Consortium to provide technologies to allow the Chinese people unfettered Internet access. The Consortium also empowers individuals in other authoritarian regimes. For example, it opened up its network and technologies to individuals in Iran who were seeking to use the Internet during protests in 2008 and played a significant role in getting news of the events out of Iran.[45] Microblogs, the 140-character messages called "weibo," are the newest means of disseminating opinions and debating issues in China. A step ahead of the government censors, an associate professor at Peking University, Hu Yong, has noted that, "Weibo has become the most prominent place for free speech in China."[46]

EMPOWERING THE NATION STATE

Authoritarian regimes have become ever more aware of the threat that ICTs pose to their control over their populations. The U.S. State Department's 2011 annual report on human rights notes that:

Today there are more than two billion people with Internet access spread across

---

[41] Tara Bahrampour, "In Syria, a push to end decades of isolation," *The Washington Post,* Apr. 17, 2011 at A1, A16.

[42] Liz Sly, "Syria says protestors killed security forces," *The Washington Post,* June 7, 2011 at A1, A7.

[43] Abeer Allam, "Revolts inspire Qataris to take on telecom group," *Financial Times,* July 20, 2011 at 6.

[44] *Id.*

[45] John Markoff, "Iranians and Others Outwit Net Censors," *The New York Times,* Apr. 30, 2011, http://www.nytimes.com/2009/05/01/technology/01filter.html?pagewanted=all.

[46] Keith B. Richburg, "In China, microblogs serve as free-speech forum," *The Washington Post,* Mar. 28, 2011 at A6.

> most countries of the world, and around five billion mobile phone subscriptions. These numbers are projected to grow dramatically in the next 15 years. And as more people gain access to these remarkable technologies, and use them both to gather and impart information on human rights and to communicate with other activists, an increasing number of governments are spending more time, money, and attention in efforts to curtail access to these new communications outlets. More than 40 governments are now using a combination of regulatory restrictions, technical controls on access to the Internet, and technologies designed to repress speech and infringe on the personal privacy of those who use these rapidly evolving technologies.[47]

The Egyptian government used the full array of ICTs available to them in its desperate attempt to manipulate whoever would listen and remain in power. In an attempt to counter the influence of Mohamed ElBaradei, the Mubarek government launched a media campaign against him, which included putting up a Facebook page of Mr. ElBaradei's family that contained photographs of his daughter in a swimsuit—images certain to inflame many Muslims.[48] Hosni Mubarek appeared on ABC News to claim that he wanted to step down as president, but he feared that chaos would result. Thus, he argued that he should remain in office to protect the people. He then used state-owned television to try to convince the public that the protestors were foreigners acting against their interests and creating unnecessary hardships.[49]

Recognizing the role that the Internet and social media sites were playing in helping the protestors organize and spread their message, Mubarek took the extraordinary measure of shutting down the Internet for five days by leveraging the country's ownership of the lines that carry communications in and out of Egypt and by threatening to pull providers' licenses if they did not immediately shut down service.[50] He then sent pro-government text messages over Vodafone and France Telecom's networks, putting the companies at risk of reputational damage and backlash from customers and activists.[51] The government's restoration of Internet service coincided with violent and coordinated attacks on the protesters in Tahrir Square,[52] providing the Mubarek

---

[47] *2010 Country Reports on Human Rights Practices,* U.S. Department of State, at 3, http://www.state.gov/ g/drl/rls/hrrpt/2010/#

[48] Heba Saleh and Roula Khalaf, "Opposition strives to exploit anger," *Financial Times*, Jan. 28, 2011 at 6.

[49] Michael Peel, Andrew England, and Steve Negus, "Anti-regime protests regain impetus," *Financial Times,* Feb. 5-6, 2011 at 2.

[50] James Glanz and John Markoff, "Egypt's Autocracy Found Internet's 'Off' Switch," *The New York Times*, International Edition, Feb, 16, 2011 at A1, A10; Tim Bradshaw, "Telecoms blackout condemned," *Financial Times,* Jan. 29-30, 2011 at 2.

[51] Andrew Parker and Jennifer Thompson, "UK Tackles Cairo over networks abuse allegations," *Financial Times,* Feb. 5-6, 2011 at 10; Andrew Parker, "Vodafone protests over mass texting," *Financial Times,* Feb. 4, 2011 at 2.

[52] Bill Woodcock, "Overview of the Egyptian Internet Shutdown," Packet Clearing House, Presentation to the DHS Infosec Technology Transition Council, Feb. 11, 2011, www.pch.net/resources/misc/Egypt-PCH-Overview.pdf.

government with the capability to coordinate the attacks and maintain situational awareness.[53]

Finally, Mubarek stunned the world by using global media channels do an about-face and defiantly refuse to step down from power as expected. In the end, his use of the media was his undoing. The crowds in Tahrir square were inflamed by Mubarek's refusal to step down, and foreign leaders were embarrassed and dismayed.[54] A spokesman for the Egyptian army then appeared on television in a somewhat "quasi-coup" to effectively announce that the military would protect the people and the stability of the country.[55]

Mubarek's heavy-handed tactic to shutdown the Internet was not without a price, economically, politically, and personally. The Organisation of Economic Co-operation and Development (OECD) estimated that the shutdown cost the Egyptian government at least U.S.$90 million due to lost revenue from communication services. The economic hit to its thriving IT sector, tourism, and other activities central to Egypt's economy was certainly substantial.[56] On the personal front, an Egyptian administrative court fined Mubarek U.S.$33.5 million and two of his colleagues an additional $57 million for "causing damage to the national economy."[57] The political costs to the Mubarek government were even greater; Egyptians sympathetic to the protestors were furious, and there was international uproar over the attacks on protestors and journalists and the shutdown of communications.[58] The UK government lodged a complaint with Egyptian authorities about the misuse of Vodafone's mobile network to send pro-government text messages,[59] and U.S. Secretary of State Hillary Clinton warned governments that blocking the Internet would impede economic growth and cause unrest.[60]

The Libyan uprising that followed Egypt, which now has turned into a full-fledged civil war, is replete with similar anecdotes and reports of the use of ICTs by Muammar Gaddafi and his regime. Gaddafi and his sons have used state media for propaganda, threats, and rants vowing to fight to "the end" or die as a martyr.[61]

---

[53] Will Englund and Leila Fadel, "Mubarek supporters confront protestors," *The Washington Post,* Feb. 3, 2011 at A1, A10-11.

[54] Roula Khalaf and Heba Saleh, "Mubarek defies calls to resign," *Financial Times,* Feb. 11, 2011 at A1; Roula Khalaf and Andrew England, "Jubilation turns to seething anger," *Financial Times,* Feb. 11, 2011 at 4; Craig Whitlock, "Mubarek refuses to step down," *The Washington Post,* Feb. 11, 2011 at A1, A8; Ernesto Londono and Leila Fadel, "For joyous throngs, history never materializes," *The Washington Post,* Feb. 11, 2011 at A1, A9

[55] Roula Khalaf, "Dramatic twist unfolds as military steps in," *Financial Times,* Feb. 11, 2011 at 4.

[56] "The economic impact of shutting down the Internet and mobile phone services in Egypt," Organisation for Economic Co-operation and Development, Feb. 4, 2011, http://www.oecd.org/document/19/0,3746,en_2649_201185_47056659_1_1_1_1,00.html.

[57] Shaimaa Fayed, "Court orders Mubarek to pay fine of $33.5 million," *The Washington Post,* May 29, 2011 at A13.

[58] *See, e.g., Michael Peel, Andrew Englund, and Heba Saleh, "Egypt clashes intensify,"* Financial Times, Feb. 4, 2011 at 1.

[59] Andrew Parker and Jennifer Thompson, "UK Tackles Cairo over networks abuse allegations," *Financial Times,* Feb. 5-6, 2011 at 10.

[60] Mary Beth Sheridan, "Clinton warns other nations; Blocking Internet will backfire," *The Washington Post,* Feb. 16, 2011 at A11.

[61] *See, e.g.,* Sudarsan Raghavan, "Gaddafi vows to fight until 'the end,'" *The Washington Post,* Feb. 23, 2011 at A1, A8; Andrew England and Heba Saleh, "Desperate Gaddafi clings on," *Financial Times,* Feb.

WikiLeaks played a familiar role as leaked cables revealed the wealth amassed by the Gaddafi family over four decades of rule, fueling the anger of long-repressed Libyans.[62] Other cables revealed the depths of his manipulation and ability to marginalize all those around him, even his children.[63] Following Egypt's lead, Gaddafi also blocked Facebook and new websites, and then shutdown the Internet prior to broad attacks on protestors. Since Libya is a monopoly telecom provider, Gaddafi's ability to shut off communications is rather straight forward, but the effect of the shutdown was not felt as acutely as in Egypt or Tunisia because Libya's connectivity is much lower than either country.[64]

Following the entry of NATO into the Libyan conflict, numerous stories appeared regarding Gaddafi's use of television to try to convince the Libyan people that the NATO forces killing them instead of supporting them. In one instance, as journalists crowded around a baby who reportedly was killed by a NATO airstrike, a Reuters reporter was given a handwritten note that said the baby was the victim of a road accident.[65]

The Syrian uprisings most parallel those of Tunisia and Egypt. Images of Hamza al-Khateeb, a thirteen year-old boy who had been cruelly tortured at the hands of Syrian officials, were broadcast on Al-Jazeera. The resulting outrage caused him to become the face of the Syrian protests, similar to Khaled Said in Egypt and Mohammed Bouazizi in Tunisia.[66] The broadcast resulted in a call for "Children's Friday," to protest the deaths of children that had occurred since the protests began, and 50,000 protestors took to the streets. The Syrian government responded by shutting off Internet service.[67]

Bahrain's protests followed a similar pattern with Al-Jazeera coverage of the protestors followed by concessions from the government, the firing of ministers,[68] government shutdowns of the Internet,[69] and blockages of Facebook and websites.[70] Al-Jazeera showed the Saudi military entering Bahrain to help the government clamp down

---

22, 2011 at 1; Heba Saleh and Andrew England, "Gaddafi vows to fight to the death," *Financial Times*, Feb. 23, 2011 at 1.

[62] Michael Peel, "Fortune amassed in 41 years of tyranny," *Financial Times*, Feb. 23, 2011 at 1, 3.

[63] Craig Whitlock, "Behind eccentricities, Gaddafi has deft control," *The Washington Post*, Feb. 23, 2011 at A9.

[64] Steven J. Vaughan-Nichols, "Libya turns off the Internet and the Massacres begin," ZDnet.com, Feb. 20, 2011, http://www.zdnet.com/blog/networking/libya-turns-off-the-internet-and-the-massacres-begin/711.

[65] Simon Denver, "Libyan regime failing to prove its accusations against NATO," *The Washington Post*, June 7, 2011 at A8.

[66] Liz Sly, "Apparent torture of boy, 13, sparks protests in Syria," *The Washington Post*, May 30, 2011 at A12.

[67] Elizabeth Flock, Syria internet services shut down as protesters fill streets," *The Washington Post, blogPost*, June 3, 2011, http://www.washingtonpost.com/blogs/blogpost/post/syria-internet-services-shut-down-as-protesters-fill-streets/2011/06/03/AGtLwxHH_blog.html.

[68] Robin Wigglesworth and Simeon Kerr, "Bahrain rulers put 'all options' on the table," *Financial Times*, Feb. 26-27, 2011 at 3.

[69] Steven J. Vaughan-Nichols, "Libya turns off the Internet and the Massacres begin," ZDnet.com, Feb. 20, 2011, http://www.zdnet.com/blog/networking/libya-turns-off-the-internet-and-the-massacres-begin/711.

[70] "Internet Enemies 2011: Countries under surveillance – Bahrain," United Nations High Commission on Refugees, http://www.unhcr.org/refworld/country,,RSF,,BHR,,4d82268a21,0.html.
Bahrain: More websites and blogs blocked by authorities," Facebook,

on protestors,[71] but, as in Libya, much communication was cutoff prior to military assaults against peaceful demonstrators.[72]

In Yemen, the pattern was repeated, with the heightened prominence of Al-Jazeera. One protestor held up a sign written in English declaring that, "Al Jazeera is part of our revolution."[73] Foreign journalists were detained and deported,[74] and dramatic televised addresses from the president were aired, but this time from Saudi Arabia, where he had fled.[75]

Other governments are learning by watching how individuals' use of ICTs have trumped years of diplomatic efforts. Ezzedine Choukri, political science professor at the University of Cairo and former diplomat and UN advisor, succinctly noted:

> [T]he protests show the limits of foreign intervention. The regime had defeated eight years of pressure from the Bush administration (and ignored President Obama's more hesitant attempts) aimed at achieving modest reforms. Yet eight days of protests convinced Hosni Mubarek to accept far-reaching reforms and an end to his rule. International players figured in the script, but only in supporting roles.[76]

As Nobel laureate, Joseph Stiglitz recently noted, "On its own, Tunisia has done more to advance the cause of democracy than any of the west's military actions in the Middle East."[77]

The U.S. government is trying to leverage the power of ICTs to promote democracy. In early 2011, the U.S. State Department began using Twitter to send U.S. messages in Arabic and Farsi to the Middle East, calling the feed the "Department of State Arabic Media Hub."[78] It already has feeds in French and Spanish and will add Chinese, Hindi, and Russian Twitter feeds in the near future.[79] U.S. Secretary of State Hillary Clinton offered to personally answer online questions from Egyptians and received more than 6,500 responses. In responding to a fraction of the questions, she acknowledged that the U.S. government had to go beyond diplomatic exchanges and include grassroots movements in its formulation of foreign policy. If nothing else, the

---

[71] "Saudi soldiers sent into Bahrain," Al-Jazeera, Mar. 15, 2011, http://english.aljazeera.net/news/middleeast/2011/03/201131412492885 0647.html.

[72] Dan Goodin, "Internet use disrupted in Bahrain as protests turn bloody," *The Register*, Feb. 18, 2011, http://www.theregister.co.uk/2011/02/18/bahrain_internet_disruption/; Joe Weisenthal, "The Mideast Right Now: Internet Shutdown In Bahrain, Violence In Yemen, Media Blackout In Iran," *Business Insider,* Feb. 16, 2011, http://www.businessinsider.com/mideast-update-march-16-2011-2.

[73] "Yet more cracking down," *The Economist,* Mar. 19, 2011.

[74] *Id.*

[75] Michael Peel and Noah Browning, "President adds twist to Yemen crisis," *Financial Times,* July 8, 2011 at 6.

[76] Ezzedine Choukri, "All Arabs will hear my street corner chatter," *Financial Times,* Feb. 5-6, 2011 at 7.

[77] Joseph Stiglitz, "The Arab spring is at risk without aid now," *Financial Times,* May 26, 2011.

[78] Ed O'Keefe and James Buck, "State Dept. launches Arabic tweets," *The Washington Post,* Feb. 10, 2011 at A17; Mary Beth Sheridan, "Clinton warns other nations; Blocking Internet will backfire," *The Washington Post,* Feb. 16, 2011 at A11.

[79] "Hillary Clinton: Internet repression 'will fail,'" BBC, Feb. 15, 2011, http://www.bbc.co.uk/news/world-us-canada-12475829?.

online session highlighted the role that the Internet is sure to play in U.S. foreign policy.[80] More recently, President Obama held the U.S. government's first town hall meeting via Twitter on July 6, 2011, and will use the data gleaned from the feeds to better understand citizens' concerns and views.[81]

Although some local Chinese communist party officials, government propaganda personnel, and municipal police officials in China have begun to use microblogs,[82] restrictive governments like Iran and China reacted negatively to the uprisings. Iran, alarmed by the anger in Tahrir Square after Mubarek refused to step down, promptly put one of its most prominent opposition leaders and a couple of his colleagues under house arrest until the threat of a protest died down.[83] The crackdowns in Iran have been so swift and communication channels so restricted that the country has seemingly blocked individual empowerment through ICTs.

China, which has routinely blocked Facebook, YouTube, and Twitter,[84] also seemingly took preemptive actions to squelch any protests that may have been inspired by the Middle Eastern events. In response to an online call for a "jasmine revolution," Chinese authorities tracked and detained activists, disconnected some texting services, filtered related Internet searches,[85] and censored television programming.[86] Subsequently, the professional social networking site LinkedIn was blocked after a discussion group called "Jasmine Voice" was started, filtering terms were expanded and Western media chiefs were called to Beijing police headquarters and reminded of rules governing interviews.[87] A few weeks later, Google accused China of disrupting its Gmail service inside the country as part of its Jasmine blockages.[88] Virtual private network services, which are commonly used to circumvent government restrictions, also experienced blockages.[89]

As evidence that the Chinese government was rattled by the Middle Eastern uprisings, a long-time business executive and 17-year resident of China declared, "We have never seen this level of control in the time that I have been here, and I have been

---

[80] Mary Beth Sheridan, "Clinton fields online questions, criticism from young Egyptians," *The Washington Post*, Feb. 24, 2011, http://www.washingtonpost.com/wp-dyn/content/article/2011/02/23/AR2011022306812.html.

[81] Mary Beth Sheridan, "Clinton warns other nations; Blocking Internet will backfire," *The Washington Post*, Feb. 16, 2011 at A11.

[82] Keith B. Richburg, "In China, microblogs serve as free-speech forum," *The Washington Post*, Mar. 28, 2011 at A6.

[83] Thomas Erdbrink, "Security forces quell protests at several locations in Tehran," *The Washington Post*, Feb. 21, 2011 at A8; Thomas Erdbrink, "Iranian opposition leader under house arrest before possible protest," *The Washington Post*, Feb. 11, 2011 at A11.

[84] Sharon LaFraniere and David Barboza, "China Tightens Censorship of Electronic Communications," *The New York Times*, Mar. 22, 2011 at A4.

[85] "Officials crack down after protest threats," *The Washington Post*, Feb. 21, 2011 at A7.

[86] Kathrin Hille, "Beijing steps up censorship of TV," *Financial Times*, May 31, 2011 at 5.

[87] Keith B. Richburg, "Wary China tightening its Internet restrictions," *The Washington Post*, Feb. 26, 2011 at A6.

[88] Richard Waters and Kathrin Hills, "Google claims Beijing is disrupting e-mail service," *Financial Times*, Mar. 22, 2011 at 18.

[89] Loretta Chao, "Google Objects to China's Acts," *The Wall Street Journal*, Mar. 22, 2011 at B3.

here since the beginning of the Internet."[90] Chinese censorship and control of communications has reached such a peak that the China Digital Times website, run by a Chinese human rights activist out of Berkeley, California, has begun posting Chinese government directives and guidelines to the print media. Website editors receive instant messages and emails regarding censoring instructions.[91]

The path to freedom via ICTs may be rocky as more governments are likely to take a hard line at the first signs of unrest, fearful that events could escalate out of control as they did in Tunisia, Egypt, and Libya. There is no clearer evidence of this risk than BBC's report during the recent London riots that, "The [UK] government is exploring whether to turn off social networks or stop people from texting during times of social unrest."[92] Twitter refused to shut down accounts of London rioters.[93]

One of the most striking examples of government empowerment through ICTs is the indirect role played by the state-backed television channels, Al-Arabhiya (Saudi Arabia) and Al-Jazeera (Qatar). Although not as well known as Al-Jazeera, it was Al-Arabhiya that provided the first images of the Tunisian protestors and gave them a voice. The channel provided coverage throughout Tunisia and has been credited with helping to bring about the quick fall of former president Zein al-Abidine Ben Ali. Al-Jazeera was slower to report on the Egyptian uprisings than Al-Arabhiya, but as the conflict escalated, the station provided full coverage and asked the Egyptian authorities to allow it to openly cover the protests. As a consequence, Al-Jazeera's live channel was removed from the Nilesat platform used for public television to a different frequency. In addition, Al-Jazeera's offices in Cairo had its phone lines cut and experienced signal interference on its Arabic news channel.[94] This infuriated the activists, who responded by Twittering other links to watch Al-Jazeera's coverage, and Al-Jazeera used its own Twitter feed to advise viewers that it was switching to "clandestine" coverage and urged protestors to contribute to the news reporting by uploading pictures, videos, and news information to dedicated websites and phone lines.[95]

The shadow power of government-owned media is an important aspect of nation state empowerment from ICTs. It is beyond question that the Qatari government played an indirect role fueling the Arab Spring uprisings through its state-owned television channel, Al-Jazeera, and broadened global support for the protestors. The coverage has been credited with speeding the fall of the governments in Tunisia and Egypt.[96] Some claim that Al-Jazeera is a "foreign policy tool used by Qatar to expand the Gulf state's

---

[90] Sharon LaFraniere and David Barboza, "China Tightens Censorship of Electronic Communications," *The New York Times*, Mar. 22, 2011 at A4.

[91] Keith B. Richburg, "Chinese censorship, uncensored," *The Washington Post*, Apr. 13, 2011 at A6.

[92] Charlie Cooke, "Cameron's Conclusion: To Curb Speech?" *The National Review*, Aug. 11, 2011, http://www.nationalreview.com/blogs/print/274344.

[93] Emma Barnett, "London riots: Twitter says all tweets must continue to flow," *The Telegraph*, Aug. 9. 2011, http://www.telegraph.co.uk/technology/twitter/8691663/London-riots-Twitter-says-all-tweets-must-continue-to-flow.html.

[94] Abeer Allam, "Slow response from al-Jazeera signals Qatar's anxieties," *Financial Times*, Jan. 29-30, 2011 at 2.

[95] Abeer Allam, "Journalists lean on social media to gain vantage point," *Financial Times*, Jan. 31, 2011 at 2.

[96] See, e.g., Abeer Allam, "Revolts inspire Qataris to take on telecom group," *Financial Times*, July 20, 2011 at 6.

diplomatic ambitions."[97] This view is supported by U.S. diplomatic cables released by WikiLeaks indicating that Qatar uses the station to enhance its clout in regional affairs.[98] It also uses it for clout in global affairs. News reports have surmised that a recent script jointly written by the U.S., France, and Britain regarding military action against Gaddafi, was carefully drafted to keep the support of Al-Jazeera, which has supported the revolution but criticized U.S. aggression against Muslims in Afghanistan and Iraq.[99]

Saudi Arabia and Qatar are cautious, however, to limit the impact of state-owned Al-Jazeera and Al-Arabhiya to their countries' external affairs. While indirectly advancing the protestors' agendas for reform via Al-Jazeera, the Qatari government has simultaneously ensured that ICTs are not used to weaken their own government, which has no parliament, free press, or civil society.[100] Despite Al-Arabhiya's role in the Tunisian revolt, during the Egyptian uprising, Saudi Arabia repeatedly called Mubarek and expressed its "support and concern about Egypt's safety and security."[101] In addition, the head of Al-Arabhiya published a newspaper article pointing out that Mubarek did not address the concerns of his population, which included his son's succession to the presidency.[102]

THE CLASH OF IDEALISM AND REALISM

The idealism of the protesters in Tunisia, Egypt, and Libya is clashing with the realism of damaged economies, fractured diplomatic structures, and a stressed population. The youth that banked on ICTs to free them from tyranny are yearning to build lives that compare with those they view in democratic countries via the Internet. Donor organizations, however, operate in very traditional ways and funds are commonly allocated to infrastructure, health, civil society, and legal/judicial reform projects that have broad goals. Whether democracy can find root in these unstable environments and flourish remains to be seen. In part, it will depend upon whether Western nations can rise to the occasion and offer innovative forms of assistance to spur entrepreneurs, raise skill levels, incentivize changes in legal frameworks, and improve the business environment. Joseph Stiglitz has encouraged private sector companies to help ensure that the Arab Spring goes in the right direction:

> A second [idea] is enlisting companies (such as Google and Facebook, which played such an important role in sparking the revolution) to provide venture capital funds, innovation prizes, and so on to help create new small and medium enterprises.[103]

---

[97] Abeer Allam, "Slow response from al-Jazeera signals Qatar's anxieties," *Financial Times*, Jan. 29-30, 2011 at 2.
[98] *Id.*
[99] Roula Khalaf, "Al-Jazeera's support is key for allies," *Financial Times*, Mar. 21, 2011 at 3.
[100] Abeer Allam, "Revolts inspire Qataris to take on telecom group," *Financial Times*, July 20, 2011 at 6.
[101] Abeer Allam, "Journalists lean on social media to gain vantage point," *Financial Times*, Jan. 31, 2011 at 2.
[102] *Id.*
[103] Joseph Stiglitz, "The Arab spring is at risk without aid now," *Financial Times*, May 26, 2011.

Tunisia's interim prime minister was quick to realize the role of international assistance and has asked for funding to "protect the Tunisian experiment."[104] His words were prescient. Although Freedom House placed Tunisia among Iran, Myanmar, Cuba, and China as one of the most restrictive regimes for online communications, following the fall of Ben Ali, Tunisian authorities opened the Internet. Now, however, the Tunisian military has ordered censorship of Facebook pages calling for violence.[105]

The idealism of government secrets is also giving way to the realism that the Internet enables instant global dissemination. WikiLeaks played an important, if tangential, role in the Middle East uprisings. Michael Peel of the Financial Times noted the breadth of their impact when he wrote:

> The documents included claims of Saudi Arabia urging a US military strike on Iran, of Yemen's government taking the blame for American missile attacks – and in a possible catalyst of recent events, of allegations of large-scale graft around the ruling family of Tunisia.[106]

Silicon Valley is also having to sort through the idealism of a connected social network and the realism of government's power to hunt down the users of the network and prosecute them. Google has been a forerunner in calling for a free and open Internet, and Eric Schmidt publicly announced his support for Wael Ghonim.[107] Facebook has five million subscribers in Egypt, the most of an Arab country, and seemed caught off guard by the sudden limelight it caught from its role in the Egyptian and Tunisian uprisings. David Kirkpatrick, author of *The Facebook Effect,* noted that the company was internally struggling with the question of whether other authoritarian governments would impose restrictions or bans on the use of Facebook out of fear that it could be used as a tool to foment unrest. The company, which requires users to use their actual names, also is discussing whether to allow political activists some measure of anonymity. Facebook has supported activists by providing technical assistance to counter government intrusions against Facebook users, such as when Tunisian officials infected Facebook users' computers with malware to collect user IDs and passwords. The company rerouted Tunisia's Facebook traffic to a site that could not be reached by local Internet Service Providers. Facebook's chief security officer carefully walked a neutral line in discussing this assistance:

> Certainly there's a political context to the particular circumstance in Tunisia, but from Facebook's perspective, what happened was a security problem that required a technological solution: we prevented an exploit that was making Facebook accounts vulnerable and restored the integrity of the compromised accounts. We

[104] Eileen Byrne, "Tunisian leader appeals for aid to safeguard democracy," *Financial Times,* Feb. 7, 2011 at 2.

[105] Steve Stecklow, "Openness of Internet is Being Tested in Tunisia," *The Wall Street Journal,* July 8-10, 2011 at 9.

[106] Michael Peel, "Tunisia sets the tone for the year," *Financial Times,* Special Report, The World 2011, Jan. 26, 2011 at 10.

[107] Jenna Wortham, "Google Praises Executive's Role in Egypt Revolt, *The New York Times,* Feb. 15, 2011, http://www.nytimes.com/2011/02/16/world/middleeast/16google.html.

would have taken the same approach to any situation where we saw a systematic exploit.[108]

Silicon Valley companies, Twitter, Google, and SayNow, have been more proactive and offered Egyptians with an alternative when Internet service is disrupted. The companies set up a service called Tweet2Speak that enables people to call a phone number and leave a message, which is then transcribed and posted on the Internet.[109] The demographics of the Middle East are a factor that no one can afford to ignore, whether a Silicon Valley company, a government leader or policymaker, or a donor organization. Young people aged 15-29 make up the highest proportion of the population in most countries in the Middle Eastern region.[110] They are intense users of ICTs, less numb to the cruel realities of authoritarian regimes, and more idealistic than their elders.

One cannot consider individual freedoms, however, without also taking into consideration national security interests. All countries are creatively exploring the use of ICTs for purposes of national security and defense. NATO, for example, has used drones to strike Muammar Gaddafi's mobile rocket launchers when traditional weapons failed. The drones are operated remotely by Central Intelligence Agency (CIA) operatives who are located thousands of miles from their targets. Military leaders argue that the drones actually help them comply with the laws of armed conflict (LOAC) by pinpointing attacks and minimizing damage and harm to civilians. Critics call for drone operations to be transferred from the CIA to the U.S. Department of Defense (DoD) because the intelligence community is not within the military chain of command, is not used to complying with the LOAC, and are less inclined to be accountable to or compensate civilians killed in attacks.[111] There is reason to agree with both sides of the discussion, but the broader question of how far intelligence organizations should insert themselves in military operations will become increasingly important as ICTs are used as weapons. Taking a positive step forward, however, the U.S. Secretary of the Air Force recently announced a new Air Force Policy Directive that prescribes guidance and procedures for the review of Air Force weapons and cyber capabilities to ensure legality under domestic and international law, including the LOAC.[112]

The increasing role of satellites in national security is another aspect of ICTs that must be considered when evaluating individual and nation state empowerment. The U.S. military has developed a space strategy that calls for better protection of its satellites and enhanced capabilities to use them as weapons. "It's a domain like air, land and sea….It's become a critical part in every other domain," noted Gen. Kevin Chilton, who previously

---

[108] Cecelia Kang and Ian Shapira, "Facebook's Egypt conundrum," *The Washington Post,* Feb. 3, 2011 at A13.

[109] *Id.*

[110] Alicia Parlapiano and Laura Stanton, *The Washington Post,* Feb. 16, 2011 at A12.

[111] "Drones and the man," *The Economist,* July 30, 2011 at 10.

[112] "Legal Review of Weapons and Cyber Capabilities," Secretary of the U.S. Air Force, Air Force Instruction 51-402, http://www.e-publishing.af.mil/?rdoFormPub=rdoPub&txtSearchWord=51-402&btnG.x=6&btnG.y=7&client=AFPW_EPubs&proxystylesheet=AFPW_EPubs&ie=UTF-8&oe=UTF-8&output=xml_no_dtd&site=AFPW_EPubs.

led U.S. Strategic Command.[113] Secretary of State Hillary Clinton has stated that the U.S. needs to protect global positioning data, missile warning system information, and communications with fighters, unmanned drones, and other surveillance systems. General James Cartwright, former vice chairman of the Joint Chiefs of Staff, has called for a multilateral effort to develop acceptable rules for the use and movement of satellites. This is not likely to be an easy path, since China, Iran, and other countries have a significant interest in space capabilities. The world was stunned when China destroyed a satellite with a ground-based missile. It was the first country to do so.[114]

As countries draw protections around their Internet linkage points, satellites, and broadcast networks, individuals may benefit from the enhanced stability of these mediums or be more quickly shut out. The latter is more likely. The risk for individual rights cannot be pushed aside as these important issues are debated. The economic impact from government controls over media also looms large. According to a McKinsey study, the Internet has generated as much growth over the past 15 years as the Industrial Revolution did in 50 years.[115] It is also an opportunity issue. The McKinsey study also indicates that the Internet has been responsible for 21 percent of all growth in mature economies, and it has created 2.6 jobs for every one that it has eliminated.[116] As Karen Kornbluh, U.S. ambassador to the OECD, and Daniel Weitzner, deputy chief technology officer in the White House, noted in a recent editorial, "It's [the Internet's] power to generate innovation is rivaled only by its potential to help people realize their rights and democratic aspirations."[117]

Jeffrey Ghannam, author of a critical report on social media in the Arab world, correctly notes that, "The convergence of social media, satellite networks, and traditional media proved pivotal to spreading the protestors' messages."[118] If governments determine that these media must have controls to ensure national security, the same controls can be used by leaders to ensure their own power —at the expense of individual freedoms.

The U.S. Government has gone out of its way to promote freedom of expression on the Internet and to condemn China and other countries who have suppressed open access to the Internet and limited communication of their people.[119] The U.S. also, however, must consider when its assistance may indirectly contribute to nation state actions against the rights of its own people. For example, the U.S. Air Force recently solicited proposals to provide the Iraqi government with a system that will enable it to wiretap and store mobile and landline phone conversations, data transmissions, and text messages and boost Iraq's capabilities to counter "criminal organizations and

---

[113] Lolita C. Baldor, "Pentagon strategy stresses the importance of satellites," *The Washington Post*, Feb. 21, 2011 at A13.
[114] *Id.*
[115] *Internet matters: The Net's sweeping impact on jobs, growth, and prosperity,"* McKinsey Global Institute, May 2011, http://www.mckinsey.com/mgi/publications/internet_matters/index.asp.
[116] *Id.*
[117] Karen Kornbluh and Daniel J. Weitzner, "Preserving a worldwide Internet," *The Washington Post*, July 15, 2011 at A15.
[118] Jeffrey Ghannam, "Freedom, beyond 140 characters," *The Washington Post*, Feb. 20, 2011 at B3.
[119] Joby Warrick, "U.S. criticizes global limits on Web access," *The Washington Post*, Apr. 9, 2011 at A2; Mary Beth Sheridan, "Clinton warns other nations; Blocking Internet will backfire," *The Washington Post*, Feb. 16, 2011 at A11.

insurgencies." In addition, the system will have advanced tracking capabilities and be able to develop a database of targets and their relationships. The Air Force has noted that Iraq's legal system has strong surveillance laws that require law enforcement to obtain a warrant before intercepting private communications. The *Washington Post* reported that the U.S. installed a similar system in Afghanistan three years ago to target drug rings.[120] What is worrisome is how these governments will use these systems, considering that their legal frameworks are fragile at best.

Against these concerns, however, lies a bright promise of how the Internet, through individuals, can provide a channel for peace that militaries and diplomats have failed to achieve. For example, Facebook page, Facebook.com/yalaYL, set up by a former Israeli diplomat, is designed to bring Israelis, Palestinians, and other Arabs together to discuss topics of interest, including regional peace. It has welcome messages from Israeli president Shimon Peres, Palestinian Authority president Mahmoud Abbas, Tony Blair, and actress Sharon Stone. By allowing people to meet virtually and discuss issues of their choice without interference or censorship, they are learning about common problems, sharing frustrations, and yearning to learn from one another. Its first month had 91,000 views and 60% of its 22,500 active users are Arabs: Palestinians, Egyptians, Jordanians, Tunisians, Moroccans, Lebanese, and Saudis.[121] As the uprisings have demonstrated, perhaps more progress can be made and more peace can be achieved through individuals and ICTs than years of diplomatic efforts.

CONCLUSION

The use of ICTs in the Arab Spring uprisings has highlighted the ways they empower the individual and boost the capabilities of the nation state. The exploitation and shut down of Internet communications by regimes across the Middle East has raised serious questions about whether these governments have violated their citizens' right to freedom of expression and access to information. Human rights groups, providers, and governments have condemned the shutdowns and declared future shutdowns must not occur. Such statements have been made, however, without the benefit of a full analysis of the rights of both sides: individuals and nation states. These proclamations also do not take into account competing legal issues, such as the right of governments to take action curbing communications to block those that would incite lawless actions.

The urgency of this issue cannot be understated; the temptation for governments to shut off communications is great, as evidenced by threats from UK leaders to cut off social networking communications during the London riots[122] and the actual shutdown of cellular service by the San Francisco transportation authorities in an attempt to disrupt a protest.[123] The role of providers also must be examined, particularly with respect to how

---

[120] "U.S. plans to give wiretapping system to Baghdad," *The Washington Post,* July 31, 2011, at A15.

[121] "Virtual Bridge Allows Strangers in Mideast to Seem Less Strange," *The New York Times,* July 10, 2011 at 9.

[122] Doug Gross, "In wake of riots, British PM proposes social media ban," CNN, Aug. 11, 2011, http://articles.cnn.com/2011-08-11/tech/london.riots.social.media_1_social-media-facebook-and-twitter-blackberrys?_s=PM:TECH.

[123] Elinor Mills, "S.F. subway muzzles cell service during protest," cnet.com, Aug. 12, 2011, http://news.cnet.com/8301-27080_3-20091822-245/s.f-subway-muzzles-cell-service-during-protest/.

they respond to government requests for information on users, whether they allow users anonymity to protect their rights and guard against persecution, and whether they refuse or comply with shutdown orders.

Since, "armies will increasingly fight with machines, not men,"[124] and civilians will increasingly be involved in modern day conflicts through their use of ICTs, it is important that military leaders, government officials, policymakers, scientists and technologies, and human rights organizations come together to examine the array of issues that have surfaced in the spring of 2011 and develop balanced principles. We must not lose a generational opportunity to advance democracy and the freedom of people around the globe. When dealing with the Internet, minutes matter and the urgency of the current situation cannot be overstated. New policies, laws, multilateral treaties, scientific advancements, and technological solutions will be required to ensure that government and individual use of ICTs is balanced and does not upset a delicate world order.

---

[124] "Drones and the man," *The Economist,* July 30, 2011 at 10.

A SECURE INTERNET TO PROMOTE WORLD-CLASS SCIENCE IN DEVELOPING COUNTRIES

WILLIAM A. BARLETTA
U.S. Particle Accelerator School, Department of Physics, Massachusetts Institute of Technology, Cambridge, USA

My comments derive from the confluence of multiple perspectives:

1. Director of a national graduate school dedicated to international outreach,
2. Former manager of an organization developing technologies that underlie major instruments of scientific inquiry,
3. Editor-in-chief of a prestigious international journal devoted to scientific instruments and methods,
4. Active research faculty at a premier research university,
5. Firm believer that talent and inventiveness in science and technology know no boundaries.

EGYPT

On June 2, 2011, Nobel laureate Ahmed Zewail told a cheering crowd in Cairo's Tahrir Square that Egypt's interim government had approved his dream for a "Cairo Caltech"[1] costing 2B$. Would such a large sum be better spent improving the general infrastructure of science education in Egypt?

In the past three years in my introductory accelerator physics course, I have provided scholarships to and taught a dozen Egyptian students motivated by an Egyptian initiative to create a new accelerator center. These students were bright, motivated, and hard-working but handicapped by deficiencies in their preparation at Egyptian universities. What kind of government investment would serve them best?

JORDAN

A decade ago, the Jordanian government gave funds for a building to house the first light source in the Middle East, SESAME. The original plan for SESAME was to install the old BESSY-1 storage ring. Germany pledged that equipment to the project. The original plan was scrapped in favor of building a ring with new equipment that would have a longer lifetime as well as some growth potential. Implementation of the storage ring has struggled, although it has brought together funding from countries with otherwise strong animosities toward each other, such as Israel and Iran. SESAME has enthusiastic support from international scientific bodies such as IUPAP and CERN. Will this project really come to fruition? When it will it be sufficient to produce world-class science? Will there be sufficient highly trained users from the Middle East to make the best use of the facility?

---

[1] Science 15 July 2 011: 333 no. 6040 p. 280.

TURKEY

The nascent Turkish Accelerator Center (TAC) aspires to build a medical accelerator a light source and even a spallation neutron source. Yet there is almost no native expertise in accelerators, although there are some first rate developers of nuclear instrumentation. Is the TAC merely wishful thinking? Or is it squandering precious resources?

QUESTIONS

Do Jordan, Turkey, Armenia, Poland, Mexico, South Africa, Thailand—the list goes on—need a synchrotron?

Is the suggestion by the DG of a major northern European lab to power new mega-science in Europe with solar energy from North Africa an example of a new sustainable colonialism? Do Libya, Tunisia, and Algeria get to control their own beamlines built at no added cost to those governments? When would a Libyan or Algerian get to be the DG in a northern European lab?

OPPORTUNITY

The rapid development of information technologies offers new opportunities to change the paradigm of the development science and technology in developing countries by leveraging high impact potential, "university-scale" technologies such as lasers, high quality ion beams, and neutron sources. From a strategic point of view the potential is to avoid the rapid divergence of scientific capabilities (especially physical sciences) between North and South. The most insidious aspect of this divergence is the loss of young scientific talent, who seek the opportunity to contribute at the highest level to the modern scientific enterprise. An incentive for the North (U.S. Europe and Japan) to promote such opportunities would be to make available analogous opportunities for major universities in the North.

What we see in the U.S., Europe and Japan is a continuing thrust to build ever-larger scientific facilities at the one-to-twenty billion-dollar scale. Be they for neutron science, studies of astrophysical matter, fusion energy R&D, or ultra-fast dynamical processes, these facilities are already unique on a regional scale. In the U.S. such facilities are the exemplars (and even the precipitators) of the decline of the infrastructure of the physical sciences at major universities and subsequent accumulation of facilities in the Department of Energy's national laboratories. At the extreme, for high energy physics, the scale of future facilities has become so large (30—100 km) that globally only one such scientific tool can be built. It is no embarrassment for developing countries to "buy a share" in a facility located elsewhere. Even the U.S. can no longer afford a machine for first-class high energy physics. International science without borders is the only credible and viable approach.

But how is sharing done with the developing world at the human scale? How do developing countries ensure that they not irreversibly left decades behind? The approach I suggest is twofold: First, identify opportunities to create university-scale facilities with the potential to yield roughly 20% to 50% of the science at a hundredth to a thousandth the cost. This approach is essential to develop the native, human infrastructure of science.

To this end small, powerful neutron tubes such as those that were developed at Berkeley or commercial table-top lasers to make GeV electron beams become more than scientific tools for "small science" condensed matter physics and chemistry. Small facilities for accelerator mass spectrometry would support many users from a wide range of disciplines. At the few million Euro scale they enable cutting edge science in research arenas that would otherwise be the exclusive domain of the North.

Still, just as the North can afford only one few-billion dollar facility, small developing countries may only be able to afford one or two few-million-dollar facilities, the most constraining factor being the scarce human capital required.

Here is the entry point for information technologies—to develop scientific human capital and to share in the international scientific enterprise in a coherent fashion. Major universities and scientific laboratories throughout the world are creating tools such as nano-positioning robotics, computational grids, and Internet co-laboratory software and hardware that will make virtual laboratories a vibrant reality. Doing so means much more than promoting active collaboration which is enabled and greatly facilitated by digital telecommunication. It means more than the sharing and analysis of scientific data on a real time basis. It means continuing telepresence in the monitoring, control and acquisition of these data. It means the secure multicast of data to assure that precious information assets are not lost either by accident or malicious design. It means the real-time sharing of the visualization of data as they are taken.

Virtual laboratories must succeed in these tasks while assuring the centralized security and safety of scarce resources at the same level that we demand in the U.S., Japan and Europe. What are the basic elements of a security culture[2] for nascent national co-laboratories?

<u>Security Program</u>
Security Organization—to build and maintain the IT security organization design. This includes composition, reporting structures, roles and responsibilities, skills, experience, and resource management.
Security Audits—activities to ensure that, periodically, all relevant objects are to be assessed against security requirements.

<u>Security Management</u>
Ownership of information—establishes an overview of assets and the security properties of these assets.
Operations and monitoring—processes and procedures to manage and monitor security architecture, as well as incident response to ensure continuous compliance to security requirements.

<u>Security Policy</u>
Security Policies—formal, prescriptive directives, policies, standards, guidelines, and procedures for implementing security requirements throughout the organization.

---

[2] This listing of characteristics is adapted from a talk at the Remote Operations Workshop Shelter Island, 17 September 2002 By Hans Frese, DESY Hamburg.

Security Agreements—formal compliance agreements with parties to ensure secure operations.

User Management
Access security—activities to implement user privileges and duties within the IT infrastructure.
Personnel related security—activities that establish and maintain awareness and education of staff.

Information Asset Security
System and network management—activities that embed security in the IT management processes.
System development and maintenance—activities that integrate secure design principles and implement functionality into information systems.

Protection and continuity
Physical security and security of environment—activities to ensure the mitigation of physical threats to the IT facilities.
Contingency Planning—necessary plans and procedures to enable recovery from major disruptions.

As virtual laboratories reach across national boundaries, all parties must come to agreement about legal norms that govern exchange and control of information and information technology. Once a country's network of scientific activity and scientist-to-scientist links are established, buying into major facilities at 2 to 4 million Euros per year makes sense.

World-class science is international and requires means for scientists from developing countries to come together with those from more privileged countries to study together, work together, and discuss together. Italy has been at the forefront in outreach here at Erice and at the ICTP in Trieste. Information technology can now help expand these efforts of scientific outreach and development a hundredfold:

- EVO—The worldwide collaboration network hosted at Caltech
- Internet simulcast of major scientific conferences
- The Citizen Cyberscience Center at CERN
- Co-laboratory tools across time zones for collaborations that never sleep
- Web-based study tools

With such tools in hand, developing countries will be fully prepared to ask for substantial participation in the global enterprises that the U.S., Europe and Japan envision to create single large facilities which are actively controlled in an integrated manner from multiple laboratories around the world. In 10 to 20 years they can aspire to full partnership.

That is the promise. What is needed is the commitment and perseverance year after year:

1. From developing countries to identify and nurture the very best developing scientists and to make the investment in and the commitment to these scientists that will keep them in their countries;
2. From the North the willingness to share high leverage, cutting edge technologies;
3. For both to break down the sociological barriers to truly nationwide and global laboratories through telepresence.
4. For everyone to work toward a culture of global internet security reinforced by a harmonious legal framework.

DEPLOYMENT AND IMPACT OF ICTS IN DEVELOPING COUNTRIES

DR. HAMADOUN TOURÉ
Secretary General, International Telecommnications Union
Geneva, Switzerland

Presented by Jody R. Westby in Dr. Toure's absence

Every country is now critically dependent on technology for commerce, finance, healthcare, emergency services, food distribution and more. Loss of vital networks would quickly cripple any nation—and none is immune to cyber attack. The pre-eminence of information and communication technologies (ICTs) and the interconnectedness of developing technologies are thus shaping a new world order, one that calls for collaboration on new issues to ensure stability. A fundamental role of ITU following the World Summit on the Information Society (WSIS) and the 2006 ITU Plenipotentiary Conference is to build confidence and security in the use of ICTs. Heads of States and government and other global leaders participating in WSIS, as well as ITU Member States, entrusted ITU to take concrete steps towards curbing the threats and insecurities related to the information society. To fulfill this mandate, ITU Secretary-General Dr Hamadoun I. Touré launched the Global Cybersecurity Agenda (GCA) as a framework for international cooperation in 2007.

The GCA aims to enhance confidence and security in the information society. It is designed for cooperation and efficiency, encouraging collaboration between all relevant stakeholders and building on existing initiatives to avoid duplicating efforts. The GCA is the first truly global multi-stakeholder and public-private alliance against cyberthreats. In 2008, ITU and the International Multilateral Partnership Against Cyber Threats (IMPACT) formally entered into a Memorandum of Understanding, after which IMPACT's state-of-the-art headquarters in Cyberjaya, Malaysia, became the physical home of the GCA. IMPACT is an international public-private initiative dedicated to enhancing the global community's capacity to prevent, defend and respond to cyberthreats. This collaboration provides ITU's 192 Member States and others with the expertise, facilities and resources to effectively enhance the global community's capability and capacity to prevent, defend against and respond to cyberthreats. Since its launch, the GCA has attracted the support and recognition of leaders and cybersecurity experts around the world. H.E. Dr. Óscar Arias Sánchez, former President of the Republic of Costa Rica and Nobel Peace Laureate, and H.E. Blaise Compaoré, President of Burkina Faso, are both Patrons of the GCA.

The GCA has fostered initiatives such as Child Online Protection (COP), the Cybersecurity Gateway and, through its partnership with IMPACT and with the support of leading global players, is currently deploying cybersecurity solutions to countries around the world.

CYBER REPRESSION: GOING WORSE. WHAT CAN BE DONE?

Ambassador Henning Wegener

Freedom of opinion and free access to information have, throughout history, been key elements in building civilized societies. They are an indispensable part of human rights and civil liberties, and are consequently centrepieces of almost all modern constitutions. Indeed, the freedom of the individual to acquire information, hold and communicate opinions could serve as a yardstick of human progress. On the other side, the definition of the limits which this principal freedom must undergo for reasons of public security, decency and *ordre public* have always been an intrinsic element of internal political debate, a permanent and necessary effort in the quest for reconciling and optimizing both individual liberty and public interest.

Government censorship in terms of systematically overstepping these limits and exercising close control over public opinion and exchange of views, mainly in respect of printed material, is a painful but recurrent part of human history, and has again and again triggered battles for the freedom of the mind. Government Internet filtering, the limits to the restriction of freedom of expression that should be observed, the balances that should be struck, and the role of the IT industry in providing the technical underpinnings for Internet control all touch on delicate issues of national sovereignty.

Attempts to counter government secrecy and empower the individual have been embodied in disclosures of government information by Wikileaks and hackers. Protections for whistleblowers may face government opposition if national security information is at issue, even if the disclosure revealed wrongdoing. The balance between transparency of government operations and protection of national security and public interests has created another dilemma regarding the use of ICTs.

NETWORK CONVERGENCE V. INTERNET KILL SWITCH

The Honorable John G. Grimes (former Assistant Secretary of U.S. Department of Defense and CIO

As the Arab Spring has progressed, the world has witnessed the shutdown of the Internet in Egypt and Syria and highly tightened controls in other countries, such as China and Iran. The ability of the Internet architecture to withstand government shutdowns has been analyzed by a non-profit research institute that supports operations and analysis in the areas of Internet traffic exchange, routing economics, and global network development. The scientific community needs to study the capabilities of ICTs and explore avenues for ensuring minimum essential communications to all countries. The Cultural Planetary Emergency will deepen lest the global connected network remain viable and capable of supporting all forms of communication in a converged environment.

TECHNOLOGY AS EMPOWERMENT OF PERSON AND STATE

Dr. Jody R. Westby, CEO, Global Cyber Risk LLC

ICTs have empowered the individual and increased the power of the nation state. The Cultural Planetary Emergency that we are facing in the Middle East and North Africa and the role that science and technology (S&T) is playing in these uprisings is changing the cultures of these countries and others around the globe. The unrest was ignited by the young fruit merchant who set himself on fire in Tunisia following police action against him, and it spread to Egypt after a 28-year-old Egyptian man named Khaled Said died from a beating by police. The unrest "went viral" when the young Google executive, Wael Ghomein, dedicated a Facebook page to Mr. Said that spread globally in a very brief period of time and inspired the Egyptians' ouster of Mr. Mubarek.

The use of the Internet, Facebook, and Twitter has been instrumental in the free dissemination of information among protestors across the Middle East and facilitated their ability to organize. These technologies have become a fount of courage, hope, and inspiration and are fueling the determination of these people to risk their lives for freedom. Beyond these beneficial uses, however, science-derived technology has been used by ruthless leaders to attack peaceful protestors and to stop the flow of communications.

To counter communication blockages, the major media channels have relied heavily upon scientific-derived technologies, such as satellite and Google maps to broadcast and report events, even when heavy censorship has been in place. This ability to bring instant information and visual depictions to the global population helped protect the human rights of the protestors, as images of brutality against unarmed citizens has resulted in condemnation by leaders of nation states and international action in the United Nations. The interests of the nation state against the rights of the individual have collided, with ICTs being the tool of chose to assert power on both sides.

THE ROLE OF SCIENCE IN ADDRESSING CULTURAL PLANETARY EMERGENCY

Dr. William Barletta, MIT and UCLA

WFS promotes international collaboration in science and technology between scientists and researchers from all parts of the world—North, South, East and West. The Federation and its members strive towards an ideal of free exchange of information, where scientific discoveries and advances are no longer restricted to a select few. The aim is to share this knowledge among the people of all nations, so that everyone may experience the benefits of the progress of science. The scientific community can help address the cultural planetary emergency by utilizing ICTs to enable scientists from around the globe to share in usage of research laboratories and facilities and to collaborate on solving global issues.

ICTs can also be used to advance scientific curricula in developing countries, break down cultural barriers, and promote discussion and engagement on the Cultural Planetary Emergency. In the field of information security, for example, international collaboration on research and development (R&D) projects is critical, yet governments typically refuse to fund efforts that involve foreign participation and only provide minimal funding for travel and meetings. The scientific community must come together and advocate the utilization of ICTs to advance global science and global scientists who can work from any country to address the Cultural Planetary Emergency.

SESSION 9

FOOD, SOIL & MEDICINE

FOCUS: GREENHOUSE GASES CONSEQUENCES AND EVIDENCE-BASED THIRD MILLENNIUM MEDICINE

THE CHEMICAL TIME BOMB CONCEPT REVISITED IN THE CONTEXT OF CLIMATE CHANGE AND OTHER RECENT DEVELOPMENTS

WILLIAM M. STIGLIANI
UNI Center for Energy & Environmental Education, University of Northern Iowa, Cedar Falls, USA

INTRODUCTION: FLOWS OF HEAVY METALS IN THE BIOSPHERE[1]

As conceptualized by Salomons (1995), one may visualize the transport of heavy metals in the environment as a wheel in which the metal moves with relentless progression from the hub to the periphery (left side of Figure 1). Heavy metals released to the air or aqueous systems (center and inner circle of the wheel, respectively) are transferred to soils and sediments (outer circle). The lifetimes of metals in air are on the order of days, and in water they are months to years. In contrast, the residence times of heavy metals in soils and freshwater and estuarine sediments can be much longer, on the order of decades to centuries or more. Eventually, through erosion and resuspension processes, the metals are transferred from the land to marine and deep ocean sediments, their ultimate sink. One may view this transport as a natural detoxification process by which air, soils, freshwaters and their sediments purge themselves of pollutants and transfer them for burial, and out of harm's way, in the sediments of the deep oceans. This voyage, however, from the terrestrial biosphere to deposition in ocean sediments, can take thousands of years or longer. The cycle is renewed though crustal uplifting when ocean sediments are transformed into new land forms, but this process takes millions of years. Hence, on timescales important to human concerns, soils and freshwater and estuarine sediments can be viewed as the ultimate sinks for heavy metals. Typically, well under 1% of the metals released to the environment reside in air and water. The remainder accumulates in soils and sediments

Numerous anthropogenic sources contribute to flows of heavy metals to the environment, including mining activities, direct industrial discharges, atmospheric deposition, corrosion of building materials and traffic-related emissions in urban areas, erosion and storm runoff from agricultural soils and paved urban areas, and leaks from municipal landfills and toxic waste dumps. Another important, sometimes widely overlooked source is natural deposits of heavy metals, which can be disturbed and mobilized unintentionally through human activities.

[1] This introduction pertains to heavy metals, but also considered in this paper are metalloids such as arsenic, inorganic pollutants such as phosphates and nitrates, and persistent organic compounds such as PCBs.

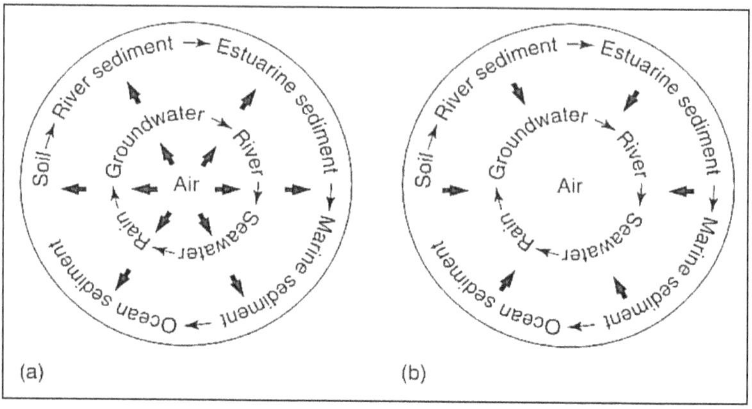

Fig. 1: Wheels represent the transport of pollutants in the environment. (a) Shows transport from air to soil and sediments; (b) shows the reverse transport from soils and sediments to mobilized pollutants in the biosphere. Source: Salomons (1995).

It is not total accumulation of a heavy metal in a soil or sediment, however, that is important, but rather its bioavailability. Bioavailability is mostly determined by the water-soluble fraction of the accumulated chemical in soil-water solution or in pore-water in sediments, while the adsorbed, insoluble phase is not available for biological uptake. In relatively unpolluted soils and sediments, the soluble phase is typically only a small fraction of total accumulation. The question raised in this paper and illustrated on the right side of Figure 1 is whether, and under what conditions, soils and sediments serving as sinks for toxic heavy metals can reverse their role and become important sources of heavy metal pollution in drinking water, crops and vegetation, and aquatic and terrestrial wildlife? We may think of such a reversal as a retoxification process in the biosphere.

THE CHEMICAL TIME BOMB CONCEPT

The Chemical Time Bomb (CTB) concept addresses the varied mechanisms by which such retoxification may occur. It was originally formulated by Stigliani (1988), and resulted in a five-year collaborative effort between the International Institute for Applied Systems Analysis (IIASA) in Laxenburg, Austria, the National Institute of Public Health and Environmental Protection in Bilthoven, the Netherlands, and the Netherlands Ministry of Housing, Physical Planning and Environment. The findings of the ensuing studies are provided in Stigliani (2002, 1996, 1995), Salomons and Stigliani (1995), Stigliani and Anderberg (1994), Stigliani et al. (1994), Stigliani (1993), Stigliani and Salomons (1993), ter Meulen et al. (1993), Stigliani and Jaffe (1993), Stigliani et al. (1993a,b), Hesterberg et al. (1992), Stigliani (1991), and Stigliani et al. (1991).

A Chemical Time Bomb is defined as *an unforeseen chain of events resulting in the delayed and sudden occurrence of harmful effects due to the mobilization or chemical transformation of chemicals stored in soils and sediments in response to saturation or alteration under certain environmental conditions* (adapted from Stigliani, 1991). The idea of chemical saturation of stored pollutants in soils is illustrated in Figure 2a. It shows that an added pollutant is partitioned between an adsorbed ("sorbed") solid state (Q) and a dissolved aqueous state (C). At an early stage of pollution, most of the pollutant remains in the sorbed phase, illustrated by the large value of the ratio of $\Delta Q_1/\Delta C_1$. As pollution continues over time, the ratio of the sorbed to mobile pollutant diminishes, as shown by the lower value of $\Delta Q_2/\Delta C_2$. Thus, a pollutant added at a constant rate over time will cause the fraction that goes into solution to become increasingly larger. The figure also shows a "maximum soil sorption capacity." This is the point at which virtually all of the input pollutant goes to the soluble phase because the soil is saturated with respect to its ability to adsorb it.

A second mode of saturation, and more important for the CTB concept, is illustrated in Figure 2b. This mechanism involves a change in a "capacity controlling property" of a polluted soil or sediment that reduces its maximum capacity from state 1 to state 2. When this occurs, a significant fraction of the sorbed chemical from state 1 (indicated by "partially filled capacity 1" on the upper curve in the figure) is desorbed as the system reaches a new equilibrium defined by state 2 (indicated by "partially filled capacity 2" on the lower curve in the figure). The amount desorbed corresponds to the downward arrow between partially filled states 1 and 2. Reaching saturation in soils via the mechanism in Figure 2a can take a long time. However, saturation can occur quickly and unexpectedly via the process indicated by Figure 2b.

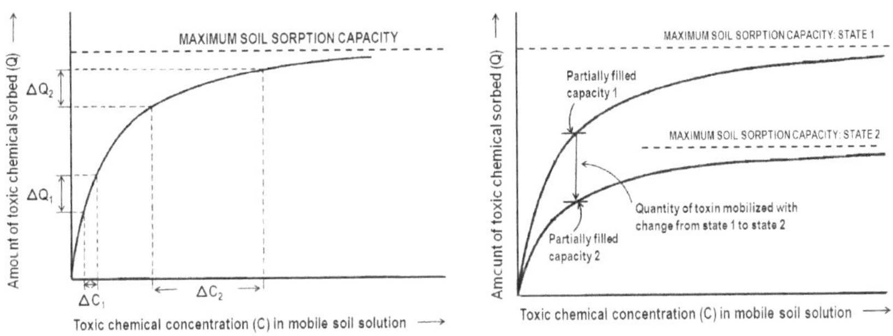

(a) **(b)**

Fig. 2: (a) Partitioning of a toxic chemical between its sorbed (Q) and insoluble (C) states; (b) effect of changing a capacity controlling property of a soil or sediment on mobilizing a stored chemical. Source: Hesterberg et al. (1992).

The story of "arsenic and old apple orchards" in Washington State, USA (Creger and Peryea, 1994; Peryea, 1991) illustrates how the second mode can operate. Acid lead arsenate ($PbHAsO_4$) was the preferred pesticide for insect control in fruit tree orchards before the introduction of DDT in 1947. Frequent high-rate applications led to substantial Pb and As accumulation in orchard topsoils. Arsenate (AsO_4^{3-}) was immobilized in the soil by adsorption to the surfaces of soil particles through ligand exchange with hydroxides (OH^-), hydrous metal-oxide minerals (Fe/Mn oxides), phyllosilicate clays, and soil carbonates. From the 1950s onwards, fruit growers began applying phosphate (PO_4^{-3}) fertilizer to their orchards to increase yield. This had the unexpected result of mobilizing AsO_4^{3-} with subsequent uptake through the roots and into the fruit. This occurred because phosphate adsorbed more strongly to the soil particles than arsenate, resulting in an exchange between these two species such that PO_4^{3-} became adsorbed and AsO_4^{3-} was mobilized; i.e.,

$$\{\text{soil particle surface}\}\text{-}AsO_4^{3-} + PO_4^{3-}(aq.) \rightarrow \{\text{soil particle surface}\}\text{-}PO_4^{3-} + AsO_4^{3-}(aq.)$$

(arsenate in immobilized form) (arsenate in mobilized form)

The mechanism at play here was a rather rapid reduction in the soil's capacity to adsorb AsO_4^{3-} and keep it safely locked in the soil. Referring to Figure 2b, the soil was in a "partially filled" state with respect to arsenate (marked by "partially filled capacity 1"). Addition of PO_4^{3-} effectively shrunk the capacity of the soil to adsorb AsO_4^{3-} by taking over the sites at the soil surfaces that had been occupied by AsO_4^{3-}. A new equilibrium was established between sorption and solubility as indicated by the curve marked by "maximum soil sorption capacity: state 2." To reach this new stage, the soil released sorbed AsO_4^{3-} to the soluble phase (in an amount marked by the arrow pointing from "partially filled capacity 1" to "partially filled capacity 2") that became available for plant uptake.

With respect to heavy metals and metalloids such as arsenic, six soil/sediment parameters are particularly important in causing the type of shift indicated in Figure 2b. These parameters, dubbed Capacity Controlling Properties (CCPs), are given in Table 1.

Thus, reducing the load of mobilizable heavy metals in the biosphere depends not only on reducing the input loads, but also on controlling human activities that affect CCPs. As noted in Table 2 such activities include land use and intensity of land use, energy production, and the numerous activities that contribute to climate change. It is particularly noteworthy that climate change may be expected to affect all the CCPs to some degree.

In the following sections I will discuss the basic chemistry of CCPs more specifically, and how changes in CCPs can trigger the mobilization of pollutants.

Table 1. Important Capacity Controlling Properties (CCPs) for Heavy Metals in Soils/Sediments

CCP	Possible Detrimental Environmental Effects from Changes in CCPs
Cation Exchange Capacity (CEC)	CEC depends on clay mineral content and type, soil organic matter, and the occurrence of manganese and iron oxides, all of which adsorb heavy metals. When CECs in soils/sediments are reduced, their capacities to retain heavy metals are also reduced.
pH	Lowering pH decreases CEC, increases heavy-metal solubility, and alters soil microbial populations.
Redox Potential (Eh)	Decreasing Eh dissolves iron and manganese oxides, resulting in the mobilization of adsorbed toxic chemicals; in mercury-contaminated sediments under low Eh, bacteria can generate highly toxic methylmercury; increasing Eh mobilizes heavy metals by dissolving insoluble metal sulfides.
Soil Organic Matter (SOM)	Decreasing SOM content reduces CEC, soil pH, buffering capacity, sorption capacity for toxic metals and organics, soil water-holding capacity; alters physical structure (e.g., increases soil erodibility), and decreases microbial activity.
Salinity	Increasing salinity solubilizes toxic chemicals by altering the ion-exchange equilibrium, increasing soluble complexation, and decreasing chemical thermodynamic activities in solution; it can also decrease microbial activity.
Microbial Activity	Altering microbial activity can affect other CCPs including SOM content, redox potential, pH; sulfate-reducing bacteria synthesize and emit methylmercury.

Source: Stigliani, 1993.

Table 2. Linkages between Human Activities and Changes in Capacity Controlling Properties (CCPs)

CCP	Human Activities Affecting CCPs
Cation Exchange Capacity (CEC)	Affected indirectly by activities that affect pH, Eh, SOM, and salinity.
pH	Atmospheric acid deposition from coal combustion; agricultural practices (liming increases pH, nitrogen fertilizer decreases pH); deforestation decreases pH; climate change altering the nitrogen cycle, and seasonal patterns of precipitation and evapotranspiration.
Redox Potential (Eh)	Drainage of wetlands (Eh increases) or flooding of dry lands (Eh declines) either as deliberate land use policy, or as a consequence of climate change.
Soil Organic Matter (SOM)	Deforestation decreases SOM; changes in vegetation; agricultural practices (recycling crop residues and manure increases SOM, intensive, high input farming without recycling decreases SOM); climate change can increase or decrease SOM.
Salinity	Irrigation with salty groundwaters; salt-water intrusion of groundwaters due to sea level rise.
Microbial Activity	Affected indirectly by activities that affect redox potential and pH, and directly by synergistic effects of multiple toxins, and climate change.

Source: Stigliani, 1993.

CATION EXCHANGE CAPACITY AS A CAPACITY CONTROLLING PROPERTY

The Role of Aluminosilicate Clays

Three constituents of soils provide Cation Exchange Capacity (CEC) for storing and immobilizing pollutants in the biosphere. One is aluminosilicate clays, a major attribute of soils in temperate regions of the world.[2] Because of their huge surface to volume ratios, and an abundance of negative charges on their surfaces, clays possess a high capacity for attracting and adsorbing positively charged "cations" (C^+) (see Figure

[2] For sake of brevity, soils containing carbonate are not discussed here in the context of the CTB phenomenon.

3a). The common natural cations found in abundance in clay soils are the four so-called "base cations" that include sodium (Na^+), potassium (K^+), magnesium (Mg^{2+}), and calcium (Ca^{2+}). These cations serve two essential roles: 1) they are nutrients for plant and animal life; and 2) they buffer the acidity of soil solutions. As shown in Figure 3a, hydrogen ions (H^+) in soil water replace base cations at the negatively charged surface sites. This step renders the solubilized base cation bioavailable, while at the same time reducing the soil water's acidity. The exchange is efficient because H^+ adsorbs more strongly to the charged clay site than the base cations. In general, cations (assuming equal concentrations) are held and displace one another in the sequence:

$$Al^{3+} > H^+ > Ca^{2+} > Mg^{2+} > K^+ > NH_4^+ > Na^+$$

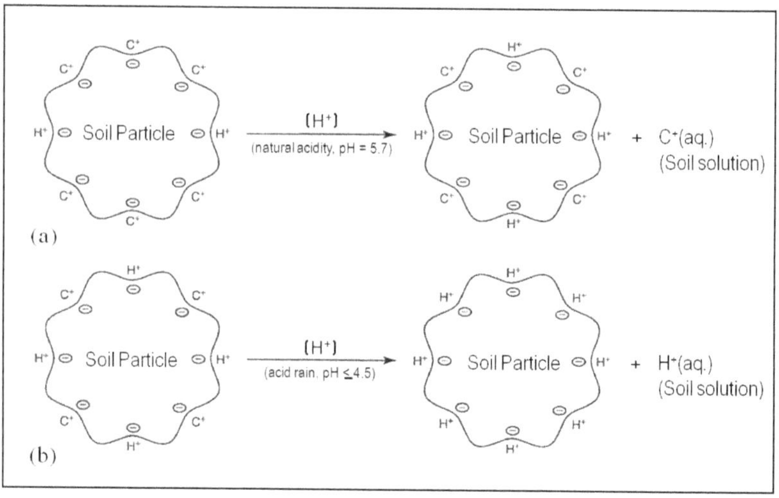

Fig. 3: (a) Natural exchange between hydrogen ion (H^+) and base cation (C^+) in a well-buffered clay soil under conditions of natural rainfall acidity (pH = 5.7); (b) exchange between H^+ and C^+ in a clay soil subjected to acid rain where pHs can be 4.5 or lower.

Unpolluted rainfall that drains through clay soils is naturally acidic (pH = 5.7), so the exchange shown in Figure 3a is an important step in the cycling of base cation nutrients in the biosphere. Problems arise when polluted rainfall (so-called "acid rain" with pHs typically 4.5 or lower) percolates through clay soils. As illustrated in Figure 3b, in this case H^+ is so concentrated that it sweeps away most of the base cations, leaving few additional surface sites available for H^+/base cation exchange. At this point, the clay has reached its saturation point with respect to its ability to buffer the acidic soil-solution water, and runoff from these soils contains acid (H^+) rather than base cations. When such soils lie within the watershed of a lake, the unbuffered surface and subsurface runoff cause lake acidification.

Soils continually replenish the base cation supply through weathering of bedrock. The supply of base cations as a component of soil (the exchangeable pool) is typically minuscule compared to the total pool that is locked in the parent rock (primary and secondary aluminosilicate minerals). While the exchangeable pool reacts rapidly with soil solution, the pool of base cations contained in mineral rocks is released relatively slowly, owing to the resistance of the parent rock to decomposition and dissolution. Weathering rates vary greatly for different soil types. The most acid-sensitive soils produce enough base cations to neutralize only 0.02 to 0.05 grams of H^+ per square meter per year ($g/m^2/yr$). In areas highly impacted by acid rain, for example Central Europe in the early 1980s, H^+ deposition rates as high as 0.60 $g/m^2/yr$ were measured.

When base cation exchange capacity in clays becomes exhausted through intensive acidification, a new buffering regime takes over, governed by reaction of acid with aluminum (Al):

$$Al(OH)_3\ (solid) + 3H^+ \rightleftarrows Al^{3+}\ (aq.) + 3H_2O$$

The pH drops to values of 4.2 or lower under the equilibrium established by this equation, where the neutralization of H^+ renders dissolution of $Al(OH)_3$ to soluble Al^{3+}. Unlike the neutralization reactions in the cation-exchange clay buffering range that generate dissolution of beneficial base cations, Al^{3+} is extremely toxic. Releases of soluble Al^{3+} from forest soils have been documented in the northeastern U.S. (Johnson et al. 1981), and in Central Europe (Schulze, 1989). Since aluminum compounds are abundant in clay soils, buffer capacity in this range is rarely depleted.

Soil Organic Matter

The second constituent of soil that adds to its CEC is "soil organic matter" (SOM). Like clays, humic material in the soil possesses a large number of negatively charged functional groups, such as dissociated carboxyl and phenolic hydroxyl groups that effectively bond to positively charged heavy metals. This high specific sorption capacity and its concentration at the soil surface make SOM a major storage compartment of anthropogenic heavy metal inputs. Even in mineral soils with less than 5% SOM content, it accounts for up to 50% of heavy metal binding (Kabata-Pendias, 2001).

Another characteristic of SOM is its strong binding capacity with organic micro-pollutants. Figure 4 shows that the concentration of stored micro-pollutants in soils correlates well with the percentage of SOM. Here is another situation, however, where a sink for pollutants can become a source of pollution. This could happen if the percentage of SOM declines due to changing physical, chemical, or climatic conditions. For example, the figure on the right shows that soils with 6% organic matter can adsorb about 4 micrograms of PCB-80 per kg of soil. If the soil concentration drops to 4%, it would only adsorb about half that amount, rendering the rest of the PCB mobilizable and biologically active.

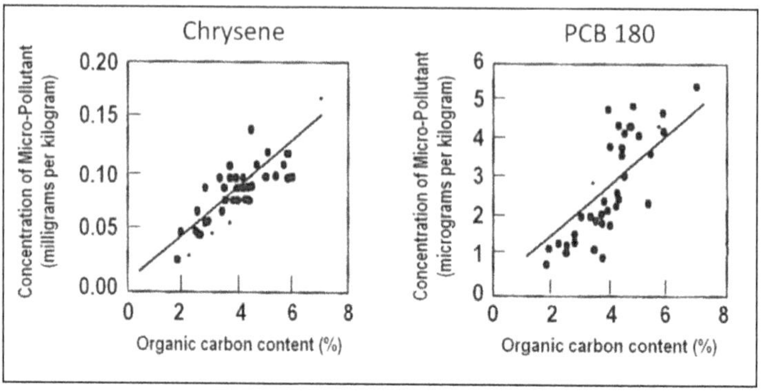

Fig. 4: Correlation between soil organic matter content and concentrations of organicmicro-pollutants in soils. Source: Salomons, 1995.

Manganese and Iron Oxides

The third constituent of soils that contributes to the CEC is an abundance of hydrous manganese and iron oxides [MnO_2, $Fe(OH)_3$]. These chemicals are ubiquitous in the environment, and coat the outer surfaces of clay minerals. Because of their negative polarity, these metal oxides provide numerous sites for binding heavy metals. They also play a critical role in mechanisms involving the Capacity Controlling Property redox potential (Eh), which is described later.

The Acidification of Big Moose Lake: a Practical Example of Depletion of CEC

Very few studies are available that have monitored the progression of soil from its natural, unpolluted state up until its loss of ability to buffer acidic inputs. The acidification of Big Moose Lake, shown in Figure 5, is one case where such information was available (Stigliani, 1988).

The lake, located in the Adirondack Mountains of New York State, is situated several hundred kilometers downwind from the industrial heartland of the U.S. Industrialization began around 1880 with the advent of coal combustion, which rapidly escalated and peaked around 1920 and remained at a more or less constant level over the next six decades. The coal contained high concentrations of sulfur, which was emitted as sulfur dioxide (SO_2) and transformed in the atmosphere into sulfuric acid. This activity constituted a major cause of acid rain in the U.S. in the 20th century (National Research Council, 1986). As depicted in the figure, the pH of the lake remained essentially constant (except for a slight decline due to natural senescence) over the entire period from 1760 to 1950. Then within the space of about 30 years (1950 to 1980) the pH declined from about 5.6 to 4.5 (corresponding to a factor of 10 increase in acidity). The figure shows that the acidification of Big Moose Lake was not synchronous with the onset of sulfur emissions in 1880. Rather, acidification occurred about 70 years after the beginning of SO_2 emissions and 30 after the sulfur emissions peaked. Accompanying the pH decline was the extinction of fish species, beginning with the most acid sensitive fish (smallmouth bass and whitefish) and ending with the extinction of lake trout, the most acid-resistant species. The delayed acidification occurred because the soils in the watershed of the lake possessed sufficient buffering capacity (from base cations Na^+, K^+, Mg^{2+}, Ca^{+2}) to neutralize the acid until the capacity was finally exhausted circa 1950.

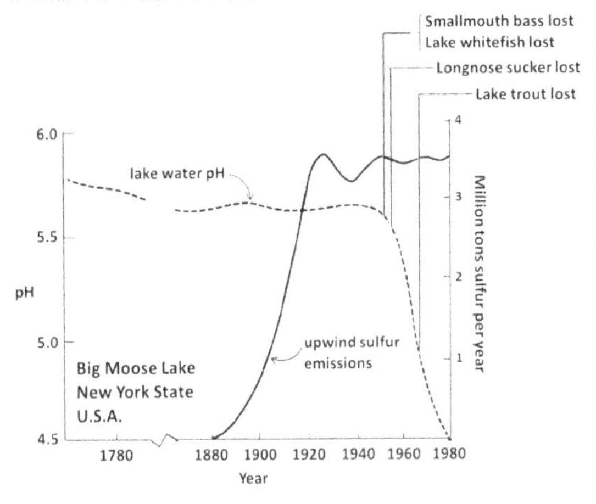

Fig. 5: History of acidification of Big Moose Lake from 1760 to 1980. Indicators include trends in pH of lakewater (dashed curve), emissions of SO_2 upwind from the lake (solid curve), and fish extinctions starting with the most acid-sensitive fish are depicted. Source: Stigliani (1988).

pH AS A CAPACITY CONTROLLING PROPERTY DETERMINING MOBILITY OF HEAVY METALS IN SOILS AND SEDIMENTS

Chemical Basis

As shown in Figure 6, another effect of soil acidification is the loss of capacity to adsorb heavy metals. This becomes a problem in soils highly contaminated with heavy metals, in which hydrogen ions (H^+) not only replace base cations, but positively-charged heavy metals as well. When this happens, a soil switches from being a sink for toxic metals such as cadmium (Cd^{2+}), zinc (Zn^{2+}), and lead (Pb^{2+}) to a source for their mobilization and release. Particularly noticeable in Figure 6 is the characteristic that the capacity of clays to adsorb the metals undergoes an abrupt change over a very narrow pH range. The inflection point of the curve occurs at pH_{50}, when 50% of the absorption capacity of the soil is exhausted. Thus, a soil contaminated with a heavy metal will quickly release the metal as its pH drops below the metal's pH_{50} value.

Table 3 shows the wide variation of values of pH_{50} for various heavy metals on two simulated clay soil substrates. Cadmium (Cd^{2+}) possesses the highest pH_{50} value, and would thus be the most susceptible to mobilization as soils acidify. In contrast, lead

(Pb^{2+}) and copper (Cu^{2+}) have the lowest pH_{50} values, and would require more strongly acidic soil conditions before they are mobilized.

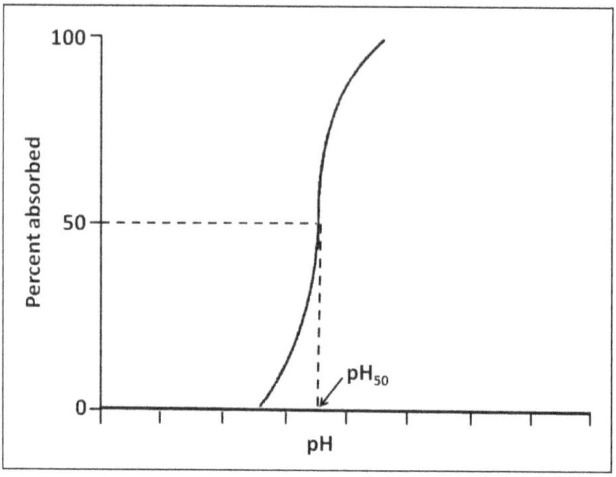

Fig. 6: Typical pH adsorption curve for divalent cations on hydrous metal oxides (pH_{50} values may range from 3 to 8). Source: National Research Council, 1985).

Table 3. Values of pH_{50} for Series of Toxic Heavy Metals on Colloidal Metal Hydroxides.

$Fe(OH)_3$ gel[a]									
Metal[b]	Pb^{+2}	<	Cu^{+2}	<	Zn^{+2}	<	Ni^{+2}	<	Cd^{+2}
pH_{50}	3.0		4.4		5.3		5.7		5.9
$Al(OH)_3$ gel									
Metal	Cu^{+2}	<	Pb^{+2}	<	Zn^{+2}	<	Ni^{+2}	<	Cd^{+2}
pH_{50}	4.8		5.2		5.6		6.3		6.6

a) Gel concentration: 9.3×10^{-2} M, Al or Fe; b) Trace metal concentration: 1.25×10^{-4} M.
Source: National Research Council, 1985.

As illustrated in Figure 7 for the case of cadmium, mobilization measured as the leaching velocity (rate at which the dissolved metal moves through the soil) increases by more than an order of magnitude as the pH declines from 7 to 4. Thus, at near-neutral pHs as found in well-limed agricultural soils, cadmium will tend to store in the soil. As soils acidify, Cd^{2+} moves increasingly into the aqueous phase, where it may migrate downward into deeper soil horizons and groundwaters, or alternatively be taken up by crops or vegetation. Such a pH decline could happen if agricultural lands are not consistently limed, or if agricultural lands are converted to forest lands.

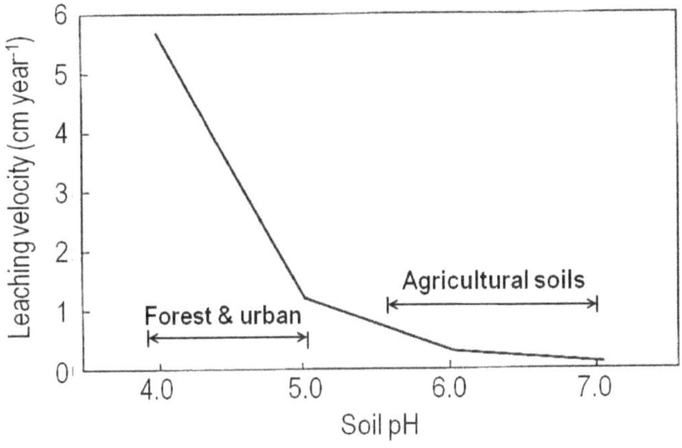

Fig. 7: Leaching velocity of cadmium as function of soil pH calculated for standard rainfall conditions in the Rhine Basin. Source: P. Jaffe and W M. Stigliani (1993) Personal communication, International Institute for Applied Systems Analysis, Laxenburg, Austria.

Practical Example of the Influence of pH and Other Capacity Controlling Parameters on Cadmium Mobility and Human Health

The impact of pH, iron and manganese oxides, carbonate, and redox potential on bioavailability and human health is obvious from viewing Table 4, which compares cadmium contamination at a site in England with one in Japan. Shipham, England was the site of zinc mining from the 16th to the 19th centuries. In Jinzu, Japan some relatively minor silver and lead mining occurred in the 16th century, but activities greatly escalated in 1905 when a large zinc refining plant came into operation, generating high tonnages of wastes that were routinely dumped into the Jinzu River, which was a source for rice irrigation. Since cadmium is present in trace amounts (up to a few percent) in virtually all zinc ores, zinc mining and refining are almost always associated with cadmium pollution.

Table 4.Impact of Cadmium Pollution on Human Health: a Comparison of Two Regions

Soil Parameter	**Shipham, England**	**Jinzu, Japan**
Maximum soil Cd concentration (includes immobilized and soluble fractions)	> 300 ppm	70 ppm
Soil pH	7.5	5.1
$CaCO_3$ Concentration	6-14 percent	0.4 percent
Mn/Fe Oxide Concentration	High	Low
Redox Potential	Moderate	Low
Health Effects	Slight	Severe

Source: Adapted from Morgan and Simms (1988).

Cadmium in the Jinzu soils was far more mobile (and thus more bioavailable) than cadmium in the Shipham soils owing to differences in CCPs. First, the lower pH in

Jinzu soils reduced their capacity to adsorb and immobilize the cadmium (see Figure 6). Second, high concentrations of manganese/iron oxides (which, as noted above, are particularly effective adsorbers of heavy metal cations such as Cd^{2+}) enhanced adsorption of cadmium in the Shipham soils, which was not the case in the Jinzu soils. The difference in redox potential was another important factor, which will be discussed in detail in the next section. The submerged soils in the rice paddy fields provided a low redox environment under which the reduced soluble forms (Mn^{2+} and Fe^{2+}) were favored over the insoluble forms [MnO_2 and $Fe(OH)_3$]. As soluble ions, Mn^{2+} and Fe^{2+} are incapable of adsorbing cadmium, as is the case for the oxidized forms MnO_2 and $Fe(OH)_3$.

Thus, even though soil cadmium levels exceeded 300 parts per million (a thousand times higher than typical rural levels) at the English site, the soil conditions rendered most of the cadmium unavailable for crop uptake. Consequently, health inventories of inhabitants ingesting food grown in this area showed only slight effects attributable to cadmium (Morgan and Simms, 1998). In contrast, soils in the Jinzu region, with cadmium levels much lower at 70 parts per million, favored the mobilization of cadmium and uptake into the rice. This resulted in the world's worst episode of cadmium poisoning, in which hundreds of persons became seriously ill from eating rice contaminated by polluted river water used for irrigation (Nogawa et al. 1999).

REDOX POTENTIAL AS A CAPACITY CONTROLLING PROPERTY IN SEDIMENTS AND AQUATIC ECOSYSTEMS

Chemical Basis

The chemical definition of an "oxidation-reduction" reaction (nicknamed "redox" reaction) is one that involves the complete or partial transfer of one or more electrons from one chemical species to another. In the biosphere, redox reactions control one of the fundamental requirements of life—the need of microbes to generate biochemical energy attained through the oxidation of organic carbon (CH_2O). In the familiar respiration reaction that sustains aerobic life (including humans), the organic carbon is oxidized by molecular oxygen (O_2) with the highest yield of metabolic energy:

$$\underset{\text{(organic carbon)}}{CH_2O} + \underset{\text{(oxidizing agent)}}{O_2} \xrightarrow[\text{(microbial mediation)}]{} \underset{\text{(reaction products)}}{CO_2 + H_2O} + \underset{\text{(derived biochemical energy)}}{502\ \text{kJ/mole}\ CH_2O}$$

Plentiful microorganisms in the aquatic and terrestrial biospheres utilize O_2 in respiration. In moist soils, waters, and sediments, however, this reaction is limited because of relatively low concentrations of O_2. But life by no means stops in anaerobic environments. Aerobic organisms die off, but others appear that utilize oxidizing agents that substitute for O_2 in the respiration reaction shown above. Nitrate (NO_3^-) serves the same purpose, albeit with less energy yield per molecule of organic carbon consumed:

$$\underset{\text{(organic carbon)}}{CH_2O} + \underset{\text{(oxidizing agent)}}{4/5NO_3^- + 4/5H^+} \xrightarrow[\text{(microbial mediation)}]{} \underset{\text{(reaction products)}}{CO_2 + 2/5N_2 + 7/5H_2O} + \underset{\text{(derived biochemical energy)}}{475\ \text{kJ/mole}\ CH_2O}$$

There are in fact six molecules (including O_2) that control the oxidizing power of the biosphere. These are, in order of decreasing energy yield: 1) oxygen (O_2); 2) nitrate (NO_3^-); 3) manganese dioxide (MnO_2); 4) ferric hydroxide [$Fe(OH)_3$]; 5) sulfate (SO_4^{2-}); and 6) carbon dioxide (CO_2). As illustrated in Figure 8, the choice of oxidizer proceeds via a preordained ordering, rather than by random utilization of all available oxidants. This ordering results from the relative oxidation potential of each molecule in its reaction with CH_2O, or, in chemical terms, its "redox" potential. The lower the redox potential, the lower the energy yield obtained from the reaction.

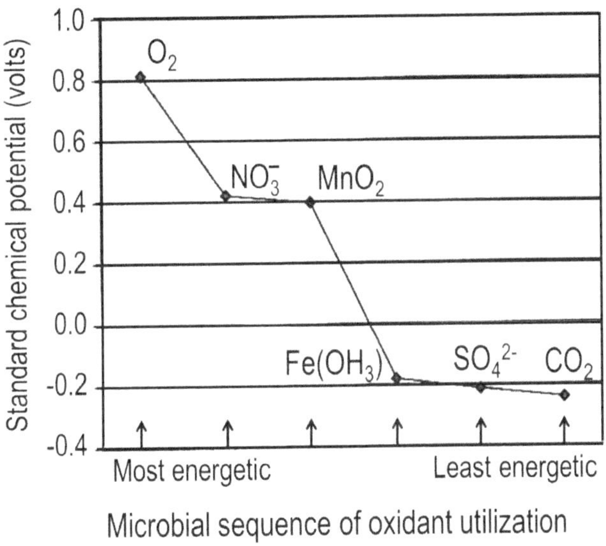

Fig. 8: Six common oxidants in the biosphere and the sequence in which they are utilized by microbes.

In an environment with a mix of oxidants, the dominant microbes are those that can tap the energy from the oxidant that yields the most energy. When that oxidant is used up, other microbes flourish that can use the oxidant with the next highest yield of energy. Thus, one can think of redox potential as a kind of "chemical switch" that determines the observed sequence by which microorganisms utilize oxidants. These reactions of course depend on a supply of organic carbon. In clean lakes with low amounts of CH_2O, microbes use O_2 as the preferred oxidant. When the amount of organic carbon exceeds about 3.4 milligrams per liter water, microbes will deplete the O_2 and nitrate, if present in the lake, becomes the preferred oxidant. When all the nitrate is depleted, then MnO_2 becomes the preferred oxidant, and so on.

As shown in Table 5, oxidants participating in redox reactions generate chemical by-products that affect the environment in different ways. With respect to nitrate, about 5% of it goes to nitrous oxide (N_2O) rather than the fully reduced product N_2. N_2O is a strong greenhouse gas (about 300 times more potent than CO_2 on a per weight basis). It also is the precursor for formation of nitric oxide (NO) in the stratosphere, which plays a key role in the nitrogen chain reactions that regulate, along with chlorine, the formation and destruction of stratospheric ozone.

The oxides of manganese (MnO_2) and iron [$Fe(OH_3)$] are insoluble. They are common environmental minerals that coat the surfaces of soil and sediment particles. They play a useful role in the environment because they efficiently adsorb deleterious pollutants such as phosphates, pesticides, and heavy metals. This situation changes, however, under anaerobic conditions when they become the preferred oxidants in the absence of nitrate and O_2. In the oxidation of CH_2O the metals are reduced to Mn^{2+} and Fe^{2+}. These are soluble forms and the pollutants that were adsorbed to the oxidized minerals are released to the water; i.e:

$$\underset{\text{[insoluble oxidized state]}}{\{MnO_2/Fe(OH)_3\}} \Leftarrow \text{(adsorbed pollutants)} \rightarrow \underset{\text{[soluble reduced state]}}{\{Mn^{2+}/Fe^{2+}\}} + \text{(soluble pollutants)}$$

Anaerobic microbes using sulfate (SO_4^{2-}) as an oxidant generate the sulfide ion (S^{2-}) as the reduced product. On the negative side, sulfide reacts with H^+ to form hydrogen sulfide (H_2S) gas, a deadly toxin. Sulfate reducing bacteria are also major players in the generation of methylmercury (Benoit et al. 2001; Benoit et al. 1999), a toxic substance posing high risks to human populations ingesting contaminated fish. On the positive side, other heavy metal ions form insoluble metal sulfides that render them unavailable to biota. Because of the natural abundance of sulfate in seawater, these reactions are particularly relevant in marine ecosystems, and in wetlands and soils that were once under the sea.

Table 5. Environmental Consequences of Microbial Redox Reactions

Oxidant	Significant Reaction Products	Impacts of Reaction Products
O_2	CO_2	Depends on source of organic carbon.[a]
NO_3^-	N_2O	A potent greenhouse gas; player in stratospheric ozone depletion.
$MnO_2/Fe(OH)_3$	Mn^{2+}/Fe^{2+}	Mobilizes phosphates, pesticides, arsenic and heavy metals that were adsorbed in oxidant forms.
SO_4^{2-}	H_2S; methylmercury; bound metal sulfides;	H_2S and methylmercury are highly toxic; metal bound sulfides are immobilized.
CO_2	CH_4	A potent greenhouse gas.

[a]CO_2 is a greenhouse gas, but will not add to the greenhouse effect if the source of organic carbon from which it is generated is renewable biomass, which is usually the case.

Methogenic bacteria that use CO_2 as the oxidant are found under extreme anaerobic conditions at the lowest redox potentials, as may be found in wetlands, flooded areas, rice paddies, and the sediments of enclosed lakes. In these conditions, bacteria produce methane (CH_4) as well as CO_2 by the disproportionation of organic carbon. Methane is a strong greenhouse gas whose potency is more than 20 times higher than CO_2 on a per weight basis.

Sudden changes in redox potential are possible, which can lead to deleterious environmental effects that appear as surprising or unpredicted occurrences. Some examples of this behavior are described in the following discussion.

Anoxia in Coastal Waters

Anoxia may occur in intermediate or deep waters of a gulf or a fjord with restricted circulation with surface waters. Figure 9 illustrates the underlying mechanism. As organic carbon is added to the deeper waters, O_2 is progressively consumed by aerobic bacterial activity, and since there is little mixing with the surface layers replenishment of O_2 from the atmosphere does not occur. (Oxygen in seawater is sufficient to oxidize only about 3.4 milligrams of carbon per liter.) Because seawater is rich in sulfate salts, the favored reaction under anaerobic conditions is sulfate reduction to hydrogen sulfide. Although H_2S is generally confined to the lower layers of seawater, during a storm event rapid mixing of the deeper, anoxic layers with the surface layer may subject marine life to high exposures of H_2S.

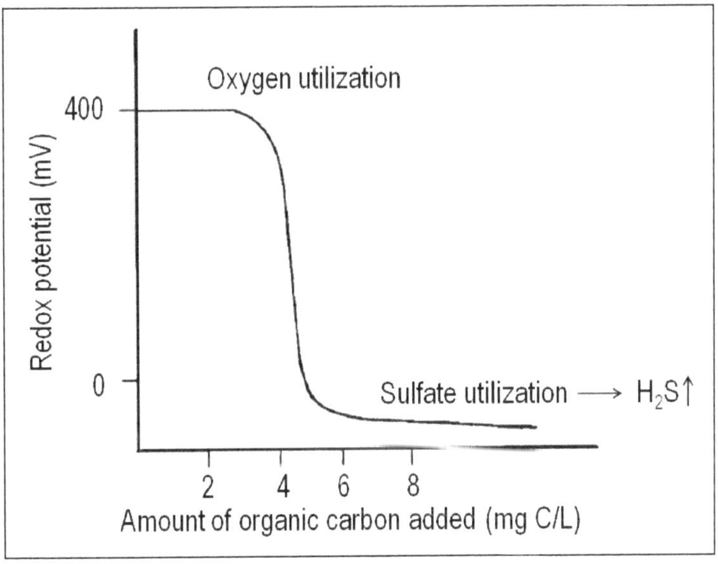

Fig. 9: Mechanism for production of toxic H_2S in marine waters with an abundance of organic carbon and restricted circulation of surface waters.

In coastal areas of Denmark in 1981 (Figure 10) and again in 1983, there were unprecedented depletions of oxygen and releases of hydrogen sulfide, causing the massive killing of fish. In a comprehensive investigation, the Danish National Agency for Environmental Protection concluded that meteorological conditions had triggered the two episodes, but that the underlying cause was the increasing over-fertilization of the coastal waters with an accompanying over-abundance of oxidizable organic carbon. The source of nutrient inputs was primarily runoff of nitrogen fertilizer from cropland (Schroder, 1985). The same phenomenon was observed in the 1980s in the Gulf of Saronik near Athens, Greece. The cause in this case was not nutrient run-off from cropland, but the direct disposal of raw sewage containing large concentrations of organic carbon.

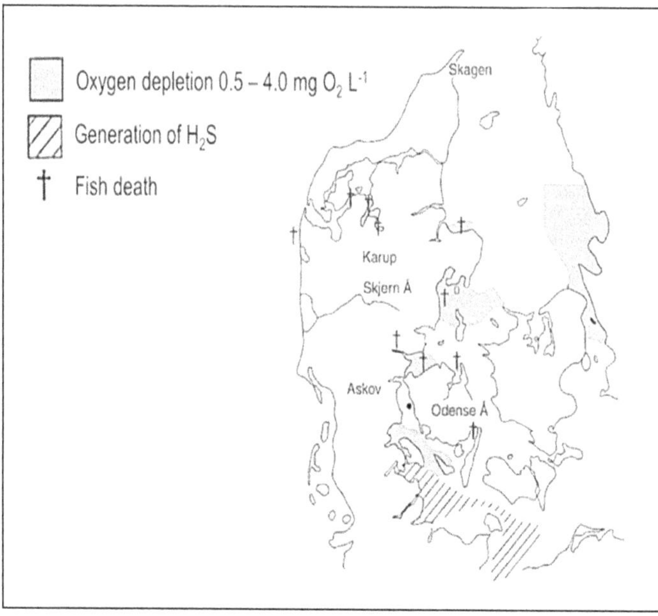

Fig. 10: Scandinavian coastal areas where oxygen depletion, fish suffocation, and generation of H_2S were recorded in September 1981. (Reproduced from Miljostyrelsen, 1984.)

Wetlands as Chemical Sinks

The episode of fish-kills depicted in Figure 10 might have been avoided if the original coastal wetlands had not been drained. Wetlands are typically anoxic and possess large amounts of organic carbon; they create natural buffer zones for nearby fresh or marine waters by trapping inflowing nitrate (Figure 11a). Nitrate enters wetlands in agricultural runoff, and bacteria use it to oxidize organic carbon via the reduction of nitrate to N_2, which is vented to the atmosphere. By depleting nitrate before it enters the estuary, wetlands limit the excessive growth of biomass and subsequent anoxic conditions in the estuary waters. Furthermore, if wetlands are of marine origin, they are likely to contain high concentrations of sulfur in the form of reduced sulfide minerals such as pyrite. Under the redox/pH conditions prevalent in wetlands, these sulfides are highly insoluble and immobilized.

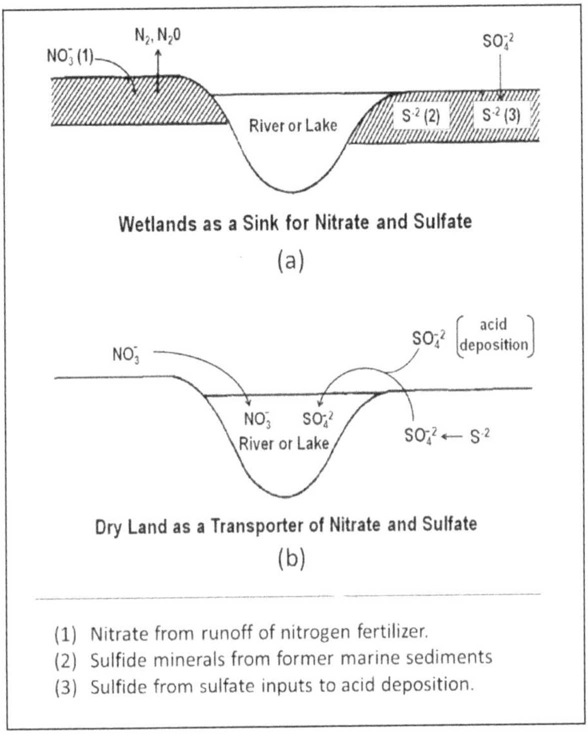

Fig. 11: (a) Ability of wetlands to buffer against nitrogen and sulfate inputs to water bodies (shaded area indicates wetland); (b) mobilization of nitrates and sulfates when wetlands are drained and exposed to atmospheric O_2. Source. Stigliani (1988).

Draining wetlands radically changes the chemistry of the landscape (compare Figure 11a to Figure 11b). Nitrates enter receiving waters unmitigated, and sulfides, upon exposure to the atmospheric O_2, are oxidized to sulfates. In contrast to sulfides, sulfate salts are appreciably soluble. Thus, draining wetlands can trigger the release of mobilized sulfuric acid (H_2SO_4), or heavy metal sulfates (e.g., $CdSO_4$, or $ZnSO_4$) that were insoluble as metal sulfides (CdS or ZnS).

An example of this phenomenon occurred in a coastal area of Sweden near the Gulf of Bothnia, where wetlands were first drained in the early 1900s and more extensively in the 1940s for conversion to agricultural lands. The draining exposed sulfides to atmospheric O_2, resulting in their oxidation to sulfuric acid. Runoff from the drained land during storm events acidified nearby lakes. The pH in one of these lakes, Lake Blamissusjon, dropped from about 5.5 in the nineteenth century to a current value of around 3. Even though agricultural activities ceased in the 1960s, the lake has not recovered; it is widely known as the most acidic lake in Sweden.

Arsenic Poisoning in Bangladesh: an Exploded Redox CTB

Prior to the 1970s, Bangladesh had one of the highest infant mortality rates in the world due to pathogens in the drinking water supply derived from polluted surface waters. At the urging of UNICEF and the World Bank, Bangladesh began a campaign in the 1970s of drilling tube wells to provide villagers with what was thought to be clean groundwater supplies. Since then about 10 million hand-pumped wells have been installed. This action reduced infant mortality and diarrheal illness by 50%. However, the plan backfired horribly when it was discovered in the 1990s that many of the wells were contaminated with arsenic arising from natural deposits in the subsurface aquifer.

A recent study in the medical journal *The Lancet* (Argos et al. 2010) found that between 35 and 77 million people have been chronically exposed to increased concentrations of arsenic. Following a cohort of nearly 12,000 Bangladeshis over the course of 10 years, the researchers found that more than 20% of deaths were attributable to arsenic. The World Health Organization has named this disaster "the largest mass poisoning in history" (WHO, 2004).

At first glance, one might wonder why this disaster happened in Bangladesh, because the amount of arsenic in the soils there is less than in many other areas in the world including Massachusetts, USA where arsenic in groundwater is not a problem. So what chain of events triggered this time bomb? The first step is to identify the reaction mechanism that caused the catastrophe. Numerous studies were conducted to elucidate the mechanism.[3] It is now clear that the **redox potential (Eh)** was the **Capacity Controlling Property** that mobilized the arsenic via the reduction of the oxidant **ferric hydroxide** [$Fe(OH)_3$] to its reduced state (Fe^{2+}) (see Figure 8 and Table 5 above). The specific reaction, mediated by microorganisms, is:

$$\underset{\text{(organic carbon)}}{CH_2O} + \underset{\text{(oxidized iron)}}{4Fe(OH)_3} + 8H^+ \longrightarrow \underset{\text{(microbial}}{4Fe^{2+}} + \underset{\text{(reduced iron)}}{11\,H_2O} + CO_2$$

Thus, the low redox conditions in the groundwater sediments favored microbial reduction of $Fe(OH)_3$ to derive metabolic energy from the oxidation of organic carbon, because of its relatively high abundance compared to other more energetically favorable oxidants (O_2, NO_3^-, and MnO_2).

The fact that $Fe(OH)_3$ was the oxidant in this reaction is extremely significant with respect to arsenic poisoning, for this form of iron appears to be particularly effective in adsorbing and immobilizing arsenic (As^{3+}) (Tufano and Fendorf, 2008). It therefore served as a large sink for the toxic chemical. From the CTB perspective, this created a potentially dangerous situation because a large toxic sink could become a large source of arsenic release under changing chemical conditions. And that is exactly what happened because the above reaction transformed $Fe(OH)_3$ to the reduced form Fe^{2+}. This form is

[3] This topic was the subject of a whole issue of the Journal of Applied Geochemistry: P. Bhattacharya et al. (eds.) (2004) Arsenic in groundwater and sedimentary aquifers. *Applied Geochemistry*, 19(2): 163-260.

water-soluble and the associated arsenic effectively desorbs and dissolves with it (Tufano and Fendorf, 2008).

The last link in the chain that needs to be solved in this CTB puzzle is the source of the organic carbon (CH_2O), without which the above reaction would not have happened. Until recently, the source of CH_2O in the groundwater was somewhat uncertain. At a typical site in Bangladesh, groundwater-irrigated rice fields and constructed ponds are the main sources of groundwater recharge. A recent study has found that recharge from the constructed ponds, but not from the rice fields, is the main source of biologically degradable CH_2O (Neumann et al. 2009). There also has to be a means of delivering CH_2O from the ponds to the depths of the groundwaters, and this was achieved by groundwater flow drawn by irrigation pumping that transported the organic-rich pond waters to a depth of around 20-30 meters. This is where most of the CH_2O reacted with the $Fe(OH)_3$, and where the peak in arsenic concentration is found (Neumann et al. 2009). This depth is also close to the average depth of the wells constructed in Bangladesh, and explains why the well waters are so contaminated.

Figure 12 shows a schematic of the mobilization process.

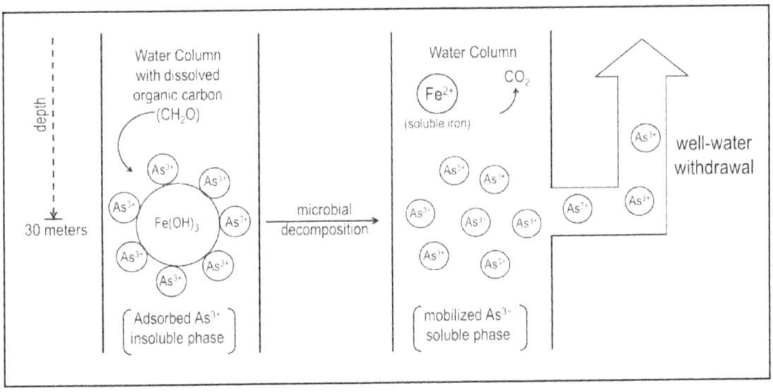

Fig. 12: Schematic of process by which arsenic is mobilized into groundwater wells in Bangladesh.

In summary, the chain of events triggering this CTB was:

1. The occurrence of natural deposits of arsenic adsorbed on the surfaces of the iron mineral $Fe(OH)_3$ in the groundwater sediments.
2. $Fe(OH)_3$ being the preferred oxidant in the redox chain due to the absence to significant concentrations of O_2, NO_3^-, and MnO_2.
3. The construction of ponds, discharge from which provided the source of CH_2O.
4. Irrigation pumping for rice production that provided groundwater flow for delivering the CH_2O to depths of 20-30 meters.
5. Microbial mediation of CH_2O oxidation via $Fe(OH)_3$ reduction, causing dissolution of $Fe(OH)_3$ to Fe^{2+} and the mobilization of As^{3+} to the groundwaters.

THE IMPLICATIONS OF CLIMATE CHANGE FOR CHEMICAL TIME BOMB OCCURRENCES

Eh/pH Ranges of Natural Aquatic Systems

Natural aquatic systems, maintained under current climatic conditions, are particularly well defined by their chemical states with regard to redox conditions (Eh) and acidity (pH). Figure 13 shows the Eh/pH diagram for major aqueous environments on Earth.

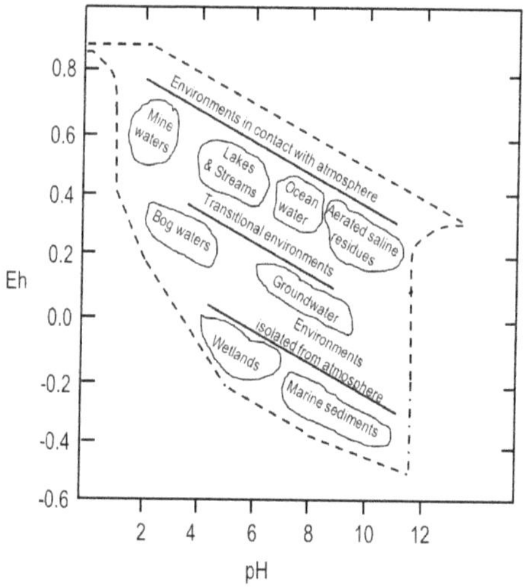

Fig. 13: Eh/pH diagram showing values of redox potential (Eh) and acidity (pH) for various natural aquatic environments on Earth. The dashed line shows the locus of measured values in these environments. Modified from Garrels and Christ (1965).

Wetlands and marine sediments characteristically possess the lowest environmental Eh values, and a wide range of pH values that do not, however, extend to extreme acidic conditions (pH values less than 4). Groundwaters and bog waters exhibit higher Eh values; the particularly low pH values in bog waters is due to the accumulation of acidic peat. Drainage from metal or coal mining (mine waters) possess even more extreme levels of acidity, resulting from the excavation of minerals under anaerobic conditions, typically containing pyrite (FeS), that become oxidized to sulfuric acid (H_2SO_4) upon exposure to the atmosphere. Well-aerated lakes, rivers, surface ocean waters, and saline residues have the highest environmental Eh values. Freshwaters possess lower pHs than marine environments, because seawater contains bicarbonate buffering salts that effectively stabilize the pH to slightly basic levels.

As noted earlier, the Eh and pH conditions of particular environments can govern whether toxic heavy metals are immobilized and biologically unavailable, or dissolved and biologically active. As shown in Figure 14, the increases in solubility for most heavy metals, indicated by the direction of the arrows, correspond to the arrow pointing simultaneously leftward and upward (i.e., toward decreasing pH and increasing Eh). Conversely, environments with high pH and low Eh values favor the burial and storage of metals in most cases. Comparing Figures 13 and 14, we can understand why one of the valuable ecosystem services provided by wetlands, owing to their low Eh and high pH values, is their ability to scavenge and immobilize toxic pollutants. We can see as well why drying out of wetlands can pose the danger of a release of toxic pollutants.

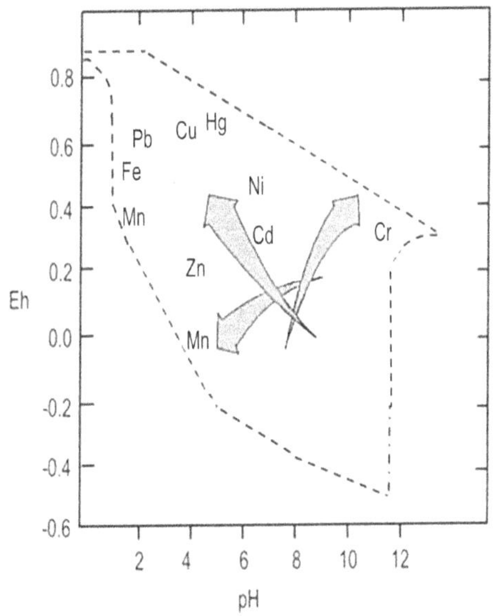

Fig. 14: How the solubility of metals is affected by changes in Eh and pH. Arrows point to increasing solubility; the solubility of most metals increases with decreasing pH and increasing Eh (assumes the absence of organic carbon). Source: Bourg (1994).

Human activities that perturb the Eh/pH diagram depicted in Figure 13 are capable of creating shifts in the biosphere's capacity to store and immobilize pollutants as shown in Figure 14. Several examples have been provided in this paper about past activities that have caused such shifts. These examples have been local and regional in scale. Climate change, however, is global in scope. It is thus reasonable to assume that the natural landscapes that evolved over the last 8,000 years of stable climate could undergo numerous shifts as they adapt to changing climatic regimes. Such shifts are likely to affect not only Eh and pH, but all the Capacity Controlling Properties discussed in this paper.

Linkages between Climatic Parameters and Capacity Controlling Properties

Progressive changes in numerous climatic indicators over recent decades leave little doubt that climate change due to human activities is already upon us. Rising global surface temperatures on land and in oceans, the increase in extreme weather events, the melting of the polar ice caps, changes in amounts and seasonal patterns of precipitation, rising sea levels, and climate-related degradation of ecosystems are well documented (IPCC, 4th assessment, 2007). The magnitude and character of future climate change is highly uncertain, because of uncertainties in the climate models, as well as in the time frame over which the global community will act to mitigate and adapt to the changes. Especially important in this regard is the pace at which carbon-based fuels are phased out and replaced by renewable energy sources (IPCC, 2011).

Figure 15 shows that major aspects of global climate change have multiple linkages with processes in the biosphere that can ultimately affect the Capacity Controlling Properties (CCPs) of soils and sediments.

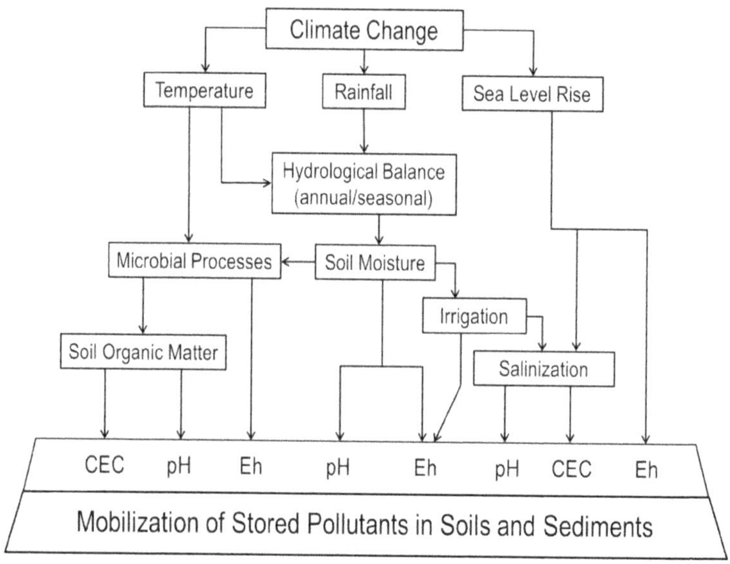

Fig. 15: Linkages between climate change and Capacity Controlling Properties regulating the mobilization of pollutants in soils and sediments.

Microbial processes and soil organic matter

A rise in temperatures increases the speed of microbial processes in soils that break down soil organic matter (SOM), a tendency that would increase CO_2 emissions to the atmosphere and decrease SOM concentrations in soils (Friedlingstein et al. 2006; Lloyd and Taylor, 1994). As discussed previously, losses of SOM will reduce the Cation Exchange Capacity (CEC) of soils, and decrease their ability to adsorb heavy metals, phosphates, and other pollutants. The buffering capacity of the soils would also be diminished, making them more prone to soil acidification (pH decline). Thus, a warming climate could lead to the desorption and mobilization of pollutants as well as soil acidification, as SOM content diminishes.

The potential degree of SOM loss, however, is quite uncertain since soils in different areas of the globe will be more prone than others to losses in organic carbon (Fung et al. 2005). Moreover, recent research suggests that global warming may alter the way microbes allocate carbon (Allison et al. 2010)—i.e., the portion of the carbon that is used to obtain microbial energy (which releases CO_2), compared to the portion that is incorporated into biomass (which stores carbon). Using model simulations, the researchers found that increasing temperatures may stimulate loss of SOM in the short term, but a shift in microbial allocation could mitigate this response over longer time periods. Also, it has been reported that warming could alter the composition of the microbial community (Balser and Wixon, 2009), which in turn could change the chemical composition of the soils; because different microbes use different soil compounds and produce different metabolites. It is currently unknown how these changes would affect SOM content.

The question of the fate of SOM during climate change has relevance far beyond the CTB concept, for it relates to the land component of the global carbon cycle and is the subject of intense research. If soils undergo a net loss of SOM globally, the land could become a net emitter of CO_2 to the atmosphere over and above any emissions caused by the burning of fossil fuels. Alternatively, if SOM increases globally, soils would be a net sink for CO_2, which would serve to mitigate climate change by offsetting emissions from burning fossil fuels. It is thus critically important for numerous reasons to adopt policies that will increase SOM by sequestering carbon into the soil.

Soil moisture

Soil moisture is an indicator of the prevailing "wetness" of a landscape, which is closely related to runoff (O'Loughlin, 1986). Runoff is part of the hydrological cycle, representing the difference between precipitation and evaporation off the landscape (including transpiration by plants). The IPCC (4th assessment, 2007) has conducted an extensive assessment of the impact of climate change on runoff based on their "SRES A1B" scenario.[4] Runoff is projected with high confidence to increase by 10 to 40% by mid-century at higher latitudes and in some wet tropical areas, including populous areas in East and South-East Asia, and decrease by 10 to 30% over some dry regions at mid-latitudes and dry tropics. There is also high confidence that many semi-arid areas (e.g., the Mediterranean Basin, western United States, southern Africa and north-eastern

[4] The SRES A1B scenario forecasts by the year 2100 a global temperature increase of 2.8°C (range: 1.7-4.4°C), and a sea level rise of 0.21-0.48 m.

Brazil) will suffer a decrease in runoff due to climate change. Drought-affected areas are projected to increase in extent, and regionally, large increases in irrigation water demand as a result of climate change are projected.

The IPCC also forecasts increases in heavy rainfall events in many regions, including some in which the mean rainfall is projected to decrease. It is likely that up to 20% of the world population will live in areas where river flood potential could increase by the 2080s. In coastal areas, sea level rise will exacerbate water resource constraints due to increased salinization of groundwater supplies.

If these climatic projections are realized, it seems likely that there will be significant changes in soil moisture, or more generally in the wetness of landscapes, in most regions of the world, with some areas becoming wetter and other areas drier. As soils become wetter, their redox potentials (Eh) will tend to decrease as they become increasingly anaerobic, particularly in the presence of organic carbon. Oxidants employed by microbes at the lower redox potentials, $Fe(OH)_3$, SO_4^{-2}, and CO_2, will be chemically reduced, and the negative impacts delineated in Table 5 could become more pronounced.

The drying out of landscapes will in many cases tend to increase both the redox potential as well as the acidity (lower pHs), especially if dried- out soils and sediments become exposed to the atmosphere. Wetland soils are rich in organic matter and microbes, ingredients that make them among the most active "biochemical reactors" in the biosphere in the processing and storing of numerous chemical pollutants. Drying out of wetlands caused by climate change shifts the Eh/pH regime to favor acidification and the mobilization of heavy metals (see Figures 11 and 14). Redox reactions can occur abiotically as well, through direct chemical reaction with atmospheric O_2. This has already been observed in the Murray-Darling river basin of Australia, one of the world's most highly regulated and stressed major river systems. An extreme drought period that has dominated the catchment since the year 2000 has lowered lake water levels, exposing thousands of hectares of sulfidic shoreline and lake-bed sediments to the atmosphere. The sediments are beginning to oxidize and acidify, and there are fears that continued drought may lead to more expansive acidification, heavy metal mobilization, anoxia, and the production of noxious gases (CSIRO, 2011).

As noted by the IPCC, the further drying out of already arid areas will tend to increase irrigation. In extensive regions of Europe, particularly in the Mediterranean Basin, and the Iberian Peninsula, Hungary, the former USSR, Bulgaria, and Romania, the ground water available for irrigation is highly saline. When the rate of evaporation is greater than the rate of precipitation, soils irrigated with water from these areas become increasing saline. Soil salinization caused by irrigation is already a serious problem in these regions (Szabolz, 1991), and as they dry out and irrigation increases, the extent and intensity of soil salinization may increase significantly. As cations from the deposited salts increase in concentration they may exchange with heavy metals adsorbed in the soil, decreasing the capacity of these soils to store the heavy metals.

Sea level rise

Rising sea levels will inundate low-lying lands, increase the salinity of rivers and bays, and salinize coastal aquifers by saltwater intrusion. Prolonged seawater inundation of previously dry, well-aerated coastal areas will cause oxygen levels to drop and the redox potential to decline, with the same consequences described above for landscapes experiencing increased wetness. This could happen in areas that are otherwise projected to become increasingly arid. Apart from the inundation effect, the chemistry of seawater itself may activate changes in the CEC and pH in surprising ways. The following example describes the mechanism by which such changes may occur.

A Practical Example: Potential Impact of Sea Level Rise on Acidity and Heavy Metal Mobilization in Coastal Lowland Acid Sulfate Soils

Wong et al. (2010) recently published a comprehensive study entitled "Seawater causes rapid trace metal mobilization in coastal lowland acid sulfate soils: implications of sea level rise for water quality." The study is important because it is one of the few studies that provide a concrete, science-based prognosis for potential linkages between sea level rise and mobilization of toxic heavy metals. The study is also interesting from the point of view of the CTB concept because it illustrates how changes in several of the Capacity Controlling Parameters described in this paper and in Figure 15 (viz., CEC, pH, and Eh) work to mobilize toxic heavy metals.

The global extent of coastal acid sulfate soils (CAS soils) is estimated to be about 170,000 km^2, with nearly half of this land (80,000 km^2) located in Australia's coastal estuaries and floodplains near where the majority of the Australian population lives, and where much of the agriculture and industry are based. These soils are characterized by high concentrations of acidity and sulfide minerals such as pyrite (FeS_2) that are released by oxidation to sulfates, and an abundance of toxic metals such as aluminum (Al), released by weathering of aluminosilicate clays. In eastern Australia, CAS soils are characterized by backswamp basins that are highly susceptible to pulses of saline water caused by saltwater intrusion, storm surges, and king tides. This risk is likely to increase in the future with predicted sea level rise and the increase in extreme events such as storm surges.

Wong et al. (2010) exposed samples of these soils to synthetic seawater with varying concentrations of sea salts, and found that desorption of heavy metals and increased acidity dominate the immediate effects of inundation. The chemical quality of seawater that makes it such a potent determinant in acidity and metal mobilization is its high "ionic strength" owing to the large salt content in the oceans, particularly sodium chloride (NaCl), but also salts of calcium and magnesium. When seawater comes in contact with CAS soils, the sea salt cations exchange with hydrogen (H^+) and heavy metal cations (C^+) at the negatively charged surface sites of clay minerals present in CAS soils. This is illustrated in the following reaction, where the sea salts, due to their high concentrations, drive the reaction strongly to the right, exhausting the CEC of the clay particles and mobilizing H^+ and C^+ in soluble form:[5]

---

[5] The prevalence of the anion Cl^- in seawater adds to the potential for mobilization of trace heavy metals by formation of soluble chloro-complexes with the metals.

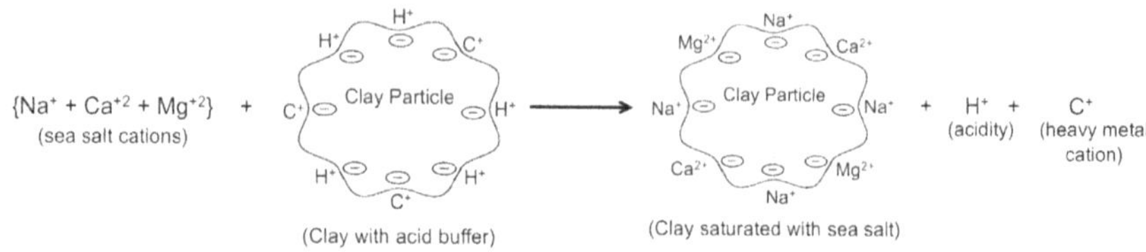

The release of H^+ decreases the pH of the soil solution, and chemical analysis revealed that highly toxic aluminum (Al^{3+}) is, by far, the dominant mobilized metal heavy cation (C^+). Lesser amounts of the heavy metals iron (Fe^{+2}), nickel (Ni^{+2}), manganese (Mn^{+2}) and zinc (Zn^{2+}) were present in all the samples studied, as well as numerous other toxic metals that were measured in some but not all the samples. After the initial drop in pH from displacement of H^+ from the clay particles, further decline in pH occurs from hydrolysis of Al^{3+} and Fe^{2+}; i.e.,

$$Al^{3+} + H_2O \longrightarrow Al(OH)^{2+} + H^+$$

$$Fe^{2+} + H_2O \longrightarrow Fe(OH)^+ + H^+$$

The enhanced acidification provided by these reactions creates a positive feedback, with lower pHs serving to induce trace heavy metal mobilization even further (see Figure 6).

The focus of this study was on the release of acidity and soluble toxic metals that would occur upon first contact between seawater and CAS soils. This "pulse" release is perhaps most relevant to tidal anomalies, storm surges and floodgate failures, all of which may increase due to climate change. Higher sea levels will also cause an upstream shift of the tidal plane, resulting in higher salinity estuarine waters further upstream. This may expand the impact on an increasing area of CAS soils. Another important finding in the study was that even low salt concentrations similar to those found in brackish waters in estuarine environments can mobilize high concentrations of aluminum and other heavy metals.

For longer periods of inundation by seawater, the CAS soils would be controlled by redox processes. Referring to Figure 13, the initial conditions under the pulse reactions are likely similar to "bog waters" where low pHs and relatively high Ehs prevail. Prolonged inundation would shift the Eh/pH conditions closer to that represented by "marine sediments" in the figure. The higher pH and low Eh values of this environment would cause the acidity to decrease, and the mobilized aluminum and iron to reprecipitate as insoluble hydroxides. There would also be a reduction in the solubility of other trace metals such as Zn, Ni, and Mn. Although this process would reverse the impacts of pollutants created by the pulse release, studies of these same CAS soils under more permanent inundation conditions have found that as the Eh level drops, dissolution of insoluble oxides of Fe^{3+} can occur resulting in the release of other trace metals and metalloids like arsenic (Burton et al. 2008; Johnston et al. 2010a, b). This is the same reaction that caused the toxic release of arsenic into drinking-water wells in Bangladesh. So it seems that either short-term pulses or long term inundation of CAS soils by seawater can have impacts on heavy metal mobilization, albeit by different mechanisms affecting different CCPs.

CONCLUSIONS

The above discussion has focused on risks arising from releases of pollutants stored in soils/sediments caused by changes in the biogeochemical conditions that favor mobilization over storage of the pollutants. The question is how to reduce the risk of such occurrences in the future? Some ideas can be suggested based on what we have learned from past experiences, current advancements in scientific understanding of soil/sediment interactions, and how well we can articulate possible scenarios of future developments relevant to these risks.

One conclusion we can make with a fair degree of certainty is that unmitigated global climate change has the potential to greatly influence all the capacity controlling properties (CCP) in soils and sediments discussed in this paper. Not all these influences will be necessarily detrimental, but others could trigger CTB-like events. Whereas such events have tended to be local or regional hotspots in the past, the concern about climate change is its potential to expand the scale of CTBs to become a global phenomenon. For example, changes in wetness of the landscape and its accompanying impacts on soils and sediments will not be confined to a few places, but will be pervasive throughout the biosphere. As shown in Figure 15, wetness (soil moisture) directly or indirectly affects all the CCPs that govern the balance between adsorption and mobilization of pollutants in the biosphere, and changes in wetness can shift this balance in ways that are likely to be extensive although not yet entirely delineated. Mitigation of climate change appears to be a critical strategy for mitigating the potential risks of CTBs, for it would significantly moderate the degree of change in wetness as well as in other climatic variables relevant to CTB phenomena. This conclusion provides yet another rationale why the international community needs to make climate change mitigation its highest priority through timely and effective actions.

Clearly, some actions would be particularly beneficial in simultaneously reducing the risks of climate change and CTBs. One would be to enact comprehensive policies that serve to maintain and increase the content of soil organic matter (SOM) in the major land use sectors. Land use practices can be applied in agriculture and forestry that would not only prevent loss of SOM, but increase it through sequestration of carbon in the soil. Such practices take carbon out of the atmosphere and help ensure that the land component of the carbon cycle will be a sink for CO_2, rather than a source of atmospheric CO_2 emissions. At the same time SOM enhances a soil's acid buffering capacity, adds sorption capacity for immobilizing pollutants, strengthens resistance to erosion, and increases fertility.

Another strategy for reducing the risks of CTBs is constructing detailed global vulnerability maps that indicate areas sensitive to mobilization of toxic metals and other pollutants. A great example is the recent study by Amini et al. (2008) who created such maps for global arsenic groundwater contamination. Beyond merely identifying regions where high arsenic concentrations in groundwater exist, they assessed the likely arsenic mobilization potential based on ambient conditions of redox potential (Eh) and pH. They constructed two global maps of vulnerability for arsenic mobilization: one under reduced groundwater conditions (low Eh); and one under high-pH/Eh conditions where arsenic is soluble in its oxidized state. Similar maps could be made for other toxic chemicals. Such maps may greatly reduce the threat of another disaster like the one experienced in Bangladesh where many millions were exposed to, and untold numbers died from, arsenic-contaminated drinking water.

As shown in Table 2, numerous land use activities and practices can affect the CCPs in soils and sediments. For example, pH is influenced by agricultural practices (liming increases pH, nitrogen fertilizer decreases pH), and deforestation (decreases pH); Eh is influenced by draining wetlands (Eh increases) and flooding dry lands (Eh decreases); and salinity is influenced by irrigation. Often these activities and practices have been conducted without consideration of their impact on CCPs and subsequent mobilization of pollutants. As a result, unintentional detrimental consequences have occurred many times. Examples cited earlier were the acidification of Big Moose Lake (USA) after seven decades of acid deposition, the acidification of Lake Lake Blamissusjon (Sweden) after the draining of bordering peat lands, cadmium poisoning in Jinzu, Japan from eating rice contaminated by effluent from a local zinc smelter, the toxification of Danish coastal waters from agricultural runoff, and the arsenic catastrophe in Bangladesh after the installation of drinking water wells.

Guided by these historical experiences and today's greatly increased scientific knowledge base, we should be capable of forecasting the impacts of our land use activities/practices as well as the implications for changes in these activities/practices. Such a forward looking "environmental impact assessment" should become standard procedure for guiding major land use policies. This work in fact has already begun, although at the moment is quite understated and in need of more articulation and intentional focus. For example, we have the forecast by the IPCC (fourth assessment, 2007), identifying the Mediterranean Basin, already a semi-arid region, as an area that will suffer further decreases in water availability due to climate change, accompanied by a likely need to increase irrigation. We know from historical experience that the groundwaters in this area are saline, and that extensive soil salinization caused by irrigation has already occurred. Combining this historical information with the prognosis of the IPCC indicates that increasing irrigation will serve to intensify and extend the areas of the Mediterranean Basin affected by soil salinization. This analysis uncovers a potentially serious vulnerability in a major European region, and decision makers should be informed about this scenario when formulating regional agricultural policies during an era of climate change. A second example is the research of Wong el al. (2010), indicating a future vulnerability in Australian coastal areas from projected sea level rise and increased storm surges. Their research showed that interaction of seawater with the coastal soils could cause extensive acidification and mobilization of toxic metals. This work contributes important new information applicable to environmental impact assessment of coastal areas subjected to sea level rise.

One of the motivations for writing this paper has been to reevaluate the significance of the idea discussed in my first paper (Stigliani, 1988), viz., that sinks for toxic chemicals could undergo changes in certain key chemical and biochemical parameters that would transform them from sinks into an emitters of toxins. I also stated that such toxic releases would tend to be time-delayed, non-linear, and manifested as surprising, unanticipated events. It seems to me that this idea is still relevant as demonstrated by the experiences of recent decades. Mostly, I hope the ideas presented in this paper will stimulate further conversation, and contribute to reducing environmental risks posed by toxic substances, especially for the sake of our children and grandchildren as we enter an era of unprecedented changes in environmental conditions.

REFERENCES

1. Allison, S.D., Wallenstein, M.D., and Bradford, M.A. (2010) "Soil-carbon response to warming dependent on microbial physiology." *Nature Geoscience*, 3: 336-340.
2. Amini, M., Abbaspour, K.C., Berg, M., Winkel, L., Hug, S.J., Hoehn, E., Yang, H., and Johnson, C.A. (2008) "Statistical modeling of global geogenic arsenic contamination in groundwater." *Environmental Science & Technology*, 42: 3669-3675.
3. Argos, M., Kalra, T., Rathouz, P.J., Chen, Y., Pierce, B., Parvez, F., Islam, T., Ahmed, A., Rakibuz-Zaman, M., Hasan, R., Sarwar, G., Slavkovich, V., van Geen, A., Graziano, J., Ahsan, H. (2010) "Arsenic exposure from drinking water, and all-cause and chronic-disease mortalities in Bangladesh (HEALS): a prospective cohort study." *The Lancet*, 376: 252-258.
4. Balser, T.C., and Wixon, D.L. (2009) "Investigating biological control over soil carbon temperature sensitivity." *Global Change Biology*, 15: 2935-2949.
5. Benoit, J.M., Gilmour, C.C., and Mason, R.P. (2001) "The influence of sulfide on solid-phase mercury bioavailability for methylation by pure cultures of *Desulfobulbus propionicus* (1 pr 3)." *Environmental Science & Technology*, 35: 121-126.
6. Benoit, J.M., Gilmour, C.C., Mason, R.P., and Heyes, A. (1999) "Sulfide controls on mercury speciation and bioavailability to methylating bacteria in sediment pore waters." *Environmental Science & Technology*, 33: 951-957.
7. Bourg, A.C.M. (1994) "Speciation of heavy metals in soils and groundwater and implications for their natural and provoked mobility." In: Salomons W, Mader P, Förstner U (eds.) *Pathways, Impact and Engineering Aspects of Metal Polluted Sites*. Springer-Verlag: Berlin.
8. Burton, E.D., Bush, R.T., Sullivan, L.A., Johnston, S.G., Hocking, R.K. (2008) "Mobility of arsenic and selected metals during re-flooding of iron- and organic-rich acidsulfate soil." *Chemical Geology*, 253: 64–73.
9. Creger, T.L., and Peryea, F.J. (1994) "Phosphate fertilizer enhances arsenic uptake by apricot liners grown in lead-arsenate-enriched soils." *HortScience*, 29: 88-92.
10. CSIRO (2011) *Acid sulfate soils along the River Murray, Lake Alexandrina and Lake Albert*. Accessed June 27, 2011 at http://www.clw.csiro.au/acidsulfatesoils/murray.html
11. Friedlingstein, P. et al. (2006) "Climate-carbon cycle feedback analysis: results from the C4MIP model intercomparison." *Journal of Climate*, 19: 3337-3353.
12. Fung, I.Y., Doney, S.C., Lindsay, K., and John, J. (2005) "Evolution of carbon sinks in a changing climate." *Proceedings of the National Academy of Sciences*, 102(32): 11201-11206.

13. Garrels, R.M. and Christ, C.L. (1965) *Solutions, Minerals, and Equilibria*. Harper & Row: New York.
14. Hesterberg, D., Stigliani, W.M., and Imeson, A.C., eds. (1992) *Chemical Time Bombs: Linkages to Socioeconomic Development*. International Institute for Applied Systems Analysis (IIASA), Executive Report 20: Laxenburg, Austria.
15. IPCC (Intergovernmental Panel on Climate Change) (2011) *Special Report on Renewable Energy Sources and Climate Change Mitigation*. IPCC: Geneva.
16. IPCC (Intergovernmental Panel on Climate Change) (2007) *Fourth Assessment Report: Climate Change 2007*. IPCC: Geneva.
17. Johnson, N.M., Driscoll, C.T., Eaton, J.S., Likens, G.E., and McDowell (1981) "'Acid rain' dissolved aluminum and chemical weathering at the Hubbard Brook Experimental Forest, New Hampshire." *Geochimica et Cosmochimica Acta*, 45: 1421-1437.
18. Johnston, S.G., Burton, E.D., Bush, R.T., Keene, A.F., Sullivan, L.A., Smith, D., McElnea, A.E., Ahern, C.R., Powell, B. (2010a) "Abundance and fractionation of Al, Fe and trace metals following tidal inundation of a tropical acid sulfate soil." *Applied Geochemistry*, 25: 323–335.
19. Johnston, S.G., Keene, A.F., Burton, E.D., Bush, R.T., Sullivan, L.A., McElnea, A.E., Ahern, C.R., Smith, C.D., Powell, B., Hocking, R.K. (2010b) "Arsenic mobilization in a seawater inundated acid sulfate soil." *Environmental Science & Technology*, 44: 1968–1973.
20. Kabata-Pendias, A. (2001) *Trace Elements in Soils and Plants* (third edition). CRC Press: Boca Raton, Florida.
21. Lloyd, J., and Taylor, J.A. (1994) "On the temperature dependence of soil respiration." *Functional Ecology*, 8: 315-323.
22. Miljostyrelsen (1984) *Oxygen Depletion and Fish Kill in 1981—Extent and Causes*. National Agency of Environmental Protection: Copenhagen.
23. Morgan, H., and Simms, D.L. (1988) *The Shipham Report: an Investigation into Cadmium Concentration and Its Implications for Human Health*. Elsevier Science Publishers: Amsterdam.
24. National Research Council (1986) *Acid Deposition: Long Term Trends*. National Academy Press: Washington, DC.
25. National Research Council (1985) *Acid Deposition: Processes of Lake Acidification*. National Academy Press: Washington, DC.
26. Neumann, R.B., Ashfaque, K.N., Badruzzaman, A.B.M, Ali, M.A., Shoemaker, J.K., and Harvey, C.F. (2010) "Anthropogenic influences on groundwater arsenic concentrations in Bangladesh." *Nature Geoscience*, 3: 46-52.
27. Nogawa, K., Kurachi, M., and Kasuya, M., eds. (1999) *Advances in the Prevention of Environmental Cadmium Pollution and Countermeasures*. Eiko Laboratory: Kanazawa, Japan.
28. O'Loughlin, E.M. (1986) "Prediction of surface saturation zones in natural catchments by topographic analysis." *Water Resources Research*, **22**: 794–804.
29. Peryea, F.J. (1991) "Phosphate-induced releases of arsenic from soils contaminated with lead arsenate." *Soil Science Society of America Journal*, 55:1301-1306.

30. Salomons, W. (1995) "Long-term strategies for handling contaminated sites and large-scale areas." In Salomons W, and Stigliani WM (eds.) *Biogeodynamics of Pollutants in Soils and Sediments, Risk Assessment of Delayed and Non-linear Responses*. Springer-Verlag: Berlin.
31. Salomons, W., and Stigliani, W.M., eds. (1995) *Biogeodynamics of Pollutants in Soils and Sediments, Risk Assessment of Delayed and Non-linear Responses*. Springer-Verlag: Berlin.
32. Schroder, H. (1985) "Nitrogen losses from Danish agriculture—trends and consequences." *Agriculture, Ecosystems and Environment*, 14: 279-289.
33. Schulze, E.D. (1989) "Air pollution and forest decline in a spruce (*Picea abies*) forest." *Science*, 244: 776-783.
34. Stigliani, W.M. (2002) "Contaminated lands and sediments: chemical time bombs?" *Encyclopedia of Global and Environmental Change*. John Wiley & Sons: Chichester, UK.
35. Stigliani, W.M. (1996) "Buffering capacity: its relevance in soil and water pollution." *New Journal of Chemistry*, 20: 205-210.
36. Stigliani, W.M. (1995) "Global perspectives and risk assessment." In Salomons, W., and Stigliani, W.M. (eds.) *Biogeodynamics of Pollutants in Soils and Sediments, Risk Assessment of Delayed and Non-linear Responses*. Springer-Verlag: Berlin.
37. Stigliani, W.M. (1993) "Overview of the chemical time bomb problem in Europe." In: ter Meulen, G.R.B., Stigliani, W.M., Salomons, W., Bridges, E.M., and Imeson, A.C. (eds.) *Proceedings of the European State-of-the-Art Conference on Delayed Effects of Chemicals in Soils and Sediments,* 2-5 September 1992, Veldhoven, The Netherlands. The Foundation for Ecodevelopment "Stichting Mondiaal Alternatief": Hoofddorp, The Netherlands
38. Stigliani, W.M., ed. (1991) *Chemical Time Bombs: Definition, Concepts, and Examples*. International Institute for Applied Systems Analysis (IIASA) Executive Report 16: Laxenburg, Austria.
39. Stigliani, W.M. (1988) "Changes in valued capacities of soils and sediments as indicators of nonlinear and time-delayed environmental effects." *International Journal of Environmental Monitoring and Assessment*, 10: 245-307.
40. Stigliani, W.M., and Anderberg, S. (1994) "Industrial metabolism at the regional level: The Rhine Basin." In: Ayres RU, and Simonis UE (eds.) *Industrial Metabolism--Restructuring for Sustainable Development*. United Nations University Press: Tokyo.
41. Stigliani, W.M., and Jaffe, P.R. (1993) *Industrial Metabolism and River Basin Studies: a New Approach for the Analysis of Chemical Pollution*. International Institute for Applied Systems Analysis (IIASA) Research Report 93: Laxenburg, Austria.
42. Stigliani, W.M., and Salomons, W. (1993) "Our father's toxic sins." *New Scientist*, 140: 38-42.
43. Stigliani, W.M., Jaffe, P., and Anderberg, S. (1994) "Metals loading of the environment." In: Socolow R, Andrews C, Berkhout F, and Thomas V (eds.) *Industrial Ecology and Global Change*. Cambridge University Press: Cambridge.

44. Stigliani, W.M., Anderberg, S., and Jaffe, P.R. (1993a) "Industrial metabolism and long-term risks from accumulated chemicals in the Rhine Basin." *Industry and Environment (UNEP)*, 16(3): 30-35.
45. Stigliani, W.M., Jaffe, P.R., and Anderberg, S. (1993b) "Heavy metal pollution in the Rhine Basin." *Environmental Science & Technology*, 27: 786-793.
46. Stigliani, W.M., Doelman, P., Salomons, W., Schulin, R., Smidt, G.R.B., and van der Zee, S.E.A.T. (1991) "Chemical time bombs: predicting the unpredictable." *Environment*, 33: 4-9/26-30.
47. Szabolz, I. (1991) "Salinization potential in European soils." In: Brouwer FM, Thomas AJ, and Chadwick MJ (eds.) *Land Use Changes in Europe, Processes of Change, Environmental Transformations and Future Patterns*. Kluwer Academic Publishers: Dordrecht.
48. ter Meulen, G.R.B., Stigliani, W.M., Salomons, W., Bridges, E.M., and Imeson, A.C., eds. (1993) *Proceedings of the European State-of-the-Art Conference on Delayed Effects of Chemicals in Soils and Sediments,* 2-5 September 1992, Veldhoven, The Netherlands. The Foundation for Ecodevelopment "Stichting Mondiaal Alternatief": Hoofddorp, The Netherlands.
49. Tufano, K.J. and Fendorf, S. (2008) "Confounding impacts of iron reduction on arsenic retention." *Environmental Science & Technology*, 42: 4777–4783.
50. WHO (World Health Organization) (2004) *Water Sanitation and Health*. World Health Organization: Geneva.
51. Wong, V., Johnston, S.G., Burton, E.D., Bush, R.T., Sullivan, L.A., and Slavich, P.G. (2010) "Seawater causes rapid trace metal mobilization in coastal lowland acid sulfate soils: implications of sea level rise for water quality." *Geoderma*, 160: 252-263.

AGRICULTURE IN A MORE POPULOUS, HIGHER ATMOSPHERIC (CO_2) WORLD–INVITING 10 BILLION TO THE BANQUET IN 2050: SOME WAYS AND MEANS

LOWELL WOOD AND PHILIP ECKHOFF
Hoover Institution, Stanford University, University of California, Lawrence Livermore National Laboratory, Stanford, CA, USA

THE CHALLENGE

The tremendous expansion in the human population's ability to produce food has allowed unprecedented population growth over the past two centuries,[1] but further increases in food production will be required over the rest of this 21st century. The dramatic increase in the food supply in some parts of the world has improved nutritional status, reduced stunting and wasting, and allowed longer, healthier, and more productive lives.[2] Authoritative estimates are for human population growth to grow to approximately 9+ billion in 2050 from the present baseline of 6+ billion,[3] peaking at approximately 10 billion before 2100. Total food production requirements are estimated to grow by 2X over the coming half-century, which is more than the population increase factor. This increased factor is due to population growth concatenated with enhanced *per capita* nutrition: more/better calories for the Developing World. For example, 35–40% of Indian children now are chronically under-nourished—possibly as many as 50% in Africa, and tens of millions of children per generation thus grow up permanently stunted, physically *and* cognitively—a huge physiological underclass that can't be rescued 'later'. Furthermore, some of the most significant population growths will occur in Africa, putting even more stress on local food supplies.[3]

Changes in the physical world over the next half-century will impact, bound, or enable human agriculture. Some geophysical features seem foredoomed. Sunlight, diurnality, continent size/shape/topography seem likely constant. Higher-than-historical atmospheric (CO_2) which is steadily and 'indefinitely' rising will have a variety of impacts. We are *de facto* "locked in" to (likely more intensive) fossil fuels usage for several decades-to-come—driven by Developing World demands for energy "created on the most practical terms," and nothing short of large-scale civilization collapse seems to "stand in their way," pre-2050±. Precipitation amounts and patterns, as well as surface water-flows, will be similar, although averages will trend upwards with larger spacetime 'patchiness' in seasonal amounts and rates. Trends in other pertinent features are more nebulous, such as global mean temperatures—and perhaps latitudinal variations. Humanity *may* "take a chance" on geoengineering, as "the least-awful alternative." This chance seems likely to occur late, possibly quite hastily and suboptimally—but *eventually* effectively, with attached basic challenges of deciding what, when, where, and how much, e.g., as sketched in Calderia and Wood, 2008.[4] Determination of winners and losers—and even determination of the *process* for provisionally determining winners and losers—will be contentious. Possible interactions of increased atmospheric (CO_2) and geoengineering with food production are represented in Figure 1.[5-6]

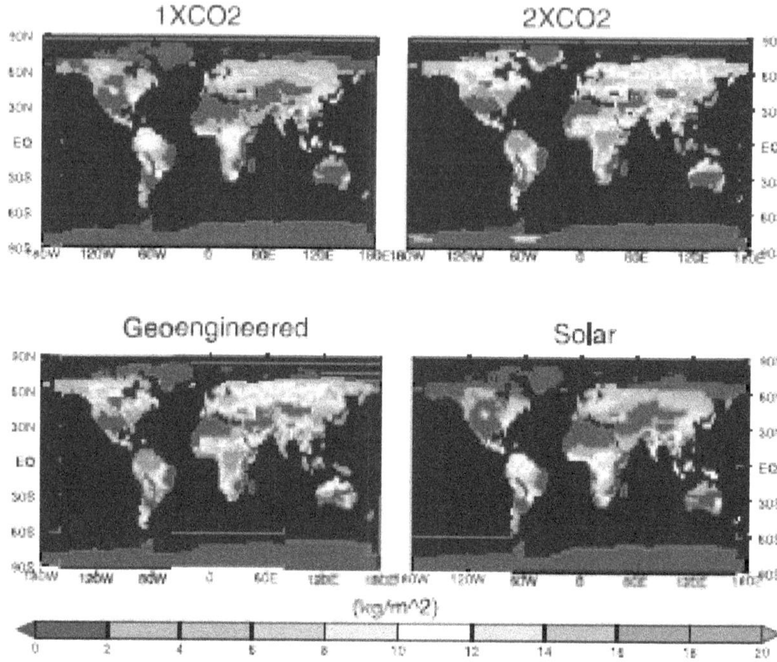

Fig. 1: Net primary (plant) productivity, in kg/m^2, of terrestrial land masses, as modeled by the IBIS code (non-N/P/K- and rootwater-constrained mode, i.e., only light-, water- and CO_2 dependent) with a slab ocean used in conjunction with the Community Climate Model (CCM3). The upper-left panel ('1XCO$_2$') shows the Earth with the preindustrial CO_2 concentration, while the lower-left panel ('Geoengineered') corresponds to twice the preindustrial CO_2 concentration and 1.8% less sunlight, resulting in the same mean global temperature; the upper-right panel ('2XCO$_2$') shows the Earth with twice the preindustrial CO_2 concentration, and the lower right ('Solar') the Earth with preindustrial CO_2 concentration together with 1.8% less sunlight. The globally-aggregated land-plant productivity in the lower-left panel is ~70% greater than in the upper-left panel, implying a global agricultural aggregated crop value difference of a large fraction of \$1 trillion/year in the enriched CO_2 and geoengineered case, although substantial cultivar-to-cultivar variations may be expected due to different sensitivities to CO_2 and ambient temperature. Much of this primary productivity gain results from just the 2X CO_2 'fertilization' effect, i.e., is indifferent to geoengineering of mean insolation and temperature, as comparison of the lower-left and upper-right panels indicates. Figure from Govindasamy et al.[6]

If present temperatures are maintained and (CO_2) doubles, large net 'fertilization' seems likely to occur: 40–70% primary productivity gains, as plants are 'fed' (far) more CO_2 per unit of water-transpired, with productivity results from commercial greenhouses with atmospheres enriched in (CO_2) being congruent with model predictions.

Even global water-supplies may become surprisingly 'malleable.' Repurposing of even modest fractions of extant riverine flows ($\leq 10^6$ m^3/sec, aggregate—1 Sverdrup) may suffice to "make all the difference" in global agriculture. A flow of 1 Sverdrup, which corresponds to approximately 3% of the Florida Gulf Stream, annually would flood the planet's currently farmed ~1.6 GHa—16 million km^2—to a depth of approximately 2 m! A third of riverine flows re-purposed, possibly from deltas, could revolutionize global agricultural productivity. Three TW of electric power suffices to desalinate a Sverdrup of seawater—or to lift this water-flow 300 meters against gravity. This 3 TWe would correspond to 25–30% of projected mid-to-late 21st Century planetary electricity supply (cf. 1 TWe in USA today). Thus water availability can be traded off against energy availability. These tradeoffs likely will be seriously considered in the face of already-present spatiotemporal water shortages and droughts.[7-9]

The starkness of the need for increased food production is enhanced by the challenges faced by current production rates. Population growth may diminish available farmlands, climate change may challenge some traditional agricultural practices, demands of richer populations for more 'costly' foodstuffs (e.g., meat) may grow, and diversion of food croplands for production of non-foodstuffs such as bio-fuels might become substantial. Ethanol creation already consumes a third of USA corn production, and food prices thus are becoming even more closely coupled to (possibly soaring) energy prices.[10] Due to these challenges and the increased needs, it is clear that *far larger* food-production advances now must be made *far faster* than ever before in history.

AVAILABLE SOLUTIONS

A wide variety of means for achieving these necessary advances are available, both 'Ordinary' approaches and 'Extraordinary' gambits. The likely path forward involves vigorously exercise of 'ordinary' ones, while concurrently developing 'extraordinary' ones as a safety net against forever-uncertain longer-range futures, a classic "belt and suspenders" strategy. In addition to increasing overall production rates, it will be necessary to dampen the troughs of food production oscillations in space and time. Humans survive for at most weeks without food, and waiting to take action until famine gnaws has never worked. Dampening these oscillations has been the historical function of granaries, one of the all-time-great inventions. Since their introduction in 9300 BC in the Jordan River valley—well before the advent of 'regular' agriculture—they have been ubiquitously popular. Their expanded and geographically extended use for ordinary *and* extraordinary oscillations and emergency situations *must* go hand-in-hand with expanded food production.

Available 'Ordinary' means involve a *concatenation* of "rather traditional" approaches. These include pervasive use of top-ranked seeds and gametes, which can be tailored for local farming environment and practices. Where and when applicable, robustness to parasites may be a key element of the ranking. In other settings, compatibility with short growing seasons allowing multiple crop-cycles annually may be preferred. In other areas, drought resistance is the most valued characteristic. The rankings will change depending on the local setting, but use of the practice of using the locally-top ranked seeds should not.

The concatenation of "rather traditional" approaches further extends "best practices" regarding farming in various locales, the improvements in which over time can be seen in Figure 2.[11] Diligent and sustained reclamation or enhancement of 'lost' and marginal lands expands available farmland and compensates for farmland lost to population pressures or climate change. Exploitation of (possibly, greatly) improved intra- and inter-seasonal weather predictions will allow time-bounding of growing seasons usefully-far-in-advance and thus support 'rational farming practices.' Such practices include rigorous suppression of weeds and herbivores—by means traditional and/or modern. Improved weather predictions will also allow better planning for local temporary food shortages.[12-13] Carefully structured and planned growing seasons allow optimized locale and time-tailored fertilizer usage,[1] which may be especially important if the land must produce multiple crops per year. Improvements require astute water supply and management, such as evaporation suppression. Low-waste harvesting, transport, storage, and distribution provide surprisingly high opportunities for gains as well, as current rates-of-waste of produced food remain high.[14] The overall system can then be rounded out with reliable and uniform distribution of foodstocks from capacious, reasonably-local stores, to *timely* substitute for local crop-failure.

Available 'Ordinary' means arguably can "do the entire job" via steady/sustained, incremental gains along each axis. All gains that are made then *multiply*: sustained average food productivity gain of ~1.75%/yr "does the 2X-by-2050 job"! This would be a distinctly non-miraculous level of agricultural accomplishment, from now to 2050, one readily exceeded throughout the 1960-80 interval of intensive agricultural R&D. Indeed, a productivity gain of 1.75%/year was nearly sustained over the past several 'mostly thinly-resourced' decades of notably sub-par agricultural investment.

Increases in gross global agricultural productivity will not suffice by itself; indeed, it will be *crucial* to extend pertinent *knowledge*, *technologies* and *practices* into essentially all Developing World settings. As this capacity is extended, comparatively large agricultural productivity gains become 'instantly available' from farmlands currently under-productive due to suboptimal practices. These gains achieved merely through lateral transfer of existing and available means not only provide overall global productivity gains, they likely provide the largest marginal gains in the areas most in need. A positive feedback cycle of economic development and farming technology advancement can then enable better-nourished Developing World populations to *leap* into Developed condition![2]

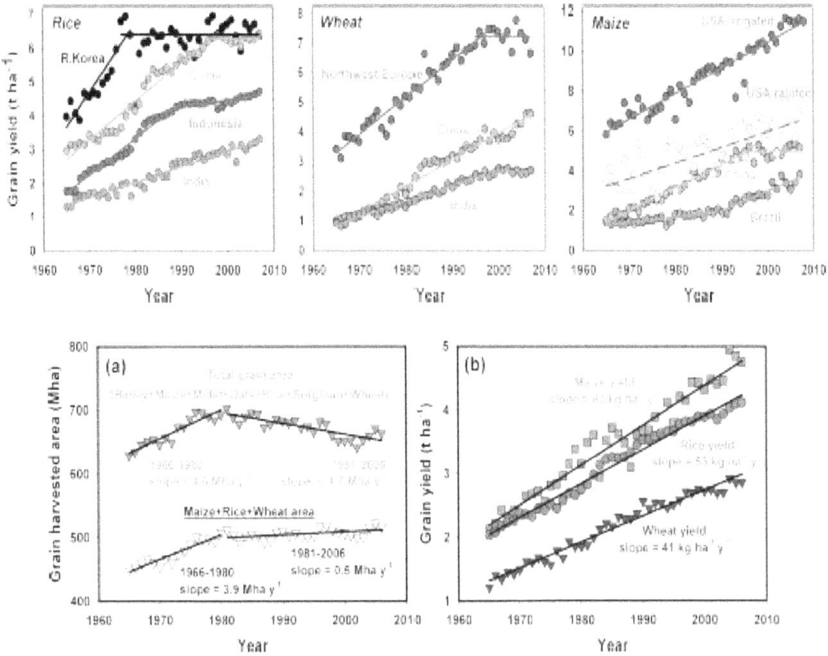

Fig. 2: Increases in grain yield, in tonnes/hectare, by grain-type, farming-type and geographical region vs. time.[11]

The alternative to establishing this productive cycle will be increased malnutrition, with its lifelong impacts on the affected and on the overall population in which they are embedded.[15] Sustained early childhood malnutrition is notorious for causing stunting and wasting[2] and is associated with poorer cognitive development[16] and educational outcomes,[17] which may have not merely lifelong but even multigenerational impacts.[18] Improved nutritional status over the past 300 years in the Developed World has been a major driver in increased average heights, extended life expectancies, and expanded lifetime work productivities, gains which typically accumulate from one generation to the next.[2] Making comparable gains in Developing World agricultural productivity relative to present average levels will involve gains *far* below those corresponding to the theoretical upper-limits on cultivar productivity.

As these gains are achieved, safety and stability of food supply against pest and cultivar diseases must be assured. More intensive monoculture—locally and globally—necessarily will increase vulnerability to threats such as wheat leaf and stem rusts, e.g., Ug99.[19] The associated risks become more severe in relatively-low resource areas with fewer available options-for-coping. These risks can be combated effectively with *sustained* anti-threats R&D *and* pervasive monitoring for advent of new threats—as well as the recrudescence of older ones.

Even with protections against diseases, it's increasingly crucial to have reliable and uniform distribution of foodstocks that are local culture-compatible. At present, food is relatively plentifully produced—*globally averaged*—but occasionally-acute shortages arise/persist in many areas. Adequate, reliable, and dependable storage and distribution infrastructures are required to 'ride through' droughts, heat-waves, and infestations in reasonably graceful manners. Challenges to infrastructure capacity and utilization will also arise from conflicts, natural disasters, or even local politics. Upgraded food storage and distribution may need to be supplemented by widely-distributed and stockpiled stores of "emergency rations." Severely malnourished infants die on one-week time-scales, but essentially all lives-at-risk could be fully protected with fully life-sustaining but low-appeal omni-nutrition packets—ones suitable even for infants. Such packets could resemble *Plumpy'nut*[12] at <1/10 the present cost/price and larger-scale versions could fully nourish a working adult indefinitely for a cost of ~$0.1/day.

EXTRAORDINARY MEANS

Beyond clearly achievable 'Ordinary' means, the potential is great for 'Extraordinary' means to provide a decisive victory over hunger and food uncertainty. These means involve a *concatenation* of steadily advancing application of molecular biological advances to food-plants and animals. These advances can improve resistance to both novel and 'traditional' pests and diseases and provide 'cooperation' with farmers' suppression of competitors/parasites. Such advances may achieve improved toleration of stresses, and thereby allow new farming techniques such as early-in-growing-season planting techniques through means of coping with wet/flooded fields.

Multiple such advances may allow broad expansion of territories supportive of two-crop-cycles-per-season farming, with different engineered stress tolerances constituting the necessary enablement in different settings. For instance, if molecular biological means could achieve growth-pausing during drought intervals, this would provide salvation of crops now (largely) lost to transient 'dry spells,' e.g., maize-during-embryogenesis. Advances on these fronts will rather inevitably result from intense, broad, and sustained R&D efforts.[20-21]

A broad frontier for extraordinary gains center on enhanced utilization of sunlight, water, fertilizers, and more to achieve net photosynthetic efficiency gains—*perhaps as high as 2X!* Long-range, high-volume transport—e.g., ~1 Sverdrup-megameter—of sustaining water would achieve greater year-to-year predictability and geographically extend agricultural productivity and the reliability thereof. High payoff goals include *in situ* fixation of nitrogen[1] or a shift from C_3 to C_4 photo-fixation of CO_2,[22] with gains as seen in Figure 3.[22] Sustained, multi-season enhancement of soil as a growth-supporting/water-retaining medium would allow reliable retention of heavier/more sporadic precipitation in plantroot-accessible forms.

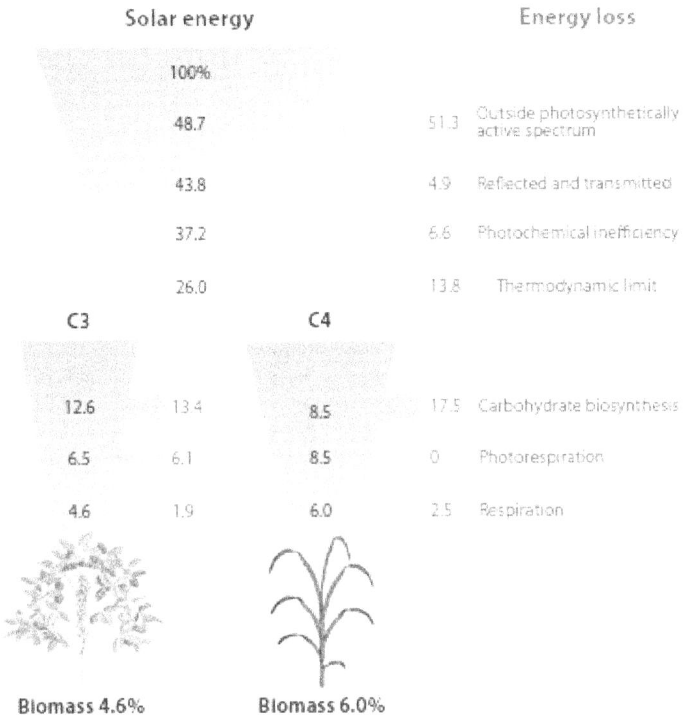

Fig 3: Effects of changing from C_3 to C_4 photo-fixation of CO_2 on overall plant biomass productivity.[22]

Further up the plant structures, one can readily envision more efficient growth and preferential production of human-desired 'products', e.g., larger, 'richer' reproductive tissues and organs. Such gains can be multiplied by improving resistance to spoilage during transport, storage, processing, marketing, or home-preparation.

These advances can be pursued via *either*-or-*both* (neo-)classical breeding and modern gene-transfer practices, with the particular selection-of-practices tailored to local preferences. For instance, it should be possible to accommodate possibly quite different 'tastes' in the PRC and the EU,[23] and for ag-animal versus human foodstuffs. This pursuit involves steady and sustained, often-incremental gains along each axis. Each achieved gain along any given axis multiplies the effects of gains along all other axes and potentially 'stack on top' of gains attained by 'ordinary means.' All together, these efforts likely will aggregate too much in excess of the requisite annual-average agricultural productivity gain of $\leq 2\%$.

DISCUSSION

Some optimizations may add margin or apply a multiplier to gains achieved. One broad arena of frequent discussion is the evolution of dietary preferences and practices. The corresponding challenge includes current and potential changes in future time driving up food requirements, but may be possible to stabilize or actually decrease food requirements over time by shifting preferences toward foodstuffs grown with higher end-to-end efficiency. Such foodstuffs include plant-fractions of higher protein content and/or higher protein efficiency as well as ag-animals such as birds that convert plant foodstuffs more efficiently into edible fractions.[1]

Changes in Developed World practices will throw long shadows onto prospects for Developing World improvements, including greater availability of raw materials and intermediate products. Changes would allow optimized use of chemical fertilizers and feed grains: e.g., should maize and soybeans be fed to cattle or to chickens? Improved pasture management is another route towards enhancement of net protein output, as is currently done in much of Europe[1] and is increasingly done in Brazil, where recent increases in cattle production have largely been driven by productivity rather than area increases.[24] Such changes and directed social evolution clearly would be the work of several decades. Peoples' food-tastes can't be 'switched' overnight but they do evolve on generational time-scales under no more than mass-marketing suasions—which time-intervals would be brief enough to correspond to longer-term food availability goals!

Another optimization focuses on effective dampening of the seasonality of food production. This can be achieved through more extensive transportation and distribution and storage infrastructures, which would help to buffer possible regional-scale perturbations, which are indeed likely to surface over sufficiently long time-scales. Such dampening would be assisted by modulation of the annual calendar of demands for food-from-abroad, enhanced local production and multi-season/-year storage, and greater production and transport of multiple crop/year foodstuffs.

Is a Second Green Revolution necessary? Quite likely *not*. More uniform 'learning of lessons' from the first one probably would suffice: use of superior seeds, indicated fertilizers, adequate water, proper preparation and maintenance of soil—all leveraged by "best farming practices." Average yields are often far below peak ones in the same area, e.g., 400^+ bushels of maize in Iowa vs. ~150 bushels regional average.[25] Indeed, world pace-setting agricultural unit productivity levels are substantially above USA-best ones, in essentially all cases, as indicted in Figure 4.[26] The basic lesson is that farmers who are "determined to win" usually do. The year 2009 saw new records in total corn production in the United States and a breaking of the yield per acre record set in 2004, showing the attainability of gains even in already-high productivity settings.[27]

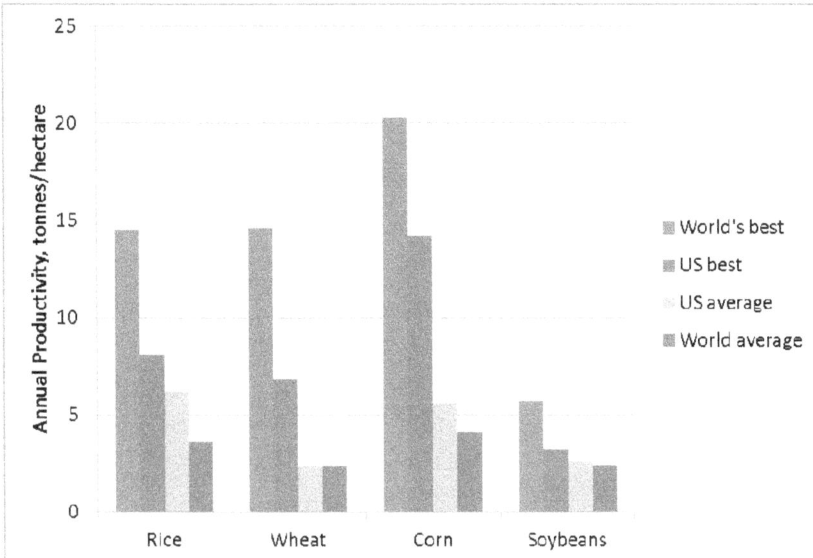

Fig. 4: Comparisons of cultivar productivities across various metrics. Clearly, average performances in both USA and global circumstances are greatly inferior to the best ones in the same geographical circumstances, suggesting the scales of average performance enhancement that are potentially available via uniform institution of "most productive farming practices." Data as quoted in Smil,. 2002.[26]

However, a Second Green Revolution would be intensely desirable for its ability to provide margins-for-shortfalls in critical areas such as Africa with its burgeoning populations, often-rudimentary farming practices and frequently-'indifferent' water supply. Not only is it desirable, it likely would be readily attainable with restoration of the agricultural R&D investment levels of a few decades ago—the more so as stunningly-large returns-on-investment seem available.

SUMMARY

Humanity "has its work cut out for it" re food production advances over the next several decades, with 2X gains in total food production being required by 2050. These are larger gains—by far—in absolute food production than during any time previously, and moreover this is to be done *after* the seemingly "easy" gains have already been attained, *and* while confronting swiftly-rising pressures on land- and water-usage, as well as competition for farmed products from non-food sectors such as fuel production!

However, even 'ordinary' means will *seemingly* suffice—*if* exercised diligently and soon, with greatly-improved food storage and distribution infrastructures—especially in imperiled areas and regions.

To be sure, greatly stepped-up R&D expenditures and infrastructure investments may *both* be required—perhaps to Green Revolution levels—to comprise adequate 'safety nets' underneath these new structures. Monitoring and evaluation of changing practices and conditions will be important components of such efforts.[28]

Some resort to 'extraordinary' means may *assure* worldwide success in greatly stepped-up food production, even in the face of multi-axially 'pressured' farmlands arising from population growth/urban expansion, bio-fuel cropping, monoculture risk-mitigations, and other presently less well-defined threats to global food productivity.

Considering these significant challenges but with a workable path forward, ***the time to act is now.*** Substantial delay will degrade modest extant margins to possibly disastrous extents. Well-thought-out, purposeful, sustained, adequately-resourced agricultural development programs must commence soon if *billions* of our fellow humans are not to be irretrievably crippled—physically *and* cognitively—by sustained poor nutrition *in our own times.*

REFERENCES

1. Smil, V. (2001) *Enriching the Earth.* Boston: Massachusetts Institute of Technology.
2. Floud, R., et al. (2011) *The Changing Body: Health, Nutrition, and Human Development in the Western World since 1700.* New York: Cambridge University Press.
3. DESA (2011) *World Population Prospects: The 2010 Revision.* New York: United Nations.
4. Caldeira, K. and L. Wood (2008) "Global and Arctic Climate Engineering: Numerical Model Studies." *Phil. Trans. R. Soc. A.* **366**: pp. 4039-4056.
5. Govindasamy, B. and K. Caldeira (2000) "Geoengineering Earth's radiation balance to mitigate CO2-induced climate change." *Geophysical Research Letters.* **27**(14): pp. 2141-2144.
6. Govindasamy, B., et al. (2002) "Impact of geoengineering schemes on the terrestrial biosphere." *Geophysical Research Letters.* **29**(22).
7. Wong, E. "Plan for China's Water Crisis Spurs Concern", in *The New York Times.* June 1, 2011: New York.
8. IEG, *Water and Development.* (2010) Washington DC: International Bank for Reconstruction and Development/ The World Bank.
9. Wong, E. "China: Drough Leaves 14 Million Chinese and Farmland Parched", in *The New York Times.* September 9, 2011: New York.
10. Polansek, T. (2011) Global Food Prices Stay High. *The Wall Street Journal.*
11. Cassman, K.G., P. Grassini, and J. van Wart (2010) *Global Food Security in a Changing Climate: Crop Yield Trends and Yield Potential,* in *Handbook of Climate Change and Agroecosystems: Impacts, Adaptation, and Mitigation.,* D. Hillel and C. Rosenzweig, Editors. Imperial College Press: London. pp. 37-51.

12. Tectonidis, M. (2006) "Crisis in Niger—Outpatient Care for Severe Acute Malnutrition." *New England Journal of Medicine*. **354**(3): pp. 224-227.
13. Verdin, J., et al. (2005) "Climate science and famine early warning." *Philosophical Transactions of the Royal Society B: Biological Sciences*. **360**(1463): pp. 2155-2168.
14. Butler, D. and N. Spencer (2010) "The Growing Problem." *Nature*. **466**: pp. 546-547.
15. Mendez, M.A. and L.S. Adair (1999) "Severity and Timing of Stunting in the First Two Years of Life Affect Performance on Cognitive Tests in Late Childhood." *The Journal of Nutrition*. **129**(8): pp. 1555-1562.
16. Martorell, R., et al. (1992) *Long-term consequences of growth retardation during early childhood*, in *Human Growth: Basic and Clinical Aspects*, M. Hernandez and J. Argente, Editors. Elsevier Science: Amsterdam.
17. Behrman, J., H. Alderman, and J. Hoddinott (2004) *Malnutrition and Hunger*, in *Global Crises, Global Solutions*, B. Lomborg, Editor. Cambridge University Press.
18. de Rooij, S.R., et al. (2010) "Prenatal undernutrition and cognitive function in late adulthood." *Proceedings of the National Academy of Sciences*. **107**(39): pp. 16881-16886, and Shulz, L. (2010) "The Dutch Hunger Winter and the developmental origins of health and disease," *Proceedings of he National Academy of Sciences*. **107**(39): pp. 16757-16758.
19. CGIAR (2010) "Virulent new strains of Ug99 stem rust, a deadly wheat pathogen." *ScienceDaily*.
20. Gilbert, N. (2010) "Inside the Hothouses of Industry." *Nature*. **466**: pp. 548-551.
21. Pennisi, E. (2011) "Decadal Plan for Plant Science Begins to Take Shape." *Science Insider*.
22. Zhu, X.-G., S.P. Long, and D.R. Ort, (2010) "Improving Photosynthetic Efficiency for Greater Yield." *Annual Review of Plant Biology*. **61**(1): pp. 235-261.
23. Henshaw, C. "China Could Sow the Seeds of GM Crops Growth," in *The Wall Street Journal*. September 9, 2011: New York.
24. Tollefson, J. (2010) "The Global Farm." *Nature*. **466**: pp. 554-556.
25. Martin, D. "Francis Chiles, 68, Dies; Sage of High Corn Yields," in *The New York Times*. 20 January 2008: New York.
26. Smil, V. (2002) *Feeding the World*. Boston: Massachusetts Institute of Technology Press.
27. USDA (2010) *Crop Production 2009 Summary*. Washington DC: United States Department of Agriculture.
28. Sachs, J., et al. (2010) "Monitoring the world's agriculture." *Nature*. **466**(7306): pp. 558-560.

EVIDENCE-BASED MEDICINE AND KNOWLEDGE TRANSLATION TYPE 2 RESEARCH

R. BRIAN HAYNES
Department of Medicine, McMaster University
Hamilton, Canada

ABSTRACT

A huge imbalance exists in what we know from science to be potentially beneficial and what we actually achieve in improving our health and preventing disease and untimely death. We need to create new knowledge on how to translate these benefits into action if we are to redress this imbalance.

Investments and advances in fundamental and applied health research are truly astounding over the past 60 years. The evidence-based medicine (EBM) movement grew from these advances, with the objective of helping policy makers, practitioners, patients and the public understand and appropriately apply the advances. But what EBM has shown to date is that the effects of research on health are minuscule compared with their potential. This low return on investment currently leads to more deaths and misery than any natural disaster in human history, but its solution requires research that is beyond the purview and abilities of traditional basic science.

Knowledge translation research (KTR) is an evolving, 21st century set of scientific strategies for not only studying the transitions between the laboratory and early clinical studies (KTR1), and between advanced clinical research and health care (KTR2, the focus of this paper), but also for overcoming the barriers that currently defeat the application of new knowledge. While it might be thought that KTR2 is a matter of "technology" (defined in the Erice statement as the study of how the power of mankind can be increased), it has recently become clear that we lack sufficient basic understanding of human learning, decision making and actions to consistently achieve health benefits. KTR2 is now seeking to comprehend and surmount the barriers to knowledge uptake, often using multidisciplinary approaches that include the scientific methodologies of epidemiology, psychology and behavioral science, geography, and others.

KTR2 and its congeners, implementation science and comparative effectiveness research, are models of Third Millennium research, all appearing in the scientific literature for the first time since the year 2000. Only if we can achieve substantive progress in this research can we expect to see the positive results of basic and applied research reflected in proportionate improvements in human health.

INTRODUCTION

A barrier, often referred to as the 'knowledge-practice gap', exists between what research has shown to be beneficial under relatively ideal circumstances and what we apply and achieve under usual circumstances. Bridging this gap will save many lives, but doing so is not simply a matter of engineering or trying harder. We need to understand more—through both fundamental and applied research—about how to translate evidence from research into practice.

Knowledge translation research (KTR) is a rapidly evolving 21st century field of science, dedicated to addressing this problem. 'Implementation science' and 'comparative effectiveness research' are closely related fields, with similar methodologies. KTR is typically conducted by multidisciplinary research teams, including epidemiologists, health services researchers, social scientists, geographers, information scientists and others. Studies can be observational or experimental, quantitative or qualitative, depending on the questions being addressed.

Advances in KTR can have profound effects on human health as they can have cross-cutting effects on the uptake of many therapeutic innovations that share common delivery routes, such as self-administered drugs and devices. For example, most efficacious treatments for acute and chronic disease conditions are delivered in the form of self-administered oral medications. If we create ways to assist patients to take one self-administered medication as prescribed, that way may have similar effects for most self-administered medications.

WHAT IS THE MAGNITUDE OF THE KNOWLEDGE TRANSLATION PROBLEM?

Substantial investments since World War II, of what is now well over U.S.$100billion/year (extrapolating from 1992 figures[1]), have created a robust biomedical and health research enterprise, with breakthroughs for many diseases. This success provides hope that healthcare services can be improved despite cutbacks in spending that are occurring in many countries. The evidence-based medicine (EBM) movement grew from the many advances in diagnostics and therapeutics, with the objective of helping all to understand and appropriately apply the genuine advances, while avoiding spurious claims.

But EBM's objectives have been largely thwarted. Simple algebra illustrates the problem. A 25% relative risk reduction in patient-important health outcomes is a moderately large effect for the efficacy of a new treatment, but typically no more than 50% of practitioners will prescribe new treatments within a generation of their validation[2] and typically no more than 50% of people who could benefit will adhere to the intervention if prescribed. Thus, a net benefit is achieved of no more than (25% x 50% x 50% =) 6% at best. For example, for many people with atrial fibrillation, oral anticoagulants can reduce the risk of stroke by over 50%,[3] but surveys show that clinicians prescribe anticoagulants for only about 50% of eligible patients,[4] and of patients for whom anticoagulants are prescribed, only about 40% take them as intended,[5] whittling a 50% relative benefit down to about 10%. About 5% of middle-aged people have atrial fibrillation and 3% these will suffer a stroke over a 15 year period as a result of the atrial fibrillation.[6] Anticoagulation, fully applied, would reduce the attributable stroke rate to 1.5% or less, but at current uptake rates, only to 2.7%. Extrapolated worldwide, annually, this amounts to a difference of over 86,000 strokes averted per year vs. over 17,000 strokes averted per year. To make matters worse, despite extensive public education, only about a third of those who suffer a stroke will arrive at a hospital soon enough to be able to benefit from the most potent treatment for reducing stroke morbidity and mortality.[7]

Overcoming these problems is inherently complex, requiring simultaneous solutions for both practitioner and patient actions to gain traction. For instance, increasing appropriate prescribing from 50% to 75%, without increasing patient adherence, would increase the benefit to only 15%, whereas increasing both prescribing and adherence to 75% would increase the net benefit to 28%. At present, efficiently, consistently and affordably solving this problem is beyond our understanding.

The knowledge-practice gap exists for virtually all health interventions. The loss that it causes for return on investment in research is staggering and assuredly leads to more deaths and suffering than any natural disaster in human history. But this is perhaps not widely appreciated when focusing on fundamental laws of nature, as it involves the interplay of many such fundamentals.

The problem of knowledge translation for health research has only recently come to the attention of research funders, who just in the first decade of the 21st century have been directing modest funds to study the problem. To make matters worse, when such funding is provided by reallocation from existing research funding, there is predictably staunch resistance from the prevailing research community.

HOW IS KNOWLEDGE TRANSLATION RESEARCH DEFINED?

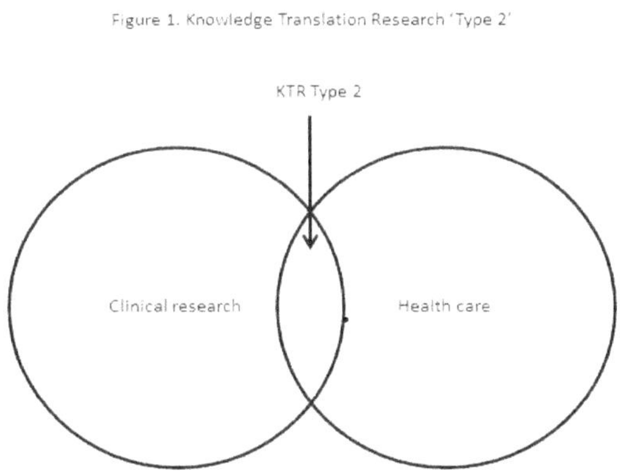

Fig. 1: Knowledge Translation Research 'Type' 2.

Knowledge translation research (KTR) attempts to understand the nature of translational problems, and to develop new methodologies for studying and overcoming them. This field is young, rapidly evolving and still being demarcated, with many attempts at definition, as chronicled by McKibbon and Lokker.[8] KTR 'type 2' (KTR2; Figure 1) is the study of the organization, retrieval, appraisal, refinement, dissemination, and timely application of knowledge from research in "real world" settings (e.g,. application of important new knowledge from research in health care). KTR2 research often involves multidisciplinary teams that include the scientific methodologies of epidemiology, psychology and behavioral science, information science, geography, and others. Largely parallel developments include *implementation science* ("...the scientific study of methods to promote the systematic uptake of clinical research findings and other evidence-based practices into routine practice, and hence to improve the quality and effectiveness of health care. It includes the study of influences on healthcare professional and organizational behavior."[9]), and *comparative effectiveness research* ("...the generation and synthesis of evidence that compares the benefits and harms of alternative methods to prevent, diagnose, treat and monitor a clinical condition, or to improve the delivery of care"[10]). While it might be thought that KTR2 research is a matter of "technology" (defined in the Erice statement as "the study of how the power of mankind can be increased"[11]), in fact, we lack basic understanding of key aspects of research synthesis, health policy making, health systems, individual and shared decision making, human health behavior, and their determinants and interactions.

WHAT IS THE NATURE OF THE KNOWLEDGE-PRACTICE GAP FOR HEALTH CARE?

Figure 2[12] depicts key steps in the path between evidence from research and appropriate use of that knowledge to promote human health. New and better research is needed at each step if use is to improve.

Figure 2. Steps between knowledge from health research evidence and appropriate use

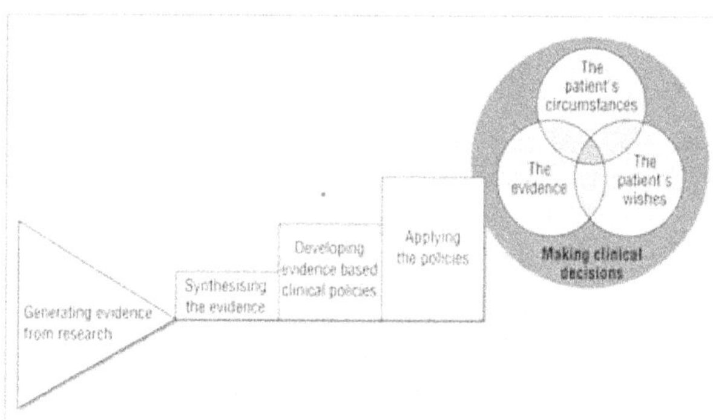

Fig. 2: Steps between knowledge from health research evidence and appropriate use.

The path begins with generating evidence from biomedical research: the shape of the wedge on the left of Figure 2 symbolizes the process of testing discoveries of potential use in health care, eliminating those that lack merit. The broad edge of the wedge represents the initial testing of innovations, which usually occurs in laboratories; many new products and processes fail and are discarded early in the testing process. Products or processes with sufficient initial merit then undergo "phase 1 and 2" human trials to assess toxicity and to estimate efficacy. Many discoveries fail, but a few merit more definitive testing in large "phase 3" controlled trials with important clinical outcomes. It is only when these advanced tests (at the point of the wedge) are successful that serious efforts at dissemination and application are warranted. Even then, the bar continues to rise for what constitutes adequate testing. For example, "efficacy" of "explanatory " testing (i.e., "Can this do more good than harm under relatively ideal circumstances?", typically in placebo-controlled trials), was once all that was needed to bring a new pharmaceutical to market, but current testing often includes "effectiveness" or "pragmatic" trials (i.e., "Does this work under usual health care circumstances?", comparing innovations against best standard care), and "efficiency" studies (i.e., "Is it worth it?", cost-effectiveness studies including patient quality of life measures). Effectiveness and efficiency studies are challenging to conduct and raise the cost of testing considerably but provide much more useful evidence for decision makers.

The steps to the right of the wedge in Figure 2 represent processes that are needed to harness research evidence for healthcare practice. These steps include synthesizing the evidence, developing clinical and health policy from the evidence, applying the policy in health care settings, and making evidence-based clinical decisions at the right place, in the right way, and at the right time. All steps must be successfully mounted to form a valid connection between evidence and practice.

Although traditional research paradigms have been contained within the wedge, KTR extends scientific principles and practice across the spectrum of translation, to include studies of evidence synthesis (e.g., to reduce bias and increase efficiency in assembling evidence[13]), health policy research (e.g., to find better ways to provide evidence-based information to policy makers[14]), health services application research (e.g., community based trials of complex interventions[15]), and studies of clinical decision-making (e.g., trials of computerized clinical decision support systems[16]). Increasingly, behavioral interventions, surgical procedures, and alternative approaches to the organization and delivery of care are being subjected to rigorous evaluation.

Three recent examples of KTR illustrate the potential. The first example is studies of "kangaroo mother care" (KMC) for premature babies. KMC includes thermal care through continuous skin-to-skin contact, support for exclusive breastfeeding or other appropriate feeding, and early recognition/response to illness. A recent systematic review of KMC trials, mainly in low- and middle-income countries, documents that KMC reduced neonatal mortality by over 50% for low birthweight babies; with over 60% reduction in serious morbidity.[17] A key question now becomes how to disseminate the practice of KMC.

My second example addresses this question. Tripathy and colleagues[18] recently published a cluster randomized trial in 2 of the poorest states in India, engaging local facilitators and women's groups to spread some old basic science results (that weren't being applied locally) about clean and safe delivery of babies, newborn care, and maternal health. The results were a striking 32% reduction in perinatal mortality and 57% reduction of moderate post-partum depression. The Society for Clinical Trials named this study as trial of the year at its meeting in May 2011.

The third example suggests that a similar approach could work in countries of much higher health status. Kaczorowski and colleagues[15] devised and tested a community-based intervention to reduce cardiovascular risk in Ontario Canada in 39 mid-sized communities. The intervention centered on volunteer-run cardiovascular risk assessment and education sessions held in community pharmacies. Statistically significant reductions favouring the intervention communities were seen in hospital admissions for acute myocardial infarction and congestive heart failure.

WHAT IS NEEDED FOR KTR TO SUCCEED (TABLE 1)?

Because of the complexity of many knowledge translation problems, it is likely that transdisciplinary teams will be needed to make progress, but creating and effectively mobilizing such teams face formidable problems in communication, culture and academic tradition. Team funding and revising tenure and promotion standards (for example, those that require single-author publications in journals of discipline) could help.

Even without such teams, many knowledge translation problems could be avoided by diagnostic and therapeutic innovations that are based on understanding of human frailties such as low adherence by practitioners and patients. I invite basic scientists to learn about the barriers to knowledge uptake and incorporate this knowledge into their research directions and undertakings.

More cures for chronic disease conditions would certainly be welcome, such as Marshall and Warren's Nobel-prize winning cure for peptic ulcer disease.[19]

Short of cures, surely we can do better than pill-shaped treatments that must be taken several times per day and in great profusion—many patients with chronic diseases take more than 20 pills per day, along with injections, not to mention 'lifestyle' changes in diet, exercise and smoking. Perhaps the behavioral demands of complex regimens can be ameliorated by such innovations as the polypill for those on multiple drugs,[20] or surgical procedures such as gastric banding for those who are unable to control their weights on diets.[21]

Table 1. Ways for knowledge translation research to succeed.

1. Foster collaborations between basic and applied researchers to understand barriers to knowledge uptake
2. Apply, in both basic and applied research planning, what we currently know about translational barriers and facilitators
3. Search for cures for chronic disease
4. Create health care interventions that minimize behavioral demands
5. Improve the methods for KT research
6. Expand KTR training programs
7. Increase funding for KT research

While awaiting such advances, we must learn to do much better with the research evidence we have. Development of new methodology has been required to make progress in KTR and more creativity in methods will be needed at each step. For example, knowledge synthesis—systematic review of all pertinent evidence concerning a given problem—is very laborious at present, with current capacity falling far short of the pace of new knowledge creation.[22] Much more efficient methods of information retrieval and updating reviews are needed, and a study led by Brian Hemens in my research group, just published,[13] provides proof of concept that efficient updating is possible, piggy-backed on an existing "knowledge refinery" and validated information retrieval processes.

Training programs for knowledge translation research are scarce, and few centres in the world have reached a "critical mass" of KT researchers to mount such programs. That said, many schools and academic departments have small programs and several larger programs exist. Many more are needed, perhaps a tenth as many as for basic science programs.

KTR requires funding just as any research, but funding agencies have just begun to trickle funds. Although there is an obvious imbalance between the rate of creation of sound research discoveries and their full utilization, no one knows how much research funding is needed to achieve acceptable balance between discovery, validation and application. At one extreme, one could solve the problem of knowledge translation in a generation by simply shutting off funds for basic research, allowing existing research to 'naturally' diffuse into practice. Or one could shift an arbitrary proportion of research funds, say, 50%, to solving application problems, but there is insufficient capacity—researchers and methods—at present for this to be feasible. A temperate approach would be to accelerate existing research funding for KTR2 by, say, 10% per annum until there is a much better balance between fundamental research and its application. As KTR2 becomes more successful, it is possible that it would indirectly generate more funds for fundamental research, because it would then be more obvious to people that the public investment in both basic and applied research reaps a high rate of return.

CONCLUSIONS

Although one might trace their origins back to Garden of Eden, knowledge translation research and its variants, implementation science and comparative effectiveness research, are truly models of science for the Third Millennium for health care, all of these names appearing in the health literature for the first time since the year 2000. KTR addresses a tragedy in science: discovery of potential value without understanding of the means of application. KTR can help science to thrive by assuring investors, both public and private, of higher return on investment. KTR can help society to promote health and ameliorate illness to full potential. KTR can help individuals to prevent or ameliorate disease to an extent that, at present, we can only wish for.

ACKNOWLEDGEMENTS

I would like to acknowledge the helpful comments of David L. Sackett, Deborah J. Cook, Juan Gabriel Ruiz Peláez and Nathalie Charpak on drafts of this manuscript.

REFERENCES

1. Michaud, C., Murray, C.J.L. Resources for health research and development in 1992: a global overview. Investing in health research and development: report of the ad hoc committee on health research relating to future intervention options. Geneva: World Health Organisation, 1996:217.
2. Antman, E.M., Lau, J., Kupelnick, B., Mosteller, F., Chalmers, T.C. (1992) "A comparison of results of meta-analyses of randomized control trials and

recommendations of clinical experts. Treatments for myocardial infarction." *JAMA*. 268(2):240-8.

3. Andersen, L.V., Vestergaard, P., Deichgraeber, P., et al. (2008) "Warfarin for the prevention of systemic embolism in patients with non-valvular atrial fibrillation: a meta-analysis." *Heart*. Dec;94(12):1607-13.
4. Bradley, B.C., et al. (2000) "Frequency of anticoagulation for atrial fibrillation and reasons for its non-use at a Veterans Affairs medical center." *Am J Cardiol.* 85(5):568-72.
5. Gialamas, A., Yelland, L.N., Ryan, P., Willson, K., Laurence, C.O., Bubner, T.K., Tideman, P., Beilby, J.J. (2009) "Does point-of-care testing lead to the same or better adherence to medication? A randomised controlled trial: the PoCT in General Practice Trial." *Med J Aust.* 191(9):487-91.
6. Conen, D., Chae, C.U., Glynn, R.J., et al. (2011) "Risk of death and cardiovascular events in initially healthy women with new-onset atrial fibrillation." *JAMA*. 305(20):2080-7.
7. Hall, R., Khan, F., O'Callaghan, C., Meyer, S., Fang, J., Hodwitz, K., Bayley, M. Ontario Stroke Evaluation Report 2011: Improving System Efficiency by Implementing Stroke Best Practices. Toronto: Institute for Clinical Evaluative Sciences; 2011.
8. http://whatiskt.wikispaces.com/Implementation+Research
9. http://www.implementationscience.com/about
10. Sox, H.C., Greenfield, S. (2009) "Comparative effectiveness research: a report from the Institute of Medicine." *Ann Intern Med.* 151(3):203-5.
11. http://www.federationofscientists.org/WfsErice.asp
12. Haynes, B., Haines, A. (1998) "Barriers and bridges to evidence-based health care practice." *BMJ* 317:273-6.
13. Hemens, B.J., Haynes, R.B. McMaster Premium LiteratUre Service (PLUS) performed well for identifying new studies for updated Cochrane Reviews. J Clin Epid (in press).
14. Lavis, J.N., Wilson, M.G., Grimshaw, J.M., Haynes, R.B., Ouimet, M., Raina, P., Gruen, R., Graham I. (2010) "Supporting the use of health technology assessments in policymaking about health systems." *Internat J Technol Assess Health Care*; 26(4):405–414.
15. Kaczorowski, J., et al. (2011) "Improving cardiovascular health at population level: 39 community cluster randomised trial of Cardiovascular Health Awareness Program (CHAP)." *BMJ*. 342:d442. doi: 10.1136/bmj.d442.
16. Haynes, R.B., Wilczynski, N.L. for the Computerized Clinical Decision Support System (CCDSS) Systematic Review Team. (2010) "Effects of computerized clinical decision support systems on practitioner performance and patient outcomes: Methods of a decision-maker-researcher partnership systematic review." *Implementation Sci*; 5:12 http://www.implementationscience.com/content/5/1/12.
17. Lawn, J.E., Mwansa-Kambafwile, J., Horta, B.L., Barros, F.C., Cousens, S. (2010) "'Kangaroo mother care' to prevent neonatal deaths due to preterm birth complications." *Internat J Epidemiol* 39(Supplement 1):i144-i154.

18. Tripathy, P., Nair, N., Barnett, S., Mahapatra, R., Borghi, J., Rath, S., Rath, S., Gope, R., Mahto, D., Sinha, R., Lakshminarayana, R., Patel, V., Pagel, C., Prost, A., Costello, A. (2010) "Effect of a participatory intervention with women's groups on birth outcomes and maternal depression in Jharkhand and Orissa, India: a cluster-randomised controlled trial." *Lancet*. 375(9721):1182-92. Epub 2010 Mar 6.
19. http://nobelprize.org/nobel_prizes/medicine/laureates/2005/press.html
20. Yusuf, S., Pais, P., Afzal, R., et al. (2009) "Effects of a polypill (Polycap) on risk factors in middle-aged individuals without cardiovascular disease (TIPS): a phase II, double-blind, randomised trial." *Lancet*. 373(9672):1341-51.
21 Picot, J., Jones, J., Colquitt, J.L., et al. (2009) "The clinical effectiveness and cost-effectiveness of bariatric (weight loss) surgery for obesity: a systematic review and economic evaluation." *Health Technol Assess*. 13(41):1-190, 215-357, iii-iv.
22. Mallett, S., Clarke, M. (2003) "How many Cochrane reviews are needed to cover existing evidence on the effects of health care interventions?" [Editorial]. *ACP J Club*. 139:A11-12.

SESSION 10

WFS GENERAL MEETING

PMP REPORTS — DEBATES AND CONCLUSIONS

AIDS AND INFECTIOUS DISEASES (AID) PMP 2011 REPORT

DR. FRANCO M. BUONAGURO
Istituto Nazionale dei Tumori, "Fondazione G. Pascale"
Napoli, Italy

In the last few years besides monitoring AIDS disease (with participation to International Conferences, including AIDS Vaccine Meeting) the AIDS and Infectious Diseases PMP focused on emerging diseases.

Since 2004, the AID PMP has been monitoring the Avian flu epidemic and a specific session was held at Erice within the context of the 2005 Planetary Emergency Seminars. More recently the PMP focused on the 2009 Flu Pandemic. The critical questions have been: Is there going to be a pandemic? Are there determinants, which one can be monitored? Can we really prevent a pandemic? Is the industry prepared for a massive vaccine production?

The epidemiological of the 2009 epidemic have been reported last year, when it was shown that the epidemic has been very mild considering the total number of infected individuals and the level of morbidity and mortality. A peculiar aspect of the pandemic H1N1 2009 has been, however, the number of children in some geographic areas, including Southern Italy. In this respect it has been reported several cases of children with cardiac involvement, which has been generally referred as the result of the flu involvement of the respiratory apparatus. In a specific case of an 11-year old young girl, instead, it has been possible to verify a direct damage of the cardiac tissue. The Flu virus was detected, by PCR technology, in pericardium and myocardium tissues along with the upper respiratory tissue, while virus was not detected in the lower respiratory tract, showing that the cardiac damage was due to a direct viral assault more that to indirect lung pathology. Furthermore, flu-related myolisis was observed also in the gastrointestinal (GI) smooth muscle cells. Such phenomenon has been reported in the scientific literature also in highly pathogenic avian-flu cases. Such mutated viruses, however, do not show any transmission advantage, given that they do not become the more frequently transmitted flu genotypes.

The key question remains on how to prevent a pandemic and the possibility to develop a more broadly protective vaccine. To this matter a specific session has been also dedicated within the European Congress of Virology, organized by the Italian Society of Virology (whose secretary is Dr. Buonaguro) at Cernobbio-IT, 7-11 April, 2010. The recent, new promising developments, mainly based on crystallographic data, have been reported and discussed in the Conference Report published on Expert Review of Vaccines (FM Buonaguro et al. 2010). Furthermore, two articles on Flu vaccine approaches in plant systems have been included within the special focus issue on Plant-based Vaccines published in the *Expert Review of Vaccines*. A fully new technology of fast-track vaccine production in tobacco plants has been recently discovered and the Bayer GmbH Industry is exploiting such strategy for preparedness to future pandemics.

Finally, members of the AID PMP are exploring further approaches to expand the WFS monitoring activity in AIDS and infectious diseases, including emerging diseases, participating to:

1. The Global Virus Network (GVN) announced at the 12th Annual International Meeting of the Institute of Human Virology, of the University of Maryland Medical School, held in Tropea-IT on 4-8 October 2010.
2. *The Infectious Agents and Cancer (IAC)* online scientific journal, published by BioMed Central, tracked by Medline, Scopus, Index Medicus and ISI Thomson. The journal is currently one of the fastest growing journals in epidemiology, oncology and Infectious diseases categories, with an unofficial impact factor (I.F.) of 2.33.

The AID PMP proposes a Workshop on Infectious Agents and Vaccine Strategies for the Erice Planetary Emergencies Session of 2012 with participation of 4 speakers, who will focus on and discuss innovative approaches to Infectious Agents monitoring and Preventive strategies. Furthermore the PMP is planning an Erice Advanced Course on *Infectious Agents: Biological Threats and Therapeutic Tools*, to discuss potential biological risks as well as recent advances in cancer therapies and vaccines based on pathogens, including oncolytic viruses.

WORLD FEDERATION OF SCIENTISTS
International Seminars on Planetary Emergencies
and Associated Meetings – 44th Session
Erice 19-24 August 2011
The Role of Science in the Third Millennium

AIDS and Infectious Diseases PMP

Franco M. Buonaguro
Member of AIDS and Infectious Diseases PMP
Director, UOC Molecular Biology & Viral Oncology
Ist. Naz. Tumori "Fond Pascale"
Napoli

AIDS and Infectious Diseases PMP

In the last few year besides

Monitoring AIDS disease
(with participation to International Conferences, including AIDS Vaccine Meeting...)

AIDS & Infectious Diseases PMP focused on
emerging diseases

The Influenza Aviaria Group Erice August - 2005

Initial Flu key questions

1. Is there going to be a pandemic?
2. Are there determinants, which one can be monitored?
3. Can we really prevent a pandemic?
4. Is the industry prepared for a massive vaccine production?

Cardiac Tamponade and Heart Failure Due to Myopericarditis as a Presentation of Infection with the Pandemic H1N1 2009 Influenza A Virus

TABLE 1. Laboratory results

Clinical sample or tissue type	Result (titer)	
	Real-time RT-PCR analysis for pandemic A/H1N1	Isolation in MDCK-SIAT1 cells (HAU)
Rhinopharyngeal swab	**Positive**	**Positive (64)**
Pharynx	Negative	Negative
Tonsil	Negative	Negative
Bronchus	**Positive**	Negative
Lung	Negative	Negative
Spleen	Negative	Negative
Heart	**Positive**	**Positive (16)**
Pericardium	**Positive**	Negative
Pericardial fluid	**Positive**	Negative

[a] Positive results are listed in bold.
[b] Real-time reverse transcription PCR (RT-PCR) analysis was performed using the CDC pandemic kit.

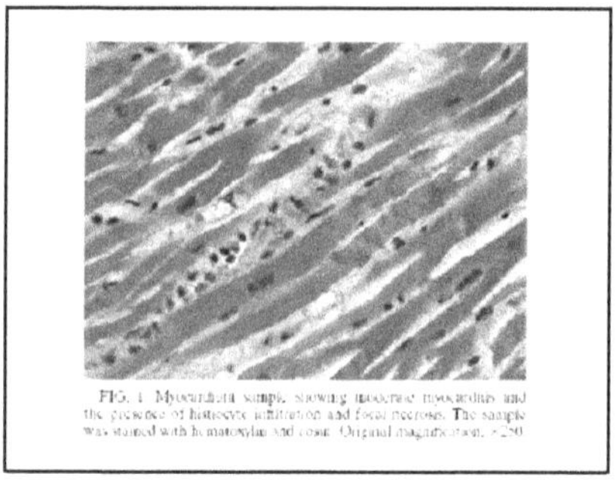
FIG. 1. Myocardium sample showing moderate myocarditis and the presence of histiocyte infiltration and focal necrosis. The sample was stained with hematoxylin and eosin. Original magnification, ×250.

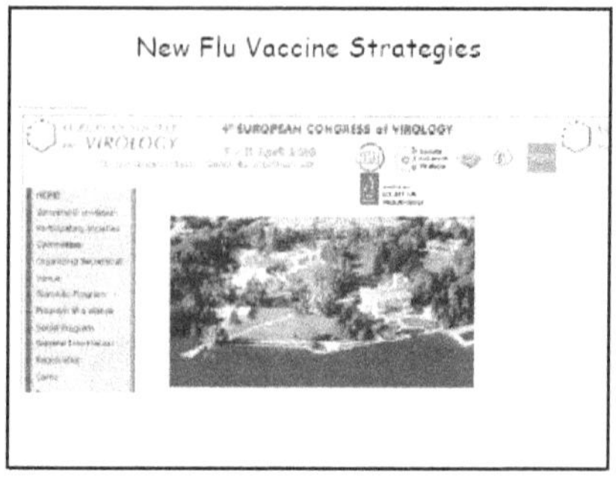
New Flu Vaccine Strategies
4th EUROPEAN CONGRESS of VIROLOGY

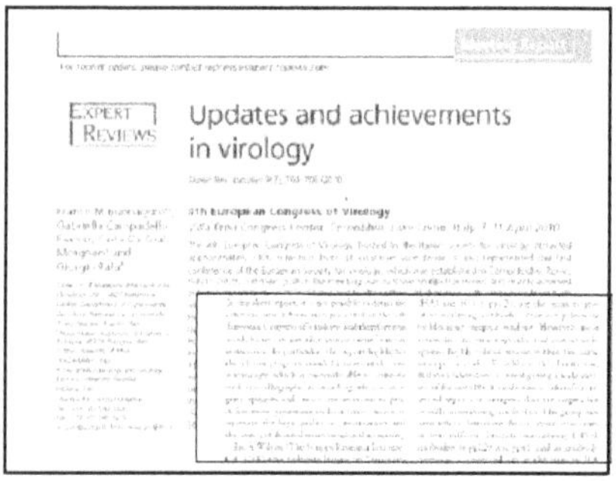
EXPERT REVIEWS
Updates and achievements in virology
4th European Congress of Virology

The INSTITUTE of HUMAN VIROLOGY
Program Information and Acknowledgements
Welcome
12th Annual International Meeting
of the Institute of Human Virology
of the University of Maryland School of Medicine
By
Robert C. Gallo, Founder and Director
Institute of Human Virology
and
Giovanni Latorre, President
University of Calabria

INSTITUTE OF HUMAN VIROLOGY
GLOBAL VIRUS NETWORK (GVN)

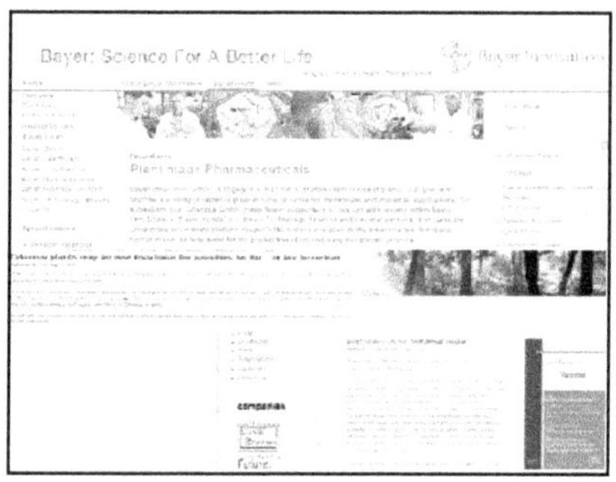
Bayer: Science For A Better Life
Plant-made Pharmaceuticals

Putting Pandemic Flu into Perspective
2001 terrorist attack with anthrax
killed five people
2002 outbreak of West Nile Virus
killed 284 people nationally in six months
2003 SARS outbreak
killed over 800 people world wide
froze Asian economies
frightened millions of people into wearing masks on the streets

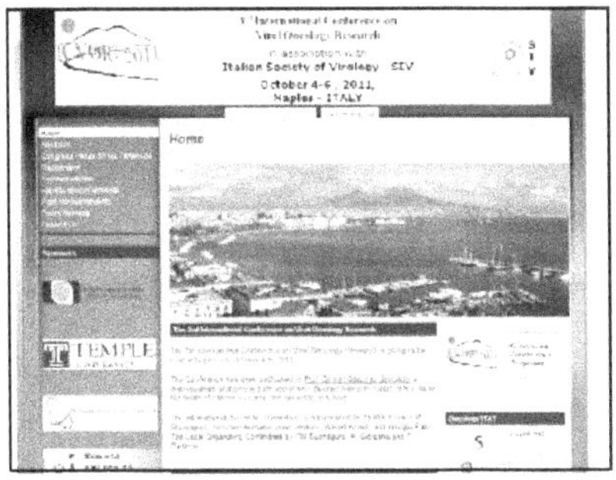
Italian Society of Virology - SIV
October 4-6, 2011,
Naples - ITALY
Home
TEMPLE

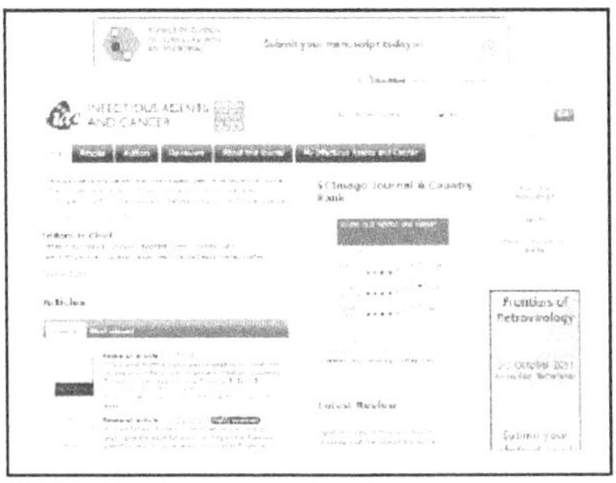
INFECTIOUS AGENTS AND CANCER
SCImago Journal & Country Rank
Frontiers of Retrovirology

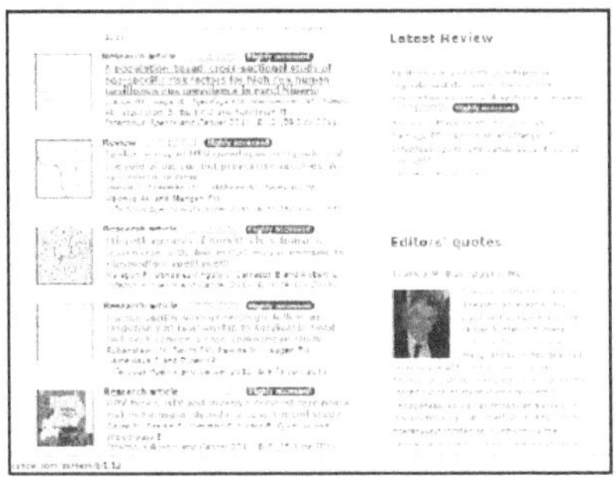
Latest Review
Editors' quotes

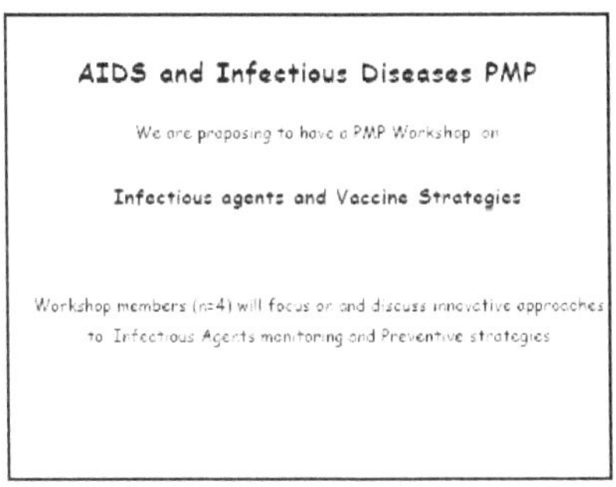
AIDS and Infectious Diseases PMP
We are proposing to have a PMP Workshop on
Infectious agents and Vaccine Strategies
Workshop members (n=4) will focus on and discuss innovative approaches to Infectious Agents monitoring and Preventive strategies

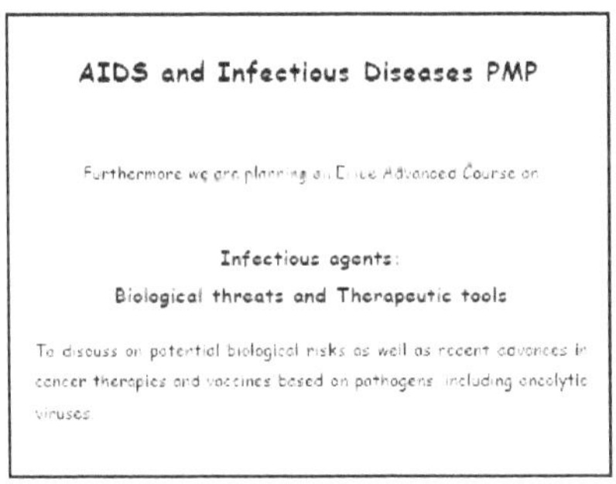
AIDS and Infectious Diseases PMP
Furthermore we are planning an Erice Advanced Course on
Infectious agents:
Biological threats and Therapeutic tools
To discuss on potential biological risks as well as recent advances in cancer therapies and vaccines based on pathogens including oncolytic viruses

MOTHER AND CHILD HEALTH PMP: 2010-2011 ACTIVITIES

DR. NATHALIE CHARPAK
Kangaroo Foundation
Bogotá, Colombia

DR. JUAN G. RUIZ
Department of Pediatrics, San Ignacio Hospital
Santafé de Bogotá, Colombia

DR. SOCORRO DE LEON-MENDOZA
Jose Fabella Memorial Hospital, Neonatology Unit
Manila, Philippines

INTRODUCTION

Creation of the Mother and Child PMP in Erice2002

PMP MANIFESTO: ***Maternal and Child mortality is a planetary emergency***

Our challenge

The Goal 4 of the millennium is to reduce by 2/3 the under-five mortality rate in developing countries before 2015. It is evident that not all the countries will reach it. In November 2010, the Kangaroo Mother Care method (KMC), a very cost effective and humane preterm infant care system, more dependent on human resources than on technology, was selected by the United Nation Development Program as an available intervention useful to reach this Goal 4 of the millennium. In our PMP, our area of expertise is KMC. We have been developing, conducting research and disseminating the method for more than 20 years. A major concern is how to overcome barriers and reach ample and effective implementation of KMC all over the world, but particularly reaching those countries and regions where its impact in infant burden of illness and quality of life would be greater. This year, our PMP focused on examining and refining the strategies for effective and timely diffusion of KMC. We invited a world leader scientist in biomedical knowledge translation and transfer, Dr. Brian Haynes, from McMaster University, to share with him this PMP challenge and to try to identify and conceptualize the best possible strategies to reach our goal.

BACKGROUND

Among the population of developing countries, children and mothers are particularly vulnerable. In addition, around 20 million Low Birth Weight infants (LBWI) are born each year because of either preterm birth or impaired prenatal growth or both, mostly in less developed countries. These births are associated with more than 30% of the mortality in the first month of life, and KMC has been proven to specifically reduce mortality and morbidity in those newborn infants. In other words, every year there are 20 million potential candidates to benefit from KMC, but only a small fraction of those infants are actually benefiting from it.

Do we have equity in neonatal mortality rates? NO

Figure 2: Variation between countries in NMRs

Source: Lawn JE, Cousens S, Zupan J. "Lancet Neonatal Survival Steering Team." *Lancet*. 2005 Mar 5-11;365(9462):891-900

The care of LBW infants represents a very high burden for health and social systems in developing countries where resources are scarce and systems are already overloaded.

Advanced neonatal care technology is effective in reducing low birth weight mortality and morbidity.

But this technology is expensive for purchase and maintenance and in resource restricted environments only a suboptimal neonatal care will be available for these fragile infants

This situation will lead to a higher rate of complications and deaths like:

- Cross-infection (in-hospital infections)
- Bronchoaspiration
- Inadequate thermal control
- Poor nutrition and growth
- Lack containment, stimulation and love
- Preventable sequels (neurological, pulmonary, blindness) not avoided

Kangaroo Mother Care (KMC) is an evidence-based technology, developed and implemented in Colombia and useful for ameliorating LBW effects on infant mortality, morbidity, nutrition and early development. KMC is a form of care for LBWI infants after transition to extra uterine life is completed, when the baby is stable. It is a complement of the technologically-based care of the newborn, not an alternative to it.

OUR ACTIVITIES DURING 2011: SCIENCE OF MEDICINE IN THE THIRD MILLENNIUM

We worked during this year on knowledge transfer and translation research applied to KMC to reach a quicker massive diffusion

What is the reality?

One of the major contributors to the gap that separates affluent and less developed countries is the inadequate access to knowledge (knowledge transfer) and when accessible, the insufficient translation of that knowledge into action

How to speed up access to KMC to all LBW in the world? What would be an effective (efficient) KMC Knowledge transfer/translation

1. Do we need to produce more specific knowledge for KMC to be accepted as an effective method of care? NO. There is a sound body of scientific knowledge sufficient to support the use of KMC. The problem is the uptake of that knowledge

   Main problem: Health care related research "consumers" are heterogeneous:

   - Low, middle and high income countries
   - Evidence-based medicine skills very heterogeneous
   - Basic quality of neonatal care very variable

   What has been the process for KMC knowledge production and translation into health care of our research group?

   - Research on safety of KMC 1989-1992
     - Observational analytical study
     - Research on efficacy and safety of KMC 1993-1996
     - Randomized controlled trial
   - Enhancement of the intervention 1996-2011
     - Explicative research (going on)

  - Pragmatic research
- Evidence based KMC guidelines with recommendations for KMC practice (to be updated)
- Translation of research results into health interventions 1994-2011
- National and international diffusion (going on)
  - Training in KMC Method (going on)
  - Resistances and quality improvement (going on)
- Economic evaluations (in course)

2. Do we need more experimental KMC evidence to convince? NO

   - What are the KMC Recent Research results?

BRAIN MOTOR FUNCTION IN ADOLESCENTS BORN VERY PRETERM AND INFLUENCE OF THE KANGAROO MOTHER CARE : A PILOT STUDY USING TRANSCRANIAL MAGNETIC STIMULATION

SCHNEIDER Cyril, PhD[1,*], CHARPAK Nathalie, MD[2], RUIZ-PELAEZ J Gabriel, MD, PhD[3], TESSIER Réjean, psy, PhD[4]

[1] Neuroscience Unit (CHUL), CHUQ Research Center, Université Laval (Rehabilitation), Québec QC, Canada; [2] Kangaroo Foundation, Bogotà DC, Colombia; [3] Javeriana University, Dept Clinical Epidemiology and Biostatistics, Bogotà DC, Colombia; [4] Pediatrics Unit (HSFA), CHUQ Research Center, Université Laval (Psychology), Québec QC, Canada

<u>Conclusion of this pilot study</u>

The KMC intervention, which generates unstressed stimulation and places the parents as the main actors, appears to have reversed and even prevented the deleterious effect of preterm birth on the integrity of the functional connectivity of the brain. The improved data in KMC adolescents were all due to conditions determined at 25–40 weeks' gestational age and not on learning during childhood. Therefore, neural plasticity in preterm children might provide the substrate to compensate for underdevelopment and improve skills during childhood and youth.

In the short- and medium-term, KMC clearly does not jeopardize the health and survival of premature infants. The results of this pilot study strongly suggest that the beneficial effects on brain rescue and development extend into adolescence. Therefore, it is appropriate to recommended use of KMC for preterm infants worldwide, as part of basic health care, to protect and rescue brain development.

- Our last research project:

Grand Challenges Canada™

Home | Grand Challenges | Integrated Innovation | Publications | News | W

Saving Brains: Unlocking Potential for Development

This program is focused on identifying and developing effective solutions for promoting cognitive development and long-term human capital formation in the face of poverty. 200 million children fail to reach their cognitive potential as a result of exposure to risk factors–malnutrition, infection, poor management of pregnancy and birth complications, and lack of stimulation and nurturing–in the first thousand days of development (from conception to 2 years of age). Effective targeted interventions may unlock the potential of the next generation of children to solve the problems burdening their societies. The complexity of the issue requires Integrated Innovation which is at the heart of the Grand Challenges Canada approach.

The key to unlocking this potential is to target the risk factor with the largest impact on cognition and human capital formation. We are looking to support assessments of the long-term effects of interventions administered in the first thousand days of development. With a common set of measurements applied across the studies, the interventions will be compared for their relative ability to promote long-term human capital formation.

Teams will be expected to re-enroll participants from studies they have previously conducted and assess the long-term effects of interventions targeted at malnutrition, infection, pregnancy and birth complications, or nurturing and stimulation that were administered during the first thousand days of development.

3. Why is it there a significant time lag between production of evidence and "official" acceptance of evidence regarding KMC?

The two largest RCT on KMC were published between 1995 and 2001 (>70% of study subjects considered in the meta-analysis). Cochrane Meta-analysis have just acknowledge KMC as been effective and safe **in 2011**

<u>The last Cochrane review:</u>

The Cochrane Collaboration, established in 1993, is an international network of more than 28,000 dedicated people from over 100 countries. They work together to help health care providers, policy makers, patients, their advocates and careers, make well-informed decisions about health care, based on the best available research evidence, by preparing, updating and promoting the accessibility of *Cochrane Reviews*—over 4,600 so far, published online in *The Cochrane Library*.

Kangaroo mother care to reduce morbidity and mortality in low birthweight infants (Review)

Conde-Agudelo A, Belizán JM, Diaz-Rossello J

THE COCHRANE COLLABORATION®

This is a reprint of a Cochrane review, prepared and maintained by The Cochrane Collaboration and published in *The Cochrane Library* 2011, Issue 3

http://www.thecochranelibrary.com

WILEY

Publishers Since 1807

Bill Gates Foundation (SNL) and Save the Children NGO: 2010

Published by Oxford University Press on behalf of the International Epidemiological Association. *International Journal of Epidemiology* 2010;39:i144–i154

doi:10.1093/ije/dyq031

'Kangaroo mother care' to prevent neonatal deaths due to preterm birth complications

Joy E Lawn,[1,2] Judith Mwansa-Kambafwile,[1,3] Bernardo L Horta,[4] Fernando C Barros[4] and Simon Cousens[5]

[1]Saving Newborn Lives/Save the Children-USA, Cape Town, South Africa, [2]Health Systems Strengthening Unit, Medical Research Council, Cape Town, South Africa, [3]Department of Public Health, Faculty of Health Sciences, University of Cape Town, Cape Town, South Africa, [4]Postgraduate Programme in Epidemiology, Universidade Federal de Pelotas, Pelotas, Brazil and [5]Infectious Diseases Epidemiology Unit, London School of Hygiene and Tropical Medicine, Keppel Street, London, UK.

- Why it happens?
  - Problems of communication between researchers and Cochrane reviewers
  - More efficient ways for evidence synthesis may be needed

4. Awareness about the existence of KMC among target health care systems has/is taken a long time.

   Despite publications in high impact medical journals, awareness among neonatologists, pediatricians, general physicians, nurses, and policy makers is suboptimal.

   <u>Why it happens?</u>

   Scientific publications and presentations at scientific events are necessary but not sufficient

   <u>What can be done about it?</u>

   - Incorporate KMC in normal health care professionals curriculum.
   - Mass media utilization?
   - Packaging KMC teaching with other widely disseminated health care interventions: oral rehydration, respiratory infection management, etc. such as IMNCI (integrated management of newborn and childhood illnesses)

Our participation

Current Women's Health Reviews

Volume 7, Number 3, August 2011

Contents

Hot Topic

Evidence-based Kangaroo Mother Care in the XXI Century

Guest Editors: Nathalie Charpak and Juan Gabriel Ruiz

Are we happy? yes But…….
$60.00 for downloading one copy of each paper!!!!
Not affordable for many of our colleagues from developing countries.

5. Do we need to change the dissemination, training and teaching KMC model?

Our model seems to produce good quality results but it is slow, time and human-resource intensive...How to evaluate the quality of the training? Cost of the training too high?

- Example of successful KMC knowledge transfer: the Kangaroo Mother Care Foundation, Philippines, Inc.
  - Registered 2008
  - Directed and operated by a multidisciplinary and multinational Board
  - Collaborative efforts with the KMC reference centre, Dr. Jose Fabella Memorial Hospital (public) and private sector (Medical Professional Organizations)
  - Three (3) additional regional KMC centres in 3 years

The Jose Fabella Memorial excellence KMC center in Manila.

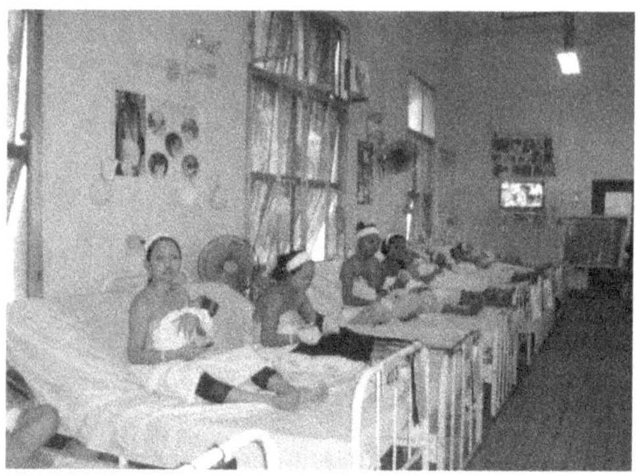

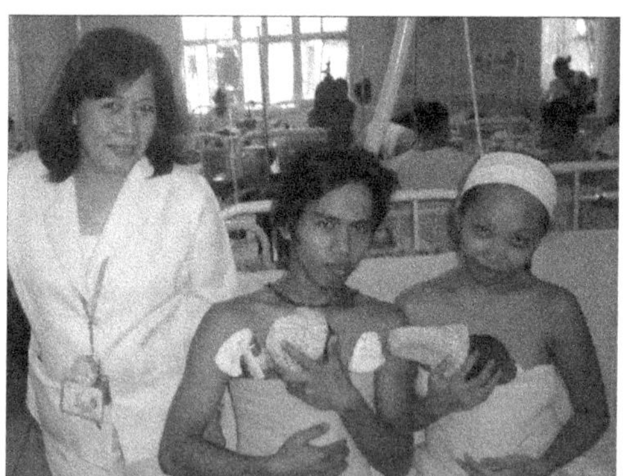

The Philippines experience in KMC training 2011: one excellence centre in Manilla and 3 regional centres in KMC.

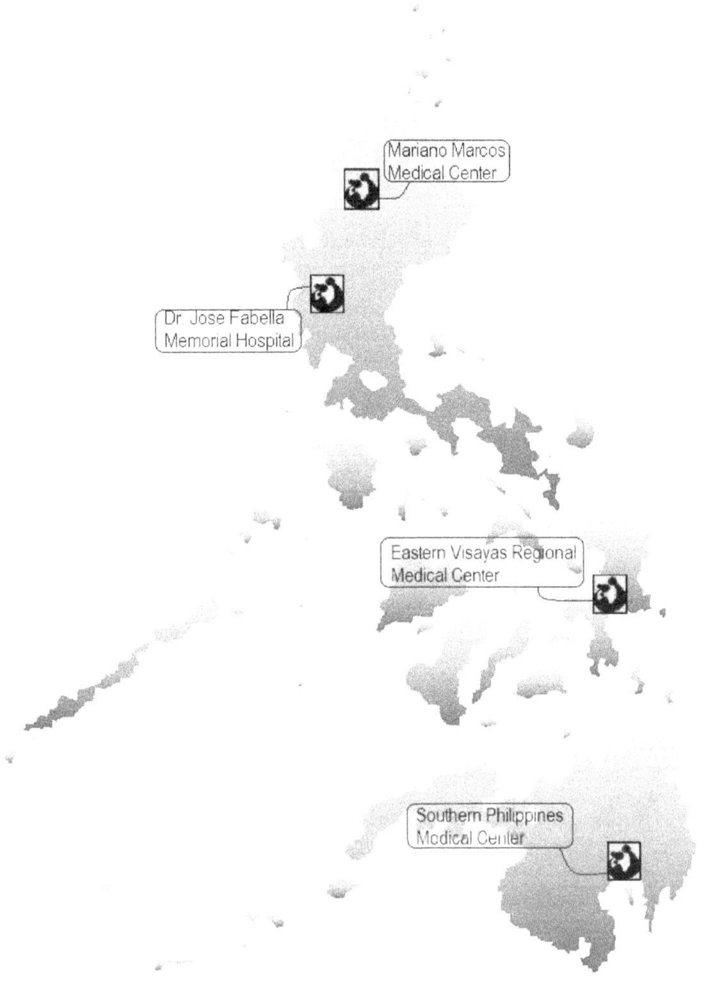

- The Colombian experience in KMC training:

  Our Model of KMC diffusion: "See one, do one, teach one".

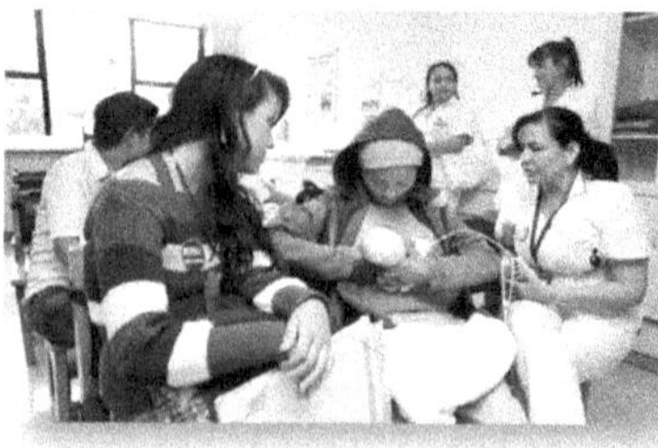

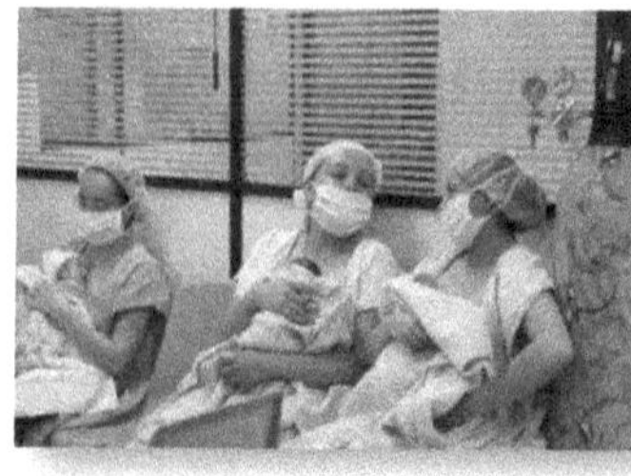

Success of knowledge transfer has been high: In 80% of national and international teams trained in Bogota, KMC implementation has been successful, despite specific needs and difficulties encountered by each program in each setting.

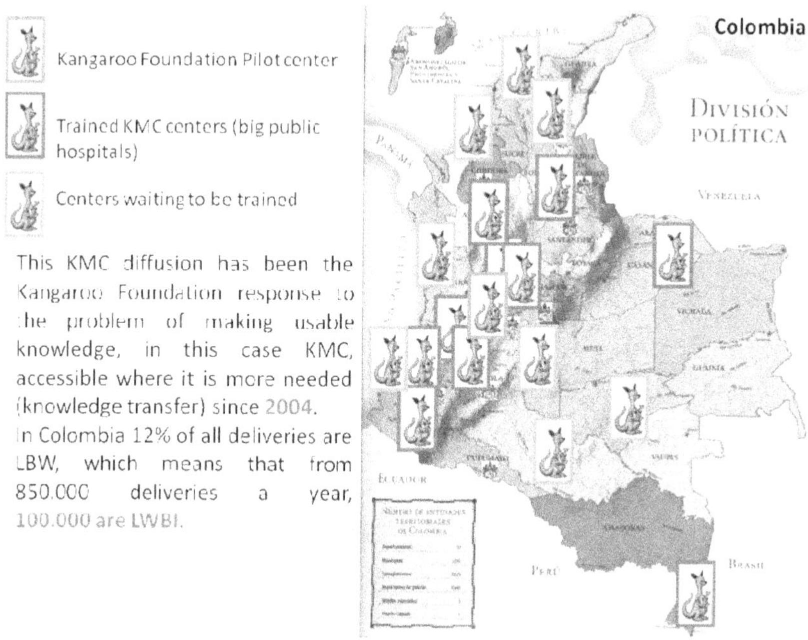

The National Training.

6. How to speed-up the KMC diffusion process and make it more efficient? THE KMC IMPLEMENTATION KIT FOR KMC TRAINERS

Libertad y Orden
Ministerio de la Protección Social
República de Colombia

OBJECTIVE

To develop an interactive platform for continuing education and virtual consultancy that help health professionals trained in KMC to:

- To strengthen their own competencies in KMC.
- To train health professionals in their hospital and their country.

WHY THIS PROJECT?

- One adequate and complete pedagogical tool does not exist presently in order to guide the KMC trainers to train and disseminate the method.
- Without a complete pedagogical tool and a virtual platform:
  - Not all trained health professionals can offer the necessary support to train and spread effectively the KMC in their own institution or country.
  - It's very difficult to stay up to date with the latest KMC events and advances.

SOME EXISTING GUIDES

- *Workbook for Kangaroo Mother Method*: South Africa, 2002, Dr. Anne-Marie Bergh.

- *Kangaroo Mother Care: a practical Guide.* WHO, 2003

- *Kangaroo Mother Care Training Manual:* United States of America, 2006, ACCESS Organization, USAID and SAVE THE CHILDREN.

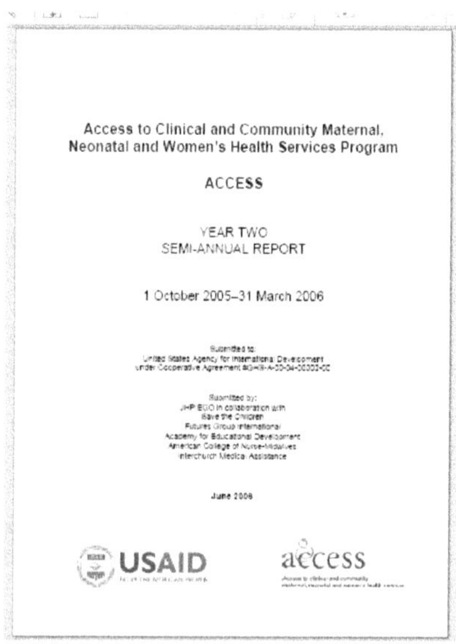

Access to Clinical and Community Maternal, Neonatal and Women's Health Services Program

ACCESS

YEAR TWO
SEMI-ANNUAL REPORT

1 October 2005–31 March 2006

Submitted to:
United States Agency for International Development
under Cooperative Agreement [illegible]

Submitted by:
JHPIEGO in collaboration with
Save the Children
Futures Group International
Academy for Educational Development
American College of Nurse-Midwives
Interchurch Medical Assistance

June 2006

USAID

access

- *KMC evidence based guidelines:* Kangaroo Foundation 2007

WHAT WE WANT TO DO...

- A specialized pedagogical tools.
- A virtual platform.
- To integrate all the already existing tools.
- To implement and disseminate KMC around the world massively.
- To have an impact on the mortality of the LBWI and to reach the millennium goals.

7. Do we need to combine several tactics and strategies to reach our goal?

- Do we need to involve politicians?
- Do we need specific national policies according to countries?
- How to take into account the variability in the quality of basic neonatal care?

We published, in January 2011, the KMC technical guidelines for Colombia in association with the Ministry of Health.

We collaborate with a new law for maternity leave that was enacted in Colombia in June 30 of 2011.

The usual maternity leave included 12 weeks beginning after delivery, 2 weeks have been added for all women and, in case of a premature delivery, the weeks of prematurity will be added to these 14 weeks and will allow the mother to stay more time with her fragile baby, first in the hospital then at home after hospital discharge.

- We have a project submitted to the Colombian research agency: Monitoring the quality of KMC care diffusion in Colombia and abroad.

Monitoring the quality of KMC care in 20 centres in Colombia

20 districts in Colombia
20 big health care centres
20 Health care teams trained

Monitoring the quality of KMC care in 10 centres in the world

10 hospitals in 10 countries:
10 large health care centres
10 Health care teams trained

Our challenge: "The burden of proof is always the responsibility of the one who is making new claims, particularly if the claim falls beyond the conventional truth, a fortiori if the claim conflicts with it, but the evidence supporting this claim must be sound".

REPORT OF THE CLIMATE PERMANENT MONITORING PANEL

CHRISTOPHER ESSEX
Department of Applied Mathematics, University of Western Ontario
London, Canada

PREAMBLE

The PMP Climate PMP meeting took place on 19th August morning, with an ad hoc committee composed of the following participants: *Mikhail Antonovsky, Christopher Essex, William Kininmonth, Ilan Koren, Richard S. Lindzen, Stephen McIntyre, Christopher Monckton, Fred Singer, Jan Szyszko and Anastasios A. Tsonis.*

The participants arrived without a standing organization for the PMP, without an agenda beyond a title "Climate", a room, and a given interval of time. Thus, preliminary discussion was needed to fill this gap. Christopher Essex was chosen to be chair by unanimous consent.

AGENDA FORMULATION

Agenda Content was decided by a round table discussion, in which all participants were invited to suggest by turn what ought to be discussed. The group present reflects some of the considerable diversity of backgrounds and viewpoints on climate without being comprehensive.

After much discussion, it was decided that both science and the interaction with society in terms of policy and politics ought to be discussed. The following agenda items were adopted.

- Scientific Questions not Adequately Explored for Major Conclusions
- Political Questions and the Interactions of Politics and Science
- Economics of Climate Mitigation

The bulk of the time was anticipated for the first two of these items.

Scientific Questions not Adequately Explored for Major Conclusions

One of the aims of this topic was not only to invite input from all participants on a variety of topics, but particularly to report on developments of a concrete project which had been invented in Erice at the 43rd Meeting in August 2010.

- The Independent General Circulation Model
  - Background: In August 2010, at the 43rd Meeting, an ad hoc committee (Essex, Kininmonth, Lindzen, McIntyre, Quirk, Shaviv) formed in the Hodgkin courtyard on the day devoted to workshops that year.
  - A major concern about modelling expressed then was lack of transparency and lack of reproducibility, two properties essential to science. Simply reporting outputs of complex empirical climate model experiments, in the form of attractive contour plots and changes, does not aid understanding of

essential computational and physical details for an outsider and even for insiders. While modelling at this level is nominally a device to grapple with behaviours of a complex system, the models themselves can become so complex that they can be as challenging to understand as the object they are intended to represent.

- Moreover, funding of modelling groups exclusively by governments tends to narrow the focus of such groups to scientific questions of interest to policy makers. A substantial amount of computational time has become mandated by governments to intercompare models from different groups in the world to produce common results.
- An open source style programming model was proposed to complement existing models, to help address some of these issues potentially putting a very different light on what models tell us. The idea was to have an independent model that would address important novel model experiments that other modelling groups would not be able or willing to address. The open source approach would utilize distributed crowd sourced expertise from volunteers thus sidestepping many of the concerns.
- Developments Since Last Year: Richard Lindzen reported to the meeting how thinking had evolved since last August. The idea of an open source alternative to big models seemed attractive, and would seem to be relatively cheap in principle (coding would rely on volunteers) the nature of the physical problem upon reflection makes a simplified modular approach necessary for crowd sourcing in the absence of appropriate physical expertise problematic.
- This kind of expertise would have to include computing expertise as well as expertise in the physics associated with climate, not to mention the specific formulations appropriate to computer climate models. This suggests that the open source aspect of the original idea would have to be scaled back if not eliminated. This puts the notion of an independent model in a different light.
- Moreover, it turns out that the source codes for most models are actually freely available, and not closed source at all. However, reverse engineering a poorly documented complex codes with unknown errors may as well be closed. In any case, distinguishing the independent model as "open source" is not strictly correct.
- Instead of crowd-sourced volunteers, at least some modelling expertise would need to be hired, which implies attractive salaries and thus a considerable sum of money. In order to make the model independent, non-governmental funds are being sought. Another development was to the decision envisage housing the hardware and staff at MIT. A volunteer board of experts overseeing the staff operations is also envisaged. Some names were suggested. A modern 256-core computer would be sufficient. The overall budget currently envisaged would be in the neighbourhood of $10 million per year.
- Fund raising and specific planning for this project is underway. More revisions can be expected.

Other items under agenda were new topics, as one should expect under the circumstances.

- Jan Szyszko gave a preview of his plenary talk, but introducing the participants to the notion of natural carbon capture using forestry techniques, which had the advantages of being relatively cheap, while exploiting natural processes, while fitting in with the goals of various significant international conventions. There was lively discussion about the biological role of carbon in the soil and there was a suggestion that a future this topic be the theme of a future session.
- Mikhail Antonovsky described various dynamical models that he and his group have used in connection to carbon dioxide amounts in time, including a point made about an improvement that ought to be made regarding a constant that he can show is not constant. This included a phase plane analysis of the carbon dioxide concentration signal, which provides an interesting independent confirmation of what satellites have been telling us about the substantial increase in biological growth due to increased levels of carbon dioxide.
- Fred Singer discussed some material for his plenary talk, which played into item i, and which was also mentioned in Richard Lindzen's plenary talk. In particular it centred around persistent differences between observations and outputs in climate models regarding tropospheric temperatures. Richard Lindzen raised the possibility that the model may actually be right. In the case of climate, not only are the models problematic, but the data base is also fraught with many problems. It can happen that a disagreement between the two may actually mean the data rather than the model is bad.
- Several participants (Kininmonth, Tsonis, Essex, and Lindzen) expressed concerns over a general lack of attention to and understanding of long-term natural variability on the multi-decadal time scale and longer. Models currently implicitly discount this and there is little further thought given to it. However this is central to being able to determine the true extent of any external forcing. It also has very significant practical importance because long-term natural variability can be every bit as significant to our lives as if it were actually climate change.

<u>Political Questions and the Interactions of Politics with Science</u>

No scientist can participate in this field without being aware of its political side. This has some implications not only for how policy is made, but it also affects what scientists feel free to ask. The participants in this session are all keenly aware that views of members of the group do not span the views held by the entire climate community. This was not a matter of our choosing, neither does that invalidate them. We make no pretence to represent the views of the entire field comprehensively. But we do maintain that we are open to scientific arguments of any kind even those with which we may disagree. It is also fair to say that we hold opinions that differ from each other, but we are all determined to discuss our differences thoughtfully nonetheless. Not asking questions and having doubts is contrary to scientific thinking.

On the matter of politics the climate problem has two aspects:

i. The nature of policy, such as regulating taxing sources of energy, reducing carbon dioxide etc.
ii. Difficulties faced by scientists who experience politically motivated mistreatment as a result of doing their jobs questioning scientific claims.

This agenda item focused on the second, rarely discussed, aspect of this issue (for a variety of reasons) not the least of which is that many scientists are untouched by this and unaware. Several participants related cases not already commonly known. Some are manifest as open processes while others are subtler, as is now manifest in the "Climategate" emails.

There was discussion of how to protect the right and obligation of scientists to ask challenging questions, to explore provocative positions, and not fear guilt by association.

This is a dangerous time for science. A recent Rasmussen poll in the United States finds 69% of respondents think that scientists have likely falsified research on global warming. The credibility of science not just climate science is at risk.

Economics of Climate Mitigation

Christopher Monckton took the session away from the physics of climate to the issue of what economic policy is appropriate given the circumstances.

Is it more cost-effective for the world to take urgent action immediately to regulate, reduce, replace, tax or trade carbon dioxide or instead to wait and see how much "global warming" occurs and then to adapt in a focused manner to any significantly adverse consequences of that warming?

In practice this question is sensitive to precisely what problematic magnitude of climate sensitivity is used, the GDP cost of measures already taken, the projected cost of putative climate-related damage that might arise in future from inaction today, the choice of pure rate-of-time-preference discount rate. There was some debate on the issue of the usefulness of discount rates for this issue without conclusion.

The quality of advice for governments from this field is at least as problematic as the state of climate science itself, but economic decisions are nonetheless central to how governments function. This is why the highly problematic Stern report was so influential, and why it is appropriate for this PMP to consider this topic.

POINT SUMMARY OF THE MEETING

On Science:

- The Independent GCM Project, which aims to help with some of the issues of models as raised, is developing.
- Long-term natural variability is a central open scientific question of both theoretical and practical importance.

On Politics:

- The right and obligation of scientists to ask questions is without contempt and reprisal is at risk, and it must be protected. Potentially important voices are being suppressed.

Suggested Future Sessions Proposed During this Meeting:

- Natural carbon capture using forestry techniques
- The costs of climate action versus climate inaction

POLLUTION PERMANENT MONITORING PANEL—2011 ANNUAL REPORT

CHAIRMAN: LORNE G. EVERETT
Chancellor Emeritus, Lakehead University, Thunder Bay, Canada, L. Everett & Associates, LLC, Santa Barbara, USA

POLLUTION PANEL ACTIVITIES 2011

- Three Plenary Sessions: Hydraulic Fracturing (HF), Water Scarcity/Pollution, Chemicals of Emerging Concern (CEC)
- 1 Workshop—Chemicals of Emerging Concern
- 11 Pollution PMP Papers and Reports, 19 August 2011
- 13 Invited Guests for Session and Workshop

CEC WORK SHOP—24 AUGUST 2011

Program organized by Lorne Everett, Stefano Parmigiani and Fred vom Saal
Speakers: Deborah Swackhamer; Jorg Drewes; Don Tillitt; Joe Braun; James Wright; Lorne Everett; Stefano Parmigiani; Fred vom Saal

Abstract for CEC Work Shop

Hundreds of contaminants of emerging concern (CECs) have been identified in our surface waters and drinking water, from plastics additives, personal care products, pesticides to pharmaceuticals, but sources are widespread rather than from easily identified 'point source" releases.

Most CECs are not new, but there is now realization that they are of emerging concern because they pose previously unrecognized risks at doses that had been considered safe. For example, recent research shows that pesticides in widespread use show adverse effects on reproduction in fish at levels below the current maximum permissible concentration in water. In addition, there is evidence from epidemiological studies suggesting that chemicals in plastic alter human fetal brain development, leading to abnormal behavior in children. New approaches for assessing risks to wildlife and humans are now being developed.

Our concern is that adverse effects of an unknown but potentially large number of chemicals can occur at doses once considered safe. Managed aquifer recharge (MAR) systems to mitigate CECs are being developed, but much remains to be learned before they can be implemented. The long-term goal involves "green chemistry" which seeks to create chemicals for use in products that do not pose environmental or human health risks.

PROPOSALS UNDER CONSIDERATION FOR 2012

1. Hydraulic Fracturing
2. Jordan Water Crisis
3. Terrorism, Cyber Attack, Water Infrastructure

4. Development of EPA Sustainability Tools
5. Regulatory Change Driven by Litigation/Media

1. **Hydraulic Fracturing—Chemical Waste Issue:** Dr. Thomas Ballestero, University of New Hampshire, USA.

Fig. 1.1.

2. **Water shortage in Jordan—Current status and potential solutions:** Dr. Muheeb Awawdeh, Department of Earth and Environmental Sciences, Yarmouk University.

Table 2. Groundwater Dramatically Declining in Jordan.

Groundwater Basin	Safe Yield (M.C.M)	Total water abstracted (M.C.M)	Balance	Abstracted from Safe Yield %
Yarmouk	40.0	43.3	-3.3	108
Azraq	24.0	59.3	-35.3	247
Amman-Zarqa	87.5	138.7	-51.2	158
Dead Sea	57.0	89.3	-32.3	157
Desi and Mudawrah	125	82.1	42.9	66
North Araba Valley	3.5	6.7	-3.2	193
Jafer	9.0	24.8	-15.8	276

- Jordan is recognized as one of the 4 most water deprived countries of the world.
- Currently the *per capita* share of water is estimated to be of the order of 140 cubic meters *per capita* per year (USA has over 9,000 cubic meters of fresh water available per year). See Tables 2.1 and 2.2.

Table 2.1

Year	Per Capita (m3)
1946	3600
1987	260
1995	210
2002	160
2015	100
2025	91

Table 2.2

Region	Domestic Water use per capita/day
Europe	130-350
Israel	250-300
Gulf State and Saudi Arabia	280-450
Egypt Syria and Iraq	>150
Jordan	80-85

3. **Terrorism, Cyber Attach and Water Infrastructure:** Distinguished Professor Frank L. Parker, Vanderbilt University, USA.

At risk, 75,000 dams, thousands of miles of pipes, aqueducts, water distributions, and sewer lines, 160,000 public drinking water facilities and 16,000 publicly owned waste water treatment facilities.

Worries are physical destruction, bio-terrorism and cyber attacks.

Risks: "raw" water systems, low water intakes, "medium" water treatment, "high" water distribution, and "very high" interlocking dependencies.

Requirements include: Updating and hardening of infrastructure, better monitoring, interconnectedness of water supply systems, protection against cyber and vandalism attacks, and multiple barriers including storage capacity, enhanced treatment systems and mutual aid.

Table 3.1. Analysis of Various Attacks and Threats.

Site of Attack/ Attack Type	Cost	Mass Casualties >1000	Skill Level Required	Cleanup and Repair Costs	Difficulty of Gaining Access and Mounting a Successful Attack	Total Risk Score > Score = > Risk
Thermonuclear/ Attack	3	10	2	10	7	4,200
Aerosol/BioAttack	7	5	5	4	6	4,200
Conventional Bomb/Attack	10	2	9	3	10	5,400
Dirty Bomb/Attack	6	7	7	7	7	14,406
Aerosol/Chemical Attack	8	7	8	5	8	17,920
Industrial Chemical Attack	9	6	7	4	8	48,384
Distribution System/ Contaminant	10	10	10	10	9	90,000

4. **Sustainability Framework:** Dr. Bernard Goldstein, University of Pittsburgh, USA.

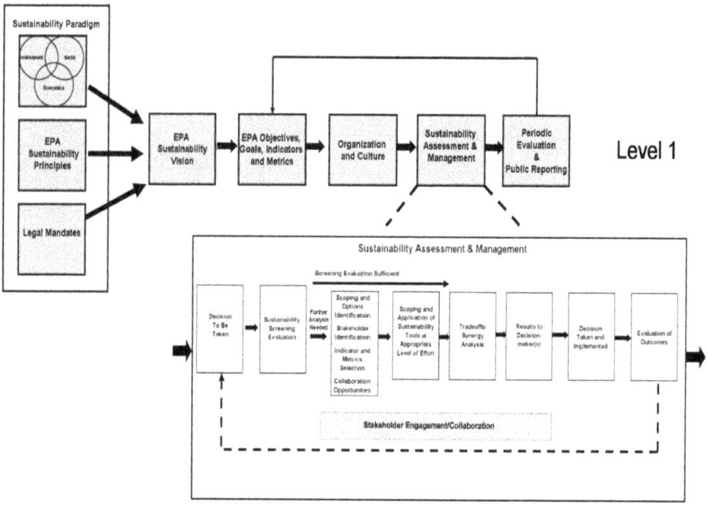

Fig. 4.1.

5. **Regulatory Changes Driven by Litigation and Media**

Examples were provided—Such as Santa Barbara California News Press, Monday 1 August 2011 (Figure 5.1).

State sets goal for limiting drinking water pollutant

Lead pollution negligence award of $320M stuns even plaintiff's attorneys

Fig. 5.1.

California Department of Public Health (CDPH) issues a Public Health Goal (PHG) January 2011 for Perchlorate in Drinking Water—Last Update: June 29, 2011. In January 2011, OEHHA released a draft technical support document for a 1-μg/L PHG for perchlorate for public comment. Perchlorate is a regulated drinking water contaminant in California, with a maximum contaminant level (MCL) of 6 micrograms per liter (μg/L). The MCL became effective October 2007. For information provided to public water systems by the CDPH Drinking Water Program about the implementation of the MCL and the scheduling of monitoring.

Perchlorate and its salts are used in solid propellant for rockets, missiles, and fireworks, and elsewhere (e.g., production of matches, flares, pyrotechnics, ordnance, and explosives). Their use can lead to releases of perchlorate into the environment.

Health Concerns—Perchlorate's interference with iodide uptake by the thyroid gland can decrease production of thyroid hormones, which are needed for prenatal and postnatal growth and development, as well as for normal metabolism and mental function in the adult. Its effects on the thyroid gland are the basis of the 6-μg/L public health goal (PHG) established in 2004 by Cal/EPA's Office of Environmental Health Hazard Assessment.

Table 1: Perchlorate Values in Lettuce Samples

Source: US FDA, CFSAN/Office of Plant & Dairy Foods, November 2004

Notes:

(1) NQ = non-quantifiable

(2) The estimated limit of quantitation (LOQ) is 1.00 ppb for lettuce

(3) For calculating the means, NQ = 0.5 ppb (1/2 LOQ)

Type of Lettuce	Location of Grower	Perchlorate (ppb)
Green Leaf Lettuce	Gadsen, Arizona	11.8
Green Leaf Lettuce	Yuma, Arizona	10.1
Green Leaf Lettuce	Yuma, Arizona	10.7
Green Leaf Lettuce	Yuma, Arizona	10.9
Green Leaf Lettuce	Yuma, Arizona	13.9
Green Leaf Lettuce	Yuma, Arizona	14.7
Green Leaf Lettuce	Yuma, Arizona	23.1
Green Leaf Lettuce	Yuma, Arizona	27.4
Iceberg Lettuce	East Yuma, Arizona	6.15
Iceberg Lettuce	Yuma, Arizona	3.71
Iceberg Lettuce	Yuma, Arizona	5.79
Iceberg Lettuce	Yuma, Arizona	6.15
Iceberg Lettuce	Yuma, Arizona	7.68
Iceberg Lettuce	Yuma, Arizona	7.73
Iceberg Lettuce	Yuma, Arizona	7.82
Iceberg Lettuce	Yuma, Arizona	10.4
Iceberg Lettuce	Yuma, Arizona	12.9
Iceberg Lettuce	Yuma, Arizona	21.4
Red Leaf Lettuce	Gadsen, Arizona	8.25
Red Leaf Lettuce	Yuma, Arizona	6.84
Red Leaf Lettuce	Yuma, Arizona	7.75
Red Leaf Lettuce	Yuma, Arizona	8.41
Red Leaf Lettuce	Yuma, Arizona	9.3
Red Leaf Lettuce	Yuma, Arizona	12.2
Red Leaf Lettuce	Yuma, Arizona	16.1
Romaine Lettuce	Yuma, Arizona	7.15
Romaine Lettuce	Yuma, Arizona	7.35
Romaine Lettuce	Yuma, Arizona	7.5
Romaine Lettuce, Organic, Romaine Hearts	Yuma, Arizona	9.17
Romaine Lettuce	Yuma, Arizona	9.46
Romaine Lettuce	Yuma, Arizona	11.5
Romaine Lettuce	Yuma, Arizona	14.6
Romaine Lettuce	Yuma, Arizona	17
Romaine Lettuce	Yuma, Arizona	24.7
Romaine Lettuce, Red	Yuma, Arizona	129

Fig. 5.2: Perchlorate in Lettuce (ppb).

POLLUTION PMP 2011 MEETING/WORKSHOP PARTICIPANTS

- Dr. Gina Calderone—ECC, USA
- Dr. Lorne Everett—L. Everett & Associates
- Distinguished Professor Frank L. Parker—Vanderbilt Univ. USA
- Dr. Fred vam Saal—University of Missouri, USA
- Dr. Stephano Parmiagiano—University of Parma, Italy
- Dr. Franco M. Buonaguro—Istituto Nazionale dei Tumori, Napoli
- Dr. Muheeb Awawdeh—Yarmouk University, Jordan
- Dr. Thomas Ballestero—University of New Hampshire, USA
- Dr. Joe Braun—Harvard University, USA
- Dr. Jorg Drewes—Colorado School of Mines, USA
- Dr. L. James Wright—University of Auckland, New Zealand
- Dr. Donald Tillitt—Columbia Environmental Research Centre, USA
- Dr. Deborah Swackhamer—University of Minnesota, USA
- Dr. Bernard Goldstein—University of Pittsburgh, USA
- Dr. Eilon Adar—Ben Gurion University, Israel
- Dr. Adalat Hasanov—National Academy of Sciences, Azerbaijan
- Dr. Rafiga Aghakishyeba—Baku State University, Azerbaijan
- J. Lauren Everett—Castaic Lake Water Agency, USA

REPORT OF THE PERMANENT MONITORING PANEL ON ENERGY, 2011

CARMINE DIFIGLIO
Office of Policy and International Affairs, U.S. Department of Energy
Washington, USA

WILLIAM FULKERSON
Institute for a Secure and Sustainable Environment, University of Tennessee
Knoxville, USA

BRUCE STRAM
Element Markets
Houston, USA

The Energy PMP organized sixteen plenary-session presentations in three topics: "Nuclear Power after Fukushima", "Unconventional Natural Gas Development" and "Energy Efficiency".

NUCLEAR POWER AFTER FUKUSHIMA

The Fukushima accidents provided a powerful backdrop to the nuclear power session. Several speakers concluded that Fukushima accidents were completely avoidable had better management and planning been in place. Japan's long record of earthquakes and tsunamis should have provided better planning benchmarks. Despite this setback, nuclear power remains an important and needed energy resource. Several speakers stressed that, with proper oversight, nuclear power can continue to provide essential energy supplies in many countries. While Fukushima has substantially clouded near-term progress on nuclear power development in certain countries, we should not assume that progress to build a new generation of nuclear power reactors will be halted. However, if nuclear power is to remain an important part of the world's energy mix, certain reforms are necessary to ensure safe operation and to restore public confidence both in the existing nuclear reactor fleet and new reactors.

Several speakers also provided valuable insights on how nuclear power issues differ among major world regions. It is necessary to discuss the future of nuclear power in a regional context as the various concerns and obstacles among regions are quite different. Consequently, different strategies are required to promote and ensure sustainable nuclear power uptake in different regions.

UNCONVENTIONAL NATURAL GAS DEVELOPMENT

The explosive growth of shale gas production in the United States could herald a similar trend in other energy-consuming regions. Large shale gas resources could substantially increase natural gas production in Europe and Asia. In particular, China could triple its natural gas production and Europe double its output by 2035 compared to an International Energy Agency reference case. With large increases in natural gas production in the major consuming regions (North America, Europe and Asia), these regions become much less dependent on natural gas imports. With this, exports of natural gas from Russia, Caspian countries and the Middle East would be much smaller. These changes in the patterns of natural gas grade would have significant energy security and geopolitical implications.

Nonetheless, environmental and other constraints could limit or slow the uptake of shale gas technology. U.S. environmental opposition to shale-gas development is growing due to incidences of environmental contamination. Many current U.S. shale gas producers need to significantly enhance their environmental practices, especially with regard to well integrity and waste-water disposal. While U.S. shale gas development is likely to continue, stronger regulatory oversight is likely. There do not appear to be geologic reasons why shale gas cannot be safely developed but more research is needed to better understand if there are formations where hydraulic fracturing should not be employed.

ENERGY EFFICIENCY

The energy efficiency session provided insights into past achievements to improve energy efficiency and messages about how to further achieve lower energy use. While OECD countries have achieved impressive improvements in energy efficiency, some of the gains can be attributed to the migration of heavy industries outside of the region. World-wide significant opportunities to improve energy use still exist and efforts need to be redoubled to realize them. For example, greater uptake of combined-cycle power plants is needed. Public policies have not successfully embraced a consistent set of requirements or incentives to exploit cost-effective efficiency opportunities.

The transportation sector merits special attention because of its high dependence on petroleum. Despite high oil prices, public policy to promote transportation energy efficiency is still needed since "market forces" produce a demonstrable under-investment in fuel efficiency.

PLANNING AND PREPARATION OF THE ENERGY PLENARY SESSIONS: ACKNOWLEDGMENTS

Nuclear: William Fulkerson and Bill Budnitz
Natural Gas: Carmine Difiglio and Lorne Everett
Energy Efficiency: Hisham Khatib
Overall Counsel and Support: Claude Manoli

<u>Energy PMP Meeting, August 2011</u>

Professor William Fulkerson announced that he would retire after serving for two years as chairman of the Energy PMP and one year as co-chairman (with Carmine Difiglio). Consequently, the Energy PMP needed to select another co-chairman. The meeting began with presentations by the two candidates who were previously nominated. William Barletta provided a talk on Inertial Fusion and Accelerators and Bob Schock discussed Policies for Sustainable Growth. After a subsequent tie vote, Professor Zichichi invited both Professors Barletta and Schock to serve as Energy PMP co-chairs with Carmine Difiglio. The remaining presentations at the Energy PMP meeting included:

- Nuclear Power after Fukushima (Tetsuo Yuhara and Hiroshi Ujita)
- China 5-Year Plan (Zhang Ziliang)
- Tokomak Fusion (Jef Ongena, Akira Miyahara)
- Cost of Climate Stabilization (Bill Fulkerson and Bruce Stram)

Regular committee business concerned the future of the Energy PMP Website (presented by Dick Wilson) and possible 2012 Activities. It should also be noted that members of the Mitigation of Terrorist Acts PMP members joined the Energy PMP meeting for the "Nuclear Power after Fukushima" discussion.

We are grateful to Bruce Stram for planning and organizing the Energy PMP meeting.

2012 Energy PMP Priorities

The Energy PMP meeting concluded with a discussion of its 2012 priorities. Three topics received particular support: energy subsidies and the net costs of low-GHG emission technologies. Energy subsidies are important in many parts of the developing world despite recent initiatives to reduce them. While reducing them is politically difficult, they discourage efficient energy investment, are a poor use of public revenues and can make countries overly dependent on fossil fuel imports. While the adverse effects of energy subsidies are not as great in OECD countries, there remain opportunities to rationalize the way different energy sources are taxed or subsidized, particularly with regard to biofuel susbsidies,

Several studies have evaluated the long term costs and benefits of energy technologies that have low greenhouse gas emissions. Some of the studies, particularly the *Energy Technology Perspectives* study of the International Energy Agency (2010) show benefits in the form of reduced fuel costs that fully reimburse the extra cost of low greenhouse gas emission technologies. It was agreed that a better understanding is needed about how the various studies reach their conclusions, how exactly to their conclusions differ and whether a better consensus would result by sponsoring a comparative analysis or holding a workshop or symposium.

Professor Schock's presentation on sustainable energy development was the third topic area. His presentation, mentioned above, focused on the currently unsustainable growth of inefficient energy services in cities. Looking forward we can anticipate billions of people moving from rural areas to cities placing an enormous pressure on energy supplies and other commodities. The Energy PMP could contribute to a better understanding of this challenge and recommend policies to improve the outlook.

There have been discussions about future activities among Energy PMP members prior to the August meeting. These discussions identified: 1) "electrification of transport and the power sector"; and 2) "energy systems and cyber security" as topics warranting special attention. They might be candidates for proposed 2012 Energy PMP plenary sessions. Another activity was suggested by Christopher Essex, Chairman of the Climate PMP. He noted that there is relatively poor representation in the IPCC Working Group III by experts on nuclear power. As Working Group III focuses on ways to mitigate greenhouse gas emissions, especially in the energy sector, a better balance of expert authors may be needed to fully consider one of the most important "zero-emission" sources of electricity. Professor Essex suggested that the Energy PMP could identify nominees for the next working group. Christopher also thought that there might be other opportunities for collaboration between the Energy and Climate PMPs.

The Energy PMP meeting adjourned agreeing that the PMP members will have some conference calls to consider these and other activities for 2012. It was noted that, very often, major events, can have a last-minute effect on our plans. Recent examples were the skyrocketing commodity prices in 2008, the Gulf of Mexico BP oil rig blowout (2010) and the Fukushima nuclear power accidents (2011).

PERMANENT MONITORING PANEL—MITIGATION OF TERRORIST ACTS

SALLY LEIVESLEY
Newrisk Limited London, Mitigation of Terrorists Acts PMP, World Federation of Scientists, London, UK

1. FUKUSHIMA-DAIICHI INCIDENT

The PMP-MTA met on 19th and 24/25 August and members presented papers across the three days in addition to utilising the 19th as a report preparation period. The Fukushima Daiichi incident which commenced on March 11, 2011 was prioritised across the three days and arrangements were made with the PMP Energy to attend lectures by the Japanese expert team to participate in the lecture and discussion session.

Additionally a special meeting was arranged between the PMP-MTA and the visiting Fukushima-Daiichi experts facilitated by Akira Miyahara, Emeritus Professor of the National Institute for Fusion Science and Nagoya University with Professor Tetsuo Yuhara, University of Tokyo and Professor Hiroshi Ujita of the Tokyo Institute of Technology and Dr. Norio Yamamoto of the Global Infrastructure Fund Research Foundation, Tokyo. The informal academic discussion focused ways in which cooperation and assistance could be maintained by the PMP-MTA on the continuing consequences suffered within Japan and for important lessons to be learned from the incident for the benefit of other countries.

These discussions will continue in communications in 2011-2012 through the liaison with PMP Energy who hosted the expert team. Attendance within some of the Workshop Sessions and contributions to the discussion by Professor Eugeni Velikhov of the Kurchatov Institute, Moscow led to many valuable insights being shared with the PMP along with informal discussions which significantly contributed to the PMPs understanding of the significance of the Japanese incident based on Chernobyl experience and the future challenges that remained to be resolved at the Fukushima Daiichi site.

2. ELECTION OF NEW CO-CHAIR/ACADEMIC LEADER AND MEMBERS ACTIVITIES 2010-2011

The PMP-MTA elected Professor Annette Sobel and passed on this recommendation to Professor Zichichi for ratification as the new Co-Chair of the PMP-MTA. Professor Sobel discussed future directions with the PMP-MTA and will provide outstanding academic leadership.

Professor Sobel is Senior Advisor for Biosecurity Engagement, Office of the Assistant Secretary of Defense for Nuclear, Chemical and Biological Defense, Assistant to the Provost for Strategic Opportunities Adjunct Professor, ECE Adjunct Professor, Community and Family Medicine, University of Missouri and Guest Scientist Los Alamos National Laboratory.

The members of the PMP-MTA requested that Dr. Sally Leivesley continue in her position as Co-Chair. Dr. A.L. Moore retired as Co-Chair following a busy period in which funding for additional meetings had been pursued. Dr. Moore contributed strategic direction and planning to the group over the past three years and is active in pursuing global linkages to extend the global reach of the outputs of the contributions that arise from the annual workshops of the PMP-MTA on global emergencies.

Throughout the past three years, the Co-Chairs of the PMP-MTA have expressed their gratitude also for the academic development and advice that has been continuously provided by the Emeritus Chair, Professor Richard Garwin (The Garwin Archive-*www.fas.org/rlg/)*. Active assistance in planning the past three years and work on web site support has also been contributed by Professor Richard Wilson who actively supports work on risk and who has, this year, published a book on *Physics is Fun* which holds many invaluable lessons for risk enthusiasts from different disciplines. (*Physics is Fun: Memoirs of a life in Physics* http://www.physicsisfun.net http://physics.harvard.edu/ ~wilson).

During the year 2010-2011, there has been active correspondence with PMP-MTA attendees including Professor Charles Ferguson of the Federation of American Scientists in order to maintain links with key research institutions and to continue a dialogue on important planetary threats such as biological terrorism. Dr. Diego Buriot within the PMP-MTA has provided a scope of activity for specialised work on biological terrorism and links to internationally recognised experts in this field for future contributions as the PMP-MTA progresses its research in the biological field.

In 2011, a number of members of the PMP-MTA were in active communication about activities and participating in consultations and discussion on terrorism events as these occurred. Some members were unable to attend the 2011 meeting. This included Professor Rajaraman who annually has provided updates on the status of terrorism and mitigation measures in India and European members, Professor Steinhausler and Lyudmilla Zaitzeva who contribute expertise across the CBRN spectrum.

In addition, the PMPs for Energy and Pollution and Water have shared discussions with the PMP-MTA to explore important questions on radiation mitigation and this was particularly important in active sharing of information from the onset day of the Fukushima Daiichi incident on March 11. Attendance at the Workshop by Bertil Galland who has contributed in past years from his expert understanding of media communications was also appreciated by the PMP.

3. 2012-2014 ACADEMIC FOCUS

Professor Sobel, in consultation with the PMP, identified a strong focus for the years 2012-2014:

The planned activities include:

- Renewed attention to biological threats, personal protection, re-emergent diseases, and opportunities to inhibit the spread of infectious diseases by the international aviation environment.
- Follow up on the science lessons-learned from the Fukushima Daiichi disaster to include an exercise using real data, to guide mitigation of terrorism and accidents involving nuclear reactors.
- Expansion of our collaboration with other PMPs, especially pollution and cyber, where there appears to be potential mutual benefit.
- In this we will make use of:

- Evolving communication and sensor network architectures for mitigation, prevention and response;
- Systems analysis approach to understanding/modeling of infectious disease re-emergence due to manmade and natural factors;
- Emphasis on infrastructure issues that must be assessed and leveraged to ensure environmental, water, animal, plant and human health despite ongoing and evolving planetary emergencies.

4. LONG TERM BRIEF OF THE PMP-MTA: CONTEXT FOR 2012-2014

Professor Richard Garwin as Emeritus Chair summarized the long term focus that has been followed by the PMP-MTA and continued in the 2011 meeting.

The PMP-MTA has a narrow brief that excludes:

- Reducing motivation.
- Interfering with execution (police work and intelligence).

From the beginning, the Panel has emphasized "high impact" terrorism:

- Terrorist-induced pandemic.
- Created substantial body of work with emphasis on mechanism of epidemics (reproduction factor R and serial interval i), feasibility of personal protective measures, and details of implementation—What, How, Messaging and Testing, Communication, Alerting, etc.
- Detonation of a stolen or improvised nuclear explosive—mitigation mostly in the avoidance of prompt (2-week) deaths due to fallout. Few-minute timeline!
- Radiological dispersal device—RDD
  - Fact vs. fear. Few prompt deaths, much economic damage from long-term denial of use of buildings and land. Calculations of spread of radioactive cloud. Lack of capability to decontaminate.
  - Ultimate RDD is attack on a nuclear reactor or spent-fuel pool.

Our work is more broadly useful, outside the realm of terrorism:

- Natural pandemics.
- Reactor accidents.
- Accidents at chemical plant.

Events and data show there is much to be done:

- Highly pathogenic avian influenza A (H5N1), also called *HPAI H5N1*.
- Pandemic H1N1 (mild)
- 11 March, 2011, Great Tohoku earthquake and tsunami
- Fukushima Daiichi multi-reactor accident; substantial comment by government of Japan (GOJ) on its failings

5. LECTURES AND ROLE PLAYING SCENARIOS 2011 WORKSHOP

A number of subjects were covered in the lectures and role play exercise scenarios that were contributed by the members in the 2011 meeting. On each of the two Workshop days, Lord Alderdice and Professor Richard Wilson provided a strategic overview of the lessons learned and ways forward. Lord Alderdice summarized for the PMP-MTA, the work of the PMP on causation of terrorist acts and he led discussion on the applications across both the PMPs so that the mitigation of terrorist acts could be consistent with knowledge that is evolving on causation. Professor Pervez Hoodbhoy and Dr. Vasiliy Krivokhizha commenced the lectures at the meeting days with strategic overviews of world events with Professor Hoodbhoy focusing on risks arising since the death of Osama bin Laden and Dr. Krivokhizha outlining the significant gaps in world order that were a risk to planetary stability and awaiting some conceptual solutions.

1. Pakistan after bin Laden
2. World overview- Libya, Afghanistan, Pakistan, Europe, Middle East
3. Biological—Pandemic studies
4. Radiological Devices—Fukushima Daiichi case study was used to identify key factors in recovery of cities post radiological distribution device ('dirty bomb') Important links with PMP Energy and Japanese experts attending the WFS Plenary session and the PMP Energy meeting enabled members of the PMP-MTA to develop further knowledge in relation to nuclear impacts on countries in the event of a terrorist attack.
5. Evaluation of technology and use of rapidly evolving communications in support of global emergencies. Specific contributions of scientific input to technology to enable rapid evacuation from contaminated zones of cities were discussed.
6. Two scenario role plays were run for members to contribute scientific advice that could be contributed by an international scientific advisory group to planetary emergencies. One scenario covered reactor threats incorporating natural disasters and terror attacks and the second scenario covered bio terror threats. Both role plays considered cyber vulnerabilities in relation to command and control systems so that the PMP members could consider the resilience of critical infrastructure under extreme threat conditions.
7. New threats to Europe's reactors from right wing extremism following the 2011 Oslo bombing and shooting by a lone activist were identified and discussed.

6. DISCUSSION POINTS FROM THE PMP-MTA WORKSHOP

<u>Professor Richard Garwin:</u>

As technology has advanced, it is now feasible to guard against loss of communications (but not necessarily of power); the latest version of the emergency communications could be routinely downloaded into the home and office PCs, so that it would be available if communication lines were down. To guard against temporary loss of power, the same capability should be and is available via smart phones and tablets.

Major questions are whether families will indeed have emergency kits and whether they will practice access to the emergency communication sites.

- 22 July 2011, Anders Breivik one-man killing operation in Norway
- 6-10 August 2011 riots, looting, and arson in London and other cities in England

Evolution of technology can aid both sides

- Wikipedia
- Web, in general
- Twitter, Facebook and other social media

There is still a need for even one good example of well-organized government capability to plan, assess, decide, and communicate with national government, local governments, and the public.

- Some progress in the United States—Center for Disease Control and Prevention (CDC), Department of Homeland Security (DHS)
    - i. e.g., *Planning Guidance for Response to a Nuclear Detonation, 2d Edition (July 2010)*
    - ii. But insufficient recognition of need for direct involvement of central government with the public

7. GENERAL DISCUSSION ON PLANETARY EMERGENCIES:

The following points were made by members:

1. Public fear was recognized as a powerful effect of planetary emergencies and scientific evidence was important in incidents to counter this public response. The risk to the world's energy supply from the fear of nuclear power which is shutting down reactors was discussed in relation to Fukushima Daiichi and also a factor that caused social consequences following an Armenian earthquake when that country shut down its reactors and suffered significant social effects from the loss of power supply. Armenia subsequently decided later to re-open their reactors but not before the fear factor had caused great deprivation.
2. New methods of risk assessment were discussed. The application of the Fukushima Daiichi incident to understanding the possible effects on cities from a radiological dispersal device attack by terrorists raised discussion on models. New strengths in complexity theory were considered and will be discussed in more detail in the 2012 meeting. Much needs to be done further to understand how we can use models and how these will assist us in modifying our approach to mitigating the risks of the consequences of a breach of a reactor.
3. Communications have been identified as critical to the mitigation of the effects of any terror attack on a reactor or on attacks on public places in order to cause mass casualties. It was recognized that panic may cause many deaths and that a

failure of the authorities to communicate to the public will contribute to the consequences. Many new methods of communication have been identified ranging from social networks to the use of sensors and these need to be explored in more detail for radiological, biological, chemical and nuclear events.
A subset of the communications and planning mechanisms for big events such as Fukushima Daiichi was explored—three phases were suggested:
- On the spot monitoring before any evacuation is considered and advised to the public
- That the level of expertise include that of Ministers of Government who will translate the scientific advice into appropriate and politically accepted measures and this means that scientific consultations should always be with the presence of the Ministers who are to carry out the political decision making.
- That there be careful preparedness with clean up teams who require psychological preparations and training in order to work in the aftermath of the nuclear incidents whether these be reactors or cities. Preparedness also covers the scientific methods of identification of the location of the nuclear fuel in the case of the Fukushima Daiichi plant. The Chernobyl experience has been that the fuel and the state of the fuel can only be found by drilling many boreholes and then by determining the best way to protect such fuel in an environment where there is an earthquake risk. In such situations there may be an option to use engineering to produce earthquake resistant covers or else there may be an option to lift the reactor vessels out of the site of the disaster and move away from populations to allow for a safe cool down period over many decades.

4. Combating the risk of social disorganization was identified as an important challenge to be pursued in 2012. Social disorganisation has been observed in social networks which have become self organizing and creating social conflict without a vision of a new social order. The many areas of international conflict in the world will add to the fragmentation of our society unless there is a new world idea that allows many different forms of social order within nations to co-exist. It was recognized that there are many different forms of government that may co-exist under a unifying vision of the new world order.
5. Consideration of fragile states such as currently experienced within Pakistan was discussed with the benefit of the lecture to the PMP by Professor Pervez Hoodbhoy. The PMP discussed the strategy of contribution of aid and assistance to NGO's that are monitored to ensure that the people are assisted with social needs and are less vulnerable to the disruption and social upheaval that is created by terrorist groups and political upheavals.

8. 2011-2012

Over the next year the PMP-MTA will continue to provide an informal hub of scientific communication between the members and in interaction with other PMPs as incidents evolve where scientific discussion may contribute knowledge to mitigation of impacts on people and the environment.

Opportunities for research and linkages across many countries are appreciated by the members of the PMP-MTA as a benefit that arises from the annual meetings in August at Erice.

Academic planning for the 2012 meeting will be led by Professor Annette Sobel and communications with Professor Sobel by members on any evolving threats or mitigation issues will be part of the planning for the 2012 Workshop.

Report compiled from PMP-MTA contributions by Co-Chair Dr Sally Leivesley October 25, 2011.

INFORMATION SECURITY PMP REPORT

HENNING WEGENER
Ambassador of Germany (ret.)
Madrid, Spain

JODY R. WESTBY
Global Cyber Risk LLC
Washington,

Continuing its work from the session in August 2010, the PMP finalized its contribution to the joint publication with the Secretary General of the International Telecommunications Union (ITU), Dr. Hamadoun Touré, *The Quest for Cyber Peace*. The book was published in January 2011 in six languages and is available free of charge on the Internet at www.itu.int/pub/S-GEN-WFS.01-1-2011. As a UN publication, the book has been distributed on a global scale, and serves as an important reference of the PMP's work for a wider audience. Working with the Secretary General and his staff has been a gratifying experience. There was, at all times, a large coincidence of views and an exemplary spirit of cooperation. This has encouraged the PMP to contemplate a further joint publication, and we will be making thematic proposals to Secretary-General Touré shortly.

In addition, selected PMP work was featured in the joint publication, *Rights and Responsibilities in Cyberspace: Balancing the Need for Security and Liberty*, between the EastWest Institute (EWI) and the WFS PMP. In contrast to the PMP's relationship with the ITU, the cooperation with the EWI, which appeared to hold promise, did not develop as hoped. The EWI was not prepared to provide a work format that would have enabled the PMP to make a contribution bringing to bear its competence and experience with appropriate recognition, so that we felt constrained to terminate the Memorandum of Understanding signed in 2010 as a basis for mutual cybersecurity activities.

Select members of the PMP participated in the annual follow-up conference to the World Summit on the Information Society at Geneva in May 2011 where Ms. Westby organized and led a mock cybercrime investigation seminar, which was very well received. The *Quest for Cyber Peace* book was also distributed in five languages.

At its working session on August 19, 2011, the PMP:

- Discussed reports on recent cybersecurity developments. Dr. Udo Helmbrecht, Director of ENISA, the European Network & Information Security Agency, reported on events in ENISA and Jody Westby reported on events in the U.S. and in the international cybersecurity R&D community.
- Worked on a new version of its Top Cybersecurity Problems List. An updated draft is now ready for final PMP review and will be distributed to the international cybersecurity R&D community for comment in fall 2011. The PMP's first Top Cybersecurity Problem List was first circulated among the international R&D community in 2008/2009.

Finally, the PMP identified and discussed future topics of interest and its work plan for 2011/2012, including appropriate topics for future collaboration with the ITU. Topics under discussion include examining risks associated with cloud computing and mobile devices, weaving multidisciplinary expertise into Computer Emergency Response Team (CERT) operations, addressing legal issues in countering the exponential growth of cybercrime, advancing international cybersecurity R&D. In addition, the PMP agreed to explore the possibility of developing a proposal for a 2012 plenary session on information security issues regarding supervisory control and data acquisition (SCADA) systems.

11

WATER & POLLUTION WORKSHOP

WATER & POLLUTION WORKSHOP

SOURCES, EFFECTS AND NEW APPROACHES TO CONTAMINANTS OF EMERGING CONCERN

AGENDA

Professor Stefano Parmigiani
Department of Evolutive and Functional Biology, University of Parma, Parma, Italy
Introduction to the Workshop

Dr. Lorne G. Everett
L. Everett & Associates, LLC, Santa Barbara, USA
Defending Low Concentrations of Toxic Chemicals in Court

Professor Deborah L. Swackhamer
Science, Technology, and Public Policy, Water Resources Center, University of Minnesota, St Paul, USA
CECs in Water: Challenges and Future Solutions

Professor Jörg E. Drewes
Environmental Science and Engineering, Colorado School of Mines, Golden, USA
Ground Water and Surface Drinking Water Sources of CECs

Professor Donald Tillitt
Biochemistry & Physiology Branch, Columbia Environmental Research Center, USGS-BRD, Columbia,
Reproductive and Endocrine-related Effects of CECs in Fish Populations

Dr. Joe M. Braun
Department of Environmental Health, Harvard University, Boston, USA
Early Life Exposure to Non-persistent Endocrine Disrupting Compounds and Childhood Neurobehavior

Professor Frederick vom Saal
Division of Biological Sciences,University of Missouri, Columbia, USA
New Approaches to Risk Evaluation of CECs

Dr. L. James Wright
Inorganic and Sustainability Cluster Department of Chemistry, University of Auckland, Auckland, New Zealand
Moving Upstream: The Green Chemistry Approach for Eliminating CECs

Chair: Dr. Lorne Everett

POLLUTION WORKSHOP
24 August 2011

NEW APPROACHES TO RISK EVALUATION FOR CHEMICALS OF EMERGING CONCERN (CECS) THAT HAVE ENDOCRINE DISRUPTING EFFECTS

FREDERICK S. VOM SAAL AND JULIA A. TAYLOR
Division of Biological Sciences, University of Missouri-Columbia, Columbia, USA

PAOLA PALANZA AND STEFANO PARMIGIANI
Dipartimento di Biologia Evolutiva e Funzionale, Universita' di Parma, Italy

ASSUMPTIONS USED IN CHEMICAL RISK ASSESSMENTS

Current assumptions used by chemical risk assessors in regulatory agencies throughout the world to assess the risks posed by environmental chemicals are based on principles laid out in the 16th century (Myers et al. 2009a). As a result, there are the flaws being identified in the regulatory system's approach to assessing the risks of posed by environmental chemicals (vom Saal and Sheehan, 1998; Myers et al. 2009a). This is slowly causing a paradigm inversion in toxicology and chemical risk assessment, but as is always the case when long-held beliefs are challenged, resistance to change is hard to overcome. The paradigm inversion is a result of the discovery that a large number (as of 2011 approximately 1000) chemicals in commerce pose a health risk due to acting as disruptors of signaling systems (intracellular, neural, endocrine) required for normal development and subsequent functioning of organ systems; these chemicals are referred to as endocrine disrupting chemicals or EDCs. The term "endocrine disrupting chemica" involves using "endocrine" in a very broad sense to encompass any chemical communication system in the body. These EDCs are an important subset of chemicals labeled as "chemicals of emerging concern or CECs". Importantly, the majority of the EDCs are not labeled as CECs because they are recently introduced chemicals. In fact, most EDCs have been in commerce for many decades and, as a result, were initially used in products without any requirement for hazard testing. As information has accumulated about the hazards associated with chemicals that can disrupt endocrine (signaling) processes, they have become prominent on lists of CECs. We will identify here the assumptions currently used in chemical risk assessments that are contradicted by published data concerning the hazards to health posed by EDCs.

A central assumption in current chemical risk assessments is that "the dose makes the poison". This assumption led to the development of testing procedures in the 1950s that relied on only examining a few very high doses of a chemical and then estimating the effects of much lower doses based on the results observed at the few very high doses (Bird, 2005; vom Saal, 2005). Thus, in experiments to assess health effects of chemicals, the current approach used by toxicologists is to only test a few very high doses of a chemical and then extrapolate (using linear models) from high-dose effects to predict a threshold daily exposure dose below which chemicals are declared "safe" (vom Saal and

Sheehan, 1998). Since the assumption that high-dose testing predicts low-dose effects is accepted as true by risk assessors, there is no requirement to verify that the dose of a chemical that is estimated to be "safe" for daily human exposure actually poses no health hazard.

As outdated as the current risk assessment hazard testing approaches are, these issues regarding flaws in the approaches used to test chemicals for health effects actually only apply to a relatively small percentage of chemicals in commerce. In the USA, for example, for the majority of chemicals in commerce, there are no hazard data at all because the regulatory laws allow products to be made from chemicals that are not tested for health effects prior to being used. The public is then assured by not only the chemical corporations but by the regulatory agencies that the chemicals are "safe". Since no pre-market testing is required, the burden of proof that a chemical poses a hazard to human health is almost impossible to overcome. For example, the EPA tried but failed to regulate asbestos, and has only banned 5 chemicals (Chameides, 2011) in the 45 years since the passage of the Toxic Substance Control Act (TSCA). Chemicals that are not tested are declared "Generally Regarded as Safe (GRAS)". In these cases, the "absence of evidence of harm" (based on no information being available) has been falsely presented to the public as "evidence for the absence of harm", implying that safety assessment has occurred. The reality is that for no chemical in commerce, whether because of GRAS approval of the chemical or flawed approaches to testing for the relatively few chemicals that have been tested, can the public have confidence of safety as a result of exposure to chemicals at low chronic levels due to using products from which the chemicals can leach.

In summary, when the current law governing environmental chemicals in the USA was written, all of the chemicals already in commerce (about 60,000 chemicals) were exempted from regulation and declared "safe" (Chameides, 2011). The laws governing the labeling of chemicals as "generally regarded as safe" have provided a method for corporations and regulatory agencies to declare chemicals that have been in commerce for many years "safe" simply because they have been used for a long time. Consequently, the reality is that this approach, that does not involve analysis of any data, results in using people as "lab rats" in a massive uncontrolled experiment.

EVIDENCE FOR HARM CAUSED BY AN EDC, BISPHENOL A (BPA)

A common question regarding the presumed safety of a chemical that has been in use for some time is: would there not be some evidence for adverse effects if the specific chemical were actually harmful? The answer is clear and disturbing regarding effects caused by EDCs and human health trends. For just one EDC, the chemical bisphenol A (BPA), Table 1 presents findings from hundreds of studies conducted since 1997 with experimental animals in relation to human health trends.

BPA is just one of approximately 100,000 chemicals in commerce, but it has been selected for intensive study because it is one of the highest volume chemicals in commerce (Bailin et al. 2008) that is a known EDC. Virtually everyone everywhere is exposed to measurable amounts of BPA that are above levels that cause harm in animals (Taylor et al. 2011), and humans everywhere are likely exposed to higher levels of BPA (by a variety of routes) relative to the low exposure levels proposed by regulatory agencies (Vandenberg et al. 2010a; Vandenberg et al. 2010b). As shown in Table 1, based on animal research as well as some recent epidemiological studies, BPA is related to a myriad of adverse health effects (Richter et al. 2007), which of course results in scientists being interested in examining its effects on the system that they specialize in studying. Thus, rather than having only toxicologists use traditional toxicological testing approaches to study BPA, this chemical has attracted the attention of ethologists, endocrinologists, neurobiologists, immunologists, reproductive biologists cancer biologists, cell biologists, molecular biologists, developmental biologists, etc. For example, we have collaborated on studies showing a variety of adverse effects on neurobehavioral outcomes in mice. One of the most interesting features of these findings is that BPA exposure during the prenatal/neonatal period of sexual differentiation of the brain has markedly different effects on males and females (Palanza et al. 2002; Laviola et al. 2005; Razzoli et al. 2005; Gioiosa et al. 2007; Palanza et al. 2008; Martini et al. 2010). Similar findings have now been reported relating behavior of children to the levels of BPA in their mothers during pregnancy (Braun et al. 2009).

Table 1. Effects observed in studies with laboratory animals in relation to human health trends.

BPA EFFECTS IN MICE & RATS	HUMAN HEALTH TRENDS
CANCER	
Prostate hyperplasia & cancer	Prostate cancer increase
Mammary hyperplasia & cancer	Breast cancer increase
MALE AND FEMALE REPRODUCTIVE SYSTEM	
Abnormal urethra / Obstruction	Hypospadias
Sperm count decrease	Sperm count decrease
Early puberty in females	Early sexual maturation
Ovarian cysts / Uterine fibroids	PCOS / Uterine fibroids
Abnormal oocyte chromosomes	Miscarriage
METABOLIC DISEASE	
Body weight increase	Obesity increase
Insulin resistance	Type 2 diabetes
Heart disease	Heart disease
BRAIN AND BEHAVIOR	
Hyperactivity / Impaired learning	ADHD
Abnormal play behavior	Autism
Abnormal gender-related behavior	

BPA is a monomer used in the manufacture of polycarbonate plastic and resins, and as an additive in polyvinyl chloride (PVC) plastic and many other products (Bailin et al. 2008); it is also used as the coating (developer) in thermal paper, from which it leaches at very high levels (Biedermann et al. 2010). While it has been known for decades that BPA mimics the activity of endogenous estrogen, since it was shown to have the efficacy on the reproductive system in female rats of the endogenous estrogen estrone (Dodds and Lawson, 1936), it is only over the last 15 years BPA has become one of the primary EDCs being chosen for study by researchers in many different disciplines. The basis for this new high level of interest among researchers in the cell, endocrine and neuroscience fields is that BPA acts as a ligand for not only the classical nuclear estrogen receptors (estrogen receptor alpha and beta), but also by more recently discovered non-genomic (rapid) cell membrane associated estrogen-receptor signaling pathways. Via these rapid estrogen receptor signaling pathways, BPA and the endogenous estrogen, estradiol, are equally potent (Watson et al. 2010).

GOVERNMENT AGENCIES HAVE NOT REGULATED BPA IN RESPONSE TO PUBLISHED FINDINGS OF HARM

The chemical BPA has become the primary example of how badly the regulatory system has failed with regard to identifying the health hazards associated with EDCs. However, it is important to again emphasize that these issues only apply to the small percentage of chemicals in commerce that have actually undergone testing for toxicity. There are a number of facts about EDCs that lead to dramatically different approaches to assessing their risks. First EDCs have entirely different effects on the fetus, infant or child than on the adult, so it is not just the dose, but the time in life of exposure that is critical; this is an issue that has not been recognized in chemical risk assessments. Second, although a brief exposure may occur only during a critical period in organ development, obvious manifestations of that exposure may not occur until adulthood or old age. In risk assessments for chemicals listed as systemic toxicants, this is missed because animals are examined in young adulthood. Third, because all hormonal systems show non-monotonic dose-response relationships, high doses of EDCs do not predict outcomes at low doses, which will be discussed in more detail below (Myers et al. 2009a), and fourth, there can be no threshold (Sheehan et al. 1999; Sheehan, 2006) for systems that are already "on" due to the presence of endogenous hormones (for example, the endogenous estrogen, estradiol) and being altered by doses of chemicals such as BPA far below the range capable of being measured by the most sensitive analytical methods currently available; for example: a 10^{-14} M dose of BPA (2.3 femtorgams/ml or 2.3 parts per quadrillion) caused a change in calcium uptake and stimulation of a rapid signaling system mediated by receptors associated with the cell membrane (Alyea and Watson, 2009; Kochukov et al. 2009). This reality stands in stark contrast to the often-stated argument that the only reason that there are chemicals of emerging concern is that our ability to detect them has gotten better, not that evidence that they can cause harm has increased. All of the issues identified here are considered to be heretical statements from the perspective of traditional toxicology. In sharp contrast, they are all readily accepted by any scientist trained in studying compounds that interact with receptors for endogenous compounds (ligands) that can bind to cellular receptors and cause responses at very low concentrations; examples of these signaling molecules are secreted hormones and growth factors.

Non-monotonic dose-response relationships are common.

One of the most contentious issues has been that current methods of risk assessment are based on the assumption that the shape of the dose-response curve is always monotonic. This is identified in toxicology as "the dose makes the poison", which refers to the prediction that the greater the exposure, the greater the response (Myers et al. 2009a). However, in experiments with hormones, drugs and other chemicals that act via hormonal mechanisms, it is very common for the dose-response curve to be non-monotonic and form an inverted U, which in endocrinology is called a "biphasic" dose-response curve (these curves have ascending and descending phases). In contrast, there appears to be a lack of awareness of this phenomenon by some toxicologists, particularly those associated with corporations that make chemicals identified as EDCs. A strange consequence of this belief that dose-response curves are always monotonic is that in many toxicological studies in which effects only occur in a restricted "low-dose" range, while the effect is not seen at higher doses, the authors conclude that there is no relationship between dose and response rather than that the relationship is non-monotonic. While not all dose-response relationships for hormones or EDCs are non-monotonic, the fact that non-monotonic dose-response relationships do commonly occur in endocrinology (Welshons et al. 2003; Myers et al. 2009a) has not been incorporated into the process of assessing risk of exposure to environmental chemicals. Thus, the assessment of risk in response to chronic low-dose exposure to chemicals is determined by examining acute toxic effects of only a few very high doses.

Guideline compliant versus academic research.

In contrast to these findings, based on traditional toxicological testing conducted for risk assessments using "guideline compliant" protocols, BPA had been incorrectly considered to be a very weak environmental estrogen. The basis for this confusion is that if a hormone or hormone-mimicking chemical or drug is only tested at a very high dose, the ability of the chemical to alter cell function at very low doses is dramatically underestimated. The molecular mechanisms that mediate this error are well understood to endocrinologists (Welshons et al. 2003; Welshons et al. 2006), but have been a source of great confusion and anxiety to risk assessors who are unwilling to acknowledge the flaws in current risk assessment methods (Myers et al. 2009a; Myers et al. 2009bb). There is an enormous discrepancy between the doses that have been found to cause harm in animals in a large number of peer-reviewed studies by independent academic scientists, which includes doses in the nanogram/kg and microgram/kg body weight/day range (Richter et al. 2007), and the no-effect dose based on testing the chemical using "guideline compliant" protocols. Thus, research using 21st century testing methods has identified adverse effects at doses far below the dose considered to cause no effect (the no effect level or NOEL) by regulatory agencies such as the European Food Safety Authority (EFSA), which as of 2011 remains at 5 milligrams/kg body weight/day (EFSA 2006). It is noteworthy that the opinion of EFSA has been subjected to harsh criticism and was rejected by the European Commission, which voted to ban the use of BPA in baby bottles in spite of assurances by EFSA of the safety of BPA.

Age at which exposure occurs is critical.

One basis for the high level of concern about BPA is that it is well known that fetuses and infants have a limited capacity to metabolize chemicals and drugs, and the glucuronidating enzymes responsible for conjugating (and largely inactivating) BPA are not present in the liver of fetuses and infants at adult levels in rodents or primates, including humans (Taylor et al. 2008 and 2011). Astonishingly, EFSA issued an opinion that infants could metabolize very high doses of BPA, far higher than doses that they might possibly be exposed to. However, this prediction contained no citation of data that would justify an opinion that contradicted every study ever conducted comparing infant and adult metabolism of BPA or other similar chemicals (EFSA, 2008). This action has led to concern regarding conflicts of interest among the members of the EFSA panel and to rejection of the position that BPA was completely safe by the European Commission (Hickman, 2010) and 9 state legislatures in the USA.

Arguments concerning the "safe" dose of BPA and the issue of "good laboratory practices".

There are now hundreds of peer-reviewed publications showing that BPA can both activate and interfere with a broad range of biological responses at doses far below those previously predicted to not cause any effects by EFSA. The explosion of research articles published in peer reviewed journals on BPA demonstrate that the models used by regulatory agencies, such as the EPA, FDA and EFSA, to estimate the safe daily intake dose failed to predict the adverse effects that are being reported at doses thousands of times lower than the estimated no effect level for BPA. These new findings have been reviewed by Richter et al. (Richter et al. 2007). This was accompanied by a consensus statement concerning the hazards posed by BPA co-signed by 38 experts (vom Saal et al. 2007), based on a U.S. National Institute of Environmental Health Sciences sponsored meeting. The scientific experts' consensus statement shows that competent scientists believe that BPA poses a hazard to human health and that the process of assessing human risk due to exposure to BPA failed to predict the actual range of doses that would cause adverse effects in animal studies and thus, potentially, also in humans.

The rationale identified by both the U.S. FDA and EFSA for rejecting the hundreds of peer-reviewed academic studies showing adverse health effects of BPA in laboratory animals at doses below the presumed no effect level was that they did not follow "guideline compliant" protocols, as indicated previously. The reason that independent academic studies do not follow these protocols is that they were developed in the middle of the 20th century. No scientist conducting research that has gone through the rigorous peer-review system required to receive government funding in any country would be awarded a grant based on proposing to use approaches that are more than a half-a-century out of date. However, it is mandated that toxicologists conducting guideline compliant studies do exactly that. These guideline compliant protocols are conducted under standards of record keeping referred to as "Good Laboratory Practices or GLP". While this might suggest that the protocols and methods used in such research must achieve a high standard, in fact this is not the case (Tweedale, 2010). GLP was instituted as a requirement for commercial laboratories conducting toxicological studies funded by chemical corporations because it was found that fraudulent data were being submitted to regulatory agencies. The equation of GLP with "good science" by regulatory agencies is thus clearly unjustified, and examples of profoundly flawed GLP studies on the health effects of BPA that were funded by chemical corporations have been identified (Myers et al. 2009b).

The lack of appropriate positive controls in industry-funded studies of BPA.

One of the most egregious flaws in industry-funded research is either the lack of an appropriate dose of a positive control in relation to the doses of BPA examined [for example, (Tyl et al. 2008)] or the lack of inclusion of any positive control compound (Tyl et al. 2002; Stump et al. 2010). A positive control "reference" chemical or drug is required to demonstrate the lack of contamination of the test system and that the investigators are competent to find any effects for a control compound that has been extensively studied for effects on the outcomes being examined; for detailed criticisms see (vom Saal and Welshons, 2006; Myers et al. 2009b). The response by representatives of industry is that: "A positive control group is neither required nor routinely employed in guideline-compliant studies, because the study design and end points are validated (i.e., known to result in adverse consequences from an appropriate chemical by appropriate routes and doses)." (Tyl, 2009). This statement is stunning in its flawed logic on a number of counts: 1. The maxim in experimental biology that drawing strong positive conclusions from entirely negative findings is unwise appears to be unknown to this investigator; 2. This statement violates the basic approach in science that in testing the null hypothesis (that there is no effect), one can only reject the hypothesis of no effect with some level of confidence. It is not possible to prove that there is no effect. Inclusion of a positive control that shows a positive effect does allow one to conclude (with some level of confidence) that the effect found for the positive control is not found for the test chemical. Thus, if, in fact, the positive control and BPA were to show different effects, which is typically not the case (vom Saal and Welshons, 2006; Myers et al. 2009b), at least if an effect occurred at an appropriate low dose of a positive control then there would be evidence that the investigators were competent to find effects on that system; 3. Finally, the idea that findings from a study are valid because at some point in the past a significant effect was found in other studies of other compounds, rendering all subsequent findings "valid", is incredible to read in a peer-reviewed commentary. The reason why concurrent positive controls are conducted is that changes from one experiment to another can occur due to a myriad of reasons, with batch-to-batch variability in contents of animal feed being just one well studied example (Heindel and vom Saal, 2008).

What is interesting is that the author (R. Tyl) of the above statement conducted a chemical-industry funded study (Tyl et al. 2008) to examine the effects of BPA in CD-1 mice, which is the mouse that we use in our laboratories. In a study conducted at the same commercial laboratory in 1995 with CD-1 mice, the control male mice had prostate weights of about 46 mg in 16-week-old males (Heindel et al. 1995), within the prostate weight range that we (Ruhlen et al. 2008), and others (Gupta, 2000), have reported using a soy-based feed. In the BPA study by Tyl, untreated (control) males were reported to have prostates that weighed over 70 mg when examined at 14 weeks old (Tyl et al. 2008), although these same males were subsequently reported to be between 16 – 19 weeks old (Tyl, 2009) and in testimony presented to the FDA Science Review Board in September 2008 were stated by Tyl to have enlarged prostates because they were 24 weeks old. If GLP studies are the "gold standard" in toxicology because GLP requires detailed record keeping, it should not be possible that the age of animals at sacrifice could be reported as occurring at three different ages in three different presentations of the same data. The clearly abnormal organ weights in the Tyl et al. (2008) study is just one example of the flawed approach to determining "validity" in GLP studies conducted for chemical risk assessments. While it is clear to experts in the field that the reproductive organs in control males in the Tyl et al. (2008) BPA study were abnormal, this was not recognized by those chosen to review the results of the study. Also, if one believes everything one does in a

GLP study is valid regardless of how discordant the data are from historic published values, then even a gross discrepancy in the data can be ignored. That EFSA persists in using this clearly flawed study as one of only two published studies on BPA that it accepts as "valid" is unacceptable.

Rejection of FDA draft risk assessment by the FDA Science Review Board.

Given the above information, it is not surprising that the U.S. FDA's science advisory board charged with determining the accuracy of the 2008 draft FDA risk assessment on BPA rejected the FDA's risk assessment of BPA and stated: "The Subcommittee finds that the draft assessment conclusions are not supported by the available data and science unless the arguments for excluding all work on effects of BPA except the Tyl studies are accepted." (FDA, 2008a). The FDA had rejected findings from all peer-reviewed publications by academic and government scientists, and following the lead of EFSA, the FDA conclusion that BPA was safe was made based on only two industry-funded studies conducted by Tyl et al. (2002; 2008) that conformed to GLP standards (FDA 2008b). Independent scientists published peer-reviewed commentaries that these two industry-funded studies were profoundly flawed (vom Saal and Hughes, 2005; Myers et al. 2009b). Subsequently, in January of 2010 the FDA issued a revised statement indicating "some concern" about the health effects of BPA and urged the public to limit exposure to BPA. However, the FDA took no regulatory action regarding BPA. In their January, 2010 press conference (FDA 2010), the FDA assistant commissioner noted:

> Current BPA food contact uses were approved under food additive regulations issued more than 40 years ago. This regulatory structure limits the oversight and flexibility of FDA. Once a food additive is approved, any manufacturer of food or food packaging may use the food additive in accordance with the regulation. There is no requirement to notify FDA of that use. For example, today there exist hundreds of different formulations for BPA-containing epoxy linings, which have varying characteristics. As currently regulated, manufacturers are not required to disclose to FDA the existence or nature of these formulations. Furthermore, if FDA were to decide to revoke one or more approved uses, FDA would need to undertake what could be a lengthy process of rulemaking to accomplish this goal.

The FDA thus publicly acknowledged that it did not have the authority to easily regulate the uses of BPA, because BPA was one of the approximately 60,000 chemicals "grandfathered" as "generally regarded as safe" when the laws governing the use of BPA and other chemicals were passed by the U.S. Congress.

In contrast to the United States, the European Union has a new chemical regulation law called "REACH—Registration, Evaluation and Authorization of Chemicals". While the law mandates testing of chemicals prior to using them in products (something that is not done in the USA), it allows chemical corporations to design the studies, analyze the data and report conclusions concerning safety to regulatory agencies. There is now irrefutable evidence that research controlled by chemical corporations is biased (Fagin and Lavelle, 1999; Markowitz and Rosner, 2000; Ong and Glantz, 2001; Michaels, 2008; Oreskes and Conway, 2010). Most relevant to this discussion, 100% of industry-funded studies of BPA have concluded that BPA is completely safe, regardless of whether the data should have led to the opposite conclusion (Haseman et al. 2001; vom Saal and Hughes, 2005; vom Saal and Welshons, 2006). There is thus a dramatic effect of source of funding on the outcome of research involving environmental chemicals, and there is also evidence of fraud by chemical corporations in an attempt to mislead the public regarding the hazards of chemicals used in products containing these chemicals.

CONCLUSIONS

Chemical manufacturers have worked with government regulatory agencies to develop a complex set of assumptions governing the decisions reached by members of the chemical risk assessment community. These assumptions have become accepted as true even though they are falsified by a huge amount of published data. When assumptions, which are essentially unverified guesses, are falsified by data, a scientist would develop new assumptions based on models that are consistent with existing data. The blatant rejection of data in favor of models by industry employees and risk assessors clearly cannot be sustained and has provoked strong responses demanding that risk assessment models that are contradicted by an extensive amount of published data be abandoned (vom Saal and Sheehan, 1998; Gies et al. 2009; Myers et al. 2009b; Vandenberg et al. 2010b).

The chemical industry has developed an array of approaches for testing BPA and other chemicals that lead to the results they want. It is logical to conclude that it is not by chance that industry-funded studies of BPA always lead to the conclusion of "no harm" and that considerable effort is expended determining the conditions necessary to achieve the desired result. One finding we have reported is that using certain types of animal feed can mask the effects of estrogenic chemicals, including potent estrogenic drugs, in mice (Ruhlen et al. 2008; Ruhlen et al. 2011). Similarly, in aquatic animals manipulation of environmental conditions, such as water temperature, can lead to false negative findings of "no harm" (Oehlmann et al. 2008). It is time for regulatory agencies to reject claims by chemical corporations that the public can be assured of the safety of chemicals based on a few industry-funded studies reporting that a chemical poses no hazard, when a large number of studies by independent scientists lead to the opposite conclusion. The clear evidence of blatant fraud by the tobacco (Ong and Glantz, 2001; Oreskes and Conway, 2010) and lead paint (Markowitz and Rosner, 2000) industries that occurred over many decades has now been extensively documented. Given the adverse human health trends related to BPA (Table 1), we cannot afford to allow few chemical corporations that manufacture BPA to continue to successfully use the same approaches that the tobacco and lead paint industries employed to hold off regulation for decades.

ACKNOWLEDGEMENTS

Funding during the preparation of this manuscript was provided to FvS by NIEHS grant ES018764 and to SP and PP by the University of Parma.

REFERENCES

1. Alyea, R.A. and C.S. Watson (2009) "Differential regulation of dopamine

transporter function and location by low concentrations of environmental estrogens and 17beta-estradiol." *Environ Health Perspect* 117:778-783.

2. Bailin, P.D., M. Byrne, S. Lewis and R. Liroff (2008) "Public awareness drives market for safer alternatives: bisphenol A market analysis report." http://www.iehn.org/publications.reports.bpa.php. Access date: April 20, 2011.
3. Biedermann, S., P. Tschudin and K. Grob (2010) "Transfer of bisphenol A from thermal printer paper to the skin." *Anal Bioanal Chem* 398:571-576.
4. Bird, J. (2005) "Hyperbole or common sense?" *Chemistry and Industry* 7 (March):14-15.
5. Braun, J.M., K. Yolton, K.N. Dietrich, R. Hornung, X. Ye, A.M. Calafat, et al. (2009) "Prenatal bisphenol A exposure and early childhood behavior." *Environ Health Perspect* 117:1945-1952.
6. Chameides, B. (2011) "In search of the toxic 5." June 13, 2011. http://www.nicholas.duke.edu/thegreengrok/insearchoftsca5. Access date: June 28, 2011.
7. Dodds, E.C. and W. Lawson (1936) "Synthetic oestrogenic agents without the phenanthrene nucleus." *Nature* 137:996.
8. EFSA (2006) "Opinion of the Scientific Panel on Food Additives, Flavourings, Processing Aids and Materials in Contact with Food on a request from the Commission related to 2,2-BIS(4-HYDROXYPHENYL)PROPANE (Bisphenol A)." *The EFSA (European Food Safety Authority) Journal* 428:1-76.
9. EFSA (2008) "Toxicokinetics of Bisphenol A - Scientific Opinion of the Panel on Food additives, Flavourings, Processing aids and Materials in Contact with Food (AFC)." Access date: May 30, 2010. E. F. S. Authority, http://www.efsa.europa.eu/EFSA/efsa_locale-1178620753812_1211902017492.htm July, 2008.
10. Fagin, D. and M. Lavelle (1999) *Toxic Deception: How the Chemical Industry Manipulates Science, Bends the Law and Endangers Your Health*. Washington, DC, Common Courage Press,.
11. FDA (2008a) FDA Science Board Subcommittee On Bisphenol A: Scientific Peer-Review Of The Draft Assessment Of Bisphenol A For Use In Food Contact Applications. October 31, 2008.
12. FDA (2008b) "Food and Drug Administration draft assessment of bisphenol A for use in Food contact applications. August 14, 2008." http://www.fda.gov/Food/FoodIngredientsPackaging/ucm166145.htm. Accessed August 2, 2010.
13. FDA (2010) "Update on bisphenol A for use in food contact applications. January 10, 2010." http://www.fda.gov/NewsEvents/PublicHealthFocus/ucm197739.htm. Access date: June 28, 2011.
14. Gies, A., B. Heinzow, H.H. Dieter and J. Heindel (2009) "Bisphenol a workshop of the German federal environment agency - March 30-31, 2009 work group report: public health issues of bisphenol a." *Int J Hyg Environ Health* 212:693-696.
15. Gioiosa, L., E. Fissore, G. Ghirardelli, S. Parmigiani and P. Palanza (2007) "Developmental exposure to low-dose estrogenic endocrine disruptors alters sex differences in exploration and emotional responses in mice." *Horm Behav* 52:307-316.

16. Gupta, C. (2000) "Reproductive malformation of the male offspring following maternal exposure to estrogenic chemicals." *Proc Soc Exp Biol Med* 224:61-68.
17. Haseman, J.K., A.J. Bailer, R.L. Kodell, R. Morris and K. Portier (2001) "Statistical issues in the analysis of low-dose endocrine disruptor data." *Toxicol Sci* 61:201-210.
18. Heindel, J.J., R.E. Chapin, J. George, D.K. Gulati, P.A. Fail, L.H. Barnes, et al. (1995) "Assessment of the reproductive toxicity of a complex mixture of 25 groundwater contaminants in mice and rats." *Fundam Appl Toxicol* 25:9-19.
19. Heindel, J.J. and F.S. vom Saal (2008) "Meeting report: batch-to-batch variability in estrogenic activity in commercial animal diets--importance and approaches for laboratory animal research." *Environ Health Perspect* 116:389-393.
20. Hickman, M. (2010) "Europe tightens 'gender bender' chemical rules." *The Independent*. London, Saturday, November 27.
21. Kochukov, M.Y., Y.J. Jeng and C.S. Watson (2009) "Alkylphenol xenoestrogens with varying carbon chain lengths differentially and potently activate signaling and functional responses in GH3/B6/F10 somatomammotropes." *Environ Health Perspect* 117:723-730.
22. Laviola, G., L. Gioiosa, W. Adriani and P. Palanza (2005) "D-amphetamine-related reinforcing effects are reduced in mice exposed prenatally to estrogenic endocrine disruptors." *Brain Res Bull* 65:235-240.
23. Markowitz, G. and D. Rosner (2000) "'Cater to the children': The role of the lead industry in a public health tragedy, 1900-1955." *Amer. J. Public Health* 90:36-46.
24. Martini, M., D. Miceli, S. Gotti, C. Viglietti-Panzica, E. Fissore, P. Palanza, et al. (2010) "Effects of perinatal administration of Bisphenol A on the neuronal nitric oxide synthase expressing system in the hypothalamus and limbic system of CD1 mice." *J Neuroendocrinol* 22:1004-12.
25. Michaels, D. (2008) *Doubt is Their Product: How Industry's Assault on Science Threatens Your Health* New York, Oxford University Press.
26. Myers, J.P., F.S. vom Saal, B.T. Akingbemi, K. Arizono, S. Belcher, T. Colborn, et al. (2009b) "Why public health agencies cannot depend on good laboratory practices as a criterion for selecting data: the case of bisphenol A." *Environ Health Perspect* 117:309-315.
27. Myers, J.P., T.J. Zoeller and F.S. vom Saal (2009a) "A clash of old and new scientific concepts in toxicity, with important implications for public health." *Environ. Health Perspect.* 117:1652–1655.
28. Oehlmann, J., M. Oetken and U. Schulte-Oehlmann (2008) "A critical evaluation of the environmental risk assessment for plasticizers in the freshwater environment in Europe, with special emphasis on bisphenol A and endocrine disruption." *Environ Res* 108:140-149.
29. Ong, E.K. and S.A. Glantz (2001) "Constructing 'sound science' and 'good epidemiology': Tobacco, lawyers, and public relations firms." *Amer. J. Public Health* 91:1749-1756.
30. Oreskes, N. and E.M. Conway (2010) *Merchants of Doubt: How a Handful of Scientists Obscured the Truth on Issues from Tobacco Smoke to Global Warming*. New York, Bloomsbury Press.
31. Palanza, P., L. Gioiosa, F.S. vom Saal and S. Parmigiani (2008) "Effects of

developmental exposure to bisphenol A on brain and behavior in mice." *Environ Res* 108:150-157.
32. Palanza, P., K.L. Howdeshell, S. Parmigiani and F.S. vom Saal (2002) "Exposure to a low dose of bisphenol A during fetal life or in adulthood alters maternal behavior in mice." *Environ. Health Perspect.* 110:415-422.
33. Razzoli, M., P. Valsecchi and P. Palanza (2005) "Chronic exposure to low doses bisphenol A interferes with pair-bonding and exploration in female Mongolian gerbils." *Brain Res Bull* 65:249-254.
34. Richter, C.A., L.S. Birnbaum, F. Farabollini, R.R. Newbold, B.S. Rubin, C.E. Talsness, et al. (2007) "In vivo effects of bisphenol A in laboratory rodent studies." *Reprod Toxicol* 24:199-224.
35. Ruhlen, R.L., K.L. Howdeshell, J. Mao, J.A. Taylor, F.H. Bronson, R.R. Newbold, et al. (2008) "Low phytoestrogen levels in feed increase fetal serum estradiol resulting in the "fetal estrogenization syndrome" and obesity in CD-1 mice." *Environ Health Perspect* 116:322-328.
36. Ruhlen, R.L., J.A. Taylor, J. Mao, J. Kirkpatrick, W.V. Welshons and F.S. vomSaal (2011) "Choice of animal feed can alter fetal steroid levels and mask developmental effects of endocrine disrupting chemicals." *Journal of Developmental Origins of Health and Disease* 2:36-48.
37. Sheehan, D.M. (2006) "No-threshold dose-response curves for nongenotoxic chemicals: Findings and applications for risk assessment." *Environ Res* 100:93-99.
38. Sheehan, D.M., E. Willingham, D. Gaylor, J.M. Bergeron and D. Crews (1999) "No threshold dose for estradiol-induced sex reversal of turtle embryos: how little is too much?" *Environ. Health Perspect.* 107:155-159.
39. Stump, D.G., M.J. Beck, A. Radovsky, R.H. Garman, L.L. Freshwater, L.P. Sheets, et al. (2010) "Developmental neurotoxicity study of dietary bisphenol A in Sprague-Dawley rats." *Toxicol Sci* 115:167-82.
40. Taylor, J.A., F.S. Vom Saal, W.V. Welshons, B. Drury, G. Rottinghaus, P.A. Hunt, et al. (2011) "Similarity of bisphenol a pharmacokinetics in rhesus monkeys and mice: relevance for human exposure." *Environ Health Perspect* 119:422-430.
41. Taylor, J.A., W.V. Welshons and F.S. Vom Saal (2008) "No effect of route of exposure (oral; subcutaneous injection) on plasma bisphenol A throughout 24h after administration in neonatal female mice." *Reprod Toxicol* 25:169-176.
42. Tweedale, T. (2010) "Good laboratory practices and safety assessments: Another view." *Environ Health Perspect* 118:194.
43. Tyl, R.W. (2009) "Basic exploratory research versus guideline-compliant studies used for hazard evaluation and risk assessment: bisphenol A as a case study." *Environ Health Perspect* 117:1644-1651.
44. Tyl, R.W., C. Myers, M. Marr, C.S. Sloan, N. Castillo, M.M. Veselica, et al. (2008) "Two-Generation Reproductive Toxicity Study of Dietary Bisphenol A (BPA) in CD-1® (Swiss) Mice." *Toxicol. Sci.* 102:362-384.
45. Tyl, R.W., C.B. Myers, M.C. Marr, B.F. Thomas, A.R. Keimowitz, D. R. Brine, et al. (2002) "Three-generation reproductive toxicity study of dietary bisphenol A in CD Sprague-Dawley rats." *Toxicol Sci* 68:121-46.
46. Vandenberg, L.N., I. Chahoud, J.J. Heindel, V. Padmanabhan, F.J. Paumgartten

and G. Schoenfelder (2010a) "Urinary, circulating, and tissue biomonitoring studies indicate widespread exposure to bisphenol A." *Environ Health Perspect* 118:1055-70.

47. Vandenberg, L.N., I. Chahoud, V. Padmanabhan, F.J. Paumgartten and G. Schoenfelder (2010b) "Biomonitoring studies should be used by regulatory agencies to assess human exposure levels and safety of bisphenol A." *Environ Health Perspect* 118:1051-4.
48. vom Saal, F.S. (2005) "Low-dose BPA: confirmed by extensive literature." *Chemistry & Industry* 7 (April):14-15.
49. vom Saal, F.S., B.T. Akingbemi, S.M. Belcher, L.S. Birnbaum, D.A. Crain, M. Eriksen, et al. (2007) "Chapel Hill bisphenol A expert panel consensus statement: integration of mechanisms, effects in animals and potential to impact human health at current levels of exposure." *Reprod Toxicol* 24:131-138.
50. vom Saal, F.S. and C. Hughes (2005) "An extensive new literature concerning low-dose effects of bisphenol A shows the need for a new risk assessment." *Environ. Health Perspect.* 113:926-933.
51. vom Saal, F.S. and D.M. Sheehan (1998) "Challenging risk assessment." *Forum for Applied Research and Public Policy* 13:11-18.
52. vom Saal, F.S. and W.V. Welshons (2006) "Large effects from small exposures. II. The importance of positive controls in low-dose research on bisphenol A." *Environ. Res.* 100:50-76.
53. Watson, C.S., Y.J. Jeng and M.Y. Kochukov (2010) "Nongenomic signaling pathways of estrogen toxicity." *Toxicol Sci* 115:1-11.
54. Welshons, W.V., S.C. Nagel and F.S. vom Saal (2006) "Large effects from small exposures. III. Endocrine mechanisms mediating effects of bisphenol A at levels of human exposure." *Endocrinol.* 147:S56-S69.
55. Welshons, W.V., K.A. Thayer, B.M. Judy, J.A. Taylor, E.M. Curran and F.S. vom Saal (2003) "Large effects from small exposures. I. Mechanisms for endocrine-disrupting chemicals with estrogenic activity." *Environ Health Perspect* 111:994-1006.

WORKSHOP PARTICIPANTS

Dr. Thomas Ballestero	University of New Hampshire Department of Civil Engineering, Durham, USA
Dr. Joe Braun	Department of Environmental Health Harvard University, Boston, USA
Professor Jörg E. Drewes	Environmental Science and Engineering Colorado School of Mines, Golden, USA
Dr. Lorne Everett Chairman	Chancellor Emeritus, Lakehead University Thunder Bay, Canada and L. Everett & Associates, LLC, Santa Barbara, USA
Dr. Bernard Goldstein	Dept of Environmental and Occupational Health, Graduate School of Public Health, Pittsburgh, USA
Professor Frank Leon Parker	Environmental Engineering Vanderbilt University, Nashville, USA
Professor Stefano Parmigiani	Evolutive and Functional Biology Università di Parma, Parma, Italy
Dr. Alfred Abed Rabbo	Water and Soil Environmental Research Unit, Bethlehem University, West Bank (Palestine), via Israel
Professor Deborah L. Swackhamer	Science, Technology, and Public Policy, Hubert H. Humphrey Institute of Public Affairs, Environmental Health Sciences, University of Minnesota, St Paul, USA
Professor Donald Tillitt	Biochemistry & Physiology, Columbia Environmental Research Center, Columbia, USA
Dr. Henry J. Vaux, Jr.,	Professor of Resource Economics and Resource Economist, Emeritus Chair, Rosenberg International Forum on Water Policy, University of California Berkeley, USA
Professor Frederick vom Saal	University of Missouri Division of Biological Sciences, Columbia, USA
Dr. L. James Wright	Department of Chemistry University of Auckland, Auckland, New Zealand

12

MITIGATION OF TERRORIST ACTS WORKSHOP

MITIGATION OF TERRORIST ACTS WORKSHOP

SCIENTIFIC SOLUTIONS FOR CBRNE TERRORISM AND CATASTROPHIC EVENTSWITH REFLECTIONS ON FUKUSHIMA AND NORWAY SITUATIONAL AWARENESS COMMUNICATIONS EVACUATION SHELTER RISK

AGENDA

SESSION No. 1—Strategic Terrorism Papers

Vasily Krivokhizha
Current Global Strategic Situation

Diego Buriot
Special Report: Bio-Terror

Robert Duncan
Complexity

Richard Wilson
Fukushima Generic Lessons: Risk Assessment, Evacuation

SESSION No. 2—Fukushima: Reflections, Applications to CBRN Terror Mitigation, Future Risk Management

Sally Leivesley
The Fukushima Effect: Communications, Catastrophic Risk and Trust. Does the World Need a Nuclear Covenant?

Annette Sobel
Fukushima Situational Awareness: Public Health, Science and Human Behavioural Factors and Health Risk Communication during Disasters

Complexity and WMD Threat Response, Early indicators, Warnings, Mitigation and Recovery

Working Group Session

Discussion and Outcomes Documentation

SESSION No. 3

Testing Scientific Communications, Situational Awareness, Evacuation, Shelter Risk Assessment Method

Sally Leivesley

Group Scenario Biological Attack Exercise

Group Role Play Scientific Solutions/In Country Responses: Reflections on Contamination Environment Post Fukushima.

Discussion

Threats from Biological, Nuclear, Radiological, Chemical, Cyber, Explosives Terrorism: Reflections on Fukushima and Europe's Nuclear Facility Risk Profile following the Norwegian Dual Attack

Richard Wilson

Overview of Workshop Day 1

SESSION No. 4—Preparing a Report on Scientific Solutions for CBRNE Terrorism with Reflections on Fukushima

Introduction

Group working session—members to summarise one of the following

1. *Global terror threat overview CBRNE 2011-2012*
2. *Fukushima Situational Awareness*
3. *Fukushima Communications*
4. *Fukushima Iodine-131 and Caesium-137 Information Criticalities*
5. *Fukushima Evacuation and Shelter—future decision making*
6. *Risk and Catastrophic Risk Assessment Methodologies for Nuclear Reactors and Critical Infrastructure Taking Into Account Fukushima and the Norwegian Attack*
7. *Complexity*

SESSION No. 5

Robert Duncan

Discussion: Novel contributions of technology to science in the mitigation of terrorist acts

Annette Sobel

Exercises and Red Team Experience

Panel documentation of Report Part 1

SESSION No. 6

Group Discussion on Report Structure Part 2

Final Report Drafting in Group

Planning: Nominated Chair for 2012-2014

Richard Garwin
Overview Workshop and Strategic Challenges

Chair: Dr. Sally Leivesley

PERMANENT MONITORING PANEL—MITIGATION OF TERRORIST ACTS, REPORT FOR THE YEAR 2010-2011

REPORT AUTHOR: PMP-MTA CO-CHAIR, DR. SALLY LEIVESLEY

An overview of the work of the PMP for 2010-2011 and in the preceding two years is discussed, in brief, in this report. Details of the work for the past three years by members of the Panel will be placed on the WFS web site to share with other scientists. The WFS web site also provides a reference point for academic institutions and members of the public who have an interest in scientific methods for mitigation of terrorist acts.

A focus on high impact terrorism by the members of the PMP-MTA has produced papers, teaching presentations and scenario based exercises. This approach to the analysis of terrorist devices, tactics and the consequences of terrorist acts has extended academic analysis into practical testing of synergies for cooperation between scientists across the world.

An important series of presentations in power point format has been produced by members of the Panel. These present practical and clear explanations of the consequences of nuclear, radiological, chemical and biological threats from terrorism and analysis of the means by which governments with scientific advice may mitigate such threats.

In annual meetings at Erice, lecture rooms have become a testing ground for the members of the PMP and guests each year—nuclear, radiological, chemical or biological threats have been assessed for near real time solutions by scientists who have participated in role plays of crisis events. Panel members from different countries have jointly developed solutions and on occasions guests from other PMPs have assisted with specialised knowledge from journalism and nuclear science.

Gap analysis in international capability for identifying and responding to catastrophic threats from international terrorism has underpinned much of the analysis and discussion. The Panel has wanted to focus on particular problems identified in gaps analysis. Solutions have been established through workshops each year and solutions arising from scientific papers are then testing in high impact terror attack scenarios using the workshop time that is supported by the WFS in special two day meetings in August.

OUTPUTS

1. Outputs from the focus on high impact terrorism have included:

- Theoretical papers.
- Teaching modules from power point presentations of essential technical and scientific information.
- The definition of technological challenges and some successful innovative solutions that have been taken back to member's countries for further consideration.
- There have been further practical tests of how science can mitigate terror effects best when there is international cooperation. There has been extensive cooperation by Panel members when challenges have arisen from terror attacks or threats of terror attacks. Most recently the Fukushima Daiichi

incident, though a natural and technological disaster has been a strong focus for this cooperative discussion.

- Translation of lessons learned from natural catastrophes such as pandemic or major incidents has been an important part of the work of the PMP. Pandemic has been a strong area of work and although the international threats of some potential pandemics have receded in the past year. The PMP revived its interest in pandemic when considering mitigation of biological terror threats. Papers from earlier years under Professor Garwin's chairmanship were directly relevant for analysis of biological terror mitigation because pandemic infection control provides some basic solutions. New information on bio-engineering and citizen scientists who share information through the World Wide Web was brought to the attention of the Panel.
- The challenge of biological threats was considered to be a theme that went beyond the three year study undertaken by the PMP and which required some specialist guests to be invited into a future meeting to deliver a more detailed understanding of 'citizen' biological work. It is considered that there are extreme asymmetric threats of bio-engineering if this technology can be carried out under the radar of the authorities and at the simplest level 'in the kitchen'.

2. Outputs from strategic threat overview work by some of the members have included:

- Two or three years of joined up analysis by members who have provided a continuation of papers each year on analysis of in-country management of terrorism. There have also been reviews of strategic global to assist the Panel with a strategic overview. Strategic overviews have provided a fundamental building block for the consideration each year of scenario based terror attacks. Analysis of the consequences have been expanded through to cover the cultural, social, psychological and other factors specific to a country or to international strategies on counter terrorism.

  Some examples of the strategic subject areas have been:

  - A global view of terror threats and political realities driving the interest of countries responding to the growing threats.
  - In-country analysis on India and Pakistan. The work on India provides a backdrop for understanding one of the most populous countries in the world where practical daily challenges to maintain its population create a different reality to catastrophic events than in the western world. The work on Pakistan has provided an increasingly serious set of factors for the Panel to consider because of the fragility of the state and the inroads of radicalism into the society. There has been considerable discussion on the potential catastrophes from terror threats to nuclear reactor safety and security and increasingly turbulent international relations. These threats in turn are significant for many other countries.

- A specialised output of significance to nuclear facilities managers and for counter terrorism mitigation was provided in two annual reports on the proliferation of fissile material based on practical work with the nuclear industry in Russia and analysis from collation of an extensive EU data base. This contribution enabled practical scenario contributions to be considered in the Panel's exercises because the data was a snapshot of the nuclear materials threats identified from losses of nuclear material each year that had been reported to the authorities in Europe. The data assisted the Panel to understand the scale and capability within criminal and terrorist activity and the overlap between organised crime and terrorism that was growing each year. Data on fissile material smuggling is essential for all nuclear facilities operators who are managing terrorist risk and designing counter measures to protect their facilities.

THEMES

Themes that have dominated the work of the PMP have included:

1. Biological and pandemic studies
2. Radiological devices
3. Evaluation of technology
   - Radiological devices, known as 'dirty bombs' have been a theme throughout the three years to 2011 and have dominated the scenario work and analysis by the Panel. This has been a reflection of strategic risks from terrorism and the Panel has been working to consider future evolution of high impact threats from terrorism in order to consider means of mitigation.
   - Scenario exercises have assisted in the analysis and open source data from government sponsored work in the United States on this subject has been one of the key sources for the Panel in undertaking a scenario exercise in 2011 on a reactor threat. One of the challenges for the scientists has been to work out at what stage scientists could identify the nature of an incident and the real time information to be communicated to politicians, the scientific community and to the public. Protection of the public requires extensive analysis and testing by the scientific community if it is to be effective.
   - Evaluation of technology has been an unplanned outcome from the meetings by the Panel but there has been innovative evolution from the workshop meetings at Erice each August over the past three years. Scenario work has been built around technological solutions and theories have been tested to identify the practical significance of certain technological solutions especially in the area of warning and evacuation routes for citizens under radiological threat.

RECOMMENDATIONS FOR POLITICIANS

This has been another output objective by the Panel. Activity has remained at the level of communications within the scientific community. One of the issues which has been taken forward has been discussion on the capacity for institutions such as CERN to assist the world with scientific knowledge and analysis at the time of any mass casualty or other catastrophic nuclear or radiological incident or nuclear terror event.

The objective has been to consider how science can deliver real time solutions under catastrophic events and how communications may be made more effective than in past catastrophic incidents. CERN has a remarkable history with communications for nuclear scientists and the knowledge of the scientific solutions within the repository of experts within CERN may be important in planetary emergencies.

SCENARIO ROLE PLAYS

In 2011, and in the preceding two years, scenarios were used to develop gap analysis by using the combined capability of the Panel who worked to solve whatever challenges arose around each of the key study areas of CBRN.

Scenario role plays also included cyber threats to critical infrastructure so that multi dimensional terror threats could be considered in real time and with a holistic approach to the threats. These cyber threats were essential knowledge in areas such as control systems for critical technology including nuclear reactors and also in the support of communications to inform and protecting the public during a catastrophic incident. The communications issues were considered across many dimensions including in space satellite systems and the World Wide Web. Social media was part of the realistic scenario consideration and the Panel worked on case studies of social media incidents and observations to identify strengths in using this medium for communications as well as understanding it as a risk in some catastrophic incidents.

Other PMPs were invited to participate in the scenarios and technical lectures by the PMP. The linkage with other PMPs was considered vital to the outputs of the PMP-MTA and in 2011 this linkage delivered very practical assistance following cooperation of the PMP Energy and also the PMP Water and Pollution. This led to participation by the Panel in August 2011 in a session with Japanese experts presenting on the Fukushima Daiichi incident.

SPECIAL CASE STUDY

Panel members agreed to present papers on Fukushima Daiichi incident and the consequences of 3/11 on Japan and Rest of the World. This incident was analysed as a special case study in 2011 and cooperation by the Panel members in sharing information was a real time activity by email and telephone following the 3/11 incident onset in Japan. Communications were shared with the PMP Energy and this greatly supported the work of the Panel members.

Specialist analysis of the Fukushima Daiichi incident included the modes of communications between scientists and governments and between scientists on a worldwide basis. Most importantly the focus of the Panel was how to inform a community faced with a catastrophic nuclear incident.

Technical analyses were supplied by members of the Panel and the focus in the 2011 workshop was to fill in the gaps in information that the Japanese government and many countries were trying to establish 5-6 months after 3/11.

NEW THREATS

A new threat to nuclear facilities has emerged after an attack by an alleged sole activist on central Oslo buildings with explosives and the follow up by shooting of young people on an island at a political camp. The suspected perpetrator had previous connections with right wing groups. A document distributed widely by the suspected perpetrator on the day of the attack to many right ring activists contained a diary of activities and a section on nuclear facilities as targets with lists of facilities across Europe and technical information on the facilities. Because of the successful initiation of an explosive device in Oslo and shooting many victims, this incident has shown there is a right wing influenced capacity to plan and use explosives. This new threat has importance for nuclear facilities managers and for consideration of measures to mitigate such terrorist acts.

CONCLUSIONS

Some themes from the 2011 Workshop will be continued into 2012 and beyond.

1. Biological threats will remain a focus for the years 20120-2014.
2. An immediate task for the Panel is to follow up of the scientific lessons from the Fukushima Daiichi incident with observations that will assist other countries to understand the post-nuclear environment. The data from Fukushima Daiichi radiation release reflects a modern, complex highly industrialised society faced with levels of contamination of cities, rural areas and marine and freshwater environments. The consequences of the incident will require new scientific solutions and this is recognised by the Japanese Government and the IAEA which has offered international assistance.
3. Communications methods and processes during nuclear reactor accidents or terror attacks will remain a continuous focus of study for the Panel.

The work of the Panel is balanced by the breadth of contributions from the members which ensure that many factors are considered to mitigate catastrophic terror acts.

In summary, international terrorism is a dynamic environment and incidents in countries each year are proving that a core scientific group with extensive analytical capability can contribute to the mitigation of the consequences of events as these occur. The preparatory work in the workshops and the informal network of activity at the WFS meetings in August and communications throughout each year by the Panel all contribute to the capacity of the Panel to contribute to scientific understanding in this field of planetary emergencies.

WORKSHOP PARTICIPANTS

Dr. Diego Buriot	Consultant and former Special Advisor Communicable Diseases UN World Health Organisation Lyon, France
Professor Robert Duncan	University of Missouri Vice Chancellor for Research Columbia, USA
Professor Richard L. Garwin	Thomas J. Watson Research Center IBM Research Division Yorktown Heights, USA
Professor Pervez Hoodbhoy	Physics Department Quaid-i-Azam University Islamabad, Pakistan
Dr. Vasiliy Krivokhizha	International Department Federal Assembly of the Russian Federation Moscow, Russia
Dr. Sally Leivesley	Newrisk Limited London Mitigation of Terrorists Acts PMP World Federation of Scientists London, UK
Professor Frank Leon Parker	Environmental Engineering Vanderbilt University Nashville, USA
Professor Annette Sobel	University of Missouri Assistant to the Provost for Strategic Opportunities Columbia, USA
Professor Richard Wilson	Department of Physics Harvard University Cambridge, USA

Chair: Dr. Sally Leivesley

13

SEMINAR PARTICIPANTS

Professor Alfred Abed Rabbo	Water and Soil Environmental Research Unit Bethlehem University Palestine, Israel
Professor Sergey V. Ablameyko	Belarusian State University Minsk, Belarus
Professor Eilon Adar	Institute for Water Research J. Blaustein Institute for Desert Research Ben Gurion University of the Negev Sede-Boker Campus, Israel
Lord John Alderdice	House of Lords London, UK
Professor Mikhail J. Antonovsky	Carbon Dioxide Division Institute of Global Climate and Ecology Moscow, Russia
Dr. Muheeb Awawdeh	Dept of Earth and Environmental Sciences Yarmouk University Irbid, Jordan
Dr. Thomas Ballestero	University of New Hampshire Department of Civil Engineering Durham, USA
Professor William A. Barletta	U.S. Particle Accelerator School Department of Physics Massachusetts Institute of Technology Cambridge, USA
Dr. Joe Braun	Department of Environmental Health Harvard University Boston, USA
Dr. Vladimir B. Britkov	Institute for Systems Analysis Russian Academy of Sciences Moscow, Russia
Dr. Franco Buonaguro	Istituto Nazionale dei Tumori "Fondazione G. Pascale" Napoli, Italy

Dr. Diego Buriot	Consultant and former Special Advisor Communicable Diseases UN World Health Organisation Lyon, France
Mr. Patrick Cailleret	ThalesRaytheonSystems CSSOM Massy, France
Dr. Gina M. Calderone	Hydrogeology Department ECC Marlborough, USA
Dr. Nathalie Charpak	Kangaroo Foundation Bogotá, Colombia
Professor Luisa Cifarelli	Department of Physics University of Bologna Bologna, Italy
Dr. Thomas B. Cochran	Natural Resources Defense Council, Inc. Washington DC, USA
Professor Andrea Contin	Department of Physics University of Bologna Bologna, Italy
Professor Pierre Darriulat	Vietnam Auger Training Laboratory Institute for Nuclear Science & Technology Hanoi, Vietnam
Dr. Socorro de Leon-Mendoza	Jose Fabella Memorial Hospital Neonatology Unit Manila, Philippines
Dr. Carmine Difiglio	Office of Policy and International Affairs U.S. Department of Energy Washington, USA
Dr. Mbareck Diop	Science & Technology Advisor to the President of Senegal Dakar, Senegal

Professor Jörg E. Drewes	Environmental Science and Engineering Colorado School of Mines Golden, USA
Professor Robert Duncan	University of Missouri Vice Chancellor for Research Columbia, USA
Professor Merab Eliashvili	Theoretical Physics Department A. Razmadze Mathematical Institute Tbilisi, Georgia
Professor Christopher Essex	Department of Applied Mathematics University of Western Ontario London, Canada
Dr. Lorne Everett	Chancellor Emeritus, Lakehead University Thunder Bay, Canada and L. Everett & Associates, LLC Santa Barbara, USA
Professor William Fulkerson	Institute for a Secure and Sustainable Environment University of Tennessee Knoxville, USA
Professor Mario Fusco Girard	Physics Department Università degli Studi di Salerno Fisciano (SA), Italy
Dr. Bertil Galland	Writer and Historian Buxy, France
Professor Richard L. Garwin	Thomas J. Watson Research Center IBM Research Division Yorktown Heights, USA
Professor Bernardino Ghetti	Dept of Pathology & Laboratory Medicine Indiana University Indianapolis, USA
Dr. Bernard Goldstein	Dept of Environmental and Occupational Health, Graduate School of Public Health Pittsburgh, USA

Professor Alberto González-Pozo	Theory and Analysis Department Universidad Autónoma Metropolitana Mexico D.F., Mexico
Dr. David Greene	Center for Transportation Analysis Energy Division Oak Ridge National Laboratory Oak Ridge, USA
The Honorable John G. Grimes	Former Assistant Secretary and Chief Information Officer U.S. Department of Energy Washington, USA
Dr. Adalat Hasanov	National Academy of Sciences Baku, Azerbaijan
Professor Brian Haynes	Department of Medicine McMaster University Hamilton, Canada
Dr. Udo Helmbrecht	European Network and Information Security Agency Heraklion, Greece
Professor Pervez Hoodbhoy	Physics Department Quaid-i-Azam University Islamabad, Pakistan
Professor Michael Jefferson	International Business and Sustainability Center London Metropolitan Business School London, UK
Lady Barbara Judge	Emeritus Chairman UK Atomic Energy Authority London, UK
Dr. Hisham Khatib	World Energy Council Amman, Jordan
Dr. William Kininmonth	Australasian Climate Research Kew Victoria, Australia

Mr. Erwin Knupp	TOTH SA Geneva, Switzerland
Dr. Edward Kobal	Slovenian Scientific Foundation Ljubljana, Slovenia
Dr. Ilan Koren	Environmental Sciences and Energy Research Weizmann Institute of Science Rehovot, Israel
Professor Jón Egill Kristjánsson	Department of Geosciences University of Oslo Oslo, Norway
Dr. Vasiliy Krivokhizha	International Department Federal Assembly of the Russian Federation Moscow, Russia
Dr. Valery P. Kukhar	Ukrainian Academy of Sciences Laboratory of Fine Organic Synthesis Kiev, Ukraine
Professor Askar Kutanov	National Academy of Sciences Former Ambassador of the Kyrgyz Republic in Japan Bishkek, Kyrgyzstan
Professor Axel Lehmann	Institute for Technical Computer Sciences German Armed Forces University München Neubiberg, Germany
Dr. Sally Leivesley	Newrisk Limited London Mitigation of Terrorists Acts PMP World Federation of Scientists London, UK
Dr. Giovanni Levi	National Scientific Research Centre Evolution of Endocrine Regulation Paris, France
Professor Richard S. Lindzen	Earth, Atmospheric & Planetary Sciences Massachusetts Institute of Technology Cambridge, USA

Professor Bruno Maraviglia	Department of Physics University of Rome "La Sapienza" Rome, Italy
Professor Sergio Martellucci	Faculty of Engineering Università degli Studi di Roma "Tor Vergata" Rome, Italy
Dr. Daniel Masiga	International Centre of Insect Physiology and Ecology (ICIPE) Nairobi, Kenya
Professor Stephen McIntyre	University of Guelph Toronto, Ontario, Canada
Professor Akira Miyahara	National Institute for Fusion Science Tokyo, Japan
Lord Christopher Monckton	The Viscount Monckton of Brenchley Carie, Rannoch, UK
Dr. André Naudi	Director of Finance and Human Resources (Former) CERN, Geneva, Switzerland
Professor Joseph Niemela	International Centre for Theoretical Physics Trieste, Italy
Professor SunKun Oh	Department of Physics Konkuk University Seoul, Korea
Dr. Jef Ongena	Plasmaphysics Laboratory Ecole Royale Militaire Brussels, Belgium
The Hon. Alessandro Pagano	Finance Commission Deputy Chamber Rome, Italy
Professor Frank Leon Parker	Environmental Engineering Vanderbilt University Nashville, USA

Professor Stefano Parmigiani	Evolutive and Functional Biology Università di Parma Parma, Italy
Professor Guido Piragino	Physics Department Università degli Studi di Torino Torino, Italy
Professor Nusrat Rajabov	Department of Mechanics and Mathematics Tajik State University Dushanbe, Tajikistan
Dr. Juan Ruiz	Department of Pediatrics San Ignacio Hospital Santafé de Bogotá, Colombia
Dr. Daniel Schaffer	Third World Academy of Sciences TWAS Trieste, Italy
Dr. Robert Schock	World Energy Council Senior Advisor London, UK
Professor Hiltmar Schubert	Fraunhofer-Institut für Chemische Technologie, ICT Pfinztal, Germany
Professor Jung Uck Seo	Former Minister of Science & Technology President Emeritus of IEEK Seoul, Korea
Dr. Nir J. Shaviv	Hebrew University of Jerusalem Mount Scopus Campus Jerusalem, Israel
Dr. Adnan Shihab-Eldin	Kuwait Foundation for the Advancement of Sciences Safat, Kuwait
Professor S. Fred Singer	Science & Environmental Policy Project Arlington, USA

Professor Annette Sobel	University of Missouri Senior Advisor for Bio-security Engagement Office of the Secretary of Defense/Nuclear, Chemical, Biological Columbia, USA
Professor William Stigliani	UNI Center for Energy & Environmental Education University of Northern Iowa Cedar Falls, USA
Dr. Bruce Stram	Element Markets Houston, USA
Professor Gheorghe Stratan	National Institute of Physics and Nuclear Engineering Bucharest-Magurele, Romania
Professor Henrik Svensmark	Center for Sun-Climate Research Danish National Space Center Technical University of Denmark Copenhagen, Denmark
Professor Deborah L. Swackhamer	Science, Technology, and Public Policy Hubert H. Humphrey Institute of Public Affairs Environmental Health Sciences University of Minnesota St Paul, USA
Professor Kyle Swanson	Department of Mathematical Sciences University of Wisconsin-Milwaukee Milwaukee WI, USA
Professor Jan Szyszko	Forestry Science University of Warsaw Warsaw, Poland
Professor Michael J. Tannenbaum	Physics Department Brookhaven National Laboratory Upton, New York, USA
Professor Donald Tillitt	Biochemistry & Physiology Columbia Environmental Research Center Columbia, USA

Professor Anastasios A. Tsonis	Department of Mathematical Sciences University of Wisconsin-Milwaukee Milwaukee, USA
Professor Hiroshi Ujita	Tokyo Institute of Technology Department of Nuclear Engineering International Nuclear Human Resource Development (Hitachi) Tokyo, Japan
Professor Jan Veizer	Department of Earth Sciences University of Ottawa Ottawa, Canada
Professor Eugeni Velikhov	Russian Research Centre "Kurchatov Institute" Moscow, Russia
Professor Frederick vom Saal	University of Missouri Division of Biological Sciences Columbia, USA
Professor Yanling Wang	Department of Medical Science Beijing University Beijing, China
Dr. Henning Wegener	Ambassador of Germany (ret.) Information Security Permanent Monitoring Panel, World Federation of Scientists Madrid, Spain
Dr. Horst Wenninger	Former Director for Technology Transfer and Scientific Computing CERN Geneva, Switzerland
Dr. Jody Westby	Global Cyber Risk LLC Information Security PMP World Federation of Scientists Washington, USA
Dr. Crispin Williams	Advanced Technology for Young Scientists from Developing Countries CERN Geneva, Switzerland

Professor Richard Wilson	Department of Physics Harvard University Cambridge, USA
Dr. Lowell Wood	Hoover Institution, Stanford University University of California Lawrence Livermore National Laboratory Stanford, CA, USA
Dr. L. James Wright	Department of Chemistry University of Auckland Auckland, New Zealand
Professor Maw-Kuen Wu	Institute of Physics Academia Sinica U.S. National Academy of Sciences Nankang, Taipei, Taiwan
Dr. Norio Yamamoto	Global Infrastructure Fund Research Foundation Minato-ku, Tokyo, Japan
Dr. Jun Yin	Councellor for Science and Technology Chinese Embassy in Italy Rome, Italy
Professor Tetsuo Yuhara	University of Tokyo The Canon Institute for Global Studies Tokyo, Japan
Professor Xiliang Zhang	Institute of Energy, Environment & Economy Tsinghua University Beijing, P.R. China
Dr. Yiyan Zhang	Second Secretary Chinese Embassy in Italy Rome, Italy

www.ingramcontent.com/pod-product-compliance
Lightning Source LLC
LaVergne TN
LVHW010636300926
849161LV00005B/954
9789814415002